# FLORA OF ASSYNT

*Ajuga pyramidalis*

x 1¼

# FLORA OF ASSYNT

### FLOWERING PLANTS AND FERNS
by
P. A. EVANS AND I. M. EVANS

### BRYOPHYTES
by
G. P. ROTHERO

Part funded by grants from

Botanical Society
of the
British Isles

SCOTTISH NATURAL HERITAGE
DUALCHAS NADAIR na h-ALBA

Glasgow Natural
History Society

Frontispiece from an original painting of *Ajuga pyramidalis*
by
Jonathan Tyler F. L. S.

*To future botanists who will discover much
more about the flora of Assynt*

Copyright 2002  P. A. Evans,  I. M. Evans  and  G. P. Rothero

All rights reserved.  No part of this publication may be reproduced, stored in a retrieval system, or transmitted in any form or by any means, electronic, mechanical, photocopying, recording or otherwise, without prior permission of the publishers.

ISBN  0-9541813-0-1

Published by P. A. Evans  and  I. M. Evans
Camera-ready copy prepared by P. A. Evans
Computer Consultants: abc Computer Services
Figures 3 and 5 by Wendy Price Cartographic Services, Inverness
Printed by CIGAM,  31 Thornhill Park, Knock, Belfast BT5 7AR

# CONTENTS

List of figures ................................................................................6

Foreword (by Alex Scott)..............................................................7

Acknowledgments ........................................................................8

Description of Assynt

       Introduction ......................................................................9

       Geology ...........................................................................11

       Climate (by Peter Kohn).................................................16

       History of the landscape .................................................20

       Vegetation
            Introduction ........................................................26
            Coast and islands................................................26
            Woodland and scrub...........................................29
            Heath, mire and crags ........................................31
            Freshwater ..........................................................34
            Limestone ...........................................................37
            Upland ................................................................39
            Man-made or influenced habitats .......................41

Flowering plants and ferns

       History of recording .......................................................47

       The present survey..........................................................58

       Plan of Species Account.................................................64

       Species Account .............................................................65

Bryophytes

       Introduction ..................................................................179

       Bryophyte communities................................................180

       History of bryophyte recording ...................................187

*continued overleaf*

Bryophytes (*continued*)

    Present survey ............................................................... 189

    Plan of Species Account ................................................ 190

    Species Account ............................................................. 191

Bibliography ................................................................................ 257

Gazetteer ..................................................................................... 262

Index to flowering plants and ferns............................................ 265

Index to bryophytes..................................................................... 279

## LIST OF FIGURES

1. Assynt: location in northern Scotland................................................ 9

2. Assynt: 10km. square boundaries and major roads............................ 9

3. Assynt: major hills, lochs and rivers; parish boundary ..................... 10

4. Assynt and adjacent areas: solid geology .......................................... 12

5. Assynt: simplified geology ................................................................ 13

6. Monthly average precipitation at Duartmore, 1981-1995................. 18

7. Temperature: mean daily maxima and minima at Duartmore, 1981-1995 ........................................................................................... 18

8. Number of vascular plant taxa recorded from each tetrad ................. 61

9. Assynt: contour map ..................................................................... Insert

The **colour plates** are situated between pages 32 and 33. They illustrate a selection of the habitats represented in the parish, in a variety of seasons. The order corresponds to that in the chapter on vegetation.

Figures 1,2,4 and 9 by Ann Cook, 3 and 5 by Wendy Price, 6 and 7 by Trish Kohn and 8 by Alex Lockton. Colour plates from photographs by I.M.E. unless otherwise credited.

# FOREWORD

This comprehensive Flora covers one of the most interesting and beautiful areas of Scotland. It is the first to tackle any part of the Highlands at this level of detail. Besides the meticulous coverage of flowering plants and ferns, the bryophytes (mosses and liverworts) have been given equal treatment. Introductory chapters include clear accounts of the geology, climate, landscape history and vegetation of the area, to enable visiting botanists to appreciate the flora in a much wider context.

The straightforward style has resulted in a very readable and easy to use document. The authors are to be congratulated on their painstaking research into the historical records, and on bringing to fruition ten years of field survey.

I believe that this publication will renew interest in the Assynt flora and stimulate botanists to find new localities for its plants and try to re-discover the old sites for species that have not been recorded recently.

The care, effort and thoroughness that have gone into this book will ensure that it will remain the most significant work on the Assynt flora for many years and will hopefully increase concern for the well-being of this extremely special area.

Alex Scott
Scottish Natural Heritage
Ullapool

# ACKNOWLEDGEMENTS

We are indebted to Scottish Natural Heritage, the Botanical Society of the British Isles and the Glasgow Natural History Society for generous contributions towards the publication of this Flora.

S.N.H. promotes the care and improvement of the natural heritage of Scotland, its responsible enjoyment, understanding, appreciation and sustainable use. One of the ways in which they do this is by supporting surveys of plant and animal life in areas of interest. Their local office is at 17 Pulteney Street, Ullapool, Wester Ross IV26 2UP.

The B.S.B.I. is the national society for anyone interested in the flowering plants and ferns of Britain and Ireland. It has a network of local recorders, organises meetings and surveys, and publishes atlases, handbooks, journals and newsletters. Correspondence about its activities may be addressed to Dept. of Botany, The Natural History Museum, Cromwell Road, London SW7 5BD.

The Glasgow N.H.S. encourages the study and recording of all branches of natural history, particularly that of the west coast of Scotland. It administers the Professor Lloyd Binns Bequest Fund which supports publications and research projects. For information about the Society contact the General Secretary, c/o Art Gallery and Museum, Kelvingrove, Glasgow G3 8AG.

We much appreciate the co-operation we have had from local landowners during the survey work for this Flora. Peter Voy, Factor to the Assynt Estate, facilitated access and the use of estate boats, and keepers Andrew MacKay and Duncan Morrison ferried us to remote areas. Jim and Margaret Payne kindly provided information about previous surveys on Ardvar.

We are eternally grateful to Alex Lockton of Whild Associates for years of support in the realm of RECORDER and recording; and to Ann and Bob Cook of abc Computer Services, whose advice was always only a telephone call away.

Help with critical groups has been generously supplied by Richard Bateman (*Dactylorhiza*), Tom Cope (*Agrostis castellana*), Clive Jermy (*Isoetes*), David McCosh (*Hieracium*), R.D.Meikle (*Salix*), C.N.Page (*Equisetum*), Chris Preston (*Potamogeton*), Rev. A.L.Primavesi (*Rosa*) and Alan Silverside (*Euphrasia*).

Tim Blackstock, Tom Blockeel and David Long have provided much appreciated help with bryophyte determinations.

Early drafts of some introductory chapters were kindly read by Malcolm Bangor-Jones, Jan Breckenridge, Bill Gilmour and Robin Noble, who made many helpful suggestions. Particular thanks are due to Chris Ferreira, who gave us the benefit of his unique knowledge of the vegetation of West Sutherland. Mike Walpole has helped us with the early literature on Sutherland botany and in many other ways and Alan Morton's advice on DMAP has been much appreciated.

Finally, there are a large number of individuals whose company we have enjoyed in the field, who have shared their expertise and finds with us, transported us to islands, or have helped us in many other ways. They include: Andy Amphlett, Margaret Barron, Charlie Bateman, Claire Belshaw, Stuart Belshaw, John Blunt, Val Blunt, Ian Bonner, Ed Brown, Ann Cook, Robin Campbell, Paul Copestake, Rod Corner, Ann Daley, Alec Dickson, Lynne Farrell, Ron Fishlock, Brian Gale, Claire Geddes, Erica Gorman, John Gorman, John Gibson, Edward Grieg-Hall, Stephen Grover, David Hawksley, Viv Hawksley, Pete Hollingsworth, Helen Jackson, Robert Kinnaird, Donald King, Eileen King, Peter Kohn, Trish Kohn, Audrey Langton, Tina Lloyd-Jenkins, Neil Lloyd-Jenkins, Gill Lockie, Trevor Lockie, Diarmid Macaulay, Iain Macdonald, Mairi MacRae, Peter MacGregor, Elma MacKenzie, Non MacLeod, Betty Matheson, Douglas McKean, Helen Morrison, Sue Murray, Lady Win Murray, Jackie Muscott, Robin Noble, David Pearman, Chris Pellant, Ian Pennie, Ken Perry, Chris Pogson, Gwen Richards, Philippa Richardson, Bill Ritchie, Ingrid Ritchie, Mary Ross, Moira Roy, Malcolm Sandals, Ro Scott, Colin Scouller, Alf Slack, Alan Stirling, Andy Summers, Leslie Tucker, Alan Underwood, Agnes Walker, Anne Walpole, Mike Walpole, Chris Warwick, Claire Warwick, Anna White and Beryl Woolner.

At the British Bryological Society field meeting in 1992 records were provided by Tom Blockeel, Alan Crundwell, Mike Fletcher, Nick Hodgetts, Peter Martin, Roy Perry, Mark Pool, Ron Porley, Phil Stanley, Rod Stern and Harold Whitehouse.

We are sorry if we have missed anyone out!

# DESCRIPTION OF ASSYNT

## INTRODUCTION

The parish of Assynt is situated in the south-western corner of Sutherland. It is one of five large parishes that constitute the botanical vice-county of West Sutherland (v.c. 108), which is shown in grey below.

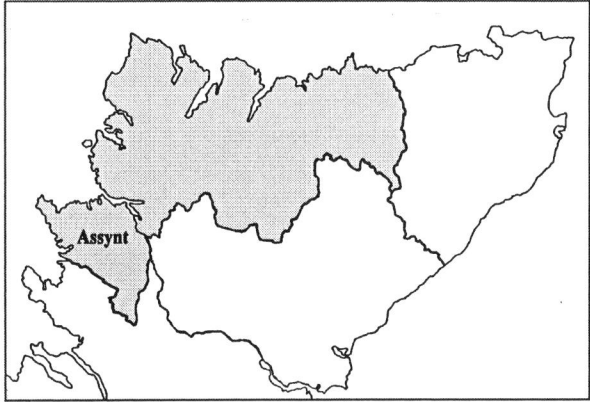

Figure 1. Assynt: location in northern Scotland.

Figure 2. Assynt: 10km. square boundaries and major roads.

Assynt stretches for some 40km. (25 miles) from Point of Stoer in the north-west to the furthest point of the Cromalt Hills (Meall a' Bhùirich Rapaig) in the south-east. It has an area of 475 sq. km. (187 sq. mls.).

In terms of the National Grid, Assynt occupies part at least of thirteen 10km. squares (hectads) in the 100km. square NC (29). However, four of the thirteen hectads (01, 10, 31 and 32) contain only tiny, albeit botanically very interesting, areas of the parish, the summit of Conival, our one 'Munro', being situated in 31. The whole of the parish is contained within the area covered by the O.S. Landranger series Sheet 15 'Loch Assynt' and parts of eight Pathfinder series (although there is only a very small area on the Strath Kanaird sheet (NC00/10).

The road pattern is relatively simple. There is a single track road from Inverkirkaig to Lochinver, and then round the coast from Lochinver to Newton. Short spurs serve the townships off this road, with a small network on the Stoer peninsula. A double track road runs from Lochinver east along Loch Assynt, with a branch at Skiag Bridge north to Unapool and Kylesku. It then continues south via Inchnadamph and Ledmore Junction to Elphin and Knockan, with a single track road branching off at Ledmore Junction down Strath Oykell.

Of the nine hectads that encompass most of Assynt, eight are traversed by roads. However, one, NC11, which includes Suilven, has neither roads nor settlements. There was one farm, Bracklach, at the western end of Cam Loch, but that has been unoccupied for well over a century. This area is accessed by tracks from Inverkirkaig and Elphin. Other remote areas may be reached by stalkers' tracks and peat roads, but do not rely on the footbridges shown on the 1:25,000 maps, many no longer exist.

Assynt had a population of 1047 at the 1991 census, spread through some 650 households. About half of these households are in the village of Lochinver, the rest in some 20 crofting townships and scattered smaller settlements and houses. The crofting townships occupy the fertile ground around the coast from Inverkirkaig to Nedd and Unapool, with outliers on the limestone at Elphin and Knockan. Those from Achmelvich to Nedd are part of the North Assynt Estate, purchased in 1993 by the Assynt Crofters' Trust. The largest landowner in the rest of the parish is the Assynt Estate.

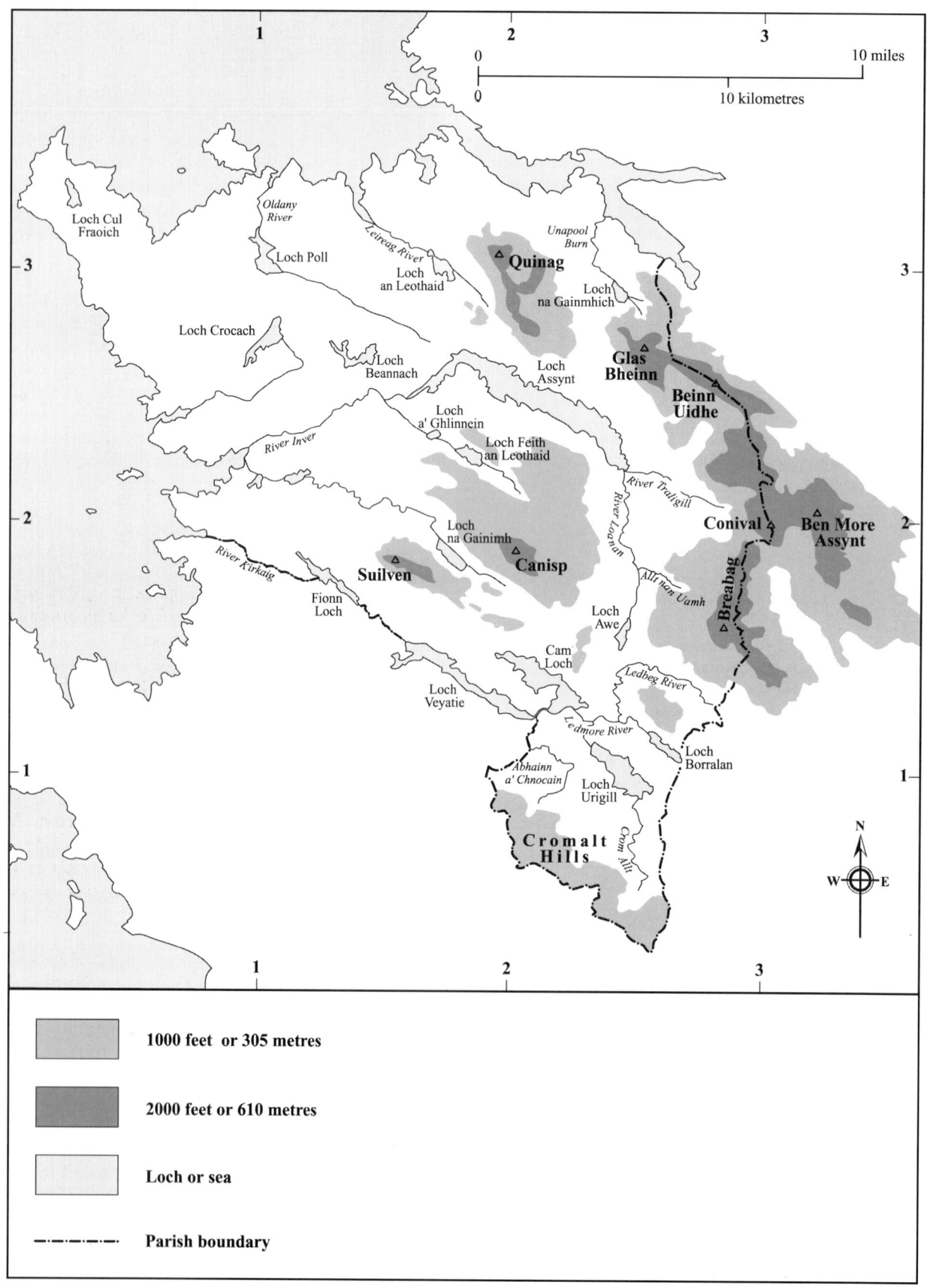

**Figure 3.** Assynt: major hills, lochs and rivers; parish boundary.

Since parish boundaries are no longer shown on readily available maps (including the most recent editions of the Pathfinder series, 1:25,000), it may be useful to describe the boundaries of Assynt (see figure 3). In the west the boundary follows the coastline, from the mouth of the River Kirkaig north to Point of Stoer, taking in the islands of A' Chleit and Soyea.

In the north it is also straightforward, following the coastline from Point of Stoer east to a point on LochGlencoul, about 300m. west of Eilean an Tuim, where a small burn enters the sea. The islands of Meall Beag and Meall Mor, although traditionally associated with Assynt, are in the parish of Eddrachillis to the north.

The eastern boundary is the obscure one. It strikes south from Loch Glencoul, over Cnoc na Creige to the saddle between Glas Bheinn and Beinn Uidhe, then south-east along the crest of Beinn Uidhe. It then turns south through rising ground to the summit of Conival. From Conival it runs south along the crest of Breabag to Meall Diamhain, then south-west to cross the A837 just east of the Altnacealgach Hotel. It continues south-west through the small hill of Gorm Cnoc, then south over Meall nan Imrichean and Meall a' Chaoruinn to the southernmost point of the parish at the summit of Meall a' Bhùirich Rapaig.

The southern boundary is fairly straightforward, following the old county boundary between Sutherland and Wester Ross along the crest of the Cromalt Hills in a north-westerly direction and crossing the A835 just north of the Knockan Visitor Centre (where it is marked by a small burn). It then follows a small burn and the river to the west of Elphin as far as Loch Veyatie, and then down that loch, Fionn Loch and the River Kirkaig to the sea.

# GEOLOGY

Since the pioneering studies of the late 19th century, Assynt has been one of the classic areas for the study of large scale geological processes. However, accounts of its geology can be couched in terminology that is a little opaque to those without an academic background in the subject.

Readers wishing to go a little deeper into the fascinating geological history of this area are referred to: *The geology of Sutherland* (Ross in Omand 1982), the introduction to *The Quaternary of Assynt and Coigach* (Lawson 1995) and *North West Highlands. A landscape fashioned by geology* (Mendum et al. 2001).

What follows is an attempt to summarise aspects of the solid geology of Assynt that appear to be significant from topographical and botanical points of view. The geology of the glacial period is covered, very briefly, in the chapter on the history of the landscape. This account draws on the sources mentioned above, amongst others, but more particularly on the introduction to his account of the main vegetation types in Ferreira's report on the *Vegetation Survey of North West Sutherland* (1995). We are grateful to Bill Gilmour and Jan Breckenridge for their comments on an early draft; they are not responsible for any continuing misapprehensions.

It may be useful, at this point, to draw attention to some terms that have been used in quite different ways by geologists and botanists. Traditionally, geologists have classified rocks, particularly igneous ones, according to their silica content: acidic rocks, with more than 63% silica by weight, intermediate rocks 52-63%, basic rocks 45-52% and ultrabasic rocks 45% or less. Nowadays, the term *silicic* (or *felsic*) is applied to light-coloured silica-rich rocks, and *mafic* to dark-coloured rocks rich in magnesium and iron.

Botanists have used the terms acidic and alkaline/basic rather more broadly, to categorise rocks, their breakdown products, groundwater, soils and vegetation. A major criterion has been the availability of bases such as calcium or magnesium, hence the terms calcareous and calcicole. Where these terms are used in this account, it is in the botanical sense. Interested readers are referred to the section in Ferreira (1995) on 'Geobotanical terms' which covers the use, and mis-use, of these and related terms in the area of overlap between geology and botany.

We would now refer you to figures 4 and 5. Figure 4 is reproduced from the 1948 edition of the Quarter Inch Geological Survey map of Northern Scotland and gives some idea of the complexity of the geology of the Assynt area. Figure 5 is a simplified version of the same map showing the five main geological regions.

These five regions are described below in approximate order of age. Most of the western part of the parish lies on Lewisian rocks, which are the oldest. Torridonian rocks constitute the Stoer peninsula and the greater part of the isolated hills for which Assynt is famous, from Quinag south to Suilven. Cambrian quartzites cap the summits of some of these hills and cover their eastern slopes. From the eastern edge of the quartzites through to the high eastern boundary of the parish is an area of great complexity, the Moine Thrust Zone, and at its southern edge a small area of the Moine proper.

Flora of Assynt

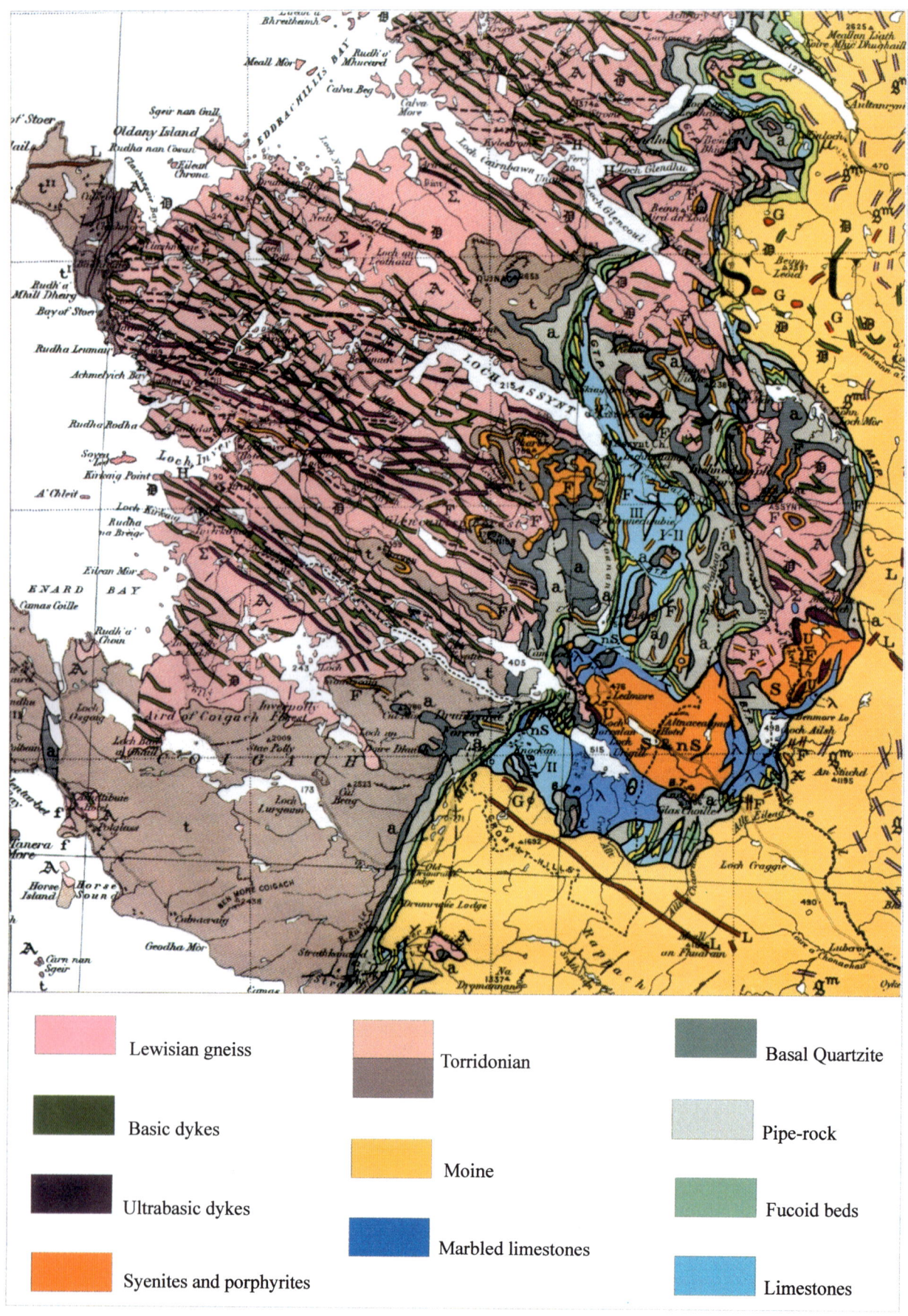

**Figure 4.** Assynt and adjacent areas: solid geology.
Reproduced from Geological Survey Quarter Inch Sheet 5 (1948).

Description of Assynt

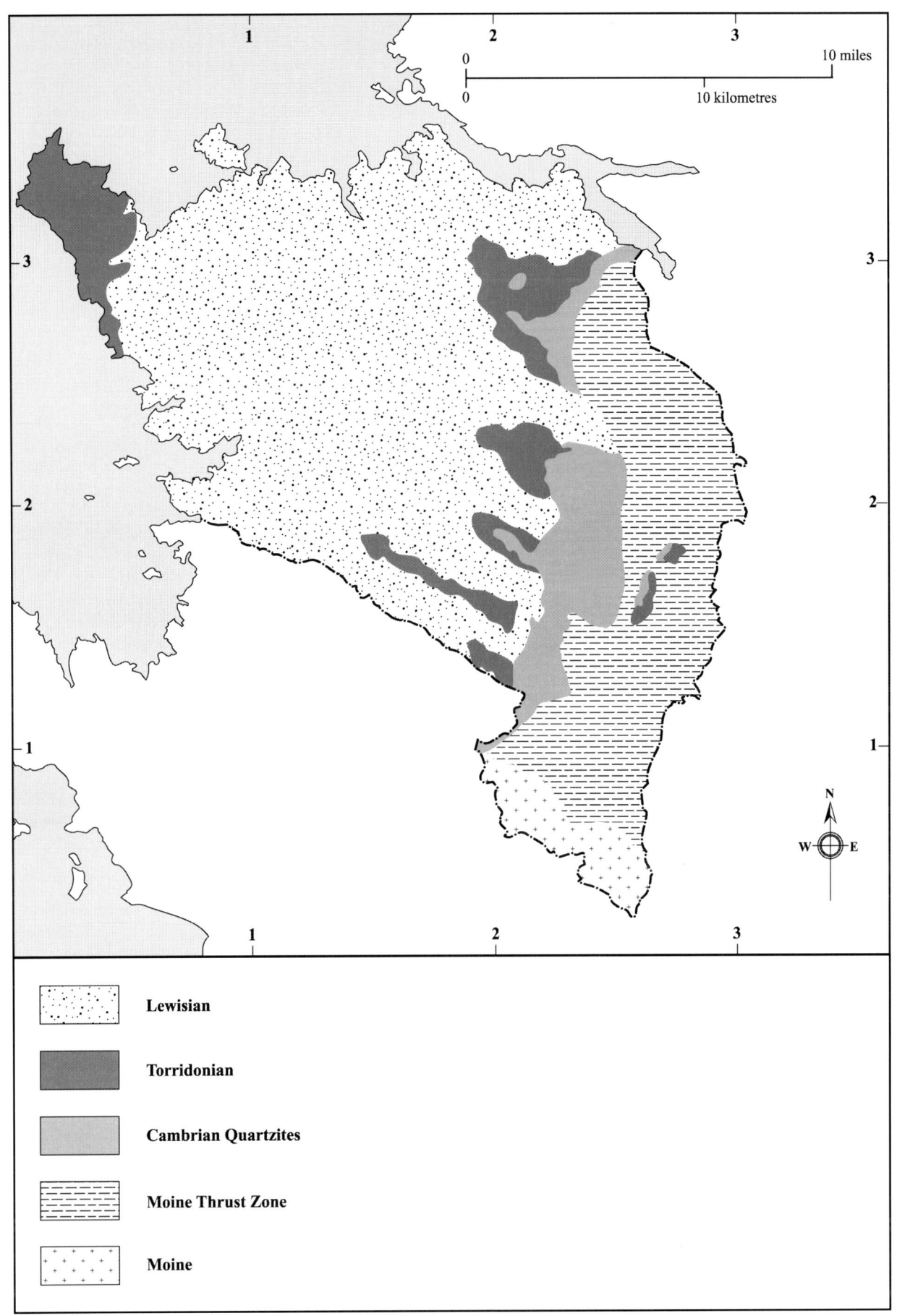

Figure 5. Assynt: simplified geology.

## Lewisian

The Lewisian gneiss comprises the oldest rocks in the area and some of the oldest on earth. They were formed deep in the earth's crust some 2900 million years ago and underwent at least two major phases of metamorphism (further change) in the following 1500 million years. Over this very long period they were also infiltrated by swarms of basic and ultrabasic dykes (narrow, linear, vertical intrusions of igneous rock), which were also subject to metamorphism. Both gneiss and dykes have been sheared and faulted by movements in the earth's crust.

The gneiss of Assynt is very varied in composition, but generally more basic than that further to the north. At its most acidic, it is a hard, pinkish, granite-like rock (granulite) that is very resistant to weathering. More typical is a grey, banded rock, with lighter silica-rich layers alternating with those of darker minerals rich in magnesium and iron (hence mafic). There are frequent inclusions, predominantly composed of these darker minerals, which are much less resistant to weathering than the lighter ones. Most important, from a botanical point of view, the mafic rocks often contain calcium and/or magnesium in a water-soluble form. It is the base-rich spring waters originating from these rocks that lead to much of the local variation in the vegetation over the gneiss.

The greater part of the gneiss lies to the west of the hills of Quinag, Canisp and Suilven, at a height ranging from sea-level to over 350m. Here, to quote Lawson (1995), 'glaciation has resulted in a deeply-dissected topography of incredible irregularity' with 'numerous steep-sided knolls with many of the intervening hollows occupied by small lochans', what has been termed the 'cnoc-and-lochan' topography. A striking feature of this topography is the alignment of many its ridges and valleys (with their lochs) in two intersecting directions, north-eastward and west-north-westward.

Once you recognize them, the basic dykes in the gneiss are a conspicuous feature of parts of the Assynt landscape; they are coloured green in figure 4. They run from east-south-east to west-north-west, and take the form of ridges of dark, bluish-black rock, liberally plastered with the white lichen *Pertusaria corallina* (which also occurs on the more mafic parts of the gneiss itself, but not in such quantity). There are good examples running parallel to the road in the townships of Drumbeg and Nedd, and others cutting across the road south-west of Oldany. The base-rich weathering products of these dykes impart a noticeable diversity to the flora in their vicinity and probably also contribute to the fertility of the soils in the townships mentioned and elsewhere.

The ultrabasic dykes are fewer in number, and run more nearly east-west; they are coloured purple in figure 4. They are often pale brown in colour, weathered in a honeycomb pattern, with narrow fissures at a variety of angles and often a dense scatter of pock-marks where some minerals have eroded out. An obvious example is on the coastal side of the road at Alltana'bradhan, cutting across the path to the mill. There are also good examples at the far end of the Clachtoll peat track (north of Creag Clais nan Cruineachd) and at the foot of the north face of Beinn Gharbh. They tend to be magnesium-rich. The dykes themselves are often almost devoid of vegetation, other than the fern *Asplenium adiantum-nigrum*, but the grassland in their vicinity has a characteristic range of species.

## Torridonian

The mudstones, sandstones and conglomerates of this series of rocks are about 1000 million years old. They were laid down under semi-arid conditions, by braided rivers and in deltas liable to seasonal flash floods, in many cases direct onto the eroded surface of the Lewisian gneiss. They occur in two main areas in Assynt. In the north-west they form the Stoer peninsula, running from near Clachtoll round to Achnacarnin. Here the gently inclined strata rise to a height of 161m. at Sìdhean Mór, providing the most impressive sea cliffs in Assynt.

Inland, the remnants of a once continuous cover of Torridonian rocks make a major contribution to the celebrated hills of Quinag, Canisp and Suilven. In places the underlying gneiss appears quite high on these hills, as on the north face of Sàil Gorm, and both the ridge of Sàil Gharbh and Canisp are capped with quartzite, but their essential character derives from the Torridonian. Features of the steep-faces of both Quinag and Suilven are rounded buttresses separated by deep chimneys ending below in stone chutes and above, round the edges of the summit ridges, terraces accessible only to those with a good head for heights.

Torridonian rocks are generally uniformly acidic and, as such, support vegetation that is floristically poor. However there thin bands of limestone in the rocks on the landward side of the Stoer peninsula, and also some thin, but distinctly calcareous, bands in places on the hills. Examples are in the eastern corrie of Quinag, the north face of Beinn Gharbh and around Meall na Braclaich, south-east of Suilven. These support a localised but very interesting calcicole flora, of which more details are given in the section on upland vegetation.

Although strictly out of sequence in terms of age, we should refer here to the igneous rocks often called 'Canisp porphyry', more generically felsites and porphyrites. They occur as sills, following the horizontal bedding planes in the Torridonian of Canisp and Suilven, and also outcrop over an area of several square kilometres in the vicinity of Beinn Gharbh and Beinn Reidh; they are coloured orange in figure 4. They date from a period of tectonic disturbance about 425 million years ago. The term *porphyry* relates to their texture, which is very distinctive, consisting of large light-coloured crystals in a finer reddish matrix. They tend to be rich in quartz and feldspar, with such minerals as they contain not readily available and thus acidic in botanical terms. The area over which they outcrop in Assynt has not been studied in sufficient detail for us to comment on the vegetation they support. However, where the sills have been faulted at the eastern end of Suilven, the resultant mineral-enrichment may be responsible for the very interesting flora recently discovered there (see the Upland section of the Vegetation chapter).

## Cambrian quartzites

These comprise the undisturbed parts of the oldest of the Cambrian rocks, the Basal or False-bedded Quartzite and Pipe Rock, laid down under marine conditions up to 600 million years ago. They are often hard white or pink quartzites, almost pure quartz in composition, and Ferreira states that they produce 'some of the most acidic and base-poor rock-origin habitats in Britain'. His opinion is borne out by the fact that tetrads 21I and J, which lie entirely on the quartzites, could only muster 80 and 79 species respectively in our survey, the least for any full tetrads. However, those totals do include two of our two rarest club-mosses.

As has been indicated, these quartzites also cap the Torridonian hills of Quinag and Canisp. On the former they are almost unvegetated apart from the moss *Racomitrium lanuginosum*, but on Canisp there is one considerable rarity associated with a very curious reddish soil-like deposit (see Upland Vegetation).

## Moine Thrust Zone

This easternmost region is where the geology of Assynt gets extremely complicated, and not only to the non-geologist, since the celebrated geologists Peach and Horne referred to it as the 'Zone of Confusion'. To quote Lawson (1995), 'this is due to the presence of large horizontally-directed dislocations or thrusts, along which masses of rock have been directed westwards'. These dislocations occurred during a period of major movement in the earth's crust around 425 million years ago. The main thrust and most easterly is the Moine Thrust, 'which carried the Moine Schists north-westwards over a complex of intensely disturbed Cambrian, Torridonian and Lewisian rocks, which themselves have moved on lower thrusts' (Ross *in* Omand 1982). The western boundary of this region in figure 5 follows the line of the lowest of these thrusts, the Sole Thrust. There is a narrow strip of unmoved Cambrian rocks lying above the quartzites (notably the Fucoid Beds) just to the west of the line of this Thrust, but for convenience we have included them in the Thrust Zone.

This region therefore contains examples of all the main rock types already mentioned, together with younger Cambrian rocks, comprising the Fucoid Beds, Salterella Grit and the lower part of the Durness Limestones. There is also in the southern part of this region a substantial intrusion of igneous rocks, centred on Loch Borralan, its most obvious feature being the 'granite' dome of Cnoc na Sròine (coloured orange in figure 4). Associated with this igneous intrusion is an area in which the limestones have been metamorphosed into marble.

The western edge of this region takes the form of a broad 'valley' extending for some 18km. from Unapool south to Elphin. Outcropping on the eastern flank of this valley, above or on the line of the road, as far south as the Ledbeg River, and again at Elphin and Knockan, is a relatively narrow band of the rusty-coloured Fucoid Beds. They are described by Ferreira as dolomitic mudstones and shales, rich in calcium, magnesium and potassium, and locally in phosphate. They are soft and easily broken down and the resultant soils are 'unusually fertile and of very considerable geobotanical importance'. Higher on the hills, they reappear on the western flanks and summit plateau of Beinn an Fhurain, in the upper part of the Traligill valley (including Bealach Traligill) and on the summit of Breabag.

Above and to the east of the Fucoid Beds is an even narrower band of the Salterella Grit, quartzite with some dolomitic (magnesium-rich) sandstone towards the top of the band. The Grit does not weather readily and has not so far been found to support a characteristic local flora.

Lying above the Salterella Grits are the massive Durness Limestones, which extend in a band, up to 3km. wide in places, from Liath Bhad on the shores of Loch Glencoul, south by Inchnadamph and Stronechrubie to Ledmore Junction. They then reappear at Elphin and Knockan, extending south-eastwards from there to the foot of the Cromalt Hills. It is unfortunate, from a botanical point of view, that much of the area in which they occur is covered with deep peat, notably at Lairig Unapool, between the valleys of the Traligill and the Allt nan Uamh, east of Beinn an Fhuarain and south-west of Loch Urigill. They vary in colour from pale to very dark grey, and in composition from pure limestone (calcium carbonate) to dolomite (calcium/magnesium carbonate), with every gradation in between.

These limestones provide some of the most striking scenery and most exciting botanising in Assynt. High points are the Traligill valley, the cliffs of Creag Sròn Chrùbaidh, the valley of the Allt nan Uamh with its famous 'Bone Caves' and the valley of the Abhainn a' Chnocain with its potholes and sinks. Set in the limestone plateau on either side of the Allt nan Uamh valley are the 'displaced' Torridonian and quartzite hills of Beinn nan Cnaimhseag and Beinn an Fhuarain; the juxtaposition of their rocks and the limestones provides sudden changes in vegetation over very short distances. The limestone rises on its eastern margin to 450m. at Cnoc Eilid Mhathain and 535m. in the Bealach Traligill, the latter being the highest that it occurs anywhere in West Sutherland.

The Bealach Traligill is the most conspicuous break in the range of hills running along the eastern boundary of the parish, from Cnoc na Creige in the north, via Glas Bheinn, Beinn Uidhe, Beinn an Fhurain and Conival, to the southern edge of Breabag at Meall Diamhain. Geologically, they tend to alternate between gneiss and and quartzite, with occasional outcrops of other Cambrian strata and igneous intrusions. A brief traverse may help set the scene for the upland section in the chapter on vegetation.

Starting in the north, Cnoc na Creige (593m.) is gneiss intersected by basic dykes, as are the lower parts of the magnificent corries on the north side of Glas Bheinn. Their upper parts and the summit of this hill (776m.) are quartzite, as is most of the ridge of Beinn Uidhe (741m.), with gneiss reappearing at its south-eastern end.

From Bealach a' Mhadaidh (pass of the wolf) west of Loch nan Cuaran (610m.), south for some four kilometres to the summit of Conival (987m.), the underlying rock is again quartzite, with Fucoid Beds (not very obvious on the ground!) and porphyritic intrusions. South of the Bealach Traligill, the Breabag ridge is again quartzites, enlivened along its western edge by Fucoid Beds and porphyritic sills.

At the southern end of Breabag, the quartzite slopes give way, along the line of the Ledbeg River, to the 'granite' of the twin rounded hills of Cnoc na Sròine (397m.). This hill is the most obvious manifestation in Assynt of the Loch Borralan Complex of igneous rocks, which also extend under deep peat on the south side of the loch. They are contemporaneous, at 425 million years, with the events that resulted in the Moine Thrust Zone. According to Ferreira, they include a suite of basic and ultrabasic rocks that show themselves locally as calcium and magnesium-rich flushes, and there are certainly signs of this on the north face of Cnoc na Sròine.

## Moine

In the south-eastern corner of Assynt there is a small area of the rocks which cover much of the rest of the northern Highlands, the Moine Schists. They run from Cnoc an t-Sasunnaich at the top of the Knockan Crags along the rounded summit ridge of the Cromalt Hills to the southernmost tip of the parish at Meall a' Bhuirich Rapaig, at a height of 400-500 m., extending north-eastwards for at most three kilometres from the ridge. They were originally deposited as sediments, derived from Lewisian-type rock, under marine conditions about 1500 years ago, and metamorphosed into crystalline schists 500 or more million years later.

In the Assynt area the schists were derived from sandstones; they are hard, acidic, and often flaggy, and the vegetation they support is fairly dour. However, there are narrow calcareous bands, which have a striking, if very local, effect on the vegetation, with abundant calcicolous mosses (Ferreira 1995). One such site is the wooded ravine and box gorge on the upper reaches of the Crom Allt, and another showing signs of base-richness is a north-facing wet crag above Loch a' Phris.

# CLIMATE

## Precipitation

Assynt is a wet and windy place. Generally in Britain the prevailing winds are south-westerly and bring moist air from the Atlantic onshore which, as it rises over the hills, cools and causes precipitation. Thus the driest parts of the parish are the coastal fringes, whilst high land towards Ben More Assynt is the wettest. Unfortunately, there are no full meteorological stations within the parish, so much of what follows has to be inferred from data for Duartmore to the north and for 'Knockanrock', which is just south of the parish boundary. Use has also been made of short-term rainfall measurements from within the parish (at Little Assynt and Inchnadamph) and recent temperature and rainfall observations at Kerrachar. These all cover comparatively short periods. Despite the limited sources of information, the wetness of the parish cannot be doubted. At Duartmore between 1981 and 1995 the rainfall averaged 1649mm. (almost 65 inches) and a very similar average was recorded at Kerrachar between 1995 and 2000. This is roughly double the rainfall of Manchester, which has the reputation of being a wet city. Even in the drier coastal parts of the parish, precipitation greatly exceeds potential losses by evaporation and transpiration.

This high rainfall seems not to be a result of any recent climate change. The Reverend William Mackenzie noted, in the *Old Statistical Account* (1794), that in Assynt 'the rain continues not only for hours but often for days: nay, for weeks, especially if the wind perseveres for a long time to blow from the west'. The Rev. Mackenzie was an accurate observer of the weather, for he also noted that 'a smart easterly wind arising and continuing for a space of twenty four hours will perfectly abate the waters, carrying off all superfluous rain from the surface and moisture from the air'. Lengthy spells of easterly winds do occur and can produce very dry periods of weather even in Assynt. In January 1996 only 25mm. (one inch) of rain fell in the entire month. However, even on the coast, uncultivated peaty land remains waterlogged through the greater part of the year and only occasionally in May or June does the surface dry out. This reflects the typical rainfall pattern. May is usually the driest month, but dry is a relative term. The long term monthly average rainfall for May at Duartmore is around 67mm. (more than two and a half inches), and for a large part of the year (September to March) monthly rainfall is around 170mm. (more than six and a half inches). Ten inches of rain in a month is by no means unusual, with 333mm. recorded at Duartmore in November 1981, and 383mm. (more than 15 inches) in February 1998 at Inchnadamph. In the latter month, only 305mm. were recorded at Little Assynt and 316mm. at Duartmore.

This pattern of a lower rainfall at Little Assynt than at Inchnadamph is repeated in most months for which records are available, but is not invariable. The limited data suggest that in really wet months Inchnadamph is appreciably wetter than Little Assynt, but in very dry months the difference is often much less and occasionally even reversed. This pattern is what might be expected since the driest months are associated with predominantly easterly winds, which deposit their rain on the eastern slopes of the hills, leaving progressively less precipitation as one moves away from the western slopes. These two recording stations lie at opposite ends of Loch Assynt and are little more than six miles apart.

Because there are no really extensive records from a single location it is difficult to be sure about long term trends. Anecdotally, recent winters have been reported as wetter than in the past, when far more snow is said to have fallen. Certainly the total precipitation recorded at Kylestrome (Omand 1982) in the winter months between 1941 and 1970 is marginally lower than that for Duartmore (1981-1995) or Kerrachar (1995-2001). Seven years is too short a period on which to base any conclusions, but the Kerrachar observations suggest that in recent years August has been drier and February wetter than earlier in the twentieth century.

As one moves inland and towards higher ground the rainfall levels undoubtedly do increase. Figures quoted by Omand (1982) for the east slope of Ben More Assynt (for an unspecified period between 1941 and 1970) give an annual rainfall of 3269mm. (130 inches), about double the rainfall on the coast. Although the pattern of rainfall is similar to that on the coast, the differences in amounts are most marked during the winter months, with December being the wettest month on Ben More, and also at 'Knockanrock'. Complete annual rainfall figures at Inchnadamph are limited to the years 1997, 1998 and 2000, when the values were 1701, 2317 and 2128mm. Comparable figures were 1640, 1821 and 1785mm. at Duartmore and 1589, 1845 and 1621mm. at Kerrachar.

All the above figures for precipitation include hail, sleet and snow. Anecdotally, snow has been less common in recent years than it was twenty or more years ago, but no hard data are available for the parish to support this suggestion. The records at Duartmore reveal that snow fell on average for nearly 25 days per year, with snow lying at 09:00 hrs. on more than 13 days per year. At sea level, lying snow may occur in any month from October to April and on the highest ground snowfalls can occur at almost any time of the year, even in July and August.

## Winds

As already mentioned, Assynt is a windy parish. The strongest winds in Britain at sea level are experienced in the north-west of Scotland and it has been suggested that this is one of the windiest permanently inhabited areas in Europe or even in the whole world! Wind is certainly a major limiting factor so far as tree growth is concerned, and the prevalence of south-westerlies is borne out by the restriction of trees, in some areas, to north and east-facing slopes. North-westerly winds are also common, and in recent years there is anecdotal evidence for an increase in the frequency of easterly winds. Wind direction can be greatly modified by the lie of the land. Although north-westerly winds are rare in Stornoway, this is the commonest direction at Shin, which is sited in a valley running from north-west to south-east.

No data on wind speeds are available for the coast of Assynt, but it is recorded that winds of 22 knots or greater blow for 14% of the time at Dounreay (in Caithness) and 21% of the time at Stornoway (on Lewis). The most exposed parts of the coast, such as the Stoer peninsula, can be expected to experience similar, or even stronger, winds than those at Stornoway.

Data on gales at Duartmore is only available for three years, when they were recorded on an average of nine days per year. A gale for this purpose is defined as a wind speed exceeding 34 knots for at least 10 minutes. Using this definition, no gales were recorded between April and July and gales were most frequent in

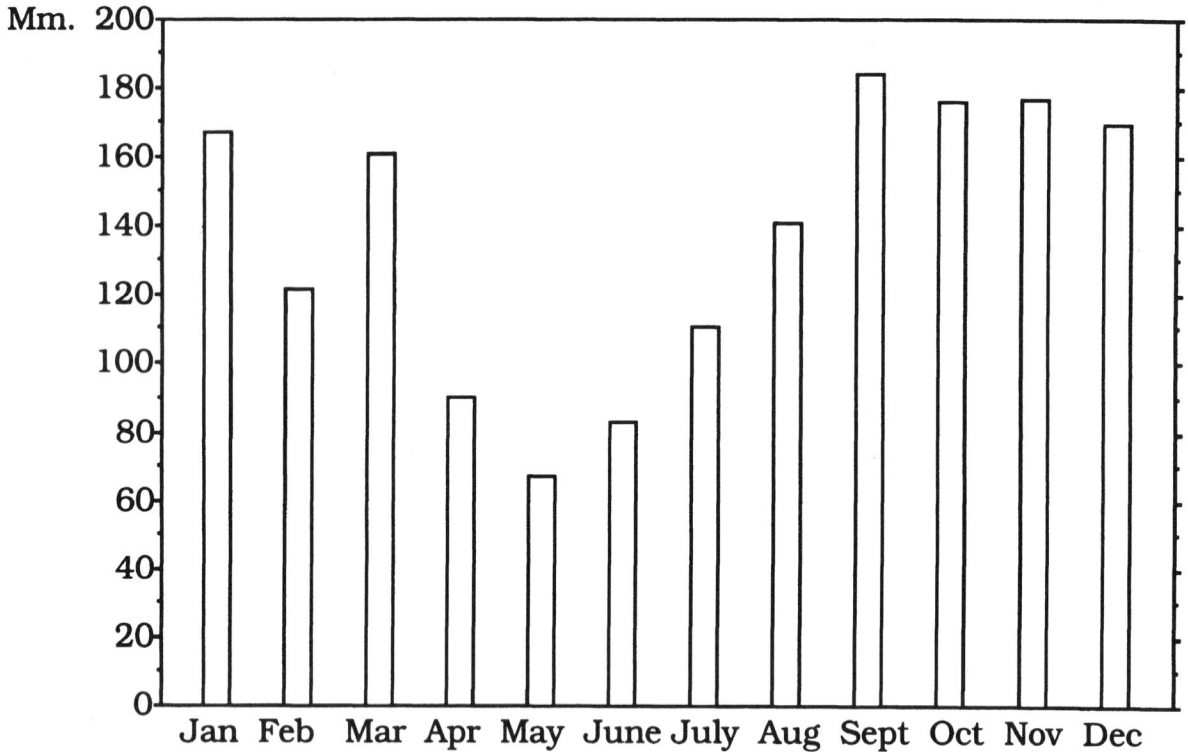

Figure 6. Monthly average precipitation at Duartmore, 1981-1995.

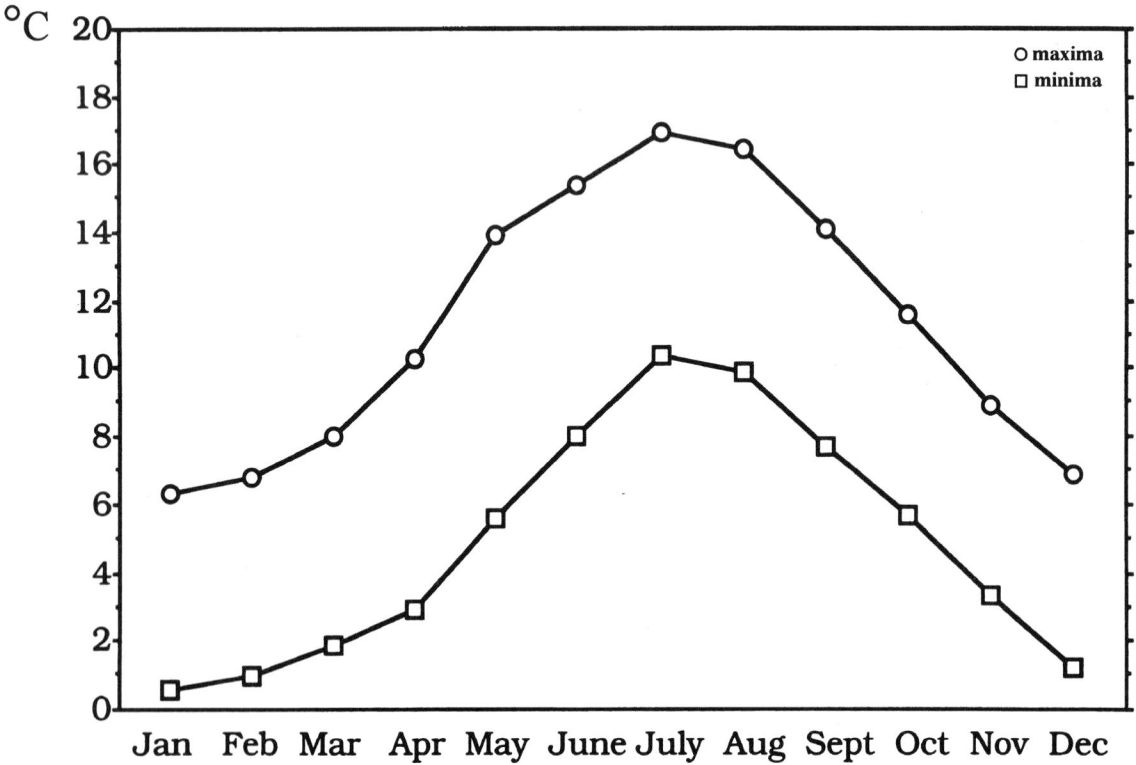

Figure 7. Temperature: mean daily maxima and minima at Duartmore, 1981-1995.

December. Over a 12 year period at Duartmore, the average wind speeds were highest in January, but November had the highest maximum wind speeds. This limited information indicates that the winter months are the windiest and May through to August are generally the quietest months. These terms are only relative. Strong winds can occur in any month and in several years spring gales have been sufficiently severe to cause leaf scorch on trees many miles inland.

## Temperature

The south-westerly winds which bring moisture from the sea also limit the temperature variations between winter and summer: summers are cool, but winters are mild. In many years the average January temperature may be little more than ten degrees cooler than the average July temperature. Even in the hottest years, the average daily maximum temperature in any month does not exceed 21°C. and is more typically nearer 17°C., with July usually the warmest month (though in 1998 the warmest month was September). Even in the coldest recent winters, the average daily maximum temperature in January and February (the coldest months) has always exceeded 4°C. These figures are for Kerrachar.

Similar temperatures are recorded at Duartmore, where the records also show that, at some point in the month, the temperature will typically rise to about 11°C. in the coldest months (December to February) and to 23 degrees in the warmest months, another indication of the small differences between the winter and summer temperatures. The night time temperatures at Duartmore typically fall no lower than -6°C. in December and January. The lowest temperature recorded there between 1981 and 1995 was -13.5°C., and the highest 28.6°C. The minimum temperatures seem surprisingly low, given that the Duartmore recording station is only 2.4km. (1.5 miles) from the sea. At Kerrachar, which lies immediately on the shores of Loch a' Chàirn Bhàin (a sea loch) and very close to the open sea, between 1995 and 2000 winter temperatures fell no lower than -3.5°C. in most winters and the lowest temperature recorded was -4.5°C. Kerrachar is only 3.5km. (just over 2 miles) from Duartmore. This marked drop in the lowest temperature even a short distance inland will undoubtedly become even more marked as one moves further from the coast. At Altnaharra, in the centre of the county and situated at only 82m.(about 300ft.) above sea level, -27°C. was recorded in the winter of 1995-96, one of the lowest temperatures recorded anywhere in Britain.

Certainly as judged by the native vegetation, Spring comes late to Assynt. This partly reflects the relatively low overall mean temperatures in the Spring months. The average overall temperature in January at Duartmore is 3.4°C. and only rises to 6.6°C. in April. There is then a sharp rise in May (9.8°), with the maximum being reached in July (13.6°) and August (13.2°). Interestingly, the temperature in November is still up at 6.1°C., only marginally cooler than April. It is probably a combination of low temperature and waterlogged substrate which delays the onset of growth in many native plants. It is noticeable that, on well drained cultivated land, growth begins much earlier and Assynt gardens can often be ahead of those much further south. This difference was particularly striking in 2001, when Assynt had an unusually dry winter and spring, whilst rainfall in southern Scotland and England was exceptionally high.

## Sunshine

The records of sunshine for Duartmore make interesting reading. Day length in December and January is very short in such northern latitudes. The maximum sunshine recorded for any single day in December is only 5 hours but this presumably represents something close to the maximum possible. On average, May is not only the driest month, but also the sunniest, with an average total of 175 hours of sunshine (compared with only 16 hours in December). The total amount of sunshine in any month ranges from as low as 3.3 hours in a very dull January to 240.6 hours in a very sunny May. Even though there are more hours with the sun above the horizon in June than in May, the maximum recorded monthly total for June is only 186.9 hours and, at 132 hours, the monthly average for June is also substantially lower than the 175 hours for May. These figures emphasise the fact that some cloud (and rain) can always be expected in June. April is the next driest and sunniest month after May and June, with an average monthly total of 113 hours of sunshine. The sunshine figures given above are taken from the period for which full data are available for Duartmore (1981 to 1995).

One can only speculate to what extent the overall pattern of comparatively mild but wet and windy winters, drier springs, and cool moist summers have influenced the composition of the native vegetation. The excess of precipitation over evaporation and transpiration and the consequent waterlogging in places must be a significant factor. Where the soil is cultivated and fertilised, most plants that can be grown in other parts of Britain can also be grown in Assynt, provided they can be sheltered from the wind. The comparative coolness of the summers and the relative lack of sunshine in high summer do limit the ripening of both seeds in late flowering herbaceous species and the wood of trees and shrubs. Even native trees such as rowan do not fruit well in many seasons and oaks only rarely set seed. The vegetation of Assynt has undergone marked changes during the last 10,000 years in response to climatic fluctuations. The current period of rapid climate change may result in further significant changes. Only time will tell.

*Peter Kohn*

# HISTORY OF THE LANDSCAPE

This chapter attempts to relate strands in the history of the landscape of Assynt to present-day patterns in its vegetation. Some aspects of the relationship have been well researched, others hardly at all, and no-one has yet attempted a synthesis. At this stage, therefore, connections are speculative and conclusions very tentative. We have, however, thought it worthwhile to bring together some of the evidence, if only to highlight aspects that require further study. In doing this, we have had the benefit of the expertise and advice of a number of experts in this area, notably Tim Lawson, Robin Noble and Malcolm Bangor-Jones. We are most grateful for their help, but absolve them from any responsibility for our interpretation of the information they have provided.

The solid geology described in the preceding chapter provides the bones of the Assynt landscape. To put some flesh on these bones, we need to consider three phases in the 'recent' history of the area, i.e. the last 15,000 years. These are glaciation, the establishment of vegetation and the arrival, increase in numbers and impact of man. Underlying all of these are broader considerations, of ongoing changes in climate and the combined effects of all the foregoing on soils and their fertility. We have tried to provide a synopsis of the three phases, but not of the underlying factors.

There are several problems for the non-specialist in the interpretation of the available information. One is dating conventions. The system used by geologists interested in the glacial period is based on radiocarbon dates ($^{14}$C), expressed as years before present (B.P.; where 'present' is A.D. 1950). It is now appreciated that 'this chronology is subject to a number of errors, not least because of long-term fluctuations in the C14 content of the earth's atmosphere'. Interested readers are referred to Edwards and Ralston (1997) for a more detailed discussion of this and related topics. Whatever its shortcomings, this system will be used here for the period up to the arrival of man, when we will switch to the more familiar B.C./A.D.

Another problem is a paucity of information from the west coast, including Assynt, for the archaeological and historic periods, at least until the 18th century. Reasons for this include the small numbers of people that have inhabited the area until comparatively recently, and its relative unimportance from an economic point of view. There has also been a lack of archaeological research, understandable given limited resources, into a part of the Highlands that does not appear to be under threat from development and is still rather remote from centres of learning. Whatever the reasons, inferences must often be drawn from what is known about a wider geographical area, and this has its pitfalls.

## Glaciation

Although there may be deficiencies in the later record, for the glacial and early post-glacial periods we are very well served by the research summarised in contributions to *The Quaternary of Assynt and Coigach* (Lawson 1995). What follows is a brief outline of the chronology of the major events and their possible effect on the local vegetation, largely based on that work.

The last ice sheet that completely covered this part of Scotland built up at at some time after 26000 years B.P. and had completely disappeared by 13000 years B.P., in a period known as the Late Devensian. There is strong evidence to suggest that, although the uppermost parts of Quinag, Canisp and Conival remained free of ice through out this period, the rest of the landscape was covered by a thick, moving ice-sheet. This ice-sheet plucked material from the hills, dropping some of it at their feet, but depositing the rest right across the landscape to the sea. There are trains of erratic boulders, most obviously of Torridonian sandstone, in well-defined bands to the west of the hills, indicating that the main direction of ice flow was to the west or west-north-west. The ice sheet also gouged out lines of weakness in the underlying gneiss, with effects on that part of our landscape that have already been mentioned.

Isolated erratics introduce an element into the local landscape that is of most interest to the lichenologist, since the lichen communities on the boulders are often subtly different from those on adjacent bedrock. Much more significant, so far as higher plants are concerned, is the nature, distribution and thickness of the rest of the material deposited by the ice (glacial till). This debris varies in size from large boulders down to rock flour. In places it was dumped as smooth sheets, in others as hummocky moraine, and much of it was later redistributed by meltwater streams, forming terraces along some of the river valleys. The finer components are the starting point for the formation of soils, and the coarser contribute to drainage, and on steep slopes form areas of scree that may be inaccessible to grazing animals. Glacial deposits are conspicuous in some places, such as the Loanan valley, but elsewhere many have since been covered by peat. The relatively small areas that are free of peat cover, too localised to have been mapped, have developed mineral soils that are highly significant for the later use of the landscape. Such areas occur both inland and, perhaps more commonly, on slopes leading down to the coast, as at the heads of sheltered sea lochs.

For some 2000 years, during a period known as the Lateglacial Interstadial, the landscape remained free of ice, although the area was probably colder than it is now. However, for about a thousand years, between 11000 and 10000 years B.P., there was a return to intensely cold conditions, with the build-up of small glaciers in corries on the north-east side of Quinag, the north side of Glas Bheinn, to the east of Canisp and also of the long ridge from Beinn Uidhe to Breabag. This period is known as the Loch Lomond Readvance or Loch Lomond Stadial. It ended with rapid warming of the climate and for the last 10,000 years the landscape has remained free of ice, although there have been marked fluctuations in temperature and precipitation.

## Establishment and development of vegetation

Evidence of changes in the vegetation cover of this area before and since the Loch Lomond Readvance comes from studies of profiles in loch sediments and peat. These were undertaken in the 1970s by Dr Winifred Pennington and associates, in Loch Sionascaig, just to the south of Assynt, supplemented by studies in Loch Borralan, Cam Loch and Loch Veyatie (see Pennington *et al.* 1972, summarised by Pennington *in* Lawson 1995). Much of the detailed information is provided by the pollen record from Loch Sionascaig.

During the 2000 years prior to the Loch Lomond Readvance, the vegetation developed from first colonisers, such as *Oxyria digyna*, through pioneer species, including willow (probably *S. herbacea*), birch (*B. nana*?), and amongst the herbs, *Artemisia* (*A. norvegica*), *Huperzia selago*, *Thalictrum*, *Rumex* and grasses. It reached its temporary climax in the form of a treeless heath dominated by crowberry and juniper.

The cold spell effectively set the clock back to zero, but at its end, about 10,000 B.P., vegetation rapidly re-established itself. After some 500 years, crowberry and juniper heath was being invaded by tree species of birch, accompanied by smaller quantities of hazel, with evidence of *Calluna* and *Sphagnum*. Pine appeared and gradually increased, and there are the first signs of oak and elm. It has been suggested that the birch/hazel woodlands still found in Assynt and adjacent areas to the south may be a relic of the first woods of this type.

Birch, hazel and ferns dominate the record from about 9000 to 7800 B.P., with pine present, but not yet spread throughout the area. Oak and elm appear to have established themselves locally. From about 7800 to 6250 B.P., there is a marked increase in the representation of pine in the record, with signs of some reduction in birch and hazel.

At about 6250 B.P., alder appears for the first time as a substantial element in the record, and at the same time there is a temporary fall in the incidence of pine, the possible disappearance of oak and an increase in *Calluna* and *Sphagnum*. These changes, together with other evidence, suggest that a deterioration of the climate resulted in increased waterlogging of catchment soils, with the start of the development of peat. At about 5000 B.P. elm disappears from the record, but there are, curiously, no signs of other significant changes in the local vegetation at this time.

There were, however, marked changes rather less than a thousand years later, with evidence, before 4000 B.P., of the steep decline of pine, not just locally, but throughout North-West Scotland. Pine found about a metre below the present surface of the peat at Achiltibuie has been radiocarbon-dated at between 4420 and 4220 B.P. and pine stumps, presumably of a similar age, have been found scattered in the peat throughout the southern part of Assynt.

Following the decline of the pine, there is evidence in the pollen record of a marked increase in *Myrica* and sedges, indicating the undoubted expansion of blanket bog communities from about 4000 B.P. Finally, a peat profile from the south-east corner of Loch Sionascaig (just over three kilometres from the modern boundaries of Assynt) contained the wood of birch dated at about 3500 B.P., together with 'clear pollen evidence for the clearance of birch. alder and pine, and charcoal from that horizon to the surface' (Pennington 1995, p.125). The appearance of charcoal in the peat has been used to support the suggestion that existing woodland was cleared by fire. Apart from any implications this might have for management of a potentially valuable resource, Noble has pointed out, that although pine will burn, mature birch and alder woodland will not, and that the charcoal could in fact be the product of heather burning or other management strategies.

So far, we have not come across a comprehensive study of the composition or relative depths of the local peat deposits, with any implications for the vegetation it bears. Ferreira (1995) has, however, some interesting comments on the considerable depth of the peat over limestone in some areas (see the chapter on the present-day vegetation).

## The arrival and establishment of man: archaeological evidence.

We are now enter the realm of the archaeologist, and for convenience change to their dating convention, the more familar one of B.C./A.D. We should bear in mind, in doing so, that the relationship between this and radiocarbon dating still has to be fully calibrated. Robin Noble has kindly provided us with notes from which much of the following summary is derived.

The first unequivocal evidence of the presence of man in Assynt is provided by substantial stone-built chambered tombs in a corridor stretching along the western edge of the limestone, from Achmore south past Inchnadamph to Altnacealgach and Elphin. These record the existence of Neolithic farming communities, who presumably chose the area for its fertility, which derived mainly from the limestone, but perhaps also because of the ease of cultivation afforded by alluvial gravels in the river valleys. These communities are believed to have been well established by 3000 B.C., and developed during a long period of climatic amelioration. There is, as yet, no evidence that any other part of Assynt was inhabited at this time. Whether this was because it was obviously so much less fertile than the limestone area, or because it was still impenetrably wooded, can only be surmised.

There is, similarly, virtually no evidence so far of any permanent human population in Assynt during the Bronze Age, which coincided in North-West Scotland with a prolonged period of climatic deterioration and the growth of blanket bog. Even if people were present, the change in climate may have rendered unusable some land formerly in cultivation, with a contraction of settlement to the most favourable ground. A flat axe from this period was found at Inchnadamph, but it may only be evidence of trade across the Highlands.

The picture changes completely during the Iron Age, from about 500 B.C., when Celtic incomers from Central Europe settled on the coast of Assynt, which up until then appears to have had, at most, a very sparse population. Evidence of settlement is provided by the imposing broch at Clachtoll, duns at Rubh' an Dunain and Loch Ardbhair and the souterrain at Glenleraig. Besides a number of variations on the idea of fortification, they almost certainly brought with them iron tools, which may have included iron-shod ploughs, allowing them to bring into cultivation heavier ground. From evidence elsewhere in Sutherland, it is thought that they occupied small farming settlements, cultivating suitable areas, but also herding animals on the hill ground and hunting. Their middens also provide evidence of the use of shellfish.

The succeeding Pictish and Early Christian periods (300-800 A.D.) have left little mark on the landscape of Assynt, but there may not have been any major changes in land use, other than the development of the practice of transhumance, i.e. the use of pastures in more remote and often higher ground during the summer months.

It seems reasonable to suggest that the fertile parts of Assynt, which are mainly on the coast and in the limestone corridor, were well settled by the time of the arrival of the Vikings in about 800-900 A.D. They presumably came by sea and most of the evidence for their presence is in coastal place-names: Kirkaig, Stoer, Oldany, Glenleraig and Unapool, but Traligill and Urigill indicate inland settlements as well. Folk legend often blames the Vikings for the destruction of 'the forest', much of which must have happened during the climatic down-turn of the Bronze Age. They also ignore the fact that Vikings needed wood quite as much as anyone else. They were, of course, farmers, and are often credited with the introduction of simple vertical-shafted watermills, a pattern used for centuries, and examples of which still exist, at for example Alltan' abradhan and Glenleraig.

What then might we surmise about the vegetation of Assynt during the period from 500 B.C. to 1000 A.D.? It seems likely that there was further reduction in the area of woodland, particularly where it occurred on level ground, but may there, perhaps, have been the beginnings of the management of this resource? Fertile ground, on the coast and limestone, would have been cultivated wherever feasible, with the arrival or increase of species we now think of as weeds of cultivation. Some of these will have been 'native' species that exploited a new niche, others may have been 'incomers', brought to the area in the seed of crop plants.

Domestic livestock, hay-making and the application of animal manure and seaweed (of which there is evidence elsewhere back to Viking times), may well have enhanced the species-diversity of the grasslands on more fertile soils. Grazing would have had some effect on accessible maritime communities, such as salt-marsh. Montane communities are unlikely to have been much affected. What then of the wet and dry heath that now occupies the greater part of the remainder of the parish? Grazing, at the levels likely, probably had little effect, but any muirburn would have begun the long-term degradation of both habitat and vegetation which has continued to the present day.

## The expanding influence of man: historical evidence.

From about A.D. 1000 there follows a period of some 700 years for which there are increasing amounts of information about land ownership and internecine strife, but precious little about the detail of land management, other than may be inferred from general trends elsewhere in Scotland. We are indebted to Malcolm Bangor-Jones for permission to use extracts from *Historic Assynt* (2000) to provide the framework for a summary of the history of this period.

After the Vikings, Assynt was successively in the hands of the MacNicols of Lewis and the MacLeods of

Lewis, a younger son of the latter family being the first of the MacLeods of Assynt. There followed some 300 years of strife, despite which, by the 17th century 'the MacLeods were actively engaged in selling cattle, the local fisheries and the manufacture of iron using ore dug from the local peat bogs'. The last is perhaps the first evidence for use of the local peat as fuel, although this may have been going on for centuries. Elsewhere in Scotland, large areas of land were regularly stripped of turf, for use as fuel, manure and in construction and there is evidence that turf manuring occurred as early as the 12th century. This may well have happened in Assynt; the local use of turf certainly continued into the 19th century.

By the early 17th century the MacKenzies, 'a land-hungry clan from Ross-shire', had acquired the feudal superiority of Assynt. During the Civil War the MacLeods suffered several raids by the Mackenzies, who, after buying up debts against the estate, in 1672 invaded Assynt. They held the estate until they went bankrupt in 1739, and in 1757 it was bought by the Sutherland family.

Assynt 'was renowned for its cattle and during the 18th century the interior of the parish became dominated by tacksmen (large farmers) who held extensive grazings'. Elsewhere, on the coast 'the population of small tenants expanded as money earned in the herring industry, and the introduction of the potato, encouraged young people to marry and start families'. The first local references to the potato are from the 1760s.

At about this time the lack of detailed information about the landscape and its people is comprehensively remedied. In 1774 John Home undertook for the Sutherland Estate a detailed survey of Assynt. His report and maps were made available to a wider public by R.J. Adam in *John Home's Survey of Assynt* (1960). Robin Adam was well acquainted with Assynt, his father having been Factor to the Estate, based at Lochinver.

The comments that follow on the landscape of the time are taken from Adam's very useful account and analysis of the original, which take into account its purpose and by whom it was commissioned. But first, we should pay tribute to Home's skills as a surveyor. He arrived at Ledbeg on 9th June and left Assynt on 16th September. In the intervening 13 weeks he surveyed on foot, with the help of two assistants and four 'lads' (two of whom failed to stand the pace) some 90,000 acres. Infield, sheiling, 'natural woods' and lochs were all chained, allowing an estimate of their area to the nearest fraction of an acre (measured in roods and falls), and the whole was mapped with a degree of detail and accuracy that allows ready comparison with Ordnance Survey maps made a century or so later.

To diverge briefly from the main theme, his accounts record that he found it necessary to lay out a good deal of money on spirits, sometimes 'for the use of self and Assistants who led the Chain to enable them to endure the fatigue of widing through Lochs and mosses from Morning early till late at night', sometimes also for 'Tenents for showing their Marches as they could not be prevaild with to do so without it' (Adam, p.xxi). He was enthusiastic about some of the natural and man-made features he came across, but daunted by an expedition on 17th July to the top of 'Braeback Hill...a continued cairn of loose stones' where 'the air is so thin and pure there is no possibility of approaching the Top of the Hill except in the heat of Summer'.

His antipathy towards the higher parts of the parish should not detract from our appreciation of the thoroughness of his survey, and the summary of areas measured is very revealing. Of the 90034 acres in all, 2203 were infield (including dwellings) and 1507 sheiling, making 3710 acres of land in any form of intensive use, just 4% of the total area. Adding in 2903 acres of 'natural woods' and 1845 of lochs, the total increases to 8458 acres, just 9.4% of the whole, leaving 81,576 acres of 'Hills, moss and rocky Muirish pasture'. (Adam does not state whether Home was using Scots or English acres, and many of the larger lochs were excluded from the calculations, but the general proportions are unaffected.)

Adam's general comments on these figures are illuminating. 'Assynt was a remote and unspecialised land', its 'population almost exclusively engaged in agriculture'. There was a clear distinction between the 'arable coastal area already heavily populated and pushing up against the limits of its resources' and a 'largely pastoral interior'. The population was estimated at 1718, of which 1175 were on the coastal farms, concentrated especially on the Point of Stoer and along the coast from Drumbeg to Unapool.

There are some clues in the survey as to the way of life of the inhabitants. Home recognized the importance of fishing as an extra source of income on the coast, and 'noticed also three instances of potato cropping there, a portent of the future' (Adam p. xlviii). Adam believed that 'the coastal farms...relied on their arable ground for subsistence and on small stocks of black cattle for the payment of rent, whilst the large inland single tacks specialised in the raising of cattle for the southern markets'. The latter formed 'one of the headstreams of the droving industry'. Once, at Achnacarnin, Home reveals that the idea of sheep-farming was in his mind, but as Adam remarks 'it was to be two decades before the great invasion began in earnest'.

There are passing references in Home's report to the agricultural techniques in use. He describes, for instance, with some incredulity, the use of the cas-chrom (hand plough), and was convinced that horse-ploughing might with advantage be introduced to the coastal farms. It is interesting to reflect, in this context, that the cas-chrom was still in use in the first half of the 20th century, alongside horse ploughing.

Although the woodlands present on the individual farms are described in some detail, there are few clues as to how they were being managed, unless his references to oak 'stools' indicates coppicing or pollarding. There is also a reference to felling, 'All the North and West of the hill above Ledbeg...is cover'd with Birch Wood very much reduced by cutting, the whole of it is green yielding very sweet grassy Pasture'.

Home's survey makes an excellent starting point for any study of the landscape of the parish over the last two centuries. Two aspects of land use have so far received special attention, although much more could be done.

R. Miller, in his paper *Land use by summer sheilings* (1967), included a detailed study of those in Assynt, based on Home's maps. He calculated that there were at least 246 sheiling sites, spread throughout the 42 farms and townships into which the land was divided, of which he examined on the ground some 200. Over three quarters were within reasonable distance (less than two miles) of settlements, and many of those on the coast were under arable cultivation. However, some were over five miles from settlements. Miller had much to say about the apparent fertility of the sheiling areas in Home's time, 'built up under treading, dunging and periodic cultivation', although evidence of the last was sparse on the inland sites. He also reflected on how much this fertility had declined since, with the use of sheiling ground probably already in decline at the time of Home's survey. His paper has done much of the ground work for a possible study of the vegetation of the sheiling areas today.

Robin Noble's recent studies on the woodlands of Assynt (1997, 2000) again make good use of Home's survey, and more on his findings may be found in the chapter on the history of recording.

To return to the late 18th century, it is a happy coincidence that Home's survey took place just seven years after the visit to Assynt by the botanist James Robertson in 1767. Much may be gathered from his account and those of succeeding generations of the botanists about the state of the landscape, especially the more fertile areas on the coast and the limestone. The chapter on the history of recording again provides an introduction to these sources of information.

Another account of the parish at about this time, anecdotal, idiosyncratic and, at times, amusing, is provided by the contribution in 1794 of the Rev. William Mackenzie to the *Old Statistical Account*. He has much to say on the yearly round of activities of the people, especially with regard to their livestock, and some good descriptions of particular areas. Of 'So-ay Island' he says, for instance, 'It is rather flat, though not wholly so. It partly abounds with heather, and sweet grass....Lambs and kids are sent there to be speaned [weaned], and taken thence in two or three weeks. If no other cattle are sent there immediately thereafter, but the pasture preserved until the beginning of November, in this event it will prove sufficient to out-winter eight or nine year olds of the cow kind'.

He is less enthusiastic about 'the extreme marches with...neighbouring landed property', where 'the surface abounds mostly with lofty mountains, hills, extensive tracts of heath, having here and there soft moss and quagmires, which often prove fatal to cows and horses'. He is rather dismissive of woodland, saying, 'There are some birch-trees thinly scattered here and there...all along from Unapool to Oldney' and 'also of the same kind in many different thin plots, far distant from one another'. However, he does acknowledge that 'the whole...though of no great value, are of considerable importance to the parish, as, in time of great storms and falls of snow, every species of cattle [i.e. livestock] resort to them for shelter, nay, they browse on the copse' [plus ça change!]. He has no specific observations on the botany of the parish, other than a passing reference to 'ivy-girt' crags at Stronechrubie, these being an innovation in the much shorter contribution, in 1837, by Rev. Charles Gordon to the *New Statistical Account*, for more details of which see the chapter on the history of recording.

Meanwhile patterns of land holding and the agricultural use of the landscape were about to become the subject of drastic change, for which we return to Bangor-Jones (2000); for a more detailed account see his work on *The Assynt Clearances* (1998). He states that 'in 1803 the Countess of Sutherland's husband succeeded as Marquis of Stafford and inherited the Bridgewater fortune. Guided by their Edinburgh lawyers, the family initiated a development programme for the estate, including the introduction of sheep and the resettlement of people on the coast, where they would be forced by the small size of their holdings to become fishermen'.

The Clearances in Assynt began in 1812 and continued into the early 1820s, during which period 'much of the parish [was reorganised] into five large sheep farms...and over 160 families were evicted, most of whom crowded into the coastal townships'.

The sheep farms covered roughly 60% of the parish and by 1820 mustered 17,000 sheep, the majority Cheviots. This number of sheep must have brought a new intensity of grazing to the hill ground, which continued through much of the 19th century.

Some measure of the extent of the crowding in coastal areas, even prior to the Clearances, is provided by figures for the population on Ru Stoer between 1774 and 1811. The number of households rose from 105 to 169 and of inhabitants from 535 to 856. This and subsequent increases must have resulted in an appreciable intensification of land use in that area.

Bangor-Jones continues, 'the estate came to exercise a very tight control over people's lives. The cultivation of holdings was encouraged, as was the building of new houses. But...the small tenants became dangerously dependant upon the potato. When that failed in the late 1840s, there was great suffering. The estate provided relief schemes, assisted emigration and reorganised the runrig townships into crofts. Recovery came in the 1860s and 1870s when slated houses began to be built by the crofters'.

Meanwhile, sheep farming had been through a difficult period after the Napoleonic Wars, but had 'started to recover in the 1830s, bringing investment in houses for farmers and shepherds, fanks and parks'. One of the forms this investment took locally was money put into planting trees on estate land, mainly around Lochinver. There is, for example, a list of nursery trees supplied to Lochinver in 1850 (Bangor-Jones, pers.comm.), which includes the following: 20,000 larch (at 8/- per 1000), 50,000 'Scotch fir' [*Pinus sylvestris*?], 10,000 spruce, and, more intriguingly, 2,000 poplar, 3,000 oak, 6,000 elm, 6,000 ash, 9,000 birch [which species?], 4,000 'sycamore or plane', 1,000 lime, 5,000 beech, 4,000 alder and 2,000 silver fir. Some of the results of these and subsequent plantings can still be seen in the mixed woodlands at Culag and beside the River Inver, but where were the ash, elm and alder set out?

'The mid-Victorian boom eventually gave way to a severe depression and from the 1880s much of the parish was made over to deer forest or stalking ground. In 1913 Assynt was sold by the Duke of Sutherland'. By an intriguing twist of fortune the purchaser was John William Stewart, later Major-General Stewart, who had been born on a croft at Nedd in 1862 and had made his fortune building railways in Canada.

The population of Assynt rose steadily from the later part of the 18th century, from 1718 in 1775, to 2395 in 1801, 2479 in 1811, 3161 in 1831, and then levelled off, being only 3178 in 1861. In the light of these figures the 19th century has been referred to as 'the high tide of settlement and land use' in Assynt. Thereafter the population began to decline and has probably now stabilised at just over 1000.

For some centuries, therefore, most of the fertile and accessible parts of Assynt now in crofting tenure were cultivated, if not continuously, then at least in rotation. When the sun is low in spring and autumn, old cultivation ridges may be seen in what is now grassland; elsewhere they have been overtaken by bracken or even woodland. Other areas were cut for hay, and well into the 20th century every house in some townships had a byre and a barn for livestock, which included sheep, cattle and horses. Most crofters had a 'tattie' patch and some grew other vegetables. In places small walled enclosures protected a clump of rhubarb and soft fruit bushes.

During the last fifty years or so, there have been substantial changes in the landscape, particularly in the crofting townships. Cultivation, even on the scale of a small-holding, is now rare, vegetable growing is uncommon, and the only stock still kept in numbers are sheep, with a few cattle. To an older generation, much of the modern landscape has the appearance of dereliction, particularly where it has colonised with birch and willow scrub.

Sheep numbers are dropping overall in crofting areas, though not in all townships, and none of the other Assynt estates now have any livestock (with the exception of a few kept by keepers). In some townships there is now a tendency to keep the sheep on the in-bye and around the villages for much of the year, rather than putting them out to the hill ground, perhaps because of a shortage of manpower to herd and gather them. Where the in-bye is wooded, particularly with hazel that was formerly coppiced, this change in practice must in the long term threaten the regeneration of the woodland. In other areas, such as the Stoer peninsula, the effects of over-grazing are evident, and the destruction wrought by muir-burn is still obvious in many places.

Crofter and community forestry schemes are introducing further changes, especially on the hill ground, where substantial areas have been deer-fenced and planted. If the acknowledged increases in the numbers of red deer over the last fifty years are also taken into account, there is now well-founded concern over the population of deer that the land can carry without degradation; substantial reductions in numbers have been advocated.

It will be interesting to see what effect these and other changes in land use have on the plant life of the parish in the future. Survey work for this Flora has provided detailed information on the distribution patterns of individual species, complementing Ferreira's vegetation survey (1995). In some cases we have suggested historical (or other) factors that may have helped determine these patterns. Any future study might most effectively concentrate on a much smaller area than the whole parish, perhaps one where there have been contrasting histories of management over the last couple of centuries (i.e. crofting and non-crofting). It will require an inter-disciplinary approach and could be fascinating.

# VEGETATION

## Introduction

The purpose of this chapter is to describe the major plant communities represented in Assynt, highlighting species and localities of interest.

We have not used the terminology of the National Vegetation Classification, for two main reasons. Firstly, neither of us is versed in its methodology. Secondly, and more important, the vegetation of the whole of North West Sutherland, including Assynt, has been the subject of an extremely detailed survey by R.E.C. Ferreira, to which further reference is made in the chapter on the history of recording. His report (1995) uses the N.V.C. terminology and may be consulted for more information.

Our appreciation of the character of local plant communities has been acquired almost subliminally in the course of the flora survey. However, we have tried to bring some order to our impressions by consulting the relevant parts of Ferreira's report. We have made extensive use of his introductory sections and site descriptions, especially in regard to the geobotanical context of communities. This brief account can only summarise the diversity and subtlety of the communities present; for more, come and see for yourself!

The communities have been grouped as follows: coast and islands; woodland and scrub; heath, mire and crags; freshwater; limestone; upland; man-made or influenced. To preserve some sense of location in each category, notes on specific features or localities are ordered from Inverkirkaig in the south-west, north to the Point of Stoer, east to Unapool, and then south along the line of the limestone to Knockan in the south-east. Others, from the centre of the parish, are slotted in where they fit best.

Before getting into detail, there are some general considerations which apply to many or all of the categories. First is the influence of our oceanic climate, with relatively equable temperatures and high rainfall. As a consequence, species largely confined to woodland further to the south and east flourish here in the open. Second is our northern latitude which, contrariwise, means that species widespread further south may be confined to our coasts. Third is the mosaic character of our vegetation, particularly on the Lewisian gneiss, where there has developed over time a landscape of frequently changing altitude, mineral richness, soil type and depth, slope and aspect. Fourth are the often subtle but nevertheless pervasive effects of the activities of man over 5,000 years or more. This is most obvious on the fertile ground of the coast and limestone areas, but may be found almost everywhere.

## Coast and islands

### Rocky shores

From Loch Kirkaig round to Lochinver, and from there north to Clachtoll, the coast, which is on Lewisian gneiss, is low and exposed. Here maritime plant communities may be compressed into a narrow band between high tide level and the lower edge of the coastal heathland. In the most exposed areas *Armeria maritima* is almost the only constant species. However, in more sheltered situations, to the leeward of headlands and in bays, a more luxuriant vegetation is developed. In crevices at the lowest level are the leathery fronds of *Asplenium marinum*, and above this, on shelves and in crevices, the bright green of *Ligusticum scoticum* and the fleshy rosettes of *Sedum rosea*, the last often abundant. Less common are the white-flowered tussocks of *Tripleurospermum maritimum* and *Silene uniflora*.

Higher, and where there is a greater depth of soil, a grassy sward develops, dominated by *Festuca rubra* and *Agrostis stolonifera*, often with *Plantago maritima* in quantity. Sheltered hollows support tall herb vegetation containing a wider variety of grasses and species such as *Angelica sylvestris*, *Luzula sylvatica* and often much *Primula vulgaris*. Flushes off mineral-rich parts of the gneiss are characterised by the grey tussocks of *Schoenus nigricans*. Stony ground on south-facing slopes has an open form of cliff grassland with much *Thymus polytrichus*, and fringing the uppermost edge of the cliffs there may be a form of maritime heath with *Empetrum nigrum* and the conspicuous silky leaves of *Salix repens*. *Sedum anglicum* is a frequent component of this community where the soil is shallow.

At Clachtoll the character of the rocky parts of coast begins to change as Torridonian sandstones take over from the gneiss. The dipping sandstones provide low cliffs from the Bay of Stoer north to Stoer Lighthouse, where there is a marked increase in height and steepness through to the Point of Stoer. Here the coast changes direction, the cliffs running south and then east to Rubh' an Dunain, this section being sheltered from the prevailing winds.

The communities of the lower sandstone cliffs are not markedly different from those on the gneiss, but where higher there are changes, although the cliffs are so steep that they can only be botanised through binoculars! In the vicinity of sea-bird colonies there occurs a nitrogen-loving community with some 'ruderal' elements. *Cochlearia officinalis* may be abundant, as also *Stellaria media* and species of *Atriplex*. Nearer the top of the cliffs, where the fulmars nest, there is lush tall herb vegetation dominated by *Rumex acetosa* and *Silene dioica*, with *Luzula sylvatica* locally abundant. At Geodh' Dearg, about half way between Point of Stoer and Rubh' an Dunain, it is possible to climb down a steep gulley to the base of the cliffs. In the shelter of this gulley a whole range of cliff communities is developed, including vertical *Salix aurita* scrub, faces covered by *Hedera helix*, and in season the conspicuous flowers of *Lonicera periclymenum*.

At Port Achnacarnin the Torridonian gives way again to gneiss, the shelter afforded by this tiny bay allowing the development of a narrow band of scrub woodland just below the cliff-top, containing birch, rowan, hazel and aspen, and a ground flora of *Allium ursinum, Hyacinthoides non-scriptus* and *Stellaria holostea*. From this point eastwards, the gneiss provides a low, indented coastline all the way to the parish boundary on Loch Glencoul. It is rugged in places, with occasional cliffs of some height, as at Culkein Drumbeg.

One species found on these rocky coasts is quite restricted in its national as well as its local distribution. *Vicia orobus* displays its pink trusses of flowers on the seacliffs from Achmelvich round to Oldany Island, and also on crags up to three km. inland. In just one place it is accompanied by *Vicia sylvatica*, which is found locally at only one other site. Another species that extends, to our initial surprise, well over two km. inland, is *Asplenium marinum*, though only on high crags, hoared over with the grey maritime lichen *Ramalina siliquosa*, which catch the full force of the south-westerlies.

At Cìrean Geardail, about a kilometre south-west of the Old Man of Stoer, a small population of the tiny twin-stemmed *Ophioglossum azoricum* occurs in short cliff-top turf dominated by *Plantago coronopus*. Just west of Rubh' an Dunain, in the short open grassy turf of cliff-top gulleys, is where we first found the tiny plants of *Anagallis minima*. We now know that it occurs widely, especially on the north coast of Assynt, but never far from the sea. An unexpected component of the cliff ledge flora nearby is *Orchis mascula*. The short sward on Rubh' an Dunain itself is a good example of vegetation shaped by a combination of exposure and heavy grazing; in late summer *Calluna vulgaris* and *Succisa pratensis* flower here at a height of only a few centimetres.

Extreme exposure is presumably also the factor that allows the tiny, normally montane, willow *Salix herbacea* to exist only 15 m. above sea level at Rubh' Dhubhard between Drumbeg and Nedd, apparently its lowest altitude on the British mainland.

Localised accumulations of shingle, sometimes quite sandy in composition, are found in small bays along these rocky coasts and also on islands such as Soyea. Characteristic of these are species of *Atriplex, Galium aparine, Potentilla anserina, Rumex crispus* and, rarely, as at Port Alltan na Bradhan, *Sonchus arvensis*. On a Torridonian storm beach at Clachtoll is our one remaining locality for *Mertensia maritima*, the few plants protected from grazing by wire mesh.

**Dunes and machair**

Assynt cannot rival in its dunes and associated habitats the huge systems elsewhere in West Sutherland, but those we have nevertheless contribute greatly to local diversity.

At Achmelvich, increasing erosion, by a combination of human pressures and great storms, as in 1983, led in 1984 to a substantial restoration scheme. Both here and at Clachtoll areas were stabilised and replanted with imported marram. There are smaller areas of dune and/or machair grassland enriched by shell sand at Stoer, Clashnessie and at Mol Bàn on the east side of Oldany Island.

The general pattern of the vegetation is similar in all of these places, but each has its own specialities. There is a sparse strandline community of *Atriplex* species, *Cakile maritima* and *Honkenya peploides*, moving up into dune proper where the sand is stabilised by *Ammophila arenaria*, fronted at Achmelvich by an impressive stand of *Leymus arenarius*. The latter species only otherwise occurs on rock, at Clachtoll and Culkein Drumbeg. *Elytrigia juncea* is sparse, but is found from Achmelvich round to Oldany Island.

In most places the areas dominated by *Ammophila* give way to semi-fixed dunes that are intensively used and grazed, and the most conspicuous elements of their vegetation are tough perennial herbs such as *Centaurea nigra* and *Heracleum sphondylium*, accompanied in places by *Daucus carota* and *Thalictrum minus*. In less stable areas, the running rhizomes of *Carex arenaria* may be seen.

At Achmelvich, Clachtoll and Stoer, these semi-fixed dunes grade almost imperceptibly into a heavily-grazed machair grassland, with *Galium verum* and *Lotus corniculatus* and, in fenced areas, *Crepis capillaris*. *Sedum acre* occurs in the sandy turf at Achmelvich and Stoer, and *Erodium cicutarium* at Stoer. *Ranunculus bulbosus*, which Ferreira mentions as a characteristic species of the Achmelvich grassland, only appears to have survived the 1980s restoration in one tiny area alongside a track, and that is the only place where it has been seen in Assynt. An unexpected discovery at Clachtoll in 1993 was *Astragalus danicus*, in an area of turf alongside the access road, to which it may have been introduced with *Ammophila* in the course of restoration work.

Where the shell-rich sand has blown up on to the surface of gneiss or Torridonian outcrops a rich community may develop, as at Achmelvich, although here it is, again, heavily grazed. Three orchids occur, *Coeloglossum viride*, *Listera ovata* and *Orchis mascula*, with *Ophioglossum vulgatum*, and often the purple flowers of *Gentaniella campestris*. Again at Achmelvich, in just such an area, alongside a lochan, is our only stand of *Scilla verna*.

Dune slacks, as such, hardly exist locally, but *Catabrosa aquatica* does occur in wet areas at Achmelvich and Clachtoll and sometimes invades the strandline at the latter locality. Two other uncommon species of the wetter areas of the machair, at Clachtoll and Stoer respectively, are *Sagina nodosa* and the brick-red orchid *Dactylorhiza incarnata* ssp. *coccinea*.

**Saltmarsh**

Small areas of saltmarsh are scattered along the coast in sheltered bays and sea-lochs, from Inverkirkaig to Unapool, the best developed examples being at Baddidarach, Lochan na Leobaig (east of Oldany Island) and around Lochs Nedd and Ardbhair.

However, we should perhaps first mention here our one wholly marine species, *Zostera marina*. Records date back to the mid-1950s, when it was found somewhere in the vicinity of Oldany Island. Washed-up leaves and rhizomes have been found by us on beaches from Achmelvich round to Port Dhrombaig, but we had not managed to locate a living stand of this elusive plant until late in 2001, when we were shown material collected just offshore at the last-named locality. The sources of the other material remain to be discovered.

Until recently our one brackish 'pondweed', *Ruppia maritima*, was only known from one locality at the head of Loch Roe. However it has now been found in quantity in the brackish loch at Duart.

The most salt-tolerant of the saltmarsh grasses, *Puccinellia maritima*, has also presented us with some difficulties locally, since the grazing pressure on our relatively small areas of such habitat means that it very rarely flowers. We have now learned to recognise it by its vegetative characters. Although almost certainly under-recorded by us, Ferreira indicates that it is the dominant species in the local saltmarsh turf, accompanied by varying amounts of *Armeria maritima*, *Glaux maritima* and *Plantago maritima*. The close-grazed turf of the *Puccinellia* zone in more sheltered bays, such as Lochan na Leobaig, has in places a golden-brown tint deriving from a diminutive fucoid seaweed referred to as *Fucus muscoides*. Two species of sedge are found at this level on the shore; *Carex viridula* ssp. *viridula* is widespread, but *Carex extensa* less so, and often best seen on the edge of runnels.

At slightly higher levels *Agrostis stolonifera* and *Festuca rubra* take over from *Puccinellia*, and *Triglochin maritima* is frequently found at this level. Higher still, the brackish grassland may have appreciable amounts of *Leontodon autumnalis*, a surprisingly adaptable species. *Hydrocotyle vulgaris* is a frequent component of wetter coastal grassland, although it also occurs for some distance inland.

Where the shore is stony, the composition of the vegetation on the upper parts of the shore may be slightly different, with *Juncus gerardii*, often accompanied by *Glaux maritima* and, less often by *Blysmus rufus* and *Eleocharis uniglumis*. *Cochlearia scotica* is also found in this sort of situation at Baddidarach and elsewhere.

Finally, in sheltered bays above normal high tide level, but where the wrack is piled up by storms, there develops a tall marsh community dominated by *Iris pseudacorus*, accompanied by *Urtica dioica*. Both of these are, of course, found further inland, but several species which grow amongst them are not, or only rarely. One is *Scutellaria galericulata*, which also occurs at the upper levels of stable shingle; a second is *Lycopus europaeus*, of which there are some fine stands at Baddidarach. Also found amongst the *Iris*, but elsewhere in crevices on rocky shores as far north as Stoer Lighthouse, is the large umbellifer *Oenanthe crocata*, here at its northernmost site in the British Isles. All three species are found far inland further south and their maritime preferences here may be more to do with January isotherms than saltwater.

**Islands**

None of the small islands off the coast of Assynt have ever been inhabited, since they are lacking in both shelter and fresh water. However, many were grazed until quite recently. Nowadays, their most conspicuous inhabitants are gulls and geese. Soyea is the largest and has a good range of communities, from stable shingle to dense *Calluna* and *Salix aurita* scrub. A large flat area in the centre of the island is completely dominated by watercress (we cannot be more precise, since on our visits none of it has been in flower or fruit); we suspect that roosting geese are responsible for its luxuriance.

Soyea also has the local distinction of at least two populations of the robust sedge *Carex otrubae*, growing at the edge of small brackish pools well within the splash zone. This is a rare and exclusively coastal species in the north of Scotland, and is known to occur in similar situations at Duartbeg, to the north of Assynt, and on Eilean Mór, just to the south, in West Ross. Of the other islands, Eilean Chrona has a reasonably diverse flora for its size, but nothing out of the ordinary.

The smallest island visited, because it was the only land in its tetrad, was Sgeir Liath, west of the mouth of Loch Nedd; it yielded just eight species of flowering plants, but, happily, one uncommon maritime moss, *Tortella flavovirens*.

Oldany, our largest island, is so close to the mainland that it can be accessed on foot at very low tides. It has a full suite of habitats, including a good range of maritime communities, both along its most exposed north-west coast and associated with the sand spit at Mol Bàn on its sheltered eastern side.

## Woodland and scrub

Most of Assynt's natural woodland is confined to a strip within five km. of the coast, with outliers along Loch Assynt and on the south side of Loch Urigill. However, there are small areas of woodland scattered across the rest of the parish, often only narrow strips along ravines or groups of trees around waterfalls. These scraps have a botanical significance out of all proportion to their size.

Two main factors appear to have determined where woodland is found today. One is exposure, especially to salt-laden south-westerly gales; most of our woodlands are in the shelter of valleys or ravines, or on the north sides of ridges or crags (with the apparent exception of some of our birch/oak woodland, see below). The other is people. The coastal strip contains most of the fertile ground in the parish, and, predictably, is also the most densely populated. All available level ground has in the past been used for growing crops. As a result, woodland is virtually confined to steep or rocky ground.

The general pattern of woodland in the parish seems to have been established by the time of Home's survey in 1774 (Adam 1960), perhaps because the remaining woodlands were recognised as a valuable resource and, until recently, managed as such. The state of our existing woodlands varies greatly. Some are now showing serious signs of senescence, others are regenerating, and the interested reader is referred to the recent study by Noble (2000) for a more detailed account of this topic.

However, few woodlands are fenced, and browsing by domestic stock and deer severely restricts regeneration, as does poorly-managed muirburn. This is demonstrated in those areas where neither grazing nor muirburn occur. Vertical crags are not ideal for the establishment of trees, but they support significant numbers, and the larger islands in lochs are often conspicuously wooded.

Planted woodland is described in the section on 'man-made' habitats; the comments that follow relate solely to natural woodlands.

### Geological factors

The structure and composition of our woodlands are largely determined by properties of the underlying rocks, in particular the degree of base-richness. The great majority are situated on the gneiss, which in Assynt, may be acidic, basic or intermediate in character, although as Ferreira points out, it is generally more basic than the gneiss further to the north.

So far as other rock types are concerned, the generally acidic Torridonian sandstones do not support much woodland locally. On the Stoer peninsula they are too exposed and intensively grazed to have anything other than occasional patches of scrub on inaccessible cliff faces, and where they occur inland, altitude and exposure both militate against woodland. The only exceptions are small areas on the south side of Cam Loch and north side of Loch Veyatie.

There are few local woodlands on Cambrian rocks, but they are of considerable interest. The largest, An Coimhleum, is situated on a sheltered north-facing slope composed of quartzite, with sills of Canisp porphyry exposed at its upper edge. It is generally acidic, but there is area of strongly basic springs in the middle of the woodland. The wooded ravine on the Allt a' Phollain in Cromalt is cut through acidic quartzite.

In contrast, the wooded ravines of the lower part of the Allt a'Chalda Mor, Allt Poll an Droighinn and Traligill River, at Inchnadamph, are cut through the Cambrian limestones, providing a substrate that is highly basic. The hanging woodland at Liath Bhad is on Fucoid Beds and limestone. The only extensive area of woodland on the limestone further south is Doire Dhubh, at the western end of Loch Urigill. There is only one tiny ravine woodland on the acidic rocks of the Moine Schists, on the upper reaches of the Crom Allt.

To give some coherence to this account of the vegetation of our woodlands we have, following Ferreira, described the general character of those on acidic substrates, which are the more numerous, and then the variations introduced by degrees of base-richness. It should be emphasised that, on the gneiss in particular, these variations in substrate may be very local, resulting in a bewildering, but very interesting, mosaic of species with differing requirements and tolerances. Canopy and ground layer have been treated separately, followed by some notes on particular woodlands.

### Canopy

The dominant tree species in almost all our woodlands is downy birch, accompanied by smaller quantities of rowan; the latter occurs as single trees on sheltered crags at higher altitudes than the birch. Silver birch is a rare planted tree in Assynt.

On acidic substrates, birch/rowan woodland may include aspen on the higher ground and hazel on the lower. Oak is also a significant component in some coastal areas and in a few places on the north side of Loch Assynt. Much of the birch/oak woodland occupies warm south-facing slopes, with particularly good examples at Baddidarach, where oaks are numerous, and at Pollachapuill. The majority of local oaks, incidentally, have characters intermediate between those of sessile and pedunculate, and they are also very heterogenous. The woodlands on the larger loch islands are also mostly acidic, with birch/rowan merging in more exposed parts into dry heath, in which large junipers may be a conspicuous feature.

On base-rich substrates there may be a much more varied canopy. Although still essentially birch/rowan or birch/hazel woodland, at times the hazel may predominate, as at Inverkirkaig and Nedd. Aspen may occur in quantity, and it is also widespread on individual crags and on the rocky margins of some sea lochs.

In places the birch/rowan/hazel/aspen woodland is further diversified by individual specimens of wych elm, goat willow and bird cherry. The best examples of this mixed woodland are on the gneiss in the vicinity of Achmelvich, Drumbeg and Nedd. Elsewhere it occurs in smaller amounts in ravines, such as those cutting through the gneiss on the Oldany River and at Creag an Spardain, and through the limestone on the Allt Poll an Droighinn and the Traligill River.

The status of ash, which might be expected to occur in such base-rich woodlands, is problematic in Assynt. It is present and regenerating freely on the Culag River and from Torbreck north to Rhicarn, but only occurs sparingly elsewhere. In contrast, holly is widespread, but usually as single trees, often on inaccessible crags, which may reflect its particular susceptibilty to browsing.

In wet hollows in woods, and on the level areas at their base, there is often swampy woodland dominated by downy birch and either common or eared willow. Willow scrub, mainly of the lower-growing eared willow, also occurs around the margins of lochs and in sheltered areas on some offshore islands. Alder is mainly confined to the margins of watercourses, but also occurs on some loch islands.

**Ground layer**

Firstly, a word of warning, which relates to our oceanic climate. As has already been outlined, species confined to woodland further south in the British Isles frequently occur here outside it, in boulder scree, on shady north-facing slopes, or under bracken and other tall herbs. Examples include *Oxalis acetosella*, almost ubiquitous wherever there is some shelter, *Anemone nemorosa*, frequently found under bracken and heather, and *Sanicula europaea*, which may occur in isolated areas of boulder scree. The latter two species may also indicate the former presence of woodland where they are now found. *Hyacinthoides non-scripta* may also occur in woodland, often in areas dominated by *Holcus mollis*, but also grows luxuriantly in the tall herb vegetation on sheltered ledges on parts of sea cliffs or inland crags that are inaccessible to grazing animals.

Likewise, luxuriant bryophyte communities occur both inside our woodlands, and elsewhere where there is a humid microclimate, even under dwarf shrub heath high on the hills. A frequent associate of the bryophytes in these situations is the tiny fern *Hymenophyllum wilsonii*.

The ground layer of the acidic woodlands varies with their slope and rockiness, but is often very species-poor. Higher areas are often dominated by *Vaccinium myrtillus* and *Deschampsia flexuosa*, with frequent *Oxalis acetosella*. Lower down, especially in the more humid woodlands, there may be fine bryophyte communities, dominated in places by hypnoid mosses, elsewhere by large oceanic liverworts. In broken ground, or on ledges inaccessible to grazing animals, a tall herb community develops, often dominated by *Teucrium scorodonia*, *Luzula sylvatica* or the larger ferns. Here also may be found some of the less common ferns, such *Dryopteris aemula*, *D. expansa* and *Gymnocarpium dryopteris*. Flushes with *Phegopteris connectilis* are signs of some slight base enrichment.

The ground layer of base-rich woodlands is, like the canopy, rather richer in species, with different communities occupying particular niches. Grazed slopes have *Ranunculus acris*, *Lysimachia nemorum* and, conspicuous in the late spring, *Primula vulgaris*. Strongly flushed areas, particularly where they are protected from grazing by boulder scree, may be occupied by sheets of *Allium ursinum*, with *Galium odoratum*, *Geum rivale* and *Sanicula europaea*. In drier more open situations, often on south-facing slopes, *Fragaria vesca* may become conspicuous, with the much rarer *Potentilla sterilis*. *Melica uniflora* is found on the basic and ultrabasic dykes which outcrop in some woodlands.

Ungrazed rocky ledges in such base-rich woodlands, whether on the gneiss or limestone, are often dominated by tall grasses such as *Brachypodium sylvaticum* or *Elymus caninus*. In wetter situations a tall herb community develops, with *Filipendula ulmaria*, *Angelica sylvestris*, *Geum rivale*, *Valeriana officinalis* and, occasionally, conspicuous patches of *Cirsium heterophyllum*. Where small burns cut through the crags the 'spiky' leaves of *Crepis paludosa* are often conspicuous.

**Areas of special interest**

All the woodlands from the Kirkaig valley round to Unapool are situated on the gneiss, and the examples that follow give some indication of the contribution that it makes to the local woodland flora.

In the Kirkaig woodlands there are small stands of *Epipactis helleborine* (most visible in the conifer plantings on the drive to Achins Bookshop), and there is also an old record for *Cephalanthera longifolia*.

The scattered woodlands stretching from Torbreck to Rhicarn and west to Ardroe and Achmelvich are probably the richest in the parish, with a high proportion of aspen and some very old hazel coppice stools. Along the roadside between Achadhantuir and Feadan there is a large population of *Cephalanthera longifolia*, here at its northernmost station in the British Isles, together with smaller quantities of *Epipactis helleborine*. Not far away, under hazel on the south side of Loch Dubh, is our only locality for the saprophytic orchid *Neottia nidus-avis*.

The next major area of woodland stretches from Oldany to Loch Dhrombaig. This contains much aspen and hazel, but the main interest is the considerable age of some of the trees, including ancient oak coppice stools at Pollachapuill and some huge goat and grey willows, some of them ancient pollards. *Ceratocapnos claviculata* occurs at the edge of the Drumbeg woodland and again in the Nedd woodlands.

The Nedd woodlands are largely on croft in-bye and contain substantial stands of hazel, much of it as large coppice stools, some huge specimens of grey willow, scattered oaks and almost as many specimens of bird cherry as there are in all the rest of the parish. There is a rich basiphile ground flora, especially in areas of large boulder scree. Their continuation up the valley of the Leireag River is noted for its riverine alders and is one of only two locations for *Carex remota*.

Further woodland extends, patchily, along the coast from Gleann Leireag to Unapool, often in ravines. It is mainly on the acidic side, with notable exceptions, such as at Creag na Spardain, where there is a fine mixed canopy together with *Festuca altissima* at its northernmost locality in the British Isles.

We now move on to a few woodlands of note on substrates other than the gneiss, bearing in mind that the woodland at Liath Bhad and the wooded ravines on the limestone are covered in the section on limestone.

An Coimhleum, on quartzite and Canisp porphyry, is a acid birchwood chiefly noteworthy for its survival in part of the parish otherwise very poorly wooded, and for its bryophytes, but the crags along its upper edge are an isolated locality for *Dryopteris expansa*.

The largest of the wooded islands in Cam Loch, Eilean na Gartaig, is a unique study in contrasts. The island is underlain by quartzite and the upper parts are a good example of ungrazed acidic woodland. Round the edge of the island, however, is a flood zone extending up to two metres above the normal summer levels of the loch, whose feeder rivers drain off limestone. This flood zone is occupied by birch/rowan woodland containing goat, grey and eared willow, beneath which there is a remarkable stand of *Allium ursinum*, accompanied by *Scrophularia nodosa*, *Orchis mascula* and *Ranunculus auricomus*, the last found in only one other locality in West Sutherland.

A tiny crag woodland on Torridonian sandstones at the edge of the western end of Cam Loch is the only locality for *Orthilia secunda*.

The extensive but elderly birch woodland on the limestone at Doire Dhubh is disappointingly poor in woodland species, probably because the limestone is here covered by a thick layer of peat. Finally, the three wooded ravines not far away in the Cromalt area are attractive, but rather acidic, noteworthy mostly for their bryophytes.

## Heath, mire and crags

The coast road from Lochinver to Unapool provides a constantly changing panorama of rocky landscapes, set with small lochs. Outwith the crofting settlements, with their close-grazed in-bye, these 'cnoc and loch' landscapes are a patchwork of colour, particularly in the late summer.

The heath and mire vegetation which makes up most of this patchwork changes abruptly with variations in aspect or slope, and is everywhere interrupted by outcrops, crags and screes of the Lewisian gneiss which underlies the area. Road cuttings reveal great variation in the gneiss itself, from a hard pinky quartz-rich type, through the typical banded form, to dark, often crumbling base-rich mineral layers. The same cuttings also show that, between the outcrops, the underlying soil-making material may be glacial till or peat, the latter often lying on the former.

Looking a little closer, we begin to notice in this apparently 'natural' landscape almost ubiquitous signs of the activities of man. Close to the road, and often also far from it, peat has been dug for centuries. Blackened stems of heather, or bleached *Sphagnum* hummocks mark areas of recent muirburn. Old walls snake across the hillsides, delineating the boundaries of ancient sheilings. More subtle are the signs of grazing; although sheep are now virtually confined to areas under crofting tenure, there is still a large red deer population.

All of these factors contribute to a small-scale mosaic of vegetation types that it is fascinating to explore, but whose character is very difficult to summarise. Perhaps the best way to give something of its flavour is to describe a short walk, in late summer, through an area not far from the footpath that runs up Gleann Leireag. It lies at an altitude of 70-140m. and was, until the early 19th century, part of the hill ground of the settlement at Glenleraig, after which it was farmed with sheep until about thirty years ago.

At first sight the rolling peaty ground appears to be covered by a rather monotonous carpet of wet heath, dominated by *Trichophorum cespitosum*, *Calluna vulgaris* and *Molinia caerulea*, with locally frequent *Myrica gale* and scattered *Erica tetralix* and *Narthecium ossifragum*. Shallow channels and seepage areas are almost solid *Molinia*, with scattered *Potentilla erecta* and *Succisa pratensis*. Steeper slopes are covered by dense *Calluna*, with occasional patches of *Erica cinerea*.

Inside the wall of an old sheiling, abandoned as such nearly two centuries ago, there is grassland composed of *Agrostis* spp., *Holcus lanatus* and *Anthoxanthum odoratum*, with scattered fronds of *Pteridium aquilinum*. On closer inspection, this grassland proves to have quite a varied herb content, including *Galium saxatile*, *Plantago lanceolata*, *Prunella vulgaris*, *Trifolium repens* and *Viola riviniana*. This is the community we call 'sweet' grassland.

In the boggy bottom of a flat valley nearby the vegetation reverts to *Trichophorum* and *Calluna*, but open runnels are edged with *Schoenus nigricans* and *Rhynchospora alba*, with the shiny leaves of *Potamogeton polygonifolius* in the water itself. Below a gneiss outcrop at the edge of this valley there is small flush with a stand of *Eleocharis quinqueflora*, occasional plants of *Carex hostiana* and *Eriophorum latifolium*, and the starfish-like rosettes of *Pinguicula vulgaris*.

Climbing up through wet heath and over a saddle, we drop down a gentle well-drained slope beneath a higher line of outcrops. Here, amongst the *Calluna* and *Erica cinerea*, there are more patches of sweet grassland matted by hypnoid mosses, with damper areas marked out by *Achillea ptarmica*, *Alchemilla vestita*, *Euphrasia* sp., *Leontodon autumnalis* and *Ranunculus acris*.

Contouring across to the nearest crag, we find shelves occupied by *Teucrium scorodonia* and little else except a few plants of *Primula vulgaris*, but running along a tight crevice under an overhang are the whorled leaves of *Galium odoratum*. In gritty grassland just at the base of the crag there is *Hypochaeris radicata* and *Thymus polytrichus*. With the *Galium*, they are indications that the gneiss of the crag, although superficially rather dour-looking, is in fact quite base-rich.

The scree immediately below the crag has two of the usual ferns, *Polypodium vulgare* and *Oreopteris limbosperma* and tucked into a damper area the neat fronds of *Phegopteris connectilis*. The base-rich character of the gneiss is confirmed by the tall herb community in the flushed lower parts of the scree, where there is *Filipendula ulmaria*, *Geum rivale*, *Lysimachia nemorum* and *Trollius europaeus*.

An open flush at the base of the scree has the tiny pink flowers of *Pinguicula lusitanica* and nearby some straw-coloured shoots of *Selaginella selaginoides*. Above us, on another, small, creviced crag, is a tightly-hugging plant of *Hedera helix* and in the rocks beneath it one scraggy sapling of *Sorbus aucuparia*. As we make our way down into the river valley below we begin to notice small bushes of *Salix aurita* that have so far escaped grazing amongst the tall *Calluna*. Alongside the river, flat areas in its flood plain are dominated by tussocky *Molinia* interspersed with the almost ubiquitous *Calluna*.

Our initial impressions of this area might have been of colourful, but run-of-the-mill wet heath and mire communities in the damper areas, with dry heath and grassland on the better drained slopes. However, the base-rich gneiss outcrops and the water draining off them add substantially to the range of communities and species. And this small area is only a sample of the overall variation and interest.

Two particular species should be mentioned here, because of the impact they can make on the landscape. Bracken, *Pteridium aquilinum*, is a frequent component of dry heath, either as scattered fronds, or forming small patches in sheltered hollows. However, on well-drained mineral soils it can dominate large areas. Areas of such soils have been prime candidates in the past for enclosure and cultivation, either as sheilings up in the hills or as croft in-bye. In the past the bracken was kept in check by cutting, livestock and ploughing. Now there are few checks to its growth and spread, although much of it does not produce spores. In its favour, it does provide, when not too dense, shelter for a number of smaller herbs.

The other species is gorse, *Ulex europaeus*, which dominates large areas of south-facing slopes in, for instance, the Kirkaig valley, and Glen Canisp. There is some suggestion that it is an introduction, originally brought in as browse, and the young shoots are certainly eaten by both sheep and deer. Its thickets may have glades into which grazing animals cannot penetrate, but its main value is to the smaller animal life that it shelters.

**Heath communities: grazing and muirburn**

Heath communities have been materially affected in most parts of Assynt under crofting tenure, and

Flora of Assynt (2002): Insert

Figure 9. Assynt: contour map.

Reproduced from Bartholomew's Half-inch Map, Sheet 58, Cape Wrath (1947).

Contours at 100ft.' 250ft., and then at 250ft. intervals.

Plate 1. Achmelvich Bay, NC0525, looking west, August 2000.
Dunes and close-grazed coastal grassland over gneiss.

Plate 2. North side of Stoer peninsula, from Rubh' an Dunain, NC0434, looking west, June 1987.
Torridonian cliffs.

Plate 3. Mouth of the Oldany River and Pollachapuill, NC1033, looking e.s.e. towards Quinag, August 2000. Saltmarsh and woodland.

Plate 4. Glenleraig, NC1431, looking north-west towards Loch Nedd, August 1993. Close-grazed saltmarsh.

Plate 5. Glenleraig, NC1431, looking north-west, August 2000.
Woodland; bracken on site of former settlement.

Plate 6. Duart, Nedd, NC1332, June 1987.
Wooded boulder scree with bryophyte mats and *Hymenophyllum wilsonii*.

Plate 7. Wooded island in Loch Poll Dhàidh, NC0729, looking north-west, May 1994. Note the lack of trees on the loch shore.

Plate 8. Sheiling at Leitir Eásaich and Loch na h-Innse Fraoich, NC1626, looking south towards Suilven, June 1995.

Plate 9. Allt an Achaidh, Cromalt, NC2306, looking south-west towards A' Chìoch, July 1994. Beautiful but dour ground!

Plate 10. Valley of the Abhainn Bad na h-Achlaise, NC1221, looking south-west, October 1989. Valley mire in autumn.

Plate 11. Gleannan a' Mhadaidh, NC1516, looking north-east, July 1999.
Site for *Asplenium septentrionale*.

Plate 12. Ultrabasic crag in ravine, Allt Caoruinn, NC1821, looking north-east, August 1997.

Plate 13. Waterfall on the Allt nan Uamh, NC2717, looking north-east, June 1995.

Plate 14. Loch na Barrach and other lochans at 300m. on north side of Suilven, NC1518, looking north-west, August 1994.

Plate 15. Loch an Ordain, NC0924, looking north-east, June 1982.
Eutrophic loch with well-developed reedswamp.

**Plate 16.** Liath Bhad, NC2530, from Loch Glencoul, looking south-west, July 1994.
Hanging woodland on Fucoid Beds and limestone.

**Plate 17.** Ardvreck peninsula from Allt a' Chalda Beag, NC2423, looking south-west, October 1993.
Limestone; site for *Astragalus danicus* and *Dactylorhiza fuchsii*.

Plate 18. View from Meallan Liath Mór, NC2218, north-east towards limestone cliffs of Creag Sròn Chrùbaidh, May 2001. Quartzite in foreground. (G.P.R.)

Plate 19. View from Meallan Liath Mór, NC2218, e.s.e., up valley of the Allt nan Uamh, May 2001. Breabag in background, with snow patches. (G.P.R.)

Plate 20. North face of Beinn Gharbh, NC2222, from across Loch Assynt, September 2000. Torridonian over gneiss.

Plate 21. Boulder scree on north-facing slope of Coire Gorm, Glas Bheinn, NC2526, May 2001. Site for many oceanic-montane hepatics. (G.P.R.)

Plate 22. View from summit of Glas Bheinn, NC2526, south-east towards Ben More Assynt and Conival, May 2001. Fell-field in foreground, late snow-lie on the hills.

Plate 23. Bad an Dìoboirich, NC1631, looking south-east towards Quinag, January 1995. 18th century sheiling with clearance cairns, and a reminder of winter conditions on the hills.

Plate 24. Long-abandoned croft and house at Rhicarn (since restored), NC0825, looking to north-east, June 1989. Site for *Fallopia japonica* and *Glechoma hederacea*.

Plate 25. Lon Ruadh and Oldany, NC0932, looking north, June 1996.
Mire, roadsides, river, disused farmhouse and garden; gorse in flower.

elsewhere, by grazing and muirburn, particularly the latter. The results of over-grazing by sheep are well-illustrated by the vegetation on either side of the fence that prevents their access to the higher parts of the cliffs west of Rubh' an Dunain. There is high *Calluna* on the seaward side, and a very short sward on the landward.

The immediate effects of muirburn are obvious enough alongside the coast road in a number of places, and can be even more striking inland. It often appears to be carried out without any regard to recommended guidelines and as a result gets out of hand and can wreak havoc. Areas of dry heath are burned back to the underlying peat, and it sweeps across mires, scorching the *Sphagna*, and up screes and crags, killing scrub and regenerating woodland and blackening the very rocks. Muirburn that got out of hand, or was accidentally or even maliciously started, has affected large areas of the parish, with hair-raising tales of fires in the past that burned the whole of the south flank of Quinag, or raged westwards from Elphin almost all the way to Lochinver. Although a few species, notably *Ajuga pyramidalis*, may become locally more abundant following such burns, the long-term damage and nutrient-impoverishment are a high price to pay for this fleeting benefit.

For these reasons the best examples of heathland vegetation in Assynt are found in the more remote areas, where muirburn has not occurred recently or so frequently, and also on loch islands. Inland from Stoer and Clachtoll for instance, in the vicinity of Loch Poll Dhaidh and Loch an Aon Aite, there is a striking number of well-grown bushes of *Juniperus communis*, a species peculiarly susceptible to muirburn, and this area also has much *Arctostaphylos uva-ursi*. *Juniperus* is also a conspicuous feature of many loch islands in the area north of Loch Assynt, such as Loch Beannach, sharing them with tall *Calluna*, *Vaccinium myrtillus* and *A. uva-ursi*.

**Mires**

The influence of base-rich elements in the gneiss on flushes and mires has been touched on in the description of part of Glenleraig given above. It is even more strikingly manifested in a series of minerotrophic mire communities (those fed by groundwater), in which *Schoenus nigricans* is a major component. Ferreira considers these to be amongst the most interesting and unusual in North-West Sutherland, and they are well represented in Assynt. He calls them the '*Schoenus* fen complex', and we give here a summary of just a few of his examples.

At Achadh Mór, *Schoenus* tussock swamp has developed over a broad terrace flushed from nearby strongly basic gneisses and at one end there are the larger tussocks of *Carex paniculata*, a species only found locally at five sites. Near Lochan Fearna (Brackloch) a *Schoenus* tussock swamp is overtopped by the saw-leaved tussocks of the large fen-sedge *Cladium mariscus*. The ribbons of *Schoenus* that snake down into a *Carex lasiocarpa* mire at Mòinteach na Dubha Chlaise house a population of the rare orchid *Dactylorhiza lapponica*, with the even rarer *D. incarnata* ssp. *cruenta* in the *C. lasiocarpa* mire itself. This is thought to be the only place in the British Isles where the orchids occur together.

Rather different is a large area of sloping ground not far from the north coast of the parish, where extensive *Schoenus* mires and stony flushes support another population of *D. lapponica*, accompanied by *Platanthera bifolia*, *Gymanadenia conopsea* and *D. incarnata* sspp. *incarnata* and *pulchella*.

This section would not be complete without mention of the ombrotrophic mires (those fed solely by rainwater) on deep peat, which are often referred to as blanket or raised bogs. The topography of the gneiss is such that large areas of deep peat have not often developed. However, there are on the north-east sides of Fionn Loch and Loch Veyatie examples of level blanket bog with some incipient raised bog, on either side of the River Inver some good raised bog and, inland of Clachtoll and Stoer, some smaller examples.

The most striking developments of this habitat in the parish are, however, over basic syenites and marbled limestones to the south of Lochs Borralan and Urigill and on the north side of Cnoc na Sròine, and also at slightly higher altitudes on the limestones east of Beinn an Fhuarain, across the plateau south-east of Inchnadamph and at Lairig Unapool. Ferreira points out that deeper peat has developed over the limestone than any of the other rock types in Assynt, and discusses the reasons for this.

The communities developed on these deep peats are not very diverse, the main one being forms of *Calluna*/*Eriophorum vaginatum* mire, with *Rubus chamaemorus* on the higher slopes, or *Trichophorum*/*E. vaginatum* mire. Pool systems are not common on these deep peats, but those in a clearing on the forested ridge between Lochs Borralan and Urigill, recently the subject of restoration measures, have an impressive development of the hummocks of the acid-loving species of *Sphagna*, *S. fuscum* and *S. austinii*.

**Cliffs and crags**

Finally, some mention should be made of the inland cliffs, crags and screes, at relatively low altitudes, which are such a feature of the gneiss landscape. These are most interesting, predictably, when the gneiss is basic or is cut by basic or ultrabasic intrusions.

The accessible lower parts of these exposures, particularly where there are damp crevices or overhangs, house a variety of small ferns. The most frequent are *Asplenium adiantum-nigrum* and *A. trichomanes*, but others may occur. A good example is on the east side of Cnoc Gorm, where a basic dyke outcrops on both sides of a small valley. On the south-west side is a shaded crevice with *A. viride*, and just across the valley, on a well-lit crag, *A. ruta-muraria*, both species that are much commoner on the limestone. The only occurrence of *Phyllitis scolopendrium* on the gneiss, in a deep crevice on the shore of Loch na h-Uidhe Doimhne, is another example.

Quite the most striking manifestation of this phenomenon, however, is in a remote ravine south of Suilven, Gleannan a' Mhadaidh, one of the magical places in Assynt! It stretches for some two kilometres, with a small boulder-choked west-flowing burn at one end and a loch at the other. According to Ferreira, the ravine follows the line of a shear zone and many of its rocks are strongly calcareous. Although not wooded, it does contain one large wych elm, and other isolated trees include goat willow and bird cherry. On the ledges in places are found all four species of *Asplenium* mentioned above, together with *Cystopteris fragilis*.

However pride of place must go to a flourishing population of *Asplenium septentrionale* on the south-facing side of the gulley. Ferreira points out that this scarce species has a most interesting distribution in the British Isles, since most of its stations are on hard, dark, mainly basic (but not calcareous) igneous and metamorphic rocks, in sites fully exposed to the sun. He suggests that it may, like *Schoenus nigricans*, have a preference for substrates with relatively high magnesium levels, or of one or more of the metal-liferous elements that are found in these rocks. It was previously found by him at one locality inland from Scourie, north of Assynt, and we have now found it in three other places in Assynt. Gleannan a' Mhadaidh is also the only place in the parish where any species of *Pyrola* has so far been found, *P. media*, under tall *Calluna* not far from the *Asplenium septentrionale*.

One especially rich and not uncommon community occurs in open dry heath on sunny south-facing rocky slopes, below base-rich gneiss crags. It is characterised by the abundance of *Thymus polytrichus*, accompanied by *Antennaria dioica*, *Hypochaeris radicata*, *Linum catharticum* and *Lotus corniculatus*, and often a range of *Euphrasia* spp. This is the preferred habitat of *Ajuga pyramidalis*, whose flat, pale green rosettes are tucked into gritty corners, often alongside the very edge of the crags.

Further interesting communities are found in the vicinity of the ultrabasic rocks which outcrop in Assynt, though much more sparingly than the basic ones. We are, once more, indebted to Ferreira for bringing them to our notice. The rocks themselves can be very distinctive; they are often brown, finely fissured and, close up, weathered into either a scatter of small pock marks or into a coarse honeycomb pattern that is exceedingly rough on the hands.

The surfaces of these ultrabasic outcrops may be strikingly devoid of higher plants, apart from fronds of *Asplenium adiantum-nigrum* in crevices with, occasionally, *Galium boreale*. Their interest is provided by the herb-rich *Calluna* heath and grassland which develops in their vicinity. These share commoner species with those mentioned above for basic crags, but have in addition *Carex flacca* and *Polygala vulgaris*, species otherwise mainly associated with the limestone.

It is perhaps no coincidence that two of the best examples of these communities are on old sheilings, at Achadh' an Ruighe Choinich on the path north of Suileag, and at Clach Airigh to the north-east of Suilven. In grassland at the former site there is a small population of the rare *Equisetum pratense*, and in flushes associated with these outcrops elsewhere in the parish there are isolated occurrences of the equally rare *Equisetum hyemale* and *E. variegatum*. *Melica nutans* is another species which seems to have an affinity for these outcrops, and the communities associated with them are, as Ferreira says, worthy of further study.

## Freshwater

### Lochs and lochans

Some 680 lochs and lochans are shown on the 1:50,000 map of Assynt. They vary in size from less than 50 m. across to the 6 km. long Loch Veyatie and 10 km. long Loch Assynt, and even more are shown at the 1:25,000 scale. As Ferreira says, 'when viewed from one of the higher hills...on a summer afternoon they provide a truly incredible spectacle of myriads of small lochans glistening in the sunlight'. Exposed shores on the five largest lochs are subject to considerable wave action and they are therefore of limited interest from a botanical point of view, except in sheltered bays. This account will therefore concentrate on the smaller ones, which we will refer to, for convenience, as lochans.

The great majority of these lochans (nearly 550) lie, for topographical reasons, on the Lewisian gneiss. According to the basicity of the rocks feeding them, they provide a full spectrum of nutrient status from strongly oligotrophic (nutrient-poor) through mildly oligotrophic and mildly eutrophic to strongly eutrophic (nutrient-rich). Most of those at lower altitudes fall into the two middle categories. Following Ferreira, the descriptions of the vegetation that follow use these categories, but it should be emphasised that some species show a wide degree of tolerance, a good example being *Littorella uniflora*. Lochans on geological substrates other than the gneiss are included in the appropriate nutrient category.

One caveat applies throughout the descriptions, and that relates to submerged aquatics. Our recording necessarily focussed on species that are identifiable from the shallows, together with evidence from washed-up material. Ferreira confined himself to floating and emergent vegetation. The Scottish Loch Survey (1988) included submerged aquatics, and we have scanned their records to supplement our own observations. However, there must remain significant gaps in our knowledge of the distribution of these species.

Strongly oligotrophic lochans

The precise composition of the emergent vegetation of these lochans, as all others, depends on their depth and the nature of the bottom, but the range of species present is fairly limited. At the margins of the stonier ones *Ranunculus flammula* is typical, with the dainty flowers of *Lobelia dortmanna* obvious in season. Submerged aquatics include both species of *Isoetes* and *Subularia aquatica*. *Littorella uniflora* is almost ubiquitous since, as mentioned above, it tolerates a wide range of nutrient status.

In slightly deeper water, there may be *Carex rostrata*, *Menyanthes trifoliata* or *Equisetum fluviatile* and, in the deepest areas, the floating aquatics *Nymphaea alba*, *Potamogeton polygonifolius* and *Sparganium angustifolium*, the strap-like leaves of the last sometimes fanning out over a wide area. In peatier lochans, *Carex rostrata* is more often accompanied by *Eriophorum angustifolium*. In high level lochans of this type the dominant species may be *Equisetum fluviatile*, with the aquatic form of *Juncus bulbosus*, or in the most extreme cases just *Eriophorum angustifolium*.

Most of the few low-level lochans that have developed on the Torridonian are in this category, while the high level corrie lochans on this substrate are often devoid of emergent vegetation. The same applies to lochans on the quartzite, often very exposed, which tend to be exceptionally acid and nutrient-poor.

Mildly oligotrophic lochans

In these, some of the ground water arises from slightly calcareous bands in the gneiss. They are particularly typical of the Assynt area and support a wide range of communities. A characteristic species, other than those already mentioned, is *Schoenoplectus lacustris*, either in pure stands, or in association with *Nymphaea*, and in deeper water, with *Potamogeton natans*. In the shallows there may be the fine leaves of *Carex lasiocarpa*, either in pure stands or with *C. rostrata*, and at the water's edge the ascending shoots and nodding heads of *Carex limosa*, although the last is also typical of miry pools.

Two species of spike-rush are found in this type of lochan, the commoner being the open stands of *Eleocharis multicaulis*, although denser stands of *E. lacustris* may also occur. In lowland examples of these lochans there may be a marginal swamp community containing *Potentilla palustris*. Another marginal species, especially on grassy shelves, is *Deschampsia setacea*, although in Assynt this seems to be confined to sites within five kilometres of the coast.

Loch an Achaidh, a shallow loch on the Torridonian rocks of the Stoer penisula, probably falls in to this category, with mats of *Menyanthes*, *Potentilla palustris* and scattered *Equisetum fluviatile*. The margins are one of the places where the vegetative characters of the two bur-reeds *Sparganium angustifolium* and *S. emersum* may be compared.

Mildly eutrophic lochans

These have, according to Ferreira, a well-marked relationship with distinctly calcareous or dolomitic bands in the gneiss. In addition to the communities found in mildly oligotrophic lochans, they may also contain *Phragmites communis*, and less commonly, as in a number of the smaller lochans north of Loch Assynt, the robust stems of *Cladium mariscus*. In two small lochans, near Camasnafriaraich and Strathcroy, the *Cladium* grades landwards through *Carex lasiocarpa* into *Schoenus nigricans*, a significant association.

One of the few lochs on Torridonian rocks that falls into this category is Loch na Claise, on the Stoer peninsula, which is probably enriched by a combination of calcareous bands in the Torridonian and wind-blown shell sand originating from Balchladich beach. It has large areas of tall reedswamp composed of *Phragmites* and *Schoenoplectus*, much *Nymphaea* and is also a site for both *Hippuris vulgaris* and *Sparganium emersum*.

Loch Awe, north of Ledmore, has little emergent vegetation other than a large bed of *Schoenoplectus* at the southern end, and from a distance looks fairly unprepossessing. However, the Fucoid Beds outcrop on its eastern margins and it also receives water from limestones to the east. The rich tall herb vegetation on its margins is an indication of raised base status, as may also be some of the species on its islands, such as *Ajuga reptans*, *Pseudorchis albida*, *Platanthera chlorantha* and an old record of *Paris quadrifolia*. It is also one of only a handful of localities on the British mainland for the Red Data Book pondweed *Potamogeton rutilus*.

Also in this category are two lochs that owe their nutrient status as much to the rivers which feed them as the rocks on which they are situated. Loch Borralan lies on syenites, but is fed by burns off marbled limestones. It has a well-developed reedswamp, and also houses one of only two populations in Assynt of the scarce aquatic fern *Pilularia globulifera*, which is locally abundant on silt in the shallows at its eastern end. The other is Cam

Loch, downstream of Loch Borralan, which has the other population of *Pilularia* and where the rich flood zone vegetation on Eilean a' Gartaig (see above in the woodland section) is also evidence of the base-richness of its waters. So also is the locally frequent occurrence of *Trollius europaeus* on its more sheltered shores.

Strongly eutrophic lochans.

These are rare, but one of the best examples is Loch an Aigeil, by the roadside between Clachtoll and Stoer, which is enriched by shell sand originating from the nearby Bay of Stoer. There is a considerable variety of emergent communities, with *Eleocharis palustris* and *Caltha palustris* close to the shore, and a locally unique stand of the pink-flowered *Persicaria amphibia*. Elsewhere in the lochan are extensive beds of *Phragmites* and *Schoenoplectus*, and another of the few local populations of *Sparganium emersum*. The loch also has a fine range of submerged aquatics including *Potamogeton alpinus*, *P. filiformis* and the rare hybrid *P. x nitens*.

Two of the few lochs in Assynt associated with the limestone come into this category, but they both lack extensive areas of emergent vegetation. Loch Urigill overlies marbled limestones, but much of it is too deep or too exposed for emergents. However, there are sheltered lagoons, bays and ox-bows around the mouth of the Crom Allt in the south-east corner, with well-developed reedswamp, consisting of *Schoenoplectus*, *Eleocharis palustris* and, unusually, *Sparganium erectum*. Other species of interest in the vicinity are *Hippuris vulgaris* and *Veronica anagallis-aquatica*; the latter is found nowhere else in the parish.

Loch Mhaolach-Coire (the Gillaroo Loch) also overlies limestone and is distinctly eutrophic. It is too high and exposed to have much emergent vegetation other than *Carex rostrata*, but has a dense population of pondweeds including *Potamogeton perfoliatus* and *P. praelongus*. In a peaty swamp on its inflow burn is one of few stands in Assynt of *Carex paniculata*.

**Burns and rivers**

The river systems of Assynt are short in Scottish terms and, taking into account their large in-stream lochs, the watercourses themselves even shorter. The two longest are the Loanan/Traligill/Inver system, containing Loch Assynt, which is some 30 km. from source to sea, and the Ledmore/Kirkaig system, containing Loch Urigill, Cam Loch, Loch Veyatie and Fionn Loch, which is about 25 km. The one other westward-draining system, out of Lochan Fada south of Canisp, is about 14 km., and the three longest on the north coast are those of the Oldany River at about 9 km., and the Leireag River and Unapool Burn at about 6km.

Although their in-stream lochs may have some ameliorating effect, the watercourses in these systems and others have steep gradients and are subject to sudden and vigorous spates. As a result they do not support substantial communities of higher plants, either aquatics or reedswamp species. This deficiency is compounded in the case of the limestone feeders to the longest system, the Allt nan Uamh and the Traligill river, since these spend much of their time underground. Characteristic species in the faster flowing stretches are *Myriophyllum alternifolium* and *Potamogeton polygonifolius*.

The exceptions to the general rule are noteworthy. At Luban Croma, on the headwaters of the Ledbeg River, there is a series of well-vegetated meanders (along the county boundary with Wester Ross), with a curious assemblage of species, including *Callitriche hamulata*, *Equisetum fluviatile*, *Glyceria fluitans*, *Sparganium natans* in quantity and the charophyte *Nitella flexilis*. *Callitriche hamulata* is particularly characteristic of the upper reaches of burns along the eastern boundaries of the parish, often forming large three-dimensional masses in pools.

More conventionally 'lowland' in character is the stretch of the Ledmore River beween Loch Borralan and Cam Loch. This has, for Assynt, quite unusually lush vegetation, with *Potamogeton alpinus* and *P. gramineus* in the watercourse, *Sparganium erectum* and *Carex aquatilis* at the edge and *Carex curta* in the flood plain marsh. The lowest part of Allt na Braclaich, meandering through a *Molinia*-dominated floodplain, also has *Carex vesicaria* in its only local station.

The lower reaches of two of the burns flowing into Loch Urigill, the Allt nam Meur and Crom Allt are exceptional in having extensive populations of *Hippuris vulgaris*, associated with *Sparganium erectum*. Na Luirgean, which drains Loch Urigill, has an initial gently graded section with *Schoenoplectus* and *Subularia*.

However, the lack of level stretches in our local watercourses is amply offset by the interest of the ravines and waterfalls on many of them, both from the point of view of higher plants and, more important, the bryophytes. Some of the wooded ravines have already been mentioned in the section on woodlands, but where waterfalls occur in open country, the vegetation associated with them can be quite lush. An example is provided by a small fall on the headwaters of the Traligill. Here the tall herb vegetation developed in the splash zone includes *Angelica sylvestris*, *Caltha palustris*, *Chamerion angustifolium*, *Epilobium alsinifolium*, *Trollius europaeus*, *Valeriana officinalis* and huge trailing stems of *Cochlearia officinalis*, at one of only two inland sites in the parish.

## Limestone

The term limestone has been used here rather loosely for two main groups of rocks, the dolomitic (magnesium-rich) mudstones and shales of the Fucoid Beds and the limestones and dolomites of the three lowest formations of the Durness Group that outcrop in Assynt. Although the vegetation associated with them is broadly similar, some species are restricted to one or the other. There are, in addition, marbled limestones at Ledbeg, around and to the south of Loch Urigill; these are very hard slow-weathering rocks and in the latter area largely overlain by peat, so their effect on the vegetation is both localised and reduced. Limestone loch communities are described in the section on aquatic habitats.

### Fucoid Beds

The most striking development of the vegetation on the Fucoid Beds is at Liath Bhad on the south-west side of Loch Glencoul, just inside the parish (continued for a short distance eastwards and also across the loch, in Eddrachillis). There is also some limestone at this site. Hanging on the crags of the outcrop is a narrow strip of mixed woodland, with birch, rowan, hazel, aspen, wych elm, goat willow and some holly.

Underneath there is a rich tall herb ledge community containing, besides commoner species, *Galium odoratum*, *Lapsana communis*, *Sanicula europaea*, *Scrophularia nodosa* and *Silene dioica*. There is also a tall grass ledge community dominated by *Brachypodium sylvaticum*, with *Bromus ramosus* and *Elymus caninus*. In crevices are *Cystopteris fragilis*, *Polystichum aculeatum* and *Scolopendrium vulgare*. *Allium ursinum* occurs in flushes and, below and around the crags, herb-rich grassland with *Coeloglossum viride*, *Listera ovata*, *Orchis mascula* and much *Trollius europaeus*.

Elsewhere the Fucoid Beds are conspicuous as rusty-coloured, lichen-encrusted low crags, surrounded by basic grassland, a good example being just to the east of the road at Lairig Unapool. *Cystopteris fragilis* is a consistent feature of these crags, as are a number of calciphile mosses. Interestingly, *Dryas octopetala* is quite absent, although locally abundant on the nearby limestone.

On a narrow rib of the Fucoid Beds crossing the Allt a' Bhealach, at the foot of Conival, occurs our largest population of *Equisetum pratense*, although it is also found in limestone flushes and, rarely, on other substrates. Alongside the headwaters of this burn, in the Bealach Traligill, the Fucoid Beds outcrop in several places. In a narrow ravine cut through them at the eastern end of the Bealach there are both the montane willowherbs *Epilobium alsinifolium* and *E. anagallidifolium*, with *Saxifraga oppositifolia*. Above the highest point of the Bealach, at about 530m., a rib of this rock slants obliquely upwards, and there supports a good stand of the arctic-alpine grass *Poa glauca*, accompanied by *Alchemilla wichurae*.

### The limestone proper

Most limestone species are found throughout the area in which it outcrops, from Lairig Unapool south through Inchnadamph and the Sròn Chrùbaidh cliffs to Allt nan Uamh, along the flanks of Beinn an Fhuarain to Ledbeg, and at Elphin and Knockan. Others are only found on the higher outcrops of Cnoc Eilid Mhathain and in the Bealach Traligill. The brief notes that follow can only give an indication of the richness of the flora of the limestone; for a delightful and more discursive account see John Raven's contribution to *Mountain Flowers* (Raven and Walters 1956).

*Dryas octopetala* is the most striking component of what have been described as calcareous heaths, accompanied by *Carex flacca*, *C. rupestris* or, less commonly and at higher altitudes, the tight hummocks of *Silene acaulis*. Off the outcrops and crags, this heath grades into calcareous grassland, which may contain much *Alchemilla alpina* or *Thymus*. Associated with these heath and grassland communities is a suite of species that has enchanted visiting botanists for over 200 years. They include the sedge *C. capillaris*, the grasses *Briza media* and *Helictotrichon pubescens*, small herbs such as *Galium sterneri*, *Persicaria vivipara*, *Polygala vulgaris* and *Viola canina*, and the orchids *Coeloglossum viride*, *Orchis mascula* and *Listera ovata*.

In shady crevices in the limestone, particularly along the usually dry watercourses, two ferns occur frequently, *Asplenium viride* and *Polystichum lonchitis*. There too may often be found the fragile stems of *Circaea* x *intermedia*, although this does turn up again nearer the coast. In drier situations on boulders and ledges are *Botrychium lunaria* and *Draba incana*, and in thinly vegetated limestone 'gravel' the tiny eyebright *Euphrasia ostenfeldii*. Flowering spikes of *Epipactis atrorubens* are found in small numbers throughout the limestone area, usually on ledges and in boulder scree where they are protected from grazing.

Springs and flushes, the latter often stony, are a predictable feature of the porous limestone. A conspicuous feature of the springs is the bright-green moss *Palustriella commutata*. The flushes are dominated by sedges and their relatives, including *Schoenus nigricans*, *Carex dioica*, *C. hostiana* and *C. oedocarpa* ssp. *brachyrhyncha*, but they are often painted yellow in season by the flowers of *Saxifraga aizoides*. *Carex pulicaris* and *Thalictrum alpinum* are associated in some flushes, and a feature of nutrient-enriched flushes at Inchnadamph and Elphin is *Veronica beccabunga*.

A characteristic shrub of the limestone is *Salix myrsinites*, which occurs widely in small ravines and other places where it has some protection from grazing. However it only shows its full potential in the older exclosure on Glac Mhór, where there are some magnificent bushes. Another shrub that appears to be virtually restricted to the limestone in Assynt in its 'wild' state is *Crataegus monogyna*; isolated and often ancient bushes are scattered along the crags from Sròn Chrùbaidh to Ardvreck Castle.

The wooded ravines on the limestone around Inchnadamph have a rich flora, which resembles that of the hanging woodland at Liath Bhad described above. The best examples are on the Traligill River and its tributary the Allt Poll an Droighinn. The canopy includes splendid examples of goat willow and wych elm, as well as hazel, birch, rowan and grey willow. On dry ledges there is the tall grass community dominated by *Brachypodium sylvaticum* and *Elymus caninus*, with some *Bromus ramosus*, but also found here is one of the Inchnadamph specialities, 'Don's Twitch'. Once thought to be a distinct species, it is now regarded only as a variety of *Elymus caninus*, with which it grows. Moist ledges support species-rich tall herb vegetation, with wet flushes dominated by *Crepis paludosa* or *Allium ursinum*.

A little further north is a rather more open ravine on the Allt a'Chalda Mor. This supports birch-hazel woodland, with rowan and wych elm, and a ground flora which includes *Epipactis helleborine* and *Listera ovata*.

These wooded ravines have other features of interest. One is a wealth of *Hieracia*, especially of the Section Cerinthoidea, another is *Ajuga reptans*, which is markedly less common in Assynt than its nationally scarce relative *A. pyramidalis*.

We conclude this section with some other specialities of particular areas of the limestone. In this context we should first mention the Inchnadamph 'zoo'. On a fissured outcrop at the eastern end of Glac Mhor there is an assemblage of species alien to the area, obviously introduced, but by whom, when, and to what end (other than to provoke unsuspecting botanists) is not known. The grykes house a flourishing population of *Phyteuma scheuchzeri*, which occurred elsewhere in the British Isles only on the walls of an Oxford college, but is there no longer. In shallow soil on shelves *Erinus alpinus* and *Campanula cochleariifolia*, both common species in cultivation, are doing well. There are also a few plants of *Silene quadrifida*, otherwise only known as an introduction on Ben Lawers, and of *Gentiana verna*, native on the limestone elsewhere in the British Isles.

Returning to our native flora, the isolated outcrop at Lairig Unapool has an impressive population of *Epipactis atrorubens*, when the deer allow it to flower. Ardvreck peninsula is the only inland station for *Astragalus danicus*, and also, rather surprisingly, the only confirmed site for *Dactylorhiza fuchsii*.

Just north of Inchnadamph, rabbit-nibbled grassland around one rib of the limestone has the rare lady's-mantle *Alchemilla glaucescens*. Off the Traligill valley, the lowest reaches of its tributary the Allt na Glaic Móire are fringed by flushed turf containing *Equisetum variegatum* and *Tofieldia pusilla*.

South of Inchnadamph are the magnificent Sròn Chrùbaidh cliffs, the upper parts of which have never been scaled by a botanist, so far as we know. However, the lowest crags and shelves are accessible with care, and support a fine range of communities and species already mentioned. In addition, the scree at their base has the hybrid fern *Polystichum* x *illyricum* in one of its few British localities, and on one shelf there is a small but healthy stand of *Gymnocarpium robertianum* in its most northerly station in the British Isles. Scattered along the lower crags, in places where they are safe from browsing, are two groups of *Sorbus rupicola*, at one of only two localities in West Sutherland.

The 'high limestone' at Cnoc Eilid Mhathain is renowned for its *Dryas*-dominated grassland, crevice and ledge communities, and has a range of less common species found sparingly, if at all, lower down, including some more generally distributed montane species. They include *Arenaria norvegica*, *Minuartia sedoides*, *Potentilla crantzii*, *Saussurea alpina* and *Silene acaulis*. In flushes east of the outcrop are *Tofieldia pusilla* and *Saxifraga oppositifolia*.

Higher still, in the Bealach Traligill, the limestone reaches its highest altitude in West Sutherland, at about 530m., and *Arenaria norvegica*, *Potentilla crantzii* and *Tofieldia* occur again in small quantity, with two of our four populations of *Saxifraga hypnoides*.

Returning to lower ground, the Allt nan Uamh valley again supports a good range of the limestone communities and species, with a number of the calcicole *Hieracia*. The crag immediately to the east of the caves has fine tall herb vegetation with some hazel scrub in its most sheltered crevices and on its summit are found *Arenaria norvegica*, *Epipactis atrorubens* and *Potentilla crantzii*.

The southernmost limestone outcrops, at Elphin and Knockan, are heavily grazed in places, and difficult to access in others. The crumbling slopes above the road are noteworthy for the amount of *Silene acaulis*, and have a conspicuous *Allium ursinum* flush. East of the townships, on the edge of the Abhainn a' Chnocain, there is flushed grassland with *Equisetum variegatum* and, in one flush nearby, with *Carex dioica* and *Eleocharis quinqueflora*, the rare hybrid horsetail *Equisetum* x *trachyodon*. Where this river enters Loch Veyatie, a few yards inside the parish (and county) boundary, we found in late summer 1996 a few flowers of *Parnassia palustris*; this has been recorded more widely in the parish, but not for some decades.

## Upland

As a working definition of 'upland', we will take land above the 300m. (1000ft) contour. On the ground this contour line marks, approximately, the lower limit of species such as *Arctostaphylos alpinus*, *Diphasiastrum alpinum* and *Salix herbacea*, though there are exceptions.

Included are all the major hills in Assynt, Quinag, Canisp and Suilven in the west, and the broken ridge from Cnoc na Creige to Breabag in the east, taking in Glas Bheinn, Beinn Uidhe and Conival. Zigzagging between these, from the eastern end of Loch Assynt southwards, are the lower hills of Beinn Gharbh and Beinn Reidh, Meallan Liath Mor, Beinn nan Cnaimhseag and Beinn Fhuarain, Cnoc na Leathaid Buidhe and Bhig, Cnoc na Sròine and the Cromalt Hills.

Within this substantial area of the parish, rock type, aspect and slope are the main factors determining the nature of the upland vegetation, together with altitude in some cases. The higher outcrops of the Fucoid Beds and limestone have already been covered in the relevant section, which leaves us, so far as the substrate is concerned, with the gneiss, Torridonian, quartzites, Moine and igneous rocks. The first three of these are major components of the single largest hill in the parish, Quinag, and it therefore serves as a good example.

One of the glories of Quinag is an area known to few other than botanists because it is well off any route to the summits. At the foot of the massive north-facing buttress of Sàil Gorm, the gneiss extends upwards to a height of 530m., forming dark crags, wet and overhanging in places, with substantial areas of boulder scree below. According to Ferreira, the gneiss at this point 'contains bands of hornblendic rock and additionally is strongly calcareous in the vicinity of the crush-zone of a fault'.

Tall herb vegetation is widespread on ledges inaccessible to deer, characterised by the presence of *Geum rivale*, *Saussurea* and *Trollius* with, in addition, *Coeloglossum viride*, *Orchis mascula* and *Rubus saxatilis*. Between the ledges may be found cushions of *Silene acaulis*, the hanging stems of *Saxifraga oppositifolia* and in flushes, *S. aizoides*. Wet crevices have *Oxyria digyna* and drier ones the ferns *Asplenium viride*, *Cystopteris fragilis*, *Dryopteris expansa*, *Polystichum aculeatum* and *P. lonchitis*. The crags and adjacent boulder scree are also very rich in bryophytes, including many species of the oceanic-montane hepatic-mat community such as *Mastigophora woodsii*.

Echoes of this species-rich basic upland vegetation occur elsewhere on Quinag, in the vicinity of narrow calcareous bands in the lower parts of the Torridonian sandstones. There is some development of these communities at the eastern side of the mouth of Bathaich Cuinneige, the large corrie between Sàil Gorm and Sàil Gharbh, but the best examples are on some east-facing crags above Lochan Bealach Cornaidh. Here in crevices may be found *Cystopteris fragilis* and *Polystichum lonchitis*, on shelves the tall herb community, with frequent *Sedum rosea* and *Trollius*, and, beneath the crags, flushes with *Thalictrum alpinum* and species-rich grassland.

The higher areas of Quinag are clothed with much more usual acid-loving upland vegetation, often species-poor. In places it is grassland, dominated by *Nardus*, or by *Agrostis* and *Festuca*, with *Alchemilla alpina*. Elsewhere there is montane heath, with *Calluna*, *Carex bigelowii*, *Empetrum hermaphroditicum* and the moss *Racomitrium lanuginosum*. In the most exposed areas, in the Bealach a' Chornaidh and elsewhere on the hill, there has developed an open fell-field community with *Juncus trifidus*, *Loiseleuria* and *Salix herbacea*. Finally, on the broad ridge of Sàil Gharbh, north of the quartzite summit, there is, on highly eroded ground, a scattered community of *Armeria maritima*, *Salix herbacea* and *Silene acaulis*, with *Luzula spicata*, occasional *Euphrasia frigida* and a very small amount of *Gnaphalium supinum*.

The Cambrian quartzites which cap the highest part of the ridge of Sàil Gharbh (808m.) lack most of these species, being almost bare fell-field on the windward side and a solid mat of *Racomitrium lanuginosum* on the leeward side. Quartzites also make up the summit of Spidean Coinich (764m.) and extend all the way down to the road on its east, and these areas are almost devoid of botanical interest, except for the very wind-pruned upland *Calluna* heath, with frequent dwarf *Juniperus* and occasional *Loiseleuria*.

None of the other hills have such a wide range of communities and species as Quinag, particularly those favouring base-rich habitats, but there are features of interest on most of them, including some truly montane species. To deal with the larger hills first, Canisp (846m.) has some similarities with Quinag. It is capped with quartzite, and the long slope between the summit and Loch Awe is an almost unrelieved slab of the same rock, encompassing the tetrad with fewest species in Assynt.

However, in an area of reddish fine-textured 'soil' just west of the summit, there is one of only two local populations of *Luzula arcuata*, accompanied by *L. spicata*. Further down the eastern slopes there is, in sheltered areas, a fine development of dwarf shrub heath with *Arctostaphylos alpinus*, *A. uva-ursi* and *Vaccinium uliginosum*, sheltering *Hymenophylum wilsonii* and a range of bryophytes; *Betula nana* has once been found in this area. Even lower down this slope, in broken fell-field on the quartzites, are a small stand of *Lycopodium annotinum* and a very much larger one of *Diphasiastrum issleri*. Finally, at the base of Torridonian crags on the north side of Canisp, there is a good development of tall herb ledge vegetation, with *Oxyria* and *Saussurea* and, on gneiss/porphyry below the crags, flushes containing frequent *Juncus triglumis*.

Suilven, the best-known of the Assynt hills, is, at 731m., not so high as either Quinag or Canisp, and does not have their mass, so it lacks some of their diversity. The path up to the bealach on the north side has gneiss faulted quite high up into the Torridonian, with frequent *Cornus suecicus*, *Epilobium anagallidifolium* in mossy flushes, and *Pseudorchis albida* on shelves with *Trollius*. The flat terrace on the summit of Caisteal Liath is almost solid *Salix herbacea* in places, with a little *Gnaphalium supinum* and *Euphrasia frigida*.

Suilven's chief claim to fame, however, is the recently-discovered flora of the steep-sided gulley between Meall Meadhonach and Meall Beag at the eastern end, where a combination of faulting and sills of Canisp porphyry has apparently resulted in a considerable degree of base-richness. Here G.P. Rothero found in 1999, in a crack on an almost inaccessible vertical face at about 600m., just three plants of *Saxifraga nivalis*, never before recorded in Assynt. In its vicinity and further down the gulley are *Cochlearia officinalis*, *Saxifraga oppositifolia*, *S. hypnoides* and a number of other basiphilous species.

We turn now to the hills down the eastern borders of the parish. Cnoc na Creige (593m.) has been little explored, except by A.G. Kenneth, who found there the alpine hawkweed named after him, *Hieracium kennethii*. In the northern corries of Glas Bheinn, the gneiss outcrops over quite an area in the lower and middle parts, although the upper parts and summit (776m.) are quartzite. There are basic bands in the gneiss, which yield good tall herb vegetation on ledges and, beneath the crags, base-rich flushes and flushed grassland. As a result the hill as a whole musters a respectable list of the upland species found locally, with some very good bryophytes. At the edges of watercourses draining two of the delightful corrie lochs, *Alchemilla glomerulans* and *A. wichurae* have each been recorded once, and in a curious stony flush on the north-western lip of Coire Dearg occurs *Deschampsia caespitosa* ssp. *alpina* in one of its two local stations.

The summit ridge of Beinn Uidhe (740m.) is perhaps the dourest hill in Assynt, for much of its length unrelieved, broken, wobbly quartzite, with a striking lack of vegetation of any kind other than cushions of *Racomitrium lanuginosum*. However in Bealach na h-Uidhe at its north-western end, where porphyry bands outcrop, *Hieracium alpinum* was discovered new to West Sutherland. *Silene uniflora* is in one of its two montane stations in Assynt on the south-western flank of the ridge, and at its south-eastern end a high outcrop of gneiss introduces a welcome element of diversity, albeit without any surprises.

From Bealach a' Mhadaidh, at the south-eastern end of Beinn Uidhe, south to the start of the ridge leading to the summit of Conival, there extends for some three kilometres an extensive area, rising from 600 to 850m. above the crags of Na Tuadhan, which we have not explored as fully as we might have done. The underlying geology is complex; on the geological map the quartzites are shown as interleaved with bands of the Fucoid Beds and Salterella Grit and also by sills of intrusive igneous rocks, although this diversity is not immediately apparent on the ground!

In gravelly patches on the ridge to the south-west of Loch nan Cuaran there there is a reasonable selection of montane species, including *Gnaphalium supinum*. Just to the south, on mossy burns flowing into Lochan nan Caorach, occur both the upland willowherbs, *Epilobium alsinifolium* and *E. anagallidifolium*, with *Veronica serpyllifolia* ssp. *humifusa* and a strikingly large-flowered form of *Cerastium fontanum*. On the edges of the burn flowing out of this lochan are *Alchemilla glomerulans* and *Persicaria vivipara*, which testify to local base-enrichment, perhaps from Imir Fada, the area draining into this lochan, So also do the presence at its northern end, in an area called 'Aeroplane Flats', of *Coeloglossum viride* and *Trollius* at an altitude of about 650m.

Conival, for all that it is the highest hill in the parish, at 987m., is disappointing so far as Assynt botany is concerned. The parish boundary runs up the northern ridge to the summit and then drops straight down into Bealach Traligill. Apart from a narrow strip of broken stony ground along the path to the summit, the western side of the hill is so steep that it has, to our knowledge, never been botanised. However, G.P. Rothero found in 1998, on both sides of the path, just north of the summit, a small population of *Luzula arcuata*, in almost certainly the same place in which it was first discovered in 1833. Nearby is a second stand of *Deschampsia cespitosa* ssp. *alpina*.

South of the Bealach Traligill, which cuts a deep gash in these eastern hills, they rise again to the 4km. long, undulating ridge known as Breabag. The northern part, as far as the headwaters of the Allt nan Uamh, is almost unrelieved quartzite and extremely dour. South of that point the quartzite is interleaved, as in Imir Fada, with bands of the Fucoid Beds and porphyroid igneous intrusive rocks. Where these latter rocks have broken down, on the gently sloping western edge of the ridge, into something approaching angular gravel, a good selection of the usual montane species reappears. In addition there are two hollows, at either end of this section of the ridge, where snow probably lies late. Here, in hummocks of moss beside the headwaters of burns, are two localities for *Sibbaldia procumbens*.

There remain the minor hills which run south from Loch Assynt to the southern boundary of the parish in the Cromalt Hills. The lowest of the Torridonian crags on the north face of Beinn Gharbh, particularly at Creagan a' Chait, have some affinities with those of Bealach a' Chornaidh on Quinag. Their basicity is revealed by the presence of *Asplenium viride*, *Coeloglossum viride*, *Orchis mascula* and the only stand of *Salix myrsinites* in Assynt off the limestone. There is in addition, cutting through the gneiss at the foot of these crags, a most imposing ultrabasic dyke, whose lava-like rock provides a foot-hold only for impressively large plants of *Asplenium adiantum-nigrum*, with *Polygala vulgaris* typically ornamenting the grassland at its foot.

The summits of Beinn Gharbh (540m.), Beinn Reidh (567m.) and the undulating plateau, at between 400 and 500m., that extends south from between them to Meallan Liath Mór, do not muster a long list of montane species, with occasional *Loiseleuria* and *Salix herbacea*. What is impressive, especially towards the southern end of this plateau, is the amount of *Arctostaphylos alpinus* on the broken Canisp porphyry and quartzites. This forms quite a luxuriant dwarf shrub heath in places, with *A. uva-ursi* and prostrate *Juniperus communis*. *A. alpinus* and *S. herbacea* reach almost their lowest altitude locally on the summit of Cnoc an Leathaid Bhuidhe, at 369m.

To the east, across the Loanan valley, are the outliers of Torridonian sandstone and quartzite of Beinn nan Cnaimhseag (568m.) and south of it, Beinn an Fhuarain (499m.). The former has, as its name suggests (*cnaimhseag* = bearberry), a large stand of *A. alpinus* on its northern flank, with in addition a small population of the delightful alpine hawkweed *Hieracium holosericeum*. The latter occurs again on the summit of Beinn an Fhuarain, as does *Silene uniflora* in one of only two inland localities, and on the northern flank of this hill is the only known local site for the hybrid willow *Salix* x *cernua*.

The twin granite hills of Cnoc na Sròine (390 and 397m.) yield no montane species except a luxuriant population of *Diphasiastrum alpinum*. There are, however, slightly basic flushes on both sides which have some species of interest.

To conclude, we should mention the exposed ridges of the Cromalt Hills, at the remote southernmost extent of Assynt. These lie on the Moine schists, which locally are rather dour, and we suspect that they have rarely before been visited by botanists. Some indication of their character is given by the name of the furthest south, Meall a' Bhuirich Rapaig (the hill of the stormy roaring). There is deep peat on the easternmost of these hills, cut back in places to bleached rock, and no upland species. At the highest point of the ridge, Meall Coire an Lochain (516m.), *Salix herbacea* makes an appearance, and below the summit, on wet north-facing crags overlooking Loch a' Phris, there is *Asplenium viride* and *Saussurea*, indicating some base-richness in the schists. Finally, at Meall an Dearcag Beag (388m.), and Cnoc an t-Sasunnaich (386m.), both *A. alpinus* and *S. herbacea* reappear, in an area with the most glorious views. And what better place for botanists from south of the border to draw this section to a close!

## Man-made or influenced habitats

The impact of man on the Assynt landscape may be appreciated at different levels of detail. The 1:50,000 map shows little more than the pattern of roads and settlements. However, antiquities such as cairns and duns, although few, remind us that use of the landscape dates back thousands of years. At a scale of 1:25,000 maps are more informative. They show, for example, the location of isolated buildings or ruins, walls and banks, sheilings and other structures concerned with the management of stock. Finally, on the ground may be seen the signs of activities that are not mapped at all. These include peat cuttings and drains, areas of recent and older muirburn and the more diffuse evidence of intensive grazing, whether by domestic stock or deer.

It should also be noted that apparently 'natural' features such as woodlands may have had their boundaries and character considerably modified by management over the centuries.

What follows are examples of species whose distribution and abundance are either dependent on, or have been substantially modified by, human activity. We have made no attempt to assign them to any hierarchy of communities, since no work has been done on this aspect of the vegetation of this area.

### Roads and tracks

Roads and tracks increase the species diversity of parts of Assynt out of all proportion to the small area they actually cover, particularly away from areas of habitation. However, since roads are mainly located on the lower more fertile ground, care should be taken in interpreting the distribution maps; the species concerned may be quite incidentally inhabiting the same corridors of land. Small roadside quarries, cuttings, embankments, drains and culverts, created when the roads were built over a century ago, all provide niches for plants. More recently, there has been extensive road widening (with some new cuts), re-haunching and the creation of many new lay-bys; these have sometimes involved the importation of material foreign to the immediate area, such as Ledmore marble or topsoil. Most of our roads are heavily salted and gritted in icy or snowy weather and piles of grit and salt are deposited in the autumn at the edge of the road in hilly places.

Double track roads have a characteristic flora in the gritty band, immediately adjacent to the carriageway, where there are raised verges (the verges on single track roads are often level with the carriageway). They include, predictably, species like *Juncus bufonius*, *Matricaria discoidea*, *Plantago major* and *Polygonum arenastrum* and, less predictably perhaps, *Gnaphalium uliginosum*, *Spergularia marina* and *S. rubra*. Salt tolerance, which probably accounts for the last two species, may also explain the fringe along many roads of *Plantago maritima* and what appear to be glaucous 'coastal' varieties of *Festuca rubra*. Other coastal species found on roadsides include *Daucus carota* at Achadhantuir, *Sagina nodosa* on a verge south of Newton, and *Carex arenaria* in a layby on Skiag, where it may have been introduced with sand. Two other more generally distributed species may also be conspicuous at the edge of the verge, *Poa humilis* and *Potentilla anserina*.

The verges themselves get a variety of treatments. Although many stretches of the double track roads are now fenced to discourage access by sheep, this has little effect on the deer. The rest of the double track and most of the single track roads are unfenced and often close-grazed by what a visitor once referred to as 'loose' sheep. They not only maintain a short sward but also dung it, so these verges are often appreciably more fertile than the adjacent land. Additionally verges on the double track roads are mown at least once a year.

These short, sometimes stony, swards have in places a distinctive and colourful flora, although it is not at all obvious why particular species occur where they do. For example, parts of the verges of the A837 east of Lochinver are yellow with *Galium verum* in summer, punctuated by occasional patches of the blue of *Campanula rotundifolia*. The former is otherwise a plant of coastal grasslands, the latter, a rare plant locally, has only once been found away from roads, on limestone. On the A894 descending from Skiag to the Drumbeg turn there are alternating stands of *Gentianella amarella* and *G. campestris*, which occur otherwise in, respectively, basic and heath grassland.

Other more robust species may occur towards the back of our few lusher verges, presumably introduced during construction or maintenance, such as *Festuca arundinacea* around Inchnadamph. *Elytrigia repens* is, happily for gardeners, an uncommon plant of such roadside verges in Assynt. A staple of verges further south, *Anthriscus sylvestris*, occurs on only one, at Elphin. Other species probably introduced by machinery are *Epilobium hirsutum*, which appeared in 1999 in the ditch of the hill leading down into Lochinver, and *Sanguisorba minor*, which turned up in a parking area and on a nearby verge at Ardvar. One final, even more bizarre introduction of this kind is *Lemna minor*, which was found in a newly cleaned-out ditch at Elphin. None of these three species had ever before been found in Assynt.

Where soil-rich spoil has imported to level verges a strikingly alien flora may be introduced. Some of it consists of short-lived weed species such as *Arabidopsis thaliana* and *Euphorbia peplus*; the tall spires of *Verbascum thapsus* which appeared in 2001 on a verge at Baddidarach probably come into this category. Another tall biennial, characteristic of disturbed rocky verges, is *Digitalis purpurea*, which can form impressive banks of colour, but disappears as the sward closes over. Other obvious imports are permanent, such as *Cirsium arvense*. It remains to be seen whether a species that recently turned up on a roadworks site at Nedd is also in this category, the sprawling var. *sativus* of *Lotus corniculatus*.

Patches of unfamiliar colour on verges brings us into the realm of garden throw-outs and other dumped material, which can introduce a further exotic element. Garden throwouts include a number of species traditionally cultivated in local gardens, such as *Alchemilla mollis*, *Crocosmia* x *crocosmiflora*, *Myrrhis odorata* and *Pentaglottis sempervirens*. A surprise was *Lamium album*, which was found amongst builders' rubble at the top of the hill leading down into Achmelvich; it only occurs elsewhere in Assynt in one local garden!

Road embankments introduce an element of well-drained scree into the local landscape and are often ornamented in spring with the golden clumps of *Dryopteris affinis*. Elsewhere they harbour dense banks of *Juncus effusus* which, although unremarkable from a floristic point of view, do provide a highly desirable micro-habitat for field voles. Some shrubby members of the rose family are also characteristic of rocky embankments, especially those with a southerly aspect. *Rubus fruticosus* appears to be near its climatic limit in Assynt and such embankments are its favoured habitat. Again, species of *Rosa* often flower and fruit more freely on road embankments than elsewhere, though why this should be so we have no idea. The most widespread of these, *R. pimpinellifolia*, can occur in impressive roadside thickets.

Most of the small roadside quarries that were excavated in the 19th century to build or extend our single track roads, have now been colonised with native species, but they do spring occasional surprises. An example is a small population of *Juncus tenuis* on the wet floor of one at Glenleraig, possibly brought in on car tyres; its only other locality in Assynt is on a track at Oldany. The quarry at Skiag, favoured parking for those climbing Quinag, also produced one year the hybrid thistle *Cirsium* x *celakovskianum*. A much larger quarry on a cut-off of the A837 west of Little Assynt, presumably opened or enlarged when this road was 'doubled' in in the 1980s, has the locally rare *Epilobium ciliatum* and one of our few weed populations of *Chamerion angustifolium*. There's excitement for you!

The peat and stalkers' tracks that were made into the interior of the parish in the late 19th or early 20th centuries were built to a standard almost equal to that of the roads, although they have never been surfaced. Their flora is essentially that of the area they are passing through, but one charactcristic plant is the curious variety *spiralis* of *Juncus effusus*, which appears to be spread by stock.

**Built-up areas, buildings and walls**

The nearest approximation to a 'built-up' area in Assynt is Lochinver, which houses about half the parish's population, much of it in the relatively new Inver Park area, and has extensive recent harbour works. Its weed flora was well documented by A.J. Wilmott and M.S. Campbell in 1943/44, and is still unusually varied for this area. *Scrophularia nodosa* is unexpectedly abundant in places, for example behind the buildings on Main Street and in the abandoned garden of the Culag Hotel. The sea wall at the south end of Main Street is rapidly being colonised by saplings of *Acer pseudoplatanus* and *Fraxinus excelsior*, and also by *Tanacetum parthenium*. Elsewhere disturbed ground supports weed populations of *Tussilago farfara*, normally a plant of river gravels in Assynt, and rare casuals such as *Alliaria petiolata*.

Older mortared walls in Lochinver have been colonised by *Asplenium trichomanes*, as have those of ruined houses elsewhere in the parish, as at Nedd. Other spleenworts are very uncommon on walls, a striking exception being the outbuildings of the old school at Drumbeg, where there is a flourishing population of *A. ruta-muraria*. The record for ferns was a long-unoccupied house at Rhicarn, whose walls bore in 1988 six species including *Phyllitis scolopendrium*; this last has only been seen on one other building, at Ardvar.

Free-standing dry-stone walls are not good habitat for higher plants, and the only frequent species is *Polypodium vulgare*. Where topped by turf, which was the practice in the past, they are often colonised by species that can withstand desiccation, such as *Aira praecox* and *Sedum anglicum*.

**Cultivated ground**

As has been pointed out in the chapter on the history of the landscape, cultivated ground, even at the scale of a 'tattie' patch, is now quite a rare sight in Assynt. Where ground is turned over, even after several decades fallow, 'traditional' arable weeds may reappear, such as *Chrysanthemum segetum*, *Euphorbia helioscopia* and *Silene alba*. Vegetable patches on the lighter soils of the coast have afforded some interesting records during the present survey. Examples are *Fumaria bastardii*, *F. officinalis* and *Polygonum boreale* at Clachtoll. Of the five species of dead-nettle recorded, only one, *L. purpureum*, is widespread. The other four have been found sparingly in gardens, *L. album* and *L. hybridum* at Achmelvich, *L. confertum* at Clachtoll and *L. amplexicaule* at Clachtoll and Culkein Stoer. Disturbance of some ground just behind the beach at the last locality also produced records of *Anchusa arvensis* and *Viola arvensis*.

The weed flora of our own garden at Nedd has intrigued us at times. We have inadvertently introduced, in sinks and pots, exotic but surprisingly persistent species such as *Geranium lucidum* and *Valerianella carinata*. *Anisantha sterilis*, *Linum bienne*, *Senecio viscosus* and *Thlaspi perfoliatum* appeared briefly, but did not last. Other local species have surprised us with their nuisance value; they include *Montia fontana* and *Stellaria uliginosa*.

All the species so far listed are annuals; some of the perennial 'thugs' commoner elsewhere are worthy of note this far north. They include all three species of *Calystegia*, of which only *C. sepium* has been recorded more than once. Their close relative *Convolvulus arvensis* has only been found once, in a garden at Baddidarach, where it had a tenuous hold on existence. Another 'thug', albeit an annual, *Impatiens glandulifera*, is however showing ominous signs of colonising ability at Clashnessie and Nedd. Another plant with this ability, though not such a menace, is *Veronica filiformis*, first recorded in 1957 and now found in mown grass in nine tetrads.

Cultivated species that have established themselves in the vicinity of gardens, sometimes long-abandoned, are another category of some interest. *Tolmiea menziesii* has spread up a wet crag at Kerrachar, and *Tropaeolum speciosum* escaped into a blackthorn thicket at Strathcroy. They may be of relatively recent origin, but an older generation of escapes includes *Fallopia japonica* which, from its occurrence in long-abandoned croft gardens, appears to have been regarded as a very desirable acquisition in the late 19th century.

Aromatic or colourful escapes include *Tanacetum vulgare*, six 'varieties' of *Mentha*, and four of *Mimulus*, the last two genera having tested our taxonomic skills, as well as those of previous generations of botanists. *Aegopodium podagraria* is so widespread that, despite its well-deserved reputation as an almost ineradicable weed, it must once have been cultivated. We wonder if the same is true of *Glechoma hederacea*, which occurs far from present-day cultivation at Rientraid. More of a mystery, though a short-lived one, was the appearance in 1995 of *Datura stramonium* in a long-disused unheated greenhouse at Glenleraig.

Two older introductions are found at either end of Loch Assynt. *Narcissus pseudonarcissus* still survives on the site of the old garden at Ardvreck Castle, which was abandoned at the end of the 18th century. More intriguing is the occurrence of the former medicinal herb *Peucedanum ostruthium* on the tiny Eilean Assynt, associated with what are thought to be medieval buildings.

**In-bye and sheilings**

Much of the richer grassland in Assynt is found on the in-bye ground of the crofting townships which extend round the coast from Inverkirkaig to Unapool and again on the limestone at Elphin and Knockan. Prior to the Second World War this ground grew crops or hay. In some places it has been abandoned to bracken or rushes, in others it is now intensively grazed, by sheep or cattle. Although we have, of course, recorded from it, we have not looked at its composition in any detail. When the grazing pressure is reduced, the grassland sometimes proves to be very rich. A good example is a field at Clashmore, which in 2000 was grazed by only four bullocks. That summer it produced at least 30 spikes of *Platanthera bifolia*, eleven of *P. chlorantha*, and two colour varieties of *Gentianella campestris*. Another is a small area of grassland alongside the former school at Unapool. After being fenced from sheep, this produced in 2001 both *P. chlorantha* and an impressively large spike of the hybrid *Dactylorhiza maculata* x *purpurella*.

Other widespread components of the local landscape that were once much more intensively used are former settlements and sheilings. They were often abandoned nearly two hundred years ago, by choice or under duress, but are still easily identifiable by a variety of features, walls, clearance cairns, cultivation ridges and drains, ruins of houses, barns, byres and sheep fanks. In their vicinity relics of grassland improvement or stock feeding such as *Lolium perenne* may persist in now remote areas, such as Dubharlainn on Loch na Loinne. Similarly, clumps of *Urtica dioica* mark the site of nutrient enrichment surely long since exhausted.

In many cases the only persistent changes to the local vegetation are rectangles of brighter green, where bracken has reclaimed the better-drained land. In places, the sweet grassland inside sheiling boundary walls contains species such as *Ophioglossum vulgatum*; it is impossible to know whether they would be present if the area had not been more intensively used in the past.

**Plantings and plantations**

There are several species of deciduous trees which although undoubtedly 'native' elsewhere in the British Isles, are usually not found far from habitation in Assynt. Obvious examples are *Salix viminalis*, planted no doubt for creel making, and *Sambucus nigra*, planted perhaps for superstitious reasons (as also was *Sorbus aucuparia* by remote houses). We just once found a bird-sown example of *Sambucus*, on scree below Creag Sròn Chrùbaidh. More localised is *Salix pentandra*, which is abundant on croft in-bye at Strathan and Badnaban and occasional elsewhere. Since all the trees appear to be male, it must have been propagated or propagate itself vegetatively. There is no one now alive who remembers the planting of these species, and the reason for the establishment of the last, other than for its decorative value, is not known.

At least two local shrubs are on the borderline of 'nativeness'. *Crataegus monogyna* is obviously, although infrequently, planted in a few places around habitation, and occurs in some numbers on the hillside behind Lochinver, presumably bird-sown. However, bushes of considerable size and obvious antiquity are also scattered along the limestone crags from Stronechrubie north to Ardvreck Castle, and the species may there be native. Similarly most of the local stands of *Prunus spinosa* are in the vicinity of past or present habitation, as at Inverkirkaig, Lochinver and on Strone Brae. However, it also occurs in scrub on the south-facing slopes of Cnoc na Sròine, where woodland is shown in Home's map for 1774, and it seems extremely unlikely that it was planted there.

Other shrubs that must originate from gardens and are no doubt bird-sown are *Cotoneaster simonsii* and all three species of *Ribes*, although *R. rubrum* appears to be a fairly long-established component of the local flora.

One of our trees, the ash, also falls into this category of borderline 'native', although it reproduces itself vigorously by seed. There is no doubt about the status of sycamore, although its saplings may occur far from any obvious plantings. Even oak, which appears to be completely native on south-facing slopes and crags from Inverkirkaig round to Unapool, may have been planted in places, such as on the in-bye at Nedd, where there is one tree on each croft.

The earliest conifer plantings, accompanied by non-native deciduous species such as beech, were made in the mid-19th century in the vicinity of Lochinver, where they occupied previously wooded areas on Cnoc na Doire Daraich (Culag Woods) and along the lower reaches of the River Inver (see the list in the section on the history of the landscape). The few silver birch that occur at the latter site are also almost certainly planted. Both areas retain elements of their native flora.

In the earlier part of the 20th century a number of small conifer plantings were established by the Assynt Estate between Inchnadamph and Ledmore, and in the 1980s substantial commercial plantings were made by the Forestry Commission on the north side of Loch Urigill. Neither of these two categories of plantation has any floristic interest, but they are important, of course, from landscape and faunistic points of view. Much more informative are the deer exclosures on the N.N.R. at Inchnadamph, the oldest dating back to the late 1950s, which show just what the native vegetation can achieve with a reduction in grazing pressure.

The purchase, in 1993, of the North Assynt Estate by the Assynt Crofters' Trust, has allowed the development of a number of crofter forestry schemes, which are, of course, fenced against deer. Located between Achmelvich and Culkein Drumbeg, and also at Knockan, they have mainly used 'native hardwoods', but also some conifers, such as *Pinus sylvestris*. More may follow; in time they will have a significant effect on the local vegetation, but so far their effect is minimal. The Assynt Estate has also fenced off, for natural regeneration and planting, two extensive areas on the south side of Loch Assynt, one at An Coimhleum and the other on a site further to the west.

**Water management, peat cutting and muirburn**

Devices for the management of water levels on our burns, rivers and lochs are spread thinly throughout Assynt. The oldest are works associated with small horizontal mills, which were abandoned during the 19th century. In the early 20th century a number of local lochs were dammed to raise the water level, presumably to improve the fishing, but these dams have broken down. In neither case does there seem to have been any long-term effects on the vegetation in their vicinity, although no doubt there was some at the time they were in use.

On a different scale is the large sluice constructed later in the 20th century to control the level of Loch Assynt. However, its only contribution to the local flora has apparently been to provide a habitat, on retaining walls, for one of our rarest native shrubs, *Viburnum opulus*. Long-term effects of the damming of several local lochs for the hydro-electric scheme below Loch Poll, which was constructed during the 1990s, remain to be seen, although the resultant fluctuations in the level of Loch na Loinne have created a substantial area of draw-down, where the former vegetation has died off.

It just remains for us to consider the effects of two long-standing activities, peat-cutting and muir-burn. Peat is still cut in a few places, but the practice was in the past very widespread, since in an area generally lacking woodland, peat was the only readily available source of fuel. The straight lines of ancient peat banks may be found in remote areas. In some townships, all the available peat within the boundaries appears to have been removed by the early part of the 20th century, if not earlier, and the inhabitants had to travel further afield for their fuel. The long-term effect on the vegetation is difficult to gauge, since the top cut was returned to the worked-over area, but there must have been some local impoverishment of the flora (see also the account of bryophyte communities).

Muirburn has been discussed in the section on heaths, mires and crags, but merits another mention. It is still being actively practised in some crofting townships, if only by a few individuals. A recent study (Hamilton *et al.* 1997) has described muirburn in the Assynt area as typically on a larger scale and more frequent than recommended in the guide-lines, poorly-controlled, and 'at least partly over areas that should be fire-free'.

Two woody species, *Arctostaphylos uva-ursi* and *Juniperus communis*, provide us with a good indication of the scale and extent of muirburn over a long period. It cannot be a coincidence that flourishing populations of both only now occur on ground that is remote from habitation. Some of the recent problems with muirburn arise from an increasing shortage of man-power. It is perhaps time that this destructive practice was abandoned, at least until it can be carried out in a properly controlled manner, and then only after due consideration of the environmental consequences.

# FLOWERING PLANTS AND FERNS

# HISTORY OF RECORDING

## Introduction

Written contributions to knowledge of the flora of Assynt date back nearly 250 years, but until the 1990s all those involved were visitors to the area, as may be the case in other remote parts of Scotland. Where published, the accounts of the earliest visits, by Scottish botanists, are to be found in journals such as the *Edinburgh New Philosophical Journal* and the *Transactions of the Botanical Society of Edinburgh*. From the 1880s, when the botanical charms of the North-West Highlands became known further afield, there are a series of papers, in particular by the Rev.E.S.Marshall and associates, in the *Journal of Botany*. In the twentieth century most of the information found its way into the publications of the Botanical Exchange Club, and its successor the Botanical Society of the British Isles. All of which means that the original sources are somewhat dispersed and many are available in the original no nearer to Assynt than Edinburgh. This is one reason why we have often quoted them at length. Another is the sheer enthusiasm for Assynt manifested in many of the accounts, its landscape as well as its botany.

Information relating to the flora of Sutherland in general, and Assynt in particular, was not finally brought together and published until 1976, with the completion by J.B.Kenworthy of *John Anthony's Flora of Sutherland*, four years after Anthony's death. The former county of Sutherland is huge, and comprises two Watsonian vice-counties, v.c. 107 (East Sutherland) and v.c. 108 (West Sutherland). Happily, Anthony had decided that, for his species account, the vice-counties would be 'further sub-divided into smaller districts - the parishes...[which are] based on the river systems and [whose] mutual boundaries are, for the most part, traced along the watersheds'. Assynt is one of the five large and ancient parishes which comprises West Sutherland, so Anthony's *Flora* gave us ready access to a summary of the available information on the flora of the parish. For detail, we were also able to consult Anthony's card indices, manuscripts and correspondence for v.c. 108, which are held by the current Vice-County Recorder (P.A.E.).

Similarly, the chapter by Anthony on the 'Botanical Exploration' of Sutherland as a whole provided a valuable framework for our account of that of Assynt. All the works listed there have been checked for specific references to the parish.

Exploration of the flora of Assynt appears to have proceeded by fits and starts, often with gaps of up to fifty years. It is therefore convenient to split the period between the first records and the present day into fifty-year periods.

## Prior to 1749

The earliest references to the plantlife of Assynt are embodied in the Gaelic place names, some of which date back at least to the 17th century. Those we have noted refer almost exclusively, and perhaps not surprisingly, to trees and shrubs, and the following examples are a selection only. The commonest element is probably *beith/e* (birch), which appears in Pollan Beithe and several Allt/an Beithe. Almost as common is *darach/daraich* (oak), which appears in Cnoc an Doire Darach, Baddidarach and Creag Dharaich.

*Fearn* (alder) is an element in the names of two lochs, on only one of which the species is still to be found. *Cnaimhseag* (bearberry) appears in Beinn na Cnaimhseag and Druim nan Cnaimhseag, on which hill and ridge alpine bearberry *Arctostaphylos alpinus* still occurs. *Aitionn/aitinn* (juniper) is an element in just one place name, Loch Ruighean an Aitinn.

Eilean na Gartaig in Cam Loch is the only name referring to a herbaceous species, *garlag* being wild garlic, which still occurs in quantity there (although there appears to have been a consonant change).

Other plant place name elements are more general, such as *driseach/drisiche*, as in Loch a' Phollain Drisich. This is derived from *dris* for a spiny shrub, either bramble or briar; given the scarcity of brambles in Assynt the reference may well be to the more frequent briar or rose. A similar case is *droigheann/droighinn*, which also refers to 'thorn, bramble, blackthorn, sloe'. Poll an Droighinn, a rocky hollow about a kilometre east of Stoer village, still has a small stand of blackthorn. However, the other place so named is an area of old sheilings at a height of about 350m. south of Beinn Uidhe; one can only speculate what spiny shrub might be involved in this case, perhaps by relationship (since it is unarmed) cloudberry *Rubus chamaemorus*?

## 1750-1799

When Anthony was summarising what was then known about the flora of Sutherland, one potentially important source of information, the earliest, was not available to him. In 1994 Henderson and Dickson published an account of the journals and other papers of James Robertson. Robertson was a protégé of John Hope, Superintendent of the Botanic Garden at Edinburgh and Professor of Botany at Edinburgh University. Hope refers to him in 1766 as 'a young man, one of the gardeners, whom I had educated myself on a botanical expedition'. Robertson undertook a brief expedition to Arran, Argyllshire etc. in 1766, but in mid-April 1767 set out on the first of five major expeditions, which were to take him throughout the Highlands, and to the Orkneys, Shetlands and Outer Isles.

For more details of his discoveries, which included a number of first records for the British Isles, see the account by Henderson and Dickson (1994). What concerns us here is the case for some of Robertson's records being the first for Assynt, which we have accepted, with the provisos noted below.

We know from his Journal for 1767 (Henderson and Dickson, pp. 55-56) that he stayed, probably towards the end of August, at Achmore, a farmhouse just half a mile north of Inchnadamph, which was occupied by the Scobie family, tacksmen on the Sutherland Estate. His general observations on his stay in Assynt are worth quoting in full, to give something of the flavour of his Journals.

'From Scowrie I went to Achmore in Assynt by the head of Khylescow & over the ...Hills. A few cultivated spots near the shore excepted, this tract is entirely composed of hill and moor. Passing over the hills I found an abundance of *Cherleria Sedoides*, Dwarf Cherleria. Here as well as on Ben Clibrick it seem'd to prefer a north aspect'.

'At Achmore there are some small cultivated fields, and large rocks of Limestone which running south-west, traverse the country for several miles. About a mile east from the Kirk of Assynt a small river [Traligill] runs near 3/4ths of a mile thro' a subterraneous passage of Lime-stone'.

'The people here assured me that a decoction of the roots of *Angelica sylvestris*/Wild Angelica, was very efficacious in removing some distempers to which their cattle were subject especially that called the black disease'.

'In the Glen above the Kirk I saw a species of Crow with red bill & feet [chough; apparently not otherwise recorded in Assynt]. It chatters like a jack-daw. Here too I saw *Ulmus campestris*/Common Elm, growing plentifully on the rocks. There is every reason to suppose this is an indigenous plant'.

''Tis said that Mariners approaching this coast observe their needles to point towards two hills in Assynt, in one or both of which 'tis probable that a magnet is contain'd'.

'From Achmore I went by Ledbeg, the crooked Loch [see below], thro' Coygach to Loch Broom'.

'The ground continues hilly, fit for pasture only. At Ledbeg there is a rock of good white marble. In a small island within the crooked Loch there is an abundance of *Ribes rubrum*/Common Currants'.

In passing, Henderson and Dickson (p.61) identified the 'crooked Loch', referred to twice in the extract above, as Loch Lurgainn, on the Achiltibuie road, at the foot of Stac Pollaidh, the Gaelic *lurgainn* referring to a crooked or unshapely leg. However, it seems much more likely that the loch referred to was in fact Cam Loch, close by the road from Ledbeg to Elphin (and on to Coigach). Not only is *cam* crooked in the Gaelic, but there is a large wooded island, Eilean na Gartaig, clearly visible and readily accessible (by boat) from the road, which may well have housed *Ribes rubrum* in Robertson's time, since it was found there again in 1959 by Alfred Slack. Taking the Achiltibuie road down Loch Lurgainn on the way to Loch Broom would have involved a considerable detour for a traveller towards the end of a very strenuous tour in 1767, and the one large island in that loch looks extremely unlikely ever to have harboured red currant.

Now to return to the issue of Robertson's Assynt records. These are contained in a manuscript, previously unknown to the botanical fraternity, in a set of papers in the Signet Library in Edinburgh (Henderson and Dickson, p.232 *et seq*.). This manuscript is headed 'Flora of Ben Hope and Ben More in...' They suggest that the missing word is most probably 'Sutherland', 'which would indicate that the mountain [referred to] was Ben More Assynt'. The manuscript takes the form of a long list of names of plants (including, at the end, bryophytes, lichens, algae and fungi), against which are tabulated the 'stations' at which they occur, shown as 1-5 over 1/4, 1/2, 3/4, 7/8 and 1. They conclude, convincingly, from an examination of which plants are shown as occurring at which stations, that this is a system, both 'original and delightfully concise', of recording the altitudinal range of the species concerned (i.e. station 1 is from the base of hill to a quarter of the way up it, etc.) In addition, some names are annotated with 'H' or 'M', which they take to imply that those plants occurred only on Ben Hope or Ben More and that those without such annotations occurred on both mountains. Again, this seems entirely reasonable.

Henderson and Dickson then go on to discuss at some length the rival claims for 'Ben More' of the hills so-named in Assynt and Coigach. Unfortunately, Robertson does not specifically mention in his Journals going up either mountain, although Achmore is within easy reach of Ben More Assynt.

We have looked at the list of species in the light of our knowledge of the flora of Assynt, and of the distribution notes in Henderson's checklist for West Ross (1991, 1992). Many of the species listed, whether from the lower or higher 'stations', do not allow us to choose between the two contenders, but there are some that seem to point unequivocally to Ben More Assynt. The most compelling is 'Scolopendrium vulgare' [*Phyllitis scolopendrium*], which is found on the limestone behind Inchnadamph (i.e. on the usual route to Ben More Assynt), but now occurs in West Ross no nearer to Coigach than the coast at Loch Ewe. Another is *Viola canina*, again on the Inchnadamph limestone, but found in West Ross only at Upper Loch Torridon. A third, higher level, species is *Saussurea alpina*, which is not recorded from Coigach, but is certainly present behind Inchnadamph.

We have therefore opted to regard Robertson's records from this table as first records for Assynt for the species concerned, with two provisos. The first is one that haunts the question of some first records for both the parish and vice-county 108 (West Sutherland). **Ben More Assynt is not and never has been within the bounds of Assynt.**

The concurrent estate, parish and vice-county boundaries run along the south-north ridge of Conival, almost two kilometres to the west of the slightly higher peak of Ben More, being connected to it by an east-west ridge. Ben More Assynt is in the parish of Creich, in vice-county 107 (East Sutherland).

The second proviso is that, although we have included Robertson's records for stations 1-3, we have, regretfully, ignored those from stations 4 and 5, which Henderson and Dickson interpret as the 'brow of the hill' and the 'summit plateau'. It is possible that parts of the brow and summit could indeed refer to Conival, rather than Ben More Assynt itself, since the latter name is often, unfortunately, used by later botanists to signify the whole massif. However, given all the 'ifs and buts' we have decided to err on the side of caution at this point.

With these provisos, what of Robertson's records from the lower stations? There are some 124 species in this category. Much of the value of the list is that it includes very common species, which were not again recorded for Assynt until the survey work for the first Atlas in the 1950s. Since we do not know how far from Achmore Robertson travelled in search of plants, his records have been described in the Species Account as from 'near Inchnadamph'.

There are a couple of apparent errors, such as *Primula veris* instead of *P. vulgaris*, and *Bunium bulbocastanum* for *Conopodium majus*. There is an anomalous record of *Plantago coronopus*, which has not since been recorded so far inland. The same applies to *Koeleria macrantha*. Then there are records of species which may well have occurred in and around Inchnadamph in the second half of the 18th century, but which have never been recorded since, and we therefore find surprising. Examples of these are *Trisetum flavescens*, *Geranium lucidum*, *Eupatorium cannabinum* and *Cruciata laevipes*. Finally, it is a tribute to Robertson's sharp eyes that he found *Ajuga reptans*, three out of four recent stations for which are at Inchnadamph, and *Parnassia palustris*, for which we only have one recent record (at Elphin).

For completeness, we should mention one further small, but usefully corroborative, source of information about plants recorded by Robertson. It is the catalogue of John Hope's Hortus Siccus (Balfour 1907), where '*Polypodium fragile*' [*Cystopteris fragilis*] is recorded as 'in Assynt J.R. 1767'. The other two local records mentioned therein are '*Ribes rubrum*' 'plentifully in an island of the Crooked Loch... Assynt' (see above) and '*Ulmus campestris* [*U. glabra*] in Sutherland and Assynt on precipices where there never had been plantations, wich elm'.

After which excellent start, only a handful of further records were made during the 18th century, although they are ones that, for reasons that Henderson and Dickson explain, reached a much wider audience than those of Robertson.

The context of this handful is in many ways more interesting than the records themselves. It is the account by Thomas Pennant of *A Tour in Scotland* made by himself and the botanist John Lightfoot in 1772 (see Bowden 1989, p.67). He describes the view, presumably from Knockan, as follows, 'I never saw a country that seemed to have been so torn and convulsed: the shock, whenever it happened, shook off all that vegetates: amongst these aspiring heaps of barreness, the sugar-loaf hill of Suil-bhein made a conspicuous figure: at their feet the blackness of the moors by no means assisted to cheer our ideas'. And this on 28th July!

However Cam Loch, 'pretty decorated [as still] with little wooded islands', may have cheered the travellers a little, as also the 'kindness and hospitality [of] the people of these parts'. Pennant and Lightfoot had apparently aspired to ascend Ben More Assynt, but having been told that the way north (from Ledmore) was impassable for horses and extremely difficult on foot, they retraced their steps to their ship at anchor at Isle Martin.

Whatever their views on the general scenery and vegetation, Pennant and Lightfoot were impressed by 'the vast limestone rock called Creg-achnocan', where they recorded *Dryas octopetala*, *Draba incana*, *Asplenium viride* and '*Polypodium* [*Polystichum*] *lonchitis* (Lightfoot 1777). All still occur there.

## 1800-1849

Although W.J.Hooker, Professor of Botany at Glasgow, 'visited many places in Scotland gathering material' for his *Flora Scotica* (1821), venturing as far afield as 'the wet moors adjoining Cape Wrath', the only records in that work that refer specifically to Assynt are repeats of Lightfoot's observations made some 50 years previously.

Botanising in the Highlands in the early part of the 19th century was a pastime for the stout-booted, not the faint-hearted. Dr Robert Graham starts his note on *Rare Scottish Plants* (Graham 1826), 'In a walk through the island of Skye, the west of Ross-shire, and Sutherland, to Caithness, in August last' [1825]. He and his companion, Mr John Home, may well have passed through Assynt, but the only record that could relate to the parish is of *Subularia aquatica*, 'in Sword Loch on the confines of Sutherland and Ross-shire', a locality yet to be identified.

In August 1827 Dr Graham was back again 'into the North of Scotland...on an excursion...with part of his pupils' (Graham 1827). They took a boat from Aberdeen to Cromarty, and proceeded thence via Tain, Bonar Bridge, Lairg, Loch Shin to Ben More Assynt, Handa, Scourie, Laxford, and points north and east, taking in Ben Hope and finishing up at Brora (Balfour 1865). There are records from 'Ben More, Assynt' of *Carex pauciflora, Utricularia intermedia, 'Apargia alpina'* [*Leontodon autumnalis*], *Poa alpina, Hieracium alpinum, Cerastium alpinum, 'Aira alpina'* [*Deschampsia cespitosa* ssp. *alpina*] and of *Drosera longifolia* from 'bog north side of Ben More, Assynt'. Regrettably, as we have seen, these records are likely to refer to the neighbouring parish of Creich.

However, there are two records in the report of this 1827 excursion which refer unambiguously to the parish. The first is '*Epipactis latifolia* on limestone rocks in Assynt'. '*E. latifolia*' is an older name for *E. helleborine* which is still found in Assynt, both on the coast and, rarely, near Inchnadamph, but the references to limestone, both in Assynt and also 'at Keoldale, parish of Durness' suggest the species seen was in fact the more frequent *E. atrorubens* (see also Marshall 1899). The second is '*Pyrus aria*', likewise 'limestone rocks, Assynt', which Anthony refers to *Sorbus rupicola*. This is still found in small quantity on the Stronechrubie cliffs, but unless it was more widespread in 1825, this was very well spotted.

Dr Graham returned to Sutherland in early August 1833 'favoured with the company of a number of friends, devoted to various branches of natural history, and some of them eminent in such pursuits'. The 'company' walked from Invergordon to Bonar Bridge, and thence 'proceeded by Oikel, Inchnadamf, Kylestrome, Scourie' and up to the north coast, returning via Strathnaver and Lairg (Graham 1833).

On this third expedition Graham records some three or four species from Assynt. The first reference is puzzling: '*Carex panicea* var. *phaeostachya*.- I found this plant on Specanconich [see below] and think there is no doubt of its specific identity with *Carex panicea*'. Despite his conviction, this 'variety' is now synonymised with *Carex vaginata*, for which there two other records for West Sutherland, Oldshoremore (1833) and Foinaven (1957). Oldshoremore now seems extremely unlikely, since the species does not normally occur below 600m. (2000ft.).

Of *Crataegus oxyacantha* [*C. monogyna*] he says 'I only saw one bush of this on a rock at Loch Assynt'; it is tempting to think that this was on the limestone at the eastern end of the loch, where some ancient examples still survive. *Luzula arcuata* was discovered 'on the ridge leading to the top of Ben More, Assynt from Inchnadamf'; this may well refer to the northern ridge of Conival, along which runs the parish and county boundary and where the species was rediscovered by G.P.Rothero in 1998.

The fourth record is of *Utricularia minor* 'found in flower only once in a small pool near the base of Speckanconick, Assynt by Mr Parnell'. In twelve years survey we too have only once found this species in flower. 'Spec(k)anconick' presents something of a problem, since the nearest locality to Assynt with that name is Speicein Coinnich on Ben Mor Coigach, and Graham's party appears not to have visited that area; is it perhaps a misnomer for Spidean Coinich on Quinag? There is some useful background information on Graham's visits in Balfour (1865).

G. and P. Anderson, in their *Guide to the Highlands and Islands of Scotland* (1834) append some plant records to their locality descriptions. They note *Hedera helix* and *Dryas octapetala* at Inchnadamph (as well as the *Sorbus rupicola* and *Ulmus glabra* already known), *Phyllitis scolopendrium* and *Hymenophyllum wilsonii* at 'Achumore' [Achmore] and *Nymphaea alba* at Kylesku.

The records for Assynt in Dr A. Murray's *The Northern Flora* (1836) are all ones published earlier by Graham.

As Anthony remarks, the botanical information contained in the accounts of the parishes in the *New Statistical Account of Scotland* (1845) varies considerably according to the knowledge of the compiler. The contribution by Rev. A. Gordon on Assynt is fairly brief, and does contain a number of possible new records, so is perhaps worth quoting in full.

'The alpine vegetation of the parish...is very similar to that which is met with in equal elevations in the greater part of the north of Scotland. As types may be mentioned, *Saussurea alpina*, *Cherleria sedoides*, *Hieracium alpinum*, *Vaccinium uliginosum*, *Asplenium viride* as plants which are not very rare in alpine districts: but less generally diffused than such as these last named, may be mentioned *Carex pulla* [*C. saxatilis*], *Carex pauciflora*, and *Arbutus alpina* [*Arctostaphylos alpinus*]'.

'The limestone districts in the parish are characterized by *Epipactis latifolia* [probably *E. atrorubens*, see above], *Dryas octopetala* - the latter in great profusion, and perhaps, in Sutherlandshire only growing on the limestone or micaceous rocks'.

'Among the rare plants found in alpine or subalpine disyltricts of the parish, may be mentioned *Pyrus aria*, *Apargia alpina* [*Leontodon autumnalis*], *Luzula alpina*, - this last found in Scotland only in three stations, of which Benmore, Assynt, is one'.

'*Silene maritima* also grows on Benmore'.

'The following may be named as yielded by the bogs in the parish:- *Carex filiformis* [*C. lasiocarpa*], *C. limosa*, *Utricularia minor*, *U. intermedia*, *Drosera anglica*, in profusion, *D. longifolia*, *D. rotundifolia*, *Sparganium fluitans* [*S. angustifolium*?], *Cladium mariscus*, in a swamp half-way between Kylestrome and Badcall, *Ligusticum scoticum* is abundant on the shores in some places'.

A number of the records are derived from earlier accounts and there is the common misapprehension that Ben More is within the bounds of the parish. *Carex saxatilis* still occurs on Ben More, and one cannot tell whether any of the other montane species may not have been recorded from there. However the list from 'bogs in the parish' does contain several first records for Assynt (if one ignores the reference to *Cladium mariscus*, from Eddrachillis!).

## 1850-1899

Despite Graham's excellent example, serious botanical exploration of Sutherland in general, and Assynt in particular, did not take off again until the 1880s. Prominent amongst those involved were Archibald Gray (1886), F.J.Hanbury (1885-1890) and the Rev. E.S.Marshall, whose visits spanned some thirty years (1887-1915). Hanbury's excursions were mainly to the north coast and he did not visit Assynt until 1890.

In the summer and autumn of 1886, Archibald Gray made 'a careful examination of the large district of Assynt, together with the western portion of the parish of Tongue' (Gray and Hinxman 1889). Gray having been 'unfortunately prevented from re-visiting Sutherland' (by 1891 he was being referred to as 'the late Mr Gray') or 'from doing full justice to the subject in a paper such as he would have liked to lay before the Society', 'his materials' were worked up by Lionel Hinxman of H.M.Geological Survey, who also contributed 'a few notes on the chief characteristics of the plant life of this part of Sutherland'. The paper was communicated to the Botanical Society of Edinburgh by the geologist Benjamin Peach and constitutes the first substantial account of the botany of the parish.

There are nearly 140 specific references to Assynt, mentioning localities ranging across the parish from Inverkirkaig, Lochinver, Achmelvich, Clachtoll and Stoer on the coast, to Unapool, Achmore, Inchnadamph, Elphin and Knockan on the eastern side. Gray apparently climbed a number of the hills, including Beinn Reidh, Beinn Gharbh, Quinag, Beinn nan Cnaimhseag and 'Coinnemheall' (Conival). He does not record any species from the north coast of the parish between Stoer and Unapool. He was the first to find *Ajuga pyramidalis*, at Torbreck, and *Arenaria norvegica*, at Inchnadamph ('Mr Gray has been the first to discover it on the mainland of Britain'). Gray recorded, in all, 474 species from West Sutherland, of which 79 were at the time of publication new to the vice-county (108), a very considerable addition to knowledge of the botany of the area.

The following year, 1887, E.S.Marshall paid a visit to Inchnadamph, with which he was very taken, 'In my opinion...the neighbourhood of Inchnadamph (108), with its varied soil and elevation, will well repay careful search', and he was as good as his word. His visit, in the late summer, was very productive of new records (Marshall 1888). Some were common enough, but not before noted, such as *Polygala serpyllifolia* 'abundant near Inchnadamph', and *Ilex aquifolium* 'clearly native about Inchnadamph, growing on the limestone cliffs, far from houses'.

Others were of difficult taxa, even in today's terms, such as 'a very curious plant...found growing in some quantity by a rill above the path from Inchnadamph to Ben More of Assynt, at about 1600ft.' This was finally diagnosed by Haussknect, the contemporary authority on the willowherbs, as the hybrid *E. anagallidifolium* x *obscurum*, which he named, after its finder, *E.* x *marshallianum*. It has only once since been recorded in the British Isles, in Stirlingshire. Marshall was the first to record *Salix myrsinites* and a number of the other limestone specialities from Inchnadamph, including *Juncus alpinoarticulatus*, which was, apparently, also collected in 1908 by 'Loch Assynt', but has not been seen since. Almost all his Assynt records on this first visit are from Inchnadamph, with just a few from Quinag, Lochan Feòir on Quinag's lower slopes, and a couple from Conival.

Marshall was back in the Highlands with F.J.Hanbury in July 1890, his enthusiasm for the Assynt limestone undiminished. 'Leaving London, we made straight for

Inchnadamph', [a] 'place...remarkable for the low altitude to which decidely alpine species descend; the high mountains near are, however, decidedly barren'. 'A day and a half at Lochinver barely gave time to explore the vegetation of the district', however. Amongst new discoveries were *Stellaria neglecta* (Inchnadamph) *Saxifraga hypnoides* (Inchnadamph), several *Hieracia*, *Hammarbya paludosa* 'in a bog near Lochinver' and *Melica nutans* on 'wooded banks of the river near Lochinver'.

In June of 1894, another of the pillars of the botanical establishment, George Claridge Druce, 'paid a visit of a few hours to Ledbeg in West Sutherland, in order to verify records made by Dr. Lightfoot in the *Flora Scotica*' (1777). The four species recorded by Lightfoot on 'Creg-achnocaen' had been included in Watson's *Topographical Botany* (1883) for West Ross (v.c.106), but Druce ascertained that the 'county boundary is marked by an iron railing near a small watercourse', and having made certain where he was recording found that 'the flora of the Sutherlandshire portion of the rocks is very interesting' (Druce 1895).

His records, from the cliffs, Cnoc an t-Sasunnaich above them, fields at Elphin and 'Knockain' and the falls at Ledbeg, paint an evocative picture of the plants of the area at that time. They include *Adoxa moschatellina* (not seen since), *Crepis virens*, *Urtica urens*, and a long list of arable weeds, many not since recorded in Assynt. He also mentions, in passing, the capture 'on the Cnoc-an rocks', by him and his photographer, who was in the carriage below, of a young golden eagle, so 'gorged from over-eating' that it could not move; this was conveyed to Ullapool, confined in a parrot cage!

The final contribution of the 19th century was by C.E. Salmon who, at the end of July 1899, spent 'a few days in Sutherland, botanizing chiefly near Inchnadamph, a small village in as wild a tract of country as is possible in Great Britain, I believe' (Salmon 1900). He was in the company of 'H.N.Dixon, W.E.Nicholson, and my brother, who were investigating the moss-flora of the district'. Salmon appears to have been the first to botanise on Canisp, although on a very unfavourable day, with 'clouds of mist and rain frequently sweeping across the mountain and leaving one with views limited to but six or eight feet!'. On an expedition up Ben More Assynt, he noted *Luzula arcuata* 'still plentiful in one compact patch on the way to the summit, where it was recorded many years ago' (see Graham 1833), and where it was re-found in 1998. He also visited Glas Bheinn and Suilven. Amongst his discoveries were *Viola lutea* at Inchnadamph, *Cerastium arcticum* at 2500ft. on Canisp, *Centaurea cyanus* in a cultivated field at Inchnadamph, several *Hieracia*, *Poa glauca* at 2000 ft. on Canisp (where it has not been seen since), and *Isoetes lacustris* in two lochs on Glas Bheinn. He was also the first to record any charophytes from Assynt.

## 1900-1949.

The first significant contribution of the 20th century was the product of a further visit by G.C.Druce, in July 1907 (Druce 1908), when he based himself at Inchnadamph, 'with its splendid range of limestone cliffs', where 'many alpines come much lower down than we are accustomed to see them, for instance *Carex capillaris*...and *Dryas octopetala*'. Although he says that 'the season was too backward to allow of any critical study of many species, the Hawthorn being in full bloom on the 14th July', he nevertheless provides records with specific Assynt localities for some 60 species, including: *Draba incana*, *Silene acaulis*, *Veronica serpyllifolia* ssp. *humifusa*, *Pseudorchis albida*, *Platanthera chlorantha* and *Luzula spicata*. *Lycopodiella inundata*, recorded by him on 'Canisp, near the base', has not yet been re-found. A useful feature of his list is that it includes Assynt records of a number of common, but previously overlooked species, such as *Caltha palustris*, *Myosotis scorpioides* and *Vicia sepium*, albeit often as varieties that no longer have any significant taxonomic status.

Given the precision with which he located the boundary between West Ross and Sutherland on the Knockan crags (see above), we have taken it, exceptionally, that when he referred to 'Ben More, Assynt 108' he knew he was in West Sutherland i.e. on the western side of Conival ridge, especially as he differentiates records from 'the eastern side 107' i.e East Sutherland.

E.S.Marshall was back at Inchnadamph for a fortnight in July 1908, with W.A.Shoolbred, this time focussing on 'critical forms, especially of *Hieracium*', of which they recorded some 18 taxa (Marshall and Shoolbred 1909). Many of these were from the limestone gorges of the Traligill River, Allt Poll an Droighinn and Allt a' Chalda Mor (still very productive), but they also collected on the north face of Canisp and Beinn Gharbh. *Hieracium shoolbredii* was later named by him (Marshall 1913) from material collected on this visit, one of the three locations in which they found it being 'a hill overlooking Loch Maol-a-Choire', i.e. Cnoc Eilid Mhathain, later to be celebrated for its limestone flora. The roses received some attention, but the taxonomy of these has changed so radically that it is difficult to equate their records with modern taxa. Other interesting records include *Equisetum variegatum* from the Loanan River and *Isoetes echinospora* from 'peaty pools, Unapool Burn', where it still occurs.

Elsewhere in Britain the three decades from 1910 are, not unsurprisingly, rather a thin time for botanical recording, and this is true also for Assynt. E.S.Marshall made one more visit to West Sutherland, in 1915, four years before he died, but confined himself to the north coast.

However the indefatigable G.C.Druce, though by now in his seventies, was still travelling extensively, presumably with a companion at the wheel of the car. In July 1923 he embarked on a tour of East Sutherland, Caithness and West Sutherland (Druce 1924a). On 12th July he motored from Bettyhill by Loch Naver to Altnaharra, then north again to Eriboll and on to Durness, explored the Cave of Smoo, and 'reached Rhiconich rather late and found the Hotel filled with fishers who are almost as dull and self-centred as golfers'! The next day he 'had to go from the west to the east coast and back, since the Scourie ferry was not working'.

All this mileage (on unsurfaced roads) was redeemed by arrival at Inchnadamph, where 'the pasture by the stream was literally full of *Habenaria bifolia*, plenty of *H. albida* and *Orchis praetermissa*' [*Dactylorhiza purpurella*?]. From Inchnadamph, he continues, 'we climbed Ben Garve where there was plenty of *Arctostaphylos alpina* and a few specimens of *Hieracium alpinum* and other alpines. On 16th they motored from Inchnadamph to Ledbeg and Knockan, where 'we saw a small pasture one mass of *Trollius. Orchis incarnata, O. praetermissa* and its var. *pulchella* [sic] and their hybrids were in great show', and 'thence through magnificent scenery...to Ullapool'. Druce continues, 'What a series of mountains between Inchnadamph and the latter place - Quinag, Canisp, Suilven, An Stack, Stack Polly, Coulmore, Coulbeg and Ben More of Coigach, most of which I have climbed but the flora of which is relatively unknown'. It was presumably on this visit that he recorded *Paris quadrifolia* 'on an islet in Loch Awe...An interesting locality', as indeed it was (Druce 1924b).

G.C.Druce paid one further brief visit to Assynt, in July 1925, when he records that 'we visited the rocks of Cnockan and obtained *Rumex arifolius* in both Ross and Sutherland' (Druce 1926). This record, of a European species, is not now accepted (Lousley and Kent 1981). Thence he proceeded to 'Leadbeg' and Inchnadamph, but his remarks more or less repeat those made about the 1923 visit. Curiously, an attempt 'to get over Scourie Ferry...after about eight miles of atrocious roads' was once again thwarted, by 'so strong a wind...that the very disagreeable Charon would not venture, so we had to come back, with only the very grand view of Quinag to repay us'.

Other distinguished botanists almost certainly visited Assynt between the two wars, but the only references to their visits are individual records in the voluminous *New County and Other Records* in the *Reports of the Botanical Exchange Club*, which we have not made the opportunity to peruse. There do not appear to be any papers of any substance relating to Assynt from this period. Perhaps the best known part of the parish, the area around Inchnadamph, was thought to have yielded all that was likely to be of interest, and in any case, a generation of possible enthusiasts had perished in the First World War, with effects on the study of natural history that occurred throughout the British Isles. Another factor may have been the new ecological approach to the study of vegetation, especially in montane areas, as exemplified on Ben Armine. further to the north (Crampton and MacGregor 1913), of which more anon.

Be that as it may, the delightfully discursive style of recording botanical expeditions had one final flowering in Assynt, a paper entitled *Autumn botanising at Lochinver (West Sutherland)*, by A.J.Wilmott of the British Museum (Natural History) and Miss M.S. Campbell (1946). This relates to a visit in the latter part of September 1943 and 'further collections' earlier in September 1944. Wartime restrictions meant that 'the collections were made on a series of short journeys by bicycle from Lochinver', although 'on a few occasions it was possible to go further afield owing to the availability of vacant seats in cars making essential journeys.' 'No attempt was made to ascend any of the mountains', but accessible habitats that looked interesting were well worked, especially since 'no note had previously been made of some common species'.

This intensive approach, from a base away from the well-worked Inchnadamph area, was very productive, and the detailed descriptions of the localities visited add greatly to the interest of the paper. They include the Culag woodlands, where *Ulex gallii* was found in what is still its only Assynt locality, Clachtoll, Stoer and as far round the coast as Drumbeg (travel was restricted further to the east, because of deep-water moorings for warships). One excellent find in 1943 was *Cephalanthera longifolia* on the road to Achmelvich.

The 1944 visit concentrated on the machair of Achmelvich Bay, where '*Gentiana septentrionalis* spangles the turf, *Ophioglossum* was occasionally seen, and *Coeloglossum* was abundant'. 'An excursion to Stoer and Balchladich led to the discovery of several *Potamogetons* at the edge of the lochan held up by the sand at Stoer Bay'. The lochan was Loch an Aigeil and amongst the discoveries were *Potamogeton alpinus, P. berchtoldii, P. gramineus, P. filiformis* and *P.* x *nitens*, all new to Assynt. *Sagina nodosa* was also noted at the edge of the loch, where it still grows.

A refreshing amount of attention was also paid to 'weeds', for instance in the garden of the Culag Hotel, which yielded *Fumaria officinalis, Geranium dissectum* and *Veronica persica*; a potato patch at Inverkirkaig and a midden at Feadan also provided useful records. One of the furthest excursions was to the lower slopes of Creag Mhór, Quinag, where Miss Campbell made the very interesting discovery, given the local geology, of *Equisetum variegatum*. Altogether about 170 taxa are listed, with detailed localities and comments. This contribution to the knowledge of Assynt plants rivals in importance that of Robertson some 170 years previously.

After the War, with access restrictions lifted, but petrol still in short supply, a trickle of botanists found their way back to Assynt, notably Mary McCallum Webster, in 1947, 1949 and on in to the 1950s, when she made a major contribution to recording the north of Scotland for the first *Atlas* (see below).

## 1950-1999: plant hunters

In this final 50-year period of our review of the botanical exploration of Assynt, something of a dichotomy opens up. The tradition of plant-hunting, by both professional and amateur botanists (in the best British tradition) continued into the 1950s, focussed from 1954 onwards by the exigencies of the 10 km. square survey of the whole of the British Isles which resulted in the *Atlas of the British Flora* (Perring and Walters 1962). This survey soaked up the energies of most botanical visitors to Assynt in the latter part of that decade. At the same time, interest was developing, in academic and conservation circles, in the ecology of the relatively unspoilt vegetation of parts of West Sutherland, in particular the moorlands, mountains and lochs. In so far as they contribute to knowledge of the flora of Assynt, these studies will be covered in a separate section of this review.

For a new generation of botanists, especially those from further afield unfamiliar with the earlier literature, the description by John Raven of the delights of Assynt (and Coigach) in the introduction to the section on North-Western Scotland in *Mountain Flowers* (Raven and Walters 1956) must have been an eye-opener. He starts, 'If there is any part of Scotland, indeed of Britain as a whole, more hauntingly lovely than the Northern tip of Skye, there is no doubt in my own mind where it lies. It is that rectangle of mountain, moorland and loch, which is bounded on the north by Kylesku and Clashnessie, on the south by Elphin and Achiltibuie'. His circuit of the area begins at Ledmore Lodge and proceeds north to Inchnadamph, of which he says 'there are few botanical centres in Britain more rewarding...nor any better suited to the indolent'.

Raven had first visited the area in 1951 in search of what has been called 'Don's Twitch', formerly *Agropyron donianum*, now, sadly, downgraded to a variety of bearded couch *Elymus caninus*. His story of the rediscovery of this grass (Raven 1952) is a fine account of botanical detective work, both in the herbarium and the field.

After extolling the manifold virtues of the Traligill valley and adjacent areas, he takes the reader up onto 'a particular limestone ridge...one of the centres of the mountain flora of the district'. This ridge is Cnoc Eilid Mhathain, above Loch Mhaolach-coire, known to aficionados as 'the high limestone', which retains its interest still, not only for limestone specialities, but also, as Raven points out, for the sudden change in the flora where the limestone abuts the Torridonian of Beinn nan Cnaimhseag.

Raven also draws attention to the rich flora of 'roadside banks...near Stoer', which 'abound with globe-flower, bitter vetch [*Vicia orobus*] and burnet rose', although here his geology lets him down, since what he took for limestone is a basic facies of the gneiss. His enthusiam for the area must have encouraged several generations of botanists to visit parts at least of Assynt. His co-author, Max Walters, who shared his enthusiasm for the Inchnadamph area, made a number of interesting records there on a joint visit with Raven in 1953 (Lipscomb and David 1981), notably *Alchemilla glaucescens* 'on the rabbit-grazed south-facing slopes of a shallow valley' not far from the Hotel, where it still occurs.

Max Walters was in 1954 appointed Director of the Distribution Maps Scheme of the Botanical Society of the British Isles, which revolutionised knowledge of the distribution of the British flora, Assynt included. Although some earlier records were incorporated into the information-gathering process, local survey work began in earnest in 1955. Some 63 record cards were completed for the nine 10 km. squares that coincide wholly or substantially with the boundaries of Assynt (i.e. NC02, 03, 11, 12, 13, 20, 21, 22, 23).

Predictably perhaps, since the contributors were, as ever, visitors to the area, there is a heavy bias in the effort expended towards areas with known botanical interest. Thus there are 17 cards for NC02 (Lochinver north to Stoer) and NC22 (Inchnadamph), 12 for NC21 (Ledmore and Allt nan Uamh), five for NC03 (Stoer round to Oldany island), three each for NC11 (Suilven), NC13 (Drumbeg) and NC23 (Unapool), two for NC20 (Cromalt) and only one for NC12 (Loch Assynt).

The total number of species recorded from each of the 10 km. squares reflects the influence of a number of factors, such as the amount of land in the square and the diversity of habitats represented, as well as the inequality of effort. On this scale NC02 leads, with 450-550 species, NC21 and NC22 follow, with 250-350 species, and the other six all scored 150-250 species.

Any survey of a relatively small area, such as Assynt, at the necessarily coarse scale of 10 km square required for a national atlas, by visiting botanists, has its shortcomings, one being the frustrating lack of detail for what we can recognise, with the benefit of hindsight, as some fascinating records. Nevertheless, the information gathered for the *Atlas* was a major contribution to our knowledge of the local flora. This is, paradoxically, as true for common species as it is for rare ones, as may be seen from the number of times that the '1950s survey' is cited under first records. Examples are *Thalictrum minus*, *Fumaria bastardii*, *Silene uniflora*, *Silene latifolia*, and there are many more.

The 63 record cards relating to Assynt were completed by some 31 botanists, major contributors being: J.Anthony, W.M.M.Baron, A.O.Chater, A.C. Crundwell, B.J.Deverall, J.H.Dickson, Miss U.K. Duncan, B.Flannigan, Miss V.Gordon, R.A.Graham, E.C.M.Haes, R.M.Harley, Mrs J.A.Harris, A.Slack, J.Ounsted, C.D.Pigott, M.E.D. and Mrs J.U.Poore, A.McG.Stirling and S.M.Walters. Other contributors included: M.E.Bradshaw, A.W.Exell, A.C.Jermy, Miss C.E.Longfield, N.M.Pritchard, M.C.F.Proctor, D.Punter, E.C. Wallace, D.Ratcliffe, G.A.M. Scott and Mrs D.L. Stewart.

John Anthony appears to have been responsible for the compilation of the master cards for the 'Assynt squares', as well as for the rest of Sutherland, and they were refereed by Mary McCallum Webster, who had an extensive knowledge of the botany of the northern Highlands. In this way some obvious misconceptions were weeded out. However, there are other records which, again with the benefit of hindsight, should not have found their way into the national database without the confirmation afforded by a voucher specimen. Many of these were thrown up by the 'Discrepancy Lists' generated in the preparation of the current *Atlas*. Since these records are in the public domain, we have included a reference to them in our species accounts, using the term 'reported' to indicate a degree of doubt.

An exception to the general rule about the scarcity of information associated with records for the first Atlas is provided by one of the botanists who recorded here. Arthur Chater, who completed two cards on a brief visit in August 1956, has kindly made available to us copies of his field notebooks for the three days he spent in the parish (together with those for the rest of his trip). They include detailed species lists, with frequencies, for the communities he examined, at precisely grid-referenced localities, and contain several very interesting records, which have been incorporated into the systematic text. They prompt us to wonder how much more irreplaceable information may be sitting in field notebooks all over the British Isles.

This is perhaps a fitting place to acknowledge the debt owed by those interested in the flora of any part of Sutherland, including Assynt, to John Anthony. J.B.Kenworthy, in the *Editorial Note* which introduces *John Anthony's Flora of Sutherland*, gives us the background to this mammoth endeavour, documenting single-handed and parish-by-parish the flora of quite the largest county in the British Isles. Anthony had been engaged on this task for nearly twenty years up to his death in 1972 at the age of 78, and his manuscripts, correspondence and card indices, now in the care of the B.S.B.I. Vice-county Recorder for West Sutherland, bear witness to the immense amount of work involved.

Most of Anthony's visits to Assynt appear to have taken place in the 1950s. He was particularly fond of the Achmelvich area and surveyed that part of NC02 in 1956. Amongst the papers in his hand is an interesting list of plants recorded in the Stoer area, headed 'Biol. Soc. 1956 in Herb. Univ. Edinburgh', which contains a reference to *Zostera* at 'Oldany'. It turns up as washed-up rhizomes and leaves on a number of Assynt beaches and has now been found growing at Port Dhrombaig, but we have so far failed to locate it in the vicinity of Oldany Island, where the habitat looks ideal. In July 1959 he found what he took to be the 'very rare arctic-alpine' *Sagina saginoides* on sea-cliffs near Stoer, and later published a note on its occurrence there and at several places on the north coast of Sutherland (Anthony 1967). His specimen from Stoer in the herbarium at Edinburgh has recently been re-examined and found to be var. *glabrata* of *S. subulata*, which still grows there.

Lest this one very understandable error might seem to diminish our appreciation of all that he did for Sutherland botany, we would refer you to his *Contribution to the Flora of Sutherland. Bettyhill Region* (Anthony 1959). With its useful map and affectionate description of the locality and its plant communities, it is an indication of how much more his *Flora of Sutherland* might have benefited from his experience of the county had he lived to complete it.

Almost thirty years elapsed between the end of fieldwork for the *Atlas* and the start of that for this Flora in 1988, followed in 1995 by fieldwork for the new *Atlas*. The Vice-county Recorder's files contain contributions, in the form of record cards, lists or notes of noteworthy finds for parts of Assynt by a number of visitors, amongst whom may be mentioned: P.Adam, Sir David Burnett, B.Burrow, J.K.Butler, R.A.R.Clarke, A.P.Conolly, M.G.Coulson (Coulson and Coulson 1969), U.K.Duncan, T.Edmondson, P.H.Gamble, G.Halliday, A.S.MacLennan, E.Norman, R.Pankhurst, Mrs M.B. Reid, J.Roberts, J.Grant Rogers, C.E.K. Scouller, A.Showler, P.J.O.Trist, and M.McC. Webster. We have cited their records wherever they provide an extra dimension to those gathered for the current flora.

However, amongst the visitors to Assynt in this period, two names stand out, those of A.G.Kenneth and A.McG.Stirling, their joint or separate visits extending from the early 1950s to the mid 1980s. Stirling reported the discovery in 1973 of the very rare hybrid fern *Polystichum* x *illyricum* on the limestone not far from Inchnadamph (Stirling 1974).

A.G.('Archie') Kenneth (1915-1989) covered a huge area of the Highlands in search of his favourite group of plants, the *Hieracia*. Furthest from his home patch of the Mull of Kintyre, but a great favourite with him, were what he saw as the under-worked hills of north-west Sutherland, in particular Foinaven, Cranstackie, Beinn Spionnaidh and the hills of the Parphe, west of the Durness road, even less frequently visited by botanists. On his way to and from these hills, he sometimes stopped off briefly in Assynt. We have had the opportunity to read his botanical notebooks, which span some 25 years, from the early 1960s to 1985, and these provide some information that didn't find its way into print.

In the very early sixties he visited the coast, recording *Ranunculus baudotii* at Achmelvich, *Tragopogon pratensis* at Clashnessie and *Carex capillaris* at Stoer. In July 1966 he paid a visit to Quinag, noting *Cryptogramma crispa* 'new for 2-2-'. perhaps somewhere in the vicinity of Lochan Bealach Cornaidh, where he also found one frond of *Polystichum lonchitis*. This is the only Assynt record for the former. He recorded *Hieracia* in visits to Quinag and Ledmore in 1971, to Conival and adjacent parts of Breabag in August 1974 (where he found 'nothing of note except hawkweeds') and to Canisp in July 1975. There is a note of a brief visit to a 'ravine near Aultnacealgach' in 1977.

In July 1983 he noted a visit to 'rocks and cliffs above Loch a' Gainmhich (nr. Quinag)'; this may well be the day he collected a *Hieracium* new to science from Cnoc na Creige, later named *H. kennethii*, 'in recognition of his fine work over many years on the hawkweeds of western Scotland' (Sell, West and Tennant 1995). In July 1984 he visited Beinn Uidhe finding it 'very poor'.

It was in 1985, apparently the year of his last visit to the north-west, that, *Hieracia* apart, he made his most notable contribution to the flora of Assynt. On 13th July he paid a second visit to Canisp, following the main ridge to the summit, if we read his notes aright. His verdict on the day was 'good day, wretched hill...not a single hawkweed seen in ground covered by me'. Nevertheless, on the way up he discovered *Diphasiastrum issleri*, in what may well be the largest population in the British Isles (Kenneth 1985). He continues, 'there is a good showing of *Arctous*, *Arctostaphylos*, and higher up, a lot of *Vaccinium uliginosum*, *V. vitis-idaea* and *Betula nana*...in a single station'. This is the only record of *B. nana* from Assynt, and a fair way from the centre of its distribution in mid and eastern Sutherland. It is possible to suggest, from the associated species, especially *V. uliginosum*, to within about a few hundred metres where it occurred, but it has not yet been refound.

Kenneth's contribution to knowledge of the *Hieracia* of Assynt is documented in the species account of that genus, but for an affectionate profile see the appreciation by Alan Stirling, companion on many field trips (Stirling 1990). The measure of the man is well summed up by an entry in his notebook for 2$^{nd}$ August 1985 when he was once again on Foinaven, 'A foul morning, with lashing rain. By return-time, the burns had risen and one was impassable. I stripped to ford it!' and that in his 70th year.

To round-off the account of 'plant-hunting' during this fifty-year period, we should mention Colin Scouller's very useful booklet *An introduction to the flowering plants and ferns of Lochbroom and Assynt* (Scouller 1988). In 1988, Graham Kay, Tim Rich, Alex Scott and Olga Stewart surveyed tetrads NC11J and 11W as a contribution to the B.S.B.I. Monitoring Survey. Scott, Bill Henderson and Terry Keatinge of Scottish Natural Heritage and its predecessors have also made some very interesting discoveries in the course of their work, notably *Asplenium septentrionale* in Gleannan a' Mhadaidh and *Ranunculus auricomus* on Eilean a' Gartaig in Cam Loch.

## 1950-1999: ecological studies

One of the earliest ecological studies of Scottish montane vegetation was that of Ben Armine in West Sutherland (Crampton and MacGregor 1913). However it was not until the 1950s that this initiative was followed up in the North-West, with work by E.A.Blake, R.E.C. Ferreira, D.N.McVean, M.E.D.Poore and D.A.Ratcliffe. Most of this work concentrated on hills outwith Assynt, such as Ben Hope, Foinaven, Meall Horn and Ben More Assynt. However, there were some observations made, incidentally, within the parish.

Poore and McVean's substantial 1957 paper on *A new approach to Scottish mountain vegetation* contains (pp.407-408) a description of 'the most remarkable area of willow scrub yet found in Scotland...on limestone pavement in Inchnadamph'. About 300 acres of a plateau lying between 700 and 900ft., where outcrops of limestone alternate with peat-filled hollows, were 'partially covered with low willow scrub, mainly of *S. myrsinites*, but also containing *S. repens*, *S. aurita* and their hybrids'. A detailed species list from five sample plots follows. The older exclosures in the vicinity of the Allt na Glaic Móire were presumably established to preserve samples of this community.

In 1962 McVean and Ratcliffe published their monograph on *Plant communities of the Scottish Highlands*. Of the 64 Floristic Tables which are a major part of this work, six contain stands recorded in Assynt, on Quinag, Meallan Liath Mór, Breabag and Conival; the lists for these stands include not only the flowering plants and ferns, but also bryophytes and lichens. In a list for Creag Liath at the southern end of Breabag occurs our first precisely localised record for *Sibbaldia procumbens*, in a place where it has been seen again recently.

In 1959 Ferreira published the first of a series of papers on montane vegetation, based on research on Ben Hope undertaken for his Ph.D. He describes his approach as geobotanical, relying as it does on a very detailed appreciation of the mineral content of the rocks involved and their weathering products. Few botanists have the expertise to adopt this approach. In 1977, having returned to the Highlands from the tropics, he offered his services and skills to the regional office of the Nature Conservancy Council at Golspie.

From this offer sprang his awe-inspiring *Vegetation survey of North West Sutherland* (1995). The fieldwork for this survey was carried out between 1977 and 1990, and we are fortunate that Assynt was part of the area covered. Our chapter on vegetation draws extensively on his findings, but he also made some significant additions to our knowledge of the flora of the parish. Examples are *Festuca altissima*, which he discovered on the wooded slopes of Creag an Spardain, above the burn which drains Loch Unapool, and *Dactylorhiza lapponica*. The report of his survey may be consulted at the local offices of Scottish Natural Heritage at Ullapool and Golspie.

Ferreira's survey specifically excluded areas with existing conservation designations, land over 488m. (1600 ft.) and submerged aquatics. So far as aquatic habitats are concerned, there are brief mentions of Loch an Aigeil and Loch an Ordain (Torbreck) in the chapter by D.H.N.Spence on *The macrophytic vegetation of freshwater lochs, swamps and associated fens* in Burnett (1964), based on studies carried out between 1958 and 1961, and plate 51 is a good photograph of Loch an Ordain.

Further interest in Assynt's lochs had to wait nearly twenty years. To complement palaeolimnological studies by Dr Winifred Pennington and co-workers (see the chapter on the history of the landscape), D.H.N.Spence and E.D.Allen of St Andrews made in 1977-78 a detailed study of the macrophytic vegetation of Loch Urigill and some 14 other lochs in what they describe as 'the Ullapool area' (Spence and Allen 1979). Five of these were in Assynt, Loch na Doire Daraich adjacent to Lochinver, Loch Mhaolach-coire, Cam Loch and Lochs Borralan and Urigill.

The study contains some very useful data on water chemistry, light penetration, substrates and exposure, with a discussion of the relationship of these factors to the development of macrophytes, significantly 'beyond depths tolerated by grazing animals'. Floristic records of interest include *Pilularia globulifera* in Loch Borralan, where it is still present in quantity, and *Hippuris vulgaris* and *Veronica scutellata* in Loch Urigill.

In 1988 some 70 Assynt lochs were surveyed in the course of the *Scottish Loch Survey* carried out by a team working for the Nature Conservancy Council. The results of the survey may, again, be consulted at local S.N.H. offices. There is, for each loch visited, a general fact sheet, sketch map of the significant communities with detailed target notes, lists of a. emergent and edge species and b. submerged and floating species, with estimates of frequency, and also some faunal notes. There is no detailed analysis of the results of the survey, nor have we had the time (or expertise) to make one, but it did come up with some important records of submerged aquatics, notably the Red Data Book species *Potamogeton rutilus* in Loch Awe, its northernmost station on the mainland of the British Isles.

So far as flowing water habitats are concerned, the only reference we have come across is to sampling stations on the Rivers Inver and Loanan for a national survey carried out in 1981 by N.T.H. Holmes. This yielded several records of interest from a floristic point of view, but we have not seen any detailed account of the findings of the survey.

The woodlands of Assynt have not received any focussed attention until very recently. McVean noted in 1958, in a paper on *Island vegetation of some Highland freshwater-lochs*, the importance to an understanding of natural forest dynamics of the woodland on such islands, including those in Loch Beannach, which he had visited in 1953-4. He describes the main island, with its one oak tree, as bearing 'a flora more closely resembling oakwood than Highland birchwood', and recorded amongst other things *Ceratocapnos claviculata* on this island. There is a passing reference in W.B.Yapp's paper on *Oaks in Scotland* (Yapp 1961) to 'a few well-grown specimens of [*Quercus*] *petraea*' by the roadside west of Loch Assynt Lodge and he suggests, on the basis of what he acknowledges are fragmentary observations, that there was formerly much oakwood in the North-West. I.D.Pennie also has observations in his thesis on *The influence of man on the vegetation of Sutherland* (1967) which may be related to the particular circumstances obtaining in Assynt.

However, we now have the benefit of the detailed studies of Assynt woodlands carried out over the last five years or more by Robin Noble (Noble 1997, 2000). In his earlier paper he compares the present location of native woodland in the parish with that mapped by Home in 1774. At first sight there appears to have been little change. He then focusses his attention on two areas, one on the coast (Drumbeg-Ardvar) and the other inland (Elphin). In the first there has been significant extension of woodland cover over the intervening two centuries. In the second the only substantial remaining area of native woodland, alongside Loch Urigill, is senescent [though currently the subject of a woodland regeneration project], and other areas mapped by Home have virtually all disappeared. Contrasting the two areas and drawing on observations made elsewhere on the coast of the North West Highlands, he suggests that 'exposure may be at least as potent a factor as grazing animals or human exploitation in determining where...woodland will or will not flourish.'

He has recently reported at much greater length on his studies, particularly of the coastal woodlands, in *The woods of Assynt* (2000), a project carried out for the Assynt Crofters' Trust. This concerns itself primarily with the woodlands from Torgawn round to the Kirkaig valley, but also refers to those in Glencanisp and on the shores of Loch Assynt. It again takes as its starting point Home's 1774 survey.

The report contains 11 very detailed case-studies, with general notes on the woods around Lochinver. It also has sections on the individual species represented in Assynt woodlands, past and present management and options for the future.

One of its most interesting revelations is the evidence of the management of individual trees as coppice or pollards stretching back over many centuries. Ancient oak coppice stools are recognised at Culkein Drumbeg, but more unusual are huge pollards of goat and grey willow. He is continuing his work and we may hopefully look forward to further insightful discussion of the relationship between the health of our woodlands and past and present management regimes.

The historical ecological approach used in this study is particularly well-suited to woodland, but would also be relevant to the future study of other aspects of the Assynt landscape, for instance the 200 or more sheilings abandoned as such during the early part of the 19th century.

# THE PRESENT SURVEY

## Origin and purpose

The plant communities of Assynt appear to represent, in microcosm, most of those found on the west coast of Sutherland. We (P.A.E and I.M.E) had been visiting the area on holiday from 1982, and after several years 'browsing', we were looking for something more challenging. To this end we started in 1988 a survey of Assynt by tetrad (two km. square).

The choice of tetrad as survey unit perhaps requires some justification, since it has not previously been attempted in the Scottish Highlands. The discipline of recording at this level of detail has proved very productive elsewhere in Britain and it was one with which we were familiar. The boundaries of Assynt include the whole or part of 164 tetrads, of which 28 may be regarded as marginal, with areas within the parish of less than 1 sq. km. Some of these marginal tetrads are offshore islands, others small areas on the coast, the edges of lochs or summits of the hills, but their botanical interest may be out of all proportion to their area.

The parish was thus considered to be the largest unit with an acknowledged geographical and historical identity that it was practicable for two people to survey in this way (it is comparable in area to the smallest Scottish and English vice-counties). Although parts of the parish had been well botanised over a period of some 250 years, it was hoped that such a survey might yield records of more than local significance. This proved to be the case.

In the event, we spent three-week holidays in June from 1988 to 1990 on the survey, before, somewhat unexpectedly, being able to move up to Assynt in the summer of 1991. There followed eight more full seasons of fieldwork (1992-1999), latterly doubling up with work over the rest of West Sutherland for the national *Atlas*. 'Topping up' and preparation of the text of the Flora has occupied a further two years. During the fieldwork we made about 400 visits to individual tetrads; some remote marginal ones were only visited once, others up to five times to cover all the potentially interesting habitats.

## Timing and technique

This far north, the useful survey season is relatively short. There is little point in starting before the beginning of June or continuing after the end of October, since only during this time are the grasses and sedges readily identifiable. Other groups, such as orchids, must be seen in flower, and in a few, such as *Callitriche*, identification requires mature fruits. Many areas are too far from the road for revisits to catch everything in its prime, so there may be some gaps in the coverage of such groups. A problem of a different kind is deer management, which begins to impose some limitations on access from August onwards.

It may be worth describing in some detail how we went about the survey, since there are substantial differences in practice from carrying out such a survey in a lowland, mainly agricultural, landscape. Time was well spent beforehand working out a sensible route to a tetrad, using tracks (where they existed), and a route around it which took in the full range of the habitats shown on the

1:25,000 map, such as lochs and burns, woodland and crags. Including travelling time, a nine hour day was usual, more if necessary. The greatest distance covered on foot in a day was 22 km., to check on a rare fern south of Suilven, but the most taxing was 20 km. round the southernmost part of the Cromalt Hills, where there are no tracks! On three occasions we hired a boat to cut down the walking time into remote areas adjacent to Lochs Assynt and Veyatie.

We estimate that, on a first visit to a tetrad, two of us could cover, reasonably comprehensively, about one sq. km. of ground spread throughout its area, leaving any other significant habitats for subsequent visits. To maximise use of the available time, especially in remote areas, we often took slightly different routes for an hour or more at a time, one of us concentrating on lochans and other water bodies, the other on crags. Some of the higher hills were visited or re-visited in the company of our co-author Gordon Rothero, taking advantage of his knowledge of montane species and his ability to reach areas that were too 'airy' for us. On these occasions we were often recording both vascular plants and bryophytes. Latterly we made more use of geological maps to identify likely areas of interest, where for instance there are basic or ultrabasic dykes cutting through the gneiss. Unfortunately there is no geological map coverage currently available for the western half of the parish.

'Common species' were recorded on an A4 recording sheet of our own devising, with columns for habitat and frequency. We usually noted the first habitat in which a species was found, but frequency referred to the tetrad as a whole and was added at the end of the visit. The 324 'common species' were ordered alphabetically as abbreviated scientific names, but for convenience in four categories: ferns and their allies; grasses; sedges and their allies; and the rest. All 'other species' were recorded with a six-figure grid reference every time we came across them. Conspicuous omissions were highlighted on the field recording sheets prior to subsequent visits and on these visits additional records were made in different colours for easy reference. Master copies for each tetrad were compiled and updated after each visit; these were never taken into the field.

The selection of species for the 'common species sheet' was based on *John Anthony's Flora of Sutherland* (1976). If that indicated that a species was present in Assynt, without any qualification or specific localities, it was included. Given that most of the information on which this text was based had been gathered by the end of the 1950s, and that there had been considerable subsequent changes in land-use locally, there were some interesting anomalies.

In the light of experience, 32 species that proved to be uncommon or rare were upgraded from 'common' to 'other'. They included *Ajuga reptans*, *Anthriscus sylvestris*, *Campanula rotundifolia*, *Dactylorhiza fuchsii*, *Fumaria officinalis*, *Koeleria macrantha*, *Lycopodium clavatum*, *Parnassia palustris*, *Trientalis europaeus* and *Viola canina*. The other side of the coin was that we finished up with a large number of precisely localised records for *Eleocharis multicaulis*, which is widespread in lochans and elsewhere. Other species for which we have numerous detailed records, such as *Hymenophyllum wilsonii* and *Trollius europaeus*, are good habitat indicators and the extra effort involved was probably not wasted.

Occurrences of the 'other species' were noted on a blank sheet, in sequence with more general notes on route, habitats, specimens collected as vouchers or for later identification, together with any other observations on the area traversed and its wildlife that took the recorders' fancy. At times, proprietary 'wet notebooks' proved their worth in the field! On return home, all material collected was identified and, if appropriate, pressed, grid references were double-checked, and the route taken was marked in highlighter on a file copy of the 1:25000 map. These maps, covered in what we came to describe as 'snail trails' in highlighter, proved very useful, both to plan further visits, and as a visible record of coverage.

Notes were made on the dominant species of discrete areas of woodland, the emergents and aquatics of almost all lochans, and on the character of the vegetation of any areas of historical interest such as sheilings. The botanical records were stored by tetrad, all other information as a daily log, with a summary in a diary. Sites and species of interest were also photographed.

Our ability to identify certain groups of plants in the field increased during the course of the survey, but material of some was always collected. An obvious example is *Utricularia* spp.; since we only once saw flowers, they had to be identified by microscopical examination of the quadrifid hairs. *Isoetes* was similarly checked, wherever possible, by examination of the megaspores. Some species of *Euphrasia* were, latterly, identified in the field, but material of other taxa was collected, as also were specimens of all *Hieracia* encountered, other than *H. vulgatum* (see account of the genus for more details). All appropriate voucher material will, in due course, be lodged with the national herbarium at the Royal Botanic Gardens, Edinburgh.

## Processing of records

This was a job for the winter. The records were entered into RECORDER, up-graded progressively to version 3.3, using a customised 'pop-up' for the 'common species'. They were mapped on DMAP, with the parish boundary parameters added where they diverged from those of the vice-county. Those with any experience of data-entry will know that this brief summary does not remotely represent the amount of work involved!

## Results

Four short and eight full-length field seasons yielded 30,676 records from the 164 tetrads. The majority of these were from our own fieldwork, but valuable contributions were made by those who attended the B.S.B.I. Field Meeting in Lochinver in 1993, local officers of Scottish Natural Heritage and its predecessors, visitors to Assynt and local naturalists. We also made use of records from the the B.S.B.I. Monitoring Survey in 1987-88 and the 1988 Scottish Loch Survey (notably of *Potamogeton* spp.). The considerable extent of our debt to particular individuals will be obvious from the acknowledgements and the species account.

The total number of taxa recorded during the survey (species, subspecies and hybrids) was 694. Of these, 100 were recorded from only one tetrad, and 23 from more than 150. *Potentilla erecta* was recorded in the greatest number of tetrads (161), closely followed by *Calluna vulgaris* (159), *Eriophorum angustifolium* (158) and *Anthoxanthum odoratum* (158); the other 19 most widespread species were, predictably, those of wet and dry heath, including one fern, *Blechnum spicant*, eight rushes, sedges and their allies, and four grasses.

There are historical records of a further 71 taxa, together with 17 that we believe may have been recorded in error. The total for Assynt since recording began in 1767 is therefore 765 taxa. To put this figure in context, the latest checklist for the whole of the British Isles (Kent 1992) lists 3087 native species, subspecies and hybrids, together with 1186 aliens in the same categories. Allowing for the fact that aliens are a small component of the Assynt flora, and also that several groups of microspecies still have to be surveyed in detail, we probably have about a fifth of the British flora within our boundaries.

## Discussion

The total numbers of taxa recorded from each tetrad are indicated in Figure 8. Totals for tetrads, other than marginal ones, range from 79 in 21J, on quartzite on the eastern slopes of Canisp, to 285 in 02W at Lochinver. The 80 taxa recorded from the adjacent tetrad 21I indicate that the figure for 21J is not the result of a 'bad day'.

51 tetrads mustered over 200 taxa. Predictably, these map out the fertile ground of the coast and the limestone, with a band across the middle connecting the two, which follows the road on the north side of Loch Assynt. A more sophisticated analysis of the records is required to determine how much roads and the settlements they serve contribute to these and other totals.

Any discussion of apparent changes in the flora of an area such as Assynt is hedged about with 'ifs, buts and maybes'. Although the parish has been visited by botanists for almost 250 years, their visits have generally been short, in high season, focussed on areas of acknowledged interest and quite sporadic (as will be seen from the chapter on the history of recording). The time and effort put into the current survey is out of all proportion to what has gone before, so we are not comparing like with like. There are also taxonomic and nomenclatural changes to take into account.

When we come to consider the broader significance of 'losses or gains', there is also the vexed question of what constitute 'native' species, which are relatively permanent features of the vegetation, as opposed to others which owe their presence, directly or indirectly, to the activities of man, which makes them less predictable. There is of course a historical dimension to this question.

We have not therefore attempted a rigorous analysis of the findings that follow. Nevertheless, some figures and comment, however speculative, may be of interest. The most readily available comparison is afforded by the taxa recorded for Assynt in *John Anthony's Flora of Sutherland* (1976). It should be emphasised that since the publication of that work a number of distinguished botanists have visited this area; any 'gains' are by no means 'all our own work'. They also include records predating that work that were not available to its author, having surfaced since from notebooks, herbaria and the national database.

### 'Losses'

In all, 71 taxa have been recorded in the past, some of them comparatively recently, but not during the present survey. About a third of these are obvious introductions, escapes from and weeds of cultivation, or casuals. For example, *Betula pendula* is thought to be an introduction here on the west coast, as are *Salix alba* and *S. fragilis*, but we cannot be sure about *S. purpurea*, which was last seen in the Kirkaig valley in 1886. *Campanula latifolia* was recorded as an introduction at Inchnadamph in 1953.

Escapes from cultivation include *Claytonia sibirica*, not seen since 1962, though probably lurking somewhere, and maybe *Symphytum officinale*, which was recorded in the 1950s.

More interesting are the weeds of cultivation, such as *Centaurea cyanus* (last seen in 1899), *Persicaria lapathifolia* (1886), *Raphanus raphanistrum* (1943) and, probably, *Silene* x *hampeana* (1955). Casuals

Flowering plants and ferns

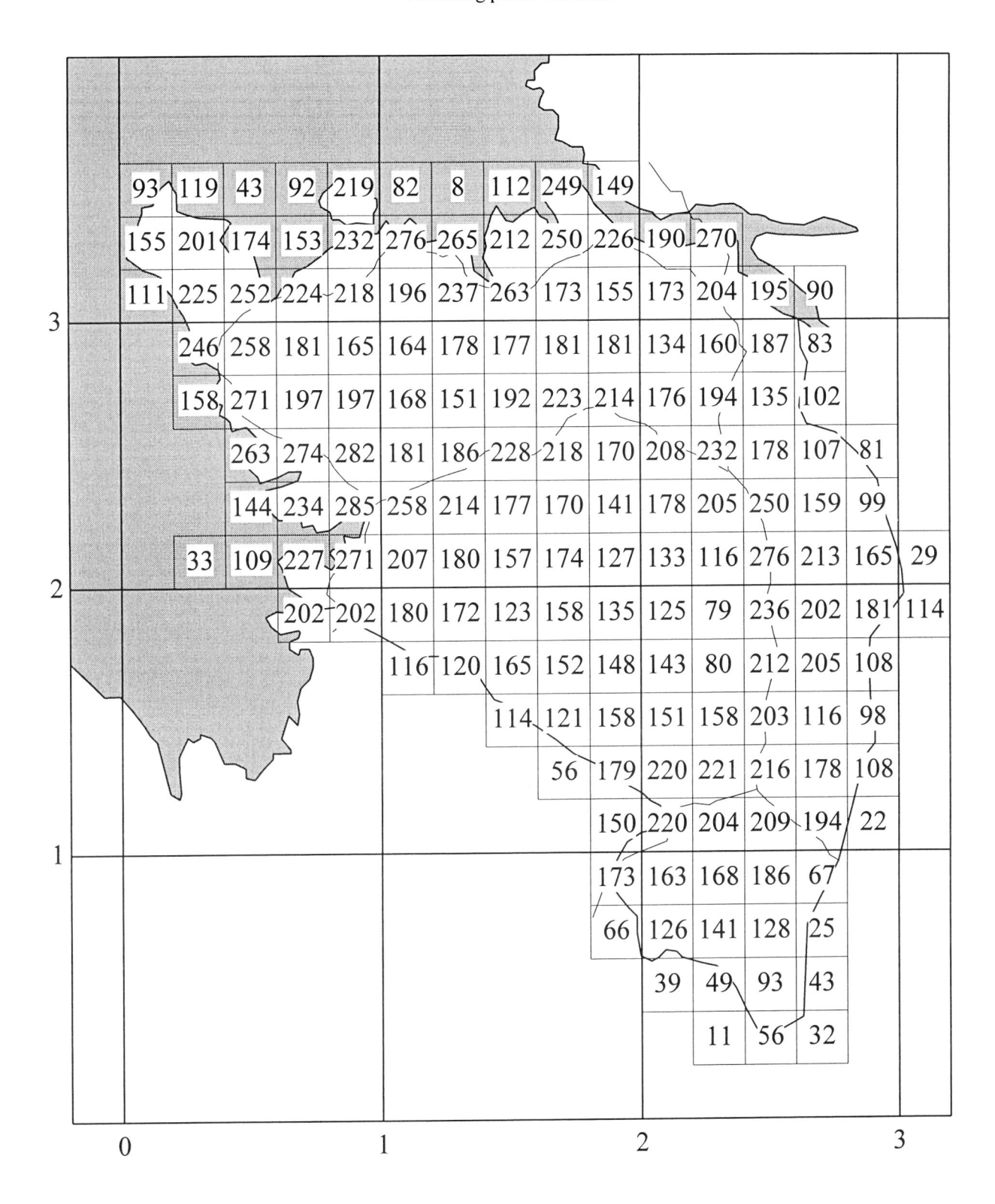

**Figure 8. Number of vascular plant taxa recorded from each tetrad.**

include: *Carduus nutans* (one plant, in 1958), *Diplotaxis muralis* (1956), *Sisybrium altissimum* (1944), *Solanum dulcamara* (1958) and *Trifolium campestre* (1886).

The remaining two thirds of the 71 species may include some genuine 'losses' from the Assynt flora; others probably survive, but have not yet been refound. Three groups are the prerogative of specialists and await expert attention (*Euphrasia* hybrids, *Rubus* and *Taraxacum*). There is material in herbaria, recently redetermined, of four roses, one subspecies and three hybrids, collected by E.S. Marshall in 1890 and 1908. He was also responsible for the discovery, in the 1880s and 1890s, of three *Epilobium* hybrids involving the montane species, which no-one has noted since. Similarly, there is recently re-verified material from Assynt of 11 species of *Hieracium*, collected between 1890 and the 1950s.

Other such species are more readily recognisable in the field, particularly if one has seen them elsewhere in the British Isles. In most cases we have no precise idea of where they were found; for some, there are approximate localities, or clues, such as associated species. Examples are perhaps most usefully listed by habitat and/or general locality, with the date when they were last reported: on the coast at Stoer, *Sagina maritima* (1953); on a crag not far from the coast at Clachtoll, *Asplenium obovatum* ssp. *lanceolatum* (1969/70); from the north-east side of Quinag, *Cryptogramma crispa* (1966); on limestone in the Inchnadamph area, *Arabis hirsuta* (1969), *Carex caryophyllea* (1961), *Juncus alpinoarticulatus* (1908) and *Viola lutea* (1889); at the base of Canisp, *Lycopodiella inundatum* (1907); further up that hill, *Betula nana* (1985!) and *Cerastium arcticum* (1899); on an island in Loch Awe, *Paris quadrifolia* (1923); at Knockan, *Adoxa moschatellina* (1894).

## 'Gains'

This Flora adds 133 taxa to those recorded for Assynt in Anthony. They are a very varied assortment. The first category consists of 26 species that appear to have been deliberately introduced, escaped from cultivation or are associated with the movement of soil or garden plants. They include the Inchnadamph 'zoo' (*Campanula cochleariifolia, Erinus alpinus, Gentiana verna, Silene alpestris* and *Phyteuma scheuchzeri*; see the species account), bird-sown species such as *Cotoneaster simonsii* and *Ribes nigrum*; and escapologists such as *Impatiens glandulifera*, three *Mentha* hybrids, three *Mimulus* hybrids and, less predictably, *Tolmeia menziesii* and *Tropaeolum speciosum*. Also presumably introduced in the more remote past, but only noted recently, are *Glechoma hederacea* and *Peucedanum ostruthium*. Garden throwouts include *Alchemilla mollis* and *Saxifraga* x *urbium*, and there are relics of cultivation such as *Phacelia tanacetifolia*.

A second category includes 19 species that are likely to have been accidentally introduced as a result of human activities. Some have probably come in with imported soil and garden plants. These are a mixture of ephemeral annual weeds and perennial 'thugs'; the former will probably disappear, the latter may well survive. They include: *Alliaria petiolata, Anisantha sterilis, Calystegia sylvatica, Convolvulus arvensis, Galeopsis bifida, Lamium album, Linum usitatissimum, Senecio viscosus, Thlaspi arvense, Valerianella carinata* and *Verbascum thapsus*.

Others have a variety of origins. *Lemna minor* appeared briefly in a newly dug ditch, and *Sanguisorba minor* after a private road was widened and re-surfaced, both presumably imported on the machinery used. *Festuca arundinacea* is only found on roadside verges in one part of the parish, and may also have been brought in during road-widening, and *Spergularia rubra* has found a niche at the edge of the widened roads. *Agrostis castellana* was found once, just off the well-used walkers' track to Suilven. *Carex laevigata* may have been introduced to the Glencanisp woodlands with young trees.

There is a small group that may well have arrived under their own steam, having efficiently dispersed seeds. These include three species of *Epilobium*.

There remain what are for the enthusiast the most exciting category of some 80 'gains', taxa 'native' to Assynt that appear not to have been previously recorded. The length of the list emphasises, once again, how many botanists have contributed to knowledge of our local flora. Some of these 'gains' reflect recently enhanced interest in difficult or critical groups and hybrids. They include *Equisetum* x *trachyodon*, eight species of *Hieracium*, several species and hybrids of *Rosa*, four hybrids of *Salix* and *Utricularia* spp.

Others appear to be so restricted in distribution locally that in the past they have just been overlooked. Examples include: *Anagallis minima, Asplenium septentrionale, Astragalus danicus, Cladium mariscus, Hippuris vulgaris, Lycopodium annotinum, Lycopus europaeus, Ophioglossum azoricum, Orthilia secunda, Pilularia globulifera, Potamogeton rutilus, Pyrola media, Ranunculus auricomus, Ruppia maritima, Saxifraga nivalis* and *Scilla verna*, together with no fewer than six sedges and eight orchids (including the four sub-species of *Dactylorhiza incarnata*, and the hybrid *D.* x *formosana*). This list includes several national rarities.

Some of these discoveries have been made in the course of organised surveys, such as ours, others were pure serendipity. Survey by tetrad allowed us to locate 142 'sites' for the nationally scarce species *Ajuga pyramidalis*, although it was of course known to occur in Assynt. Quite as pleasing, however, was the discovery, by our co-author Gordon Rothero, of two small populations of the nationally scarce montane species *Luzula arcuata*, one verifying a record over a century old, the other completely new.

## The future

No account of the flora of an area is ever complete; it can only represent a 'snapshot' of the period in which it was carried out, with the perspective provided by records from the past. It would be remiss of us to end this chapter without suggesting some aspects of the flora of Assynt that might repay further attention.

There were, inevitably, deficiencies in our geographical coverage of the parish. Some remote areas were only visited once. In the case of the Cromalt Hills, the geology and topography is such that we probably did not miss much of significance. Of the Torridonian hills, Quinag was quite fully surveyed, but further attention to the higher crags on Canisp and the 'skirts' of Suilven might be productive. Of the hills along the eastern boundary, more attention could be paid to the north face of Cnoc na Creige, but Glas Bheinn and Beinn Uidhe were reasonably well covered, especially the latter, which is almost devoid of vegetation over large areas.

Parts of the high plateau from Loch na Cuaran south to Beinn an Fhurain remain unexplored, as does the very steep south-west face of Conival. The northern two km. of Breabag were not visited, although since they appear to be broken quartzite, they may not yield much that is new. There are some crags whose steep or vertical faces defeated us and may never have been explored by any botanist, but the only substantial area of lower ground not visited by us is a stretch about a kilometre wide along the south side of Suilven, partly because of its remoteness, but also because the alignment of the tetrad boundaries, the geology and topography, made it unlikely that any significant habitat had gone unrecorded.

The logistics of obtaining the use of boats during the fishing season, with reasonable weather guaranteed, defeated our attempts to visit some of the more noteworthy islands in the larger lochs, although we were able to incorporate in the species account some interesting records made by others.

All the larger coastal islands were visited, with the exception of Eilean a' Ghamhna, but we did not get out to many of the small islands east of Oldany Island and north of Drumbeg, other than Sgeir Liath, which is the only land in tetrad 13H, and was visited, yielding just eight species.

Finally, we come to species or groups of species that would benefit from further study. We only started collecting material of the *Hieracia* seriously in 1997 and coverage is very patchy in all but the limestone areas. Similarly *Euphrasia*, other than the nine species mapped, deserves more attention, as do *Rubus* and *Taraxacum*. We were virtually defeated by the coastal species of *Atriplex*, and know that we seriously under-recorded *Puccinellia maritima*, since we have only recently learned to identify it in the vegetative state. Our recording of *Potamogeton* spp. was seriously constrained by the availability of boats, but as mentioned above we were able to make use of the records of this genus from the Scottish Loch Survey. Other aquatics were however probably reasonably recorded from the margins and washed-up material.

As has been indicated above, some species probably still await rediscovery, despite our best endeavours. Others, notably weeds of cultivation such as *Persicaria lapathifolia* and *Raphanus raphanistrum*, have not been seen for a long time, and unless there us a radical change in patterns of land-use they may only survive in the buried seed-bank. Other challenges will become apparent from perusing the species account, and we should be only too pleased to point visitors in a likely direction.

Lastly, there should be new species to discover in Assynt. An obvious example is *Orobanche alba*, which has recently been found on its usual host, *Thymus polytrichus*, on the gneiss about four km. to the south of Inverkirkaig, in West Ross; there is also an older record from Oldshoremore, to the north.

*Salix phylicifolia* may occur, since we have found two hybrids of which it is a parent. More of a challenge is presented by *Pyrola minor* and *P. rotundifolia*, which both occur elsewhere in West Sutherland, since members of this genus seem reluctant to flower this far north. Needless to say, we should like to hear if you find anything of interest. Meanwhile we shall hope to fill some of the 'holes' in the maps.

# PLAN OF SPECIES ACCOUNT

The nomenclature and sequence of the families, genera and species follows Stace's *New Flora of the British Isles* (2nd edition 1997).

Scientific names are in *italic* with the authorities in normal type. The only synonyms cited are names used in *John Anthony's Flora of Sutherland* (Kenworthy 1976); they are shown in normal type enclosed in brackets. Gaelic names are taken from *Gaelic Names of Plants* (Clark and Macdonald 1999) and are in normal type. English names are in **bold**.

The records are held on computer in RECORDER 3.3 and the distribution maps have been produced on DMAP. Dots which appear to be in the sea are tetrads containing small islands and those apparently outwith the parish boundary are marginal tetrads with very small areas in Assynt.

The number of tetrads from which each species has been recorded during the present survey is on the first line of the species account; the maximum possible is 164. A general statement of frequency follows, with an indication of habitat preference.

Details of records are only given when the number of tetrads is 6 or fewer. Where the author of the record is not quoted it is P.A. and I.M.Evans. Locality names have been taken from the O.S 1:50,000 or, if not on this, the 1:25,000. Most of the localities are named in Gaelic; we have done our best to employ the correct accents, a difficult task, as there is not total unanimity in their use. Locality names mentioned in historic records are given in quotes where the spelling differs from that used today.

For the sources of the historical records see the chapter on the history of recording and the bibliography. They are also contained in a comprehensive card index which is available for consultation.

Publications which are referred to frequently have been abbreviated thus:
*John Anthony's Flora of Sutherland* (Kenworthy 1976) as *Anthony*.
*Atlas of the British Flora* (Perring and Walters 1962) as *Atlas*.
*Critical Supplement to the Atlas of the British Flora* (Perring 1968) as *Critical Supplement*.
*Scarce Plants in Britain* (Stewart, Pearman and Preston, 1994) as *Scarce Plants*.

# SPECIES ACCOUNT

## LYCOPODIOPSIDA
**Clubmosses and Quillworts**

LYCOPODIACEAE

*Huperzia selago* (L.) Bernh. ex Schrank & Martius
(*Lycopodium selago* L.)
Garbhag an t-Slèibhe
**Fir Clubmoss**

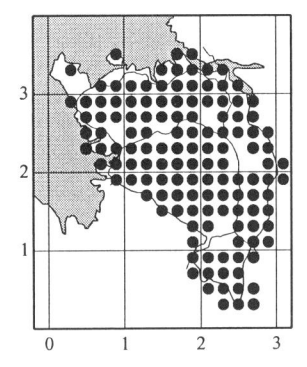

131 tetrads. Widespread and common in wet heath, stony flushes and open unshaded habitats, where the vegetation is thin. It occurs from high on the hills down almost to sea level, but is less frequent in limestone districts. First recorded in 1767 from the Inchnadamph area by J.Robertson.

*Lycopodiella inundata* (L.) Holub
(*Lycopodium inundatum* L).
Garbhag Lèana
**Marsh Clubmoss**

Not seen during the present survey. This elusive plant was first recorded in July 1907 from 'near the base' of Canisp by G.C.Druce (the date of 1903 in *Anthony* appears to be an error). It has so far defeated the efforts of a number of botanists to re-find it; some likely ground is the area of wet peat and pools immediately north of Loch Awe, at NC2416. An undated record from NC03 is mapped in *Scarce Plants in Britain* but, judging by the source, may be an error.

*Lycopodium clavatum* L.
Lus a'Mhadaidh-ruaidh
**Stag's-horn Clubmoss**

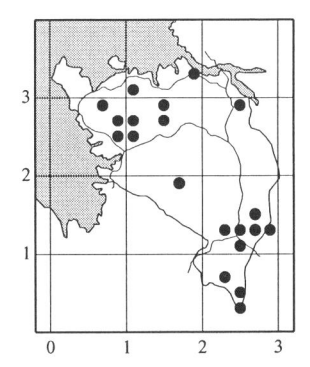

20 tetrads. Found in only a limited number of localities, although often in good stands where it does occur. Its prostrate growth, beneath heath vegetation, means that it is almost certainly under-recorded, although in an extensive population the upright fertile shoots bearing pale green cones may be visible from a considerable distance. Particularly abundant on the northern slopes of Cnoc na Sròine and along the margins of forestry rides at Ledmore. First recorded in 1886 from Elphin by A.Gray.

*Lycopodium annotinum* L.
Lus a'Bhalgair
**Interrupted Clubmoss**

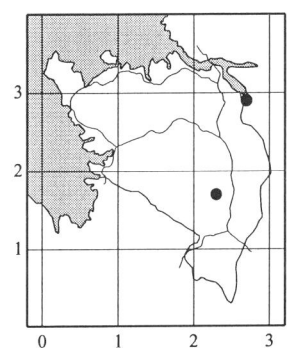

2 tetrads. Found at 440m. amongst heath vegetation on an ungrazed ledge just above water level, at Loch Bealach a'Bhuirich, by G.P.Rothero, and at 340m. on dry stony ground south of the Allt Mhic Mhurchaidh Gheir, on the south-eastern slopes of Canisp. It was reported in 1981 from NC13. First recorded in 1994 from the first of these sites.

*Diphasiastrum alpinum* (L.) Holub
(*Lycopodium alpinum* L.)
Garbhag Ailpeach
**Alpine Clubmoss**

38 tetrads. Found regularly on the higher hills, where it grows in exposed situations receiving little protection from surrounding vegetation. First recorded during the 1950s survey for the first *Atlas*.

*Diphasiastrum issleri* (Rouy) Holub
(*D. complanatum* ssp. *issleri* (Rouy) Jermy)
Garbhag Issler
**Yellow Cypress Clubmoss**

1 tetrad. The only known locality for this Red Data Book species is Canisp, where it was first recorded by A.G.Kenneth in 1985. An estimate of the extent of the site was made in 1998, during the present survey. Growing on stony, only lightly vegetated ground, the impressively large population

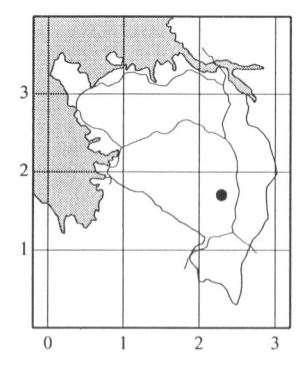

stretches for 500m. along approximately the 250m. contour, on the botanically dour Cambrian quartzite of the south-eastern slopes of Canisp. Its slightly yellow-green colour makes it easily recognisable. It is interesting that barely a kilometre away on the same hill, is one of Assynt's two localities for *Lycopodium annotinum*.

SELAGINELLACEAE

*Selaginella selaginoides* (L.) P.Beauv.
Garbhag Bheag
**Lesser Clubmoss**

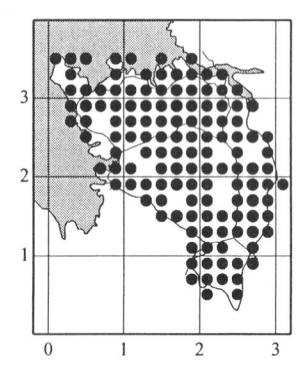

129 tetrads. Extremely widely distributed in wet flushes and seepage areas, where there is a degree of base-richness, even if very localised. It grows in open unshaded situations, and where *Pinguicula lusitanica* occurs *Selaginella* is often nearby. First recorded in 1767 from the Inchnadamph area by J.Robertson.

ISOETACEAE

*Isoetes lacustris* L.
Luibh nan Cleiteagan
**Quillwort**

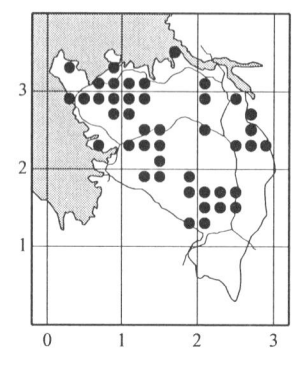

43 tetrads. This species has been found to be widespread in lochs where the bottom is composed of sand, gravel or small stones, from near sea level to over 500m. A characteristic sight is the spiky leaves curving upwards from beneath a medium-sized stone, where a germinating spore had presumably been caught. The plants were searched for on foot in shallow water and, as they may grow to a depth of 6m., the species may be under-recorded. One record is from a burn and a few were from drifted material. All determinations of this and the following species were made by microscopic examination of the megaspores. First recorded in 1899 from 'two lochs on Glas Bheinn' by C.E.Salmon.

*Isoetes* x *hickeyi* Taylor & Luebke
*I. lacustris* x *I. echinospora*

Specimens taken from Loch an Leothaid and Loch a'Choire Dhuibh in 1995 had malformed megaspores, suggesting that the plants may have been of hybrid origin.

*Isoetes echinospora* Durieu
Luibh Cleite an Earraich
**Spring Quillwort**

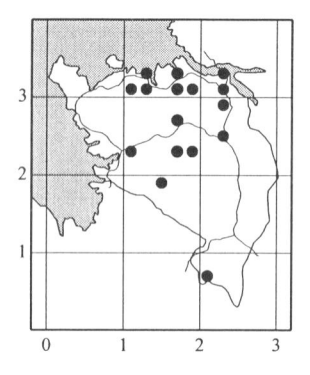

16 tetrads. Considerably less common than *I. lacustris*, but there is no noticeable difference in the character of the lochs in which the two species occur. *I. echinospora* has been found once at an altitude of 340m., but that is an isolated case and most of the records come from lower ground. It has been found once in a burn, the same locality that produced the first record, made in 1908, from 'peaty pools, Unapool Burn' by E.S.Marshall and W.A.Shoolbred.

# EQUISETOPSIDA
**Horsetails**

EQUISETACEAE

*Equisetum hyemale* L.
Biorag
**Rough Horsetail**

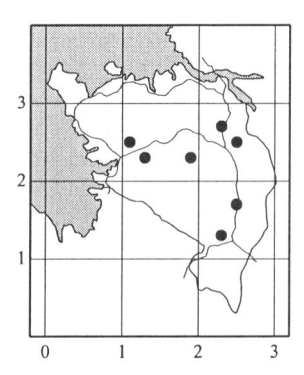

7 tetrads. Rare. Found in four instances in limestone flushes, but the other three records appear to be associated with basic or ultrabasic dykes cutting through the gneiss.

The largest population extends for 20m., growing in the riffle of a small burn. First recorded in 1886 at '*Achumore*' by A.Gray. A small stand is still present on what we judge to be that site.

*Equisetum* x *trachyodon* A.Braun
*E. hyemale* x *E. variegatum*
**Mackay's Horsetail**

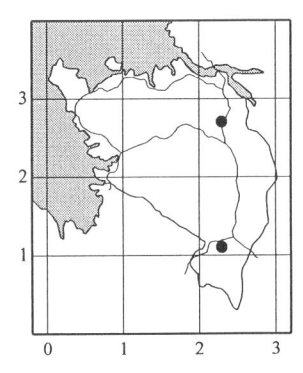

2 tetrads. This hybrid was first recorded by R.E.C.Ferreira, A.Scott and W.Henderson in 1984, growing in a basic flush near the Abhainn a'Chnocain, with *E. variegatum* nearby. Subsequently it was noted by I.A.Macdonald in a limestone flush at Lairig Unapool, at a site which also contains *E. hyemale*.

*Equisetum variegatum* Schleich. ex F.Weber & D.Mohr
Earball an Eich Caol
**Variegated Horsetail**

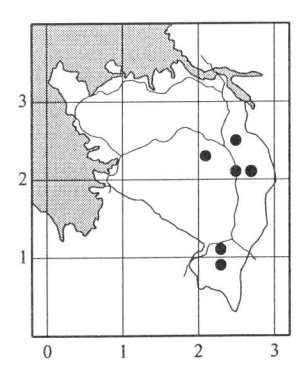

6 tetrads. Rare, but may be quite prolific where it does occur. A very characteristic habitat is the fringe of vegetation jutting out over the water, at the edge of a burn running through limestone grassland, as for example the Allt a'Chalda Beag. Five of the records are from limestone and one from the foot of Beinn Gharbh, where an ultrabasic dyke cuts through the gneiss. First recorded in 1887 'by the Traligill, Inchnadamph' by E.S.Marshall. An interesting later record is from 'a roadside bog south of Creag Mhór', on the south side of Quinag, where it was found in 1943 by M.S.Campbell. This locality was probably on the gneiss.

*Equisetum fluviatile* L.
Clois
**Water Horsetail**

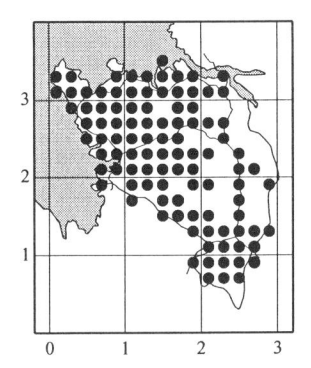

101 tetrads. The commonest horsetail in the area. Found abundantly in lochs and on the margins of rivers and burns where there is slow-moving water. Also occurs frequently in bog and wet heath where there is no open water. First recorded in 1894 'near Elphin and Ledbeg' by G.C.Druce.

*Equisetum arvense* L.
Earball an Eich
**Field Horsetail**

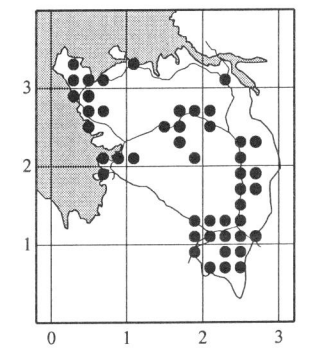

46 tetrads. On dry, disturbed ground along roadsides and in small quarries. Also along rivers and burns, on bare ground at the water's edge. First recorded in *Anthony*.

*Equisetum pratense* Ehrh.
Earball an Eich Dubharach
**Shade Horsetail**

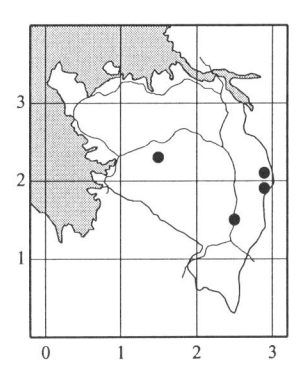

4 tetrads. Rare. On the upper reaches of the River Traligill it has colonised a slope of earthy scree created by a landslip, with a much smaller group close at hand, on shingle at the water's edge. A very good population grows on the Fucoid Beds on a rocky shelf above the Allt a'Bhealaich. In considerably smaller quantity it was found in dry heath on the edge of a burn, on the lower western slopes of Beinn an Fhuarain, in the vicinity of limestone bands. First recorded in 1955 on the north side of Quinag by M.E.D.Poore; found in 1969 on the River Traligill by M.G.Coulson and in 1983 by R.E.C.Ferreira in flushed grassland associated with ultra-basic rocks in the gneiss east of Achadh Mór. We cannot find the source of a record from Drumbeg given in *Anthony*.

*Equisetum sylvaticum* L.
Cuiridin Coille
**Wood Horsetail**

92 tetrads. A misnomer in Assynt, where we have not seen it in woodland. The species is widespread, presumably able to thrive in the open on wet and dry heath because of the high atmospheric moisture.

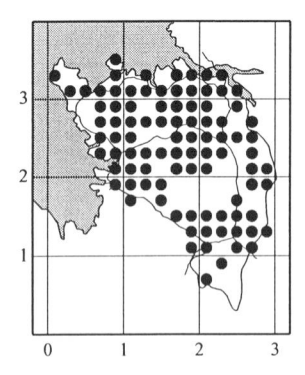

It was first recorded in 1943, in a 'rocky gulley north-east of Inverkirkaig' by A.J.Wilmott and M.S.Campbell.

*Equisetum palustre* L.
Cuiridin
**Marsh Horsetail**

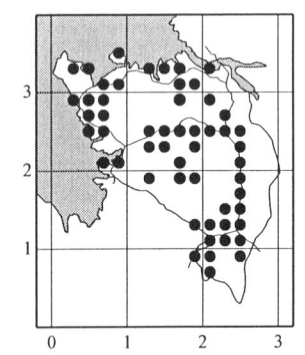

54 tetrads. Widespread in marshes, ditches, edges of burns and rivers; also in places where there may temporarily be shallow standing water. First recorded in 1894 'near Elphin and Ledbeg' by G.C.Druce.

# PTEROPSIDA
**Ferns**

OPHIOGLOSSACEAE

*Ophioglossum vulgatum* L.
Teanga na Nathrach
**Adder's-tongue**

14 tetrads. Although far from common, this small fern occupies a variety of habitats. It has been found growing under heather, in dune grassland, and in 'sweet' grassland, sometimes under light bracken cover. In the hills, such areas of grass with bracken are often indicative of earlier grazings or cultivation which, being sited on the better soils, support a richer flora than the surrounding hill ground. First recorded in 1943 from 'machair at Achmelvich' by M.S.Campbell.

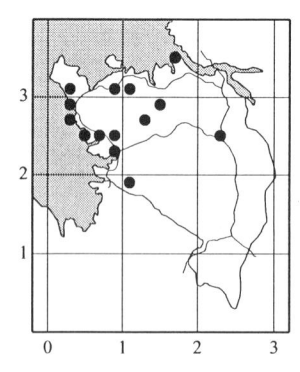

*Ophioglossum azoricum* C.Presl
Teanga na Nathrach Beag
**Small Adder's-tongue**

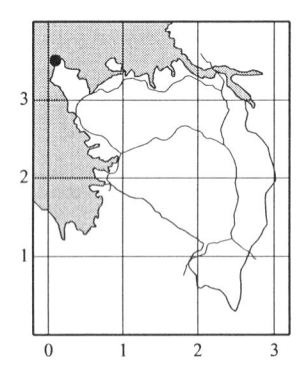

1 tetrad. The only record made during the present survey was of more than 100 plants, growing in close turf near the edge of an 80m. cliff, near Cìrean Geardail, about 1km. south-west of the Old Man of Stoer. It was first recorded by R.M.Corner in 1958 from coastal grassland near Raffin on the Stoer peninsula, but was only determined as this species in 1978; it is therefore not mentioned in *Anthony*. We have attempted to re-find the plant at this site but without success, and it may well have succumbed to grazing pressure during the intervening 40 years.

*Botrychium lunaria* (L.) Sw.
Lus nam Mìos
**Moonwort**

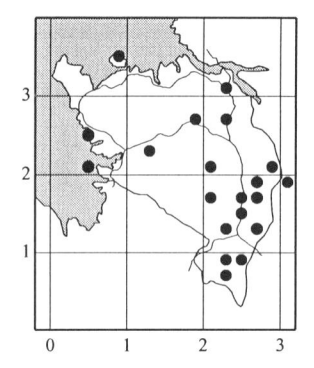

20 tetrads. The greatest concentration of records is in the limestone areas, but the plant is also found in sandy coastal grassland and dry heath grassland. At least three of the sites are on roadsides, where the plant seems to have become established during changes in the road pattern and has survived because of the continuing thinness of the vegetation on the stony ground. An easily overlooked fern, which may well be under-recorded because of its small size. First recorded in 1886 by A.Gray as 'abundant in limestone districts...Durness-Assynt'.

OSMUNDACEAE

*Osmunda regalis* L.
Raineach Rìoghail
**Royal Fern**

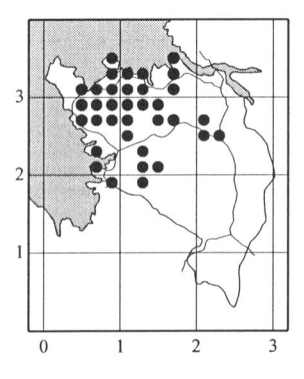

35 tetrads. Most frequent on low ground where there is an abundance of acid lochs and burns. Cover provided by their overhanging margins gives the fern

necessary protection from grazing animals, to which it is clearly very attractive. The best stands are on islands in lochs and we have never seen it growing on open unprotected ground. First recorded in 1886 by A.Gray, as 'Assynt-Stoer…very abundant in certain localities…in spite of the tourist'; presumably a reference to the Victorians' penchant for ferns.

## ADIANTACEAE

*Cryptogramma crispa* R.Br.
Raineach Pheirsill
**Parsley Fern**

Recorded in 1966 on 'Quinag NC22' by A.G.Kenneth. It has not been seen since.

## MARSILIACEAE

*Pilularia globulifera* L.
Feur a'Phiobair
**Pillwort**

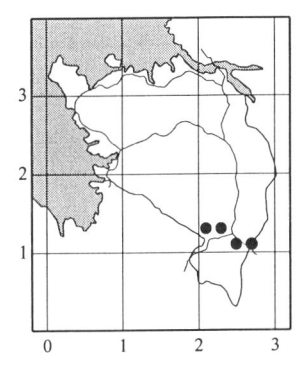

4 tetrads. This represents two populations in each of two lochs. During the present survey it was noted at the north-western end of Loch Borralan in 1988 and 1992 and at two sites on Cam Loch, one in 1992 and one in 1993. The plants are often in silty conditions in the shelter of a bay. In some instances the population continued out from the water on to the shore, but the significance of this is questionable, as loch levels can vary considerably. The Loch Borralan populations were seen again in 1999, but repeated searches failed to produce the Cam Loch plants. There are no records in *Anthony*, who regarded it as extinct in both West and East Sutherland. First recorded in 1977 by D.H.N.Spence and E.D.Allen on 'a stony loch shore' on Loch Borralan, where it was refound by E.Charter in 1983.

## HYMENOPHYLLACEAE

*Hymenophyllum wilsonii* Hook.
Raineach Còinnich
**Wilson's Filmy-fern**

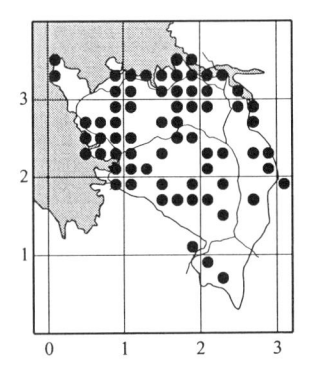

67 tetrads. The greatest density of records is in the birch woodlands around the coast, where it grows luxuriantly on large boulders and tree stumps. Less characteristic sites were found on the Stoer peninsula, where it was growing in turf in two places, firstly on the north-facing side of a shallow gully and secondly in the lee of large rocks on a very exposed slope. Also unexpectedly, it was seen on Canisp, growing beneath *Arctostaphylos alpinus* at 500m., and deep in north-facing boulder scree on Meallan Liath Mór. It is likely that there are more records to be made at altitude if only the right habitats can be identified. First recorded from '*Achumore*' in Anderson (1834), and from the gorge of Allt Poll an Droighinn by A.Gray in 1886.

[*Trichomanes speciosum* Willd.
Raineach Chill Airne
**Killarney Fern**

The gametophyte of this species has been found both north and south of Assynt. A search, by A.C.Jermy in 1993, of suitable sites near Stoer Lighthouse and also inland locations in Torridonian boulder scree, proved unsuccessful.]

## POLYPODIACEAE

*Polypodium vulgare* L.
Clach-raineach Chaol
**Common Polypody**

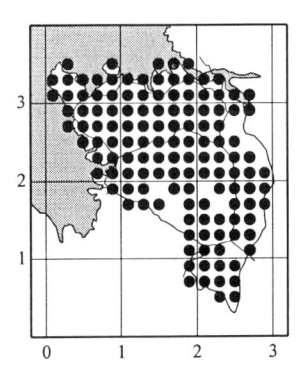

130 tetrads. An extremely common fern growing both in woodland and unshaded situations, epiphytically, on rocks or on dry ground. In the early years of the survey it was recorded only as an aggregate. Latterly we checked for the presence of *P. interjectum*, but in spite of this only one record was made. The map therefore represents all records apart from this one. First recorded in 1767 near Inchnadamph by J.Robertson.

*Polypodium interjectum* Shivas
(*P. vulgare* ssp. *prionodes* Rothm.)
Clach-raineach Mheadhanach
**Western Polypody**

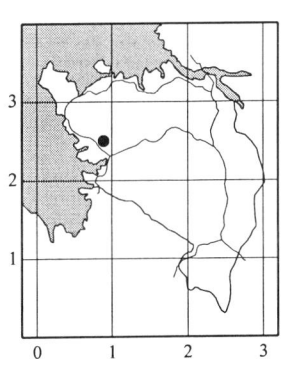

1 tetrad. Found in woodland at Rhicarn. See account of the preceding species. First confirmed by A.C.Jermy from material collected in 1970 at Torgawn by M.G.Coulson.

DENNSTAEDTIACEAE

*Pteridium aquilinum* (L.) Kuhn
Raineach Mhòr
**Bracken**

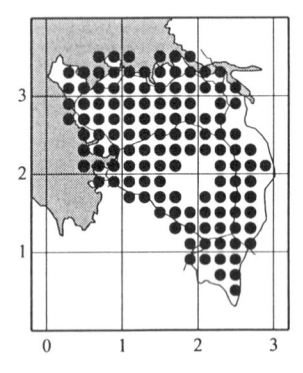

126 tetrads. Widespread and common on well-drained sites in woodland and on moorland, only being absent from the higher hills. Patches of bracken, in a landscape otherwise clothed in heather, deergrass and rock, often mark sites of former cultivation, the grassland being invaded by the fern when active use of the area diminished. As these grassy areas would have been chosen for their fertility, it follows that they are usually richer in small herbs in spite of the bracken cover. First recorded in 1767 near Inchnadamph by J.Robertson.

THELYPTERIDACEAE

*Phegopteris connectilis* (Michx.) Watt
(Thelypteris phegopteris (L.) Slosson)
Raineach Fhaidhbhile
**Beech Fern**

123 tetrads. Common and widespread in woodland and on rocky slopes, where there is both shelter and moisture. It can form quite dense and extensive colonies on wet cliff ledges and in gullies It is a little less frequent in limestone areas, but whether the reason for this is the character of the substrate or lack of suitable habitat is difficult to say. First recorded in 1943 from 'a gully near Inverkirkaig' and 'rocks at the south foot of Creag Mhór, Quinag' by A.Wilmott and M.S.Campbell.

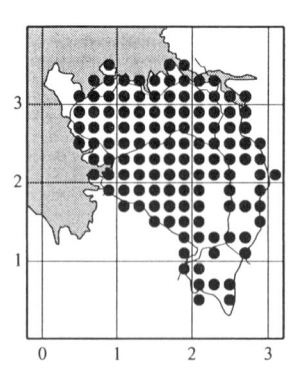

*Oreopteris limbosperma* (Bellardi ex All.) Holub
(Thelypteris oreopteris (Ehrh.) Slosson)
Crim-raineach
**Lemon-scented Fern**

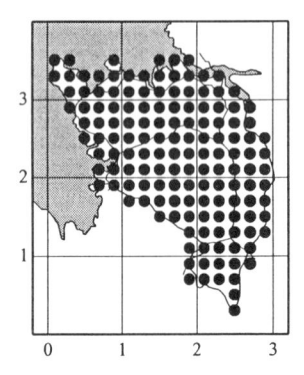

144 tetrads. One of the most widespread ferns in the parish, found in moist but well-drained places in both woodland and moorland, always making use of the shelter provided by cliffs, gorges and burn banks. First recorded in 1767 near Inchnadamph by J.Robertson as '*Acrostichum thelypteris*'.

ASPLENIACEAE

*Phyllitis scolopendrium* (L.) Newman
Teanga an Fhèidh
**Hart's-tongue**

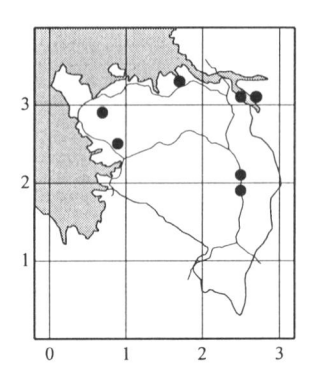

7 tetrads. An uncommon fern, not occurring anywhere in quantity. Two of the records are from the mortared stone walls of buildings at Ardvar and Rhicarn and two from limestone cliffs and screes in the Inchnadamph area. Of the remainder, two tetrads lie on a narrow outcrop of limestone/Fucoid Beds in woodland at Liath Bhad, on the south side of Loch Glencoul, and the fern also occurs abundantly on the sea-cliff just over the parish boundary on a continuation of the same rock. A surprising discovery was 15-20 plants in a 'horizontal gryke', in a basic dyke cutting through the gneiss on the shore of Loch na h-Uidhe Doimhne, less than a metre above the surface of the water. First recorded in 1767 near Inchnadamph by J.Robertson. It was also noted in 1943, on the 'wall of the Culag Hotel garden, Lochinver' by A.J.Wilmott and M.S.Campbell.

*Asplenium adiantum-nigrum* L.
Raineach Uaine
**Black Spleenwort**

86 tetrads. The most common fern on cliffs and outcrop rock, although less frequent on walls. Found mostly on the gneiss, sometimes on Torridonian sandstone but absent from many tetrads which lie on limestone.

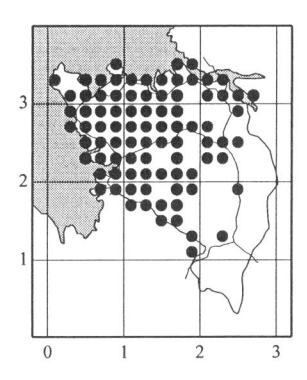

R.E.C.Ferreira has pointed out the interesting association of this species with ultrabasic dykes (Ferreira 1995). First recorded in 1943 in three localities by A.J.Wilmott and M.S.Campbell.

*Asplenium obovatum* ssp. *lanceolatum* (Fiori) P. da Silva
Lus a'Chorrain Lannach
**Lanceolate Spleenwort**

The first and only record is from Clachtoll, the most northerly station from which it has been recorded in the British Isles. It was found there on a gneiss crag, west of Cnoc Beag, by M.G.Coulson in 1969/70 and identified by A.C..Jermy, but has not been seen since.

*Asplenium marinum* L.
Raineach na Mara
**Sea Spleenwort**

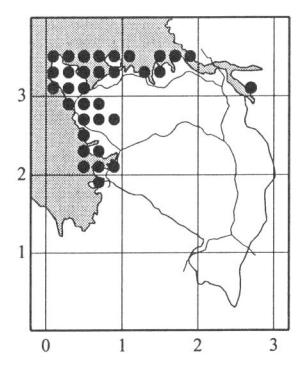

33 tetrads. Found where there is suitable cliff habitat around the coast, particularly on the western shores of the parish where exposure is greatest. It occurs in five inland sites: at Clachtoll are two sites approximately 0.5km. from the sea, one a cliff on the eastern shore of Loch an Aigeil with five plants noted and the other not far away with more than 30 plants; at Creag Liath, 0.75km. from the sea, nine plants were recorded on an inland cliff; 2.5 km. from the coast two tufts of the fern were noted on a cliff on the eastern shore of Loch a'Phollain Drisich; the most outstanding of the inland sites, Creag Clais nan Cruineachd, is nearly 3km. from the sea and the species occurs there in two tetrads, two plants in one and 26 countable ones, with others too high to assess, in the other. First recorded in 1886 at 'Lochinver' by A.Gray.

*Asplenium trichomanes* L.
Dubh-chasach
**Maidenhair Spleenwort**

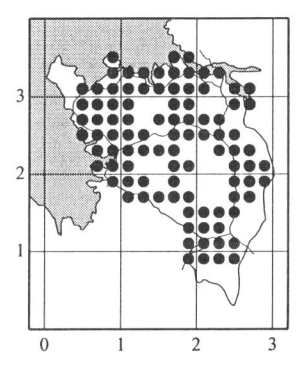

95 tetrads. Widespread in shady situations on cliffs, crags and walls. Owing, it has to be confessed, to an inconsistency in recording technique throughout the period of the survey, we are unable to map *A. trichomanes* ssp. *trichomanes*. The map above shows those records which were only identified to species level. *A. trichomanes* ssp. *quadrivalens* was noted during the second half of the survey and a map for that follows. First recorded in 1943 on 'rocks at the south foot of Creag Mhór, Quinag' by A.J.Wilmott and M.S.Campbell.

ssp. *quadrivalens* D.E.Mey.

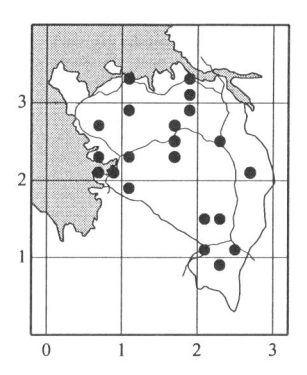

21 tetrads  For limitations in recording see under the species. Of the habitats represented on the map, two are mortared walls and four are limestone; gneiss and Torridonian sandstone are also represented.

*Asplenium viride* Huds.
Ur-thalmhainn
**Green Spleenwort**

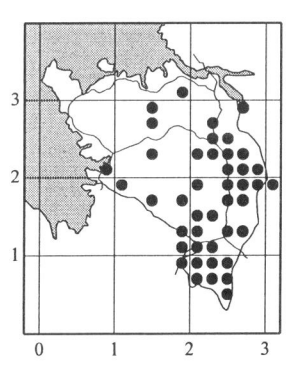

43 tetrads. A calcicole whose distribution, as will be seen from the map, corresponds to a great extent to that of the limestone areas. Other sites are at the base of the Torridonian and where basic and ultrabasic dykes cut through the gneiss. The fern requires shelter and shade and can grow luxuriantly where these conditions are met. In some parts of Britain this species favours high ground, but in the West of Scotland it is common at low levels. First recorded in 1772 at '*Creg-a-chnocain*' by T.Pennant and J.Lightfoot.

*Asplenium ruta-muraria* L.
Rù Bhallaidh
**Wall-rue**

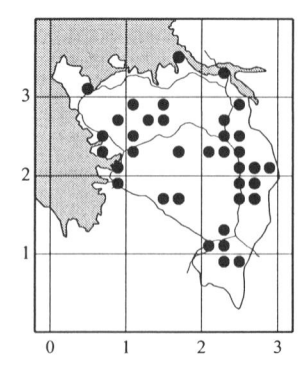

36 tetrads. Although the greatest concentration of records comes from rocks in the limestone areas, the good scatter on the west side of the parish shows that it can tolerate a variety of rock types, given sunshine and a degree of shelter provided by fissures. There are few records from walls. First recorded in 1894 from Knockan crags by G.C.Druce.

*Asplenium septentrionale* (L.) Hoffm.
Lus a'Chorrain Gòbhlach
**Forked Spleenwort**

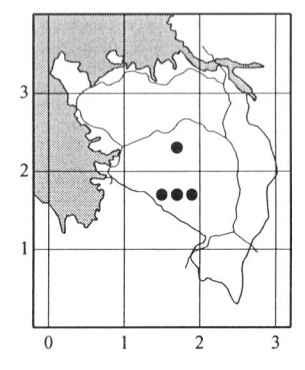

4 tetrads. Not recorded in *Anthony* for either West or East Sutherland. The first record for Assynt was made by A.Scott and W.Henderson in 1991 in Gleannan a'Mhadaidh, growing on the dark rock of a basic dyke, which shows on the south-facing wall of this narrow, steep-sided valley. Another population, of 20+ plants, was found by G.P.Rothero further along the same cliff. The other sites are in similar situations. Five plants and a single plant were found on crags in two tetrads to the east. On an outcrop north of the gorge of the Allt an Tiaghaich, G.P.Rothero found eight tufts.

WOODSIACEAE

*Athyrium filix-femina* (L.) Roth
Raineach Moire
**Lady Fern**

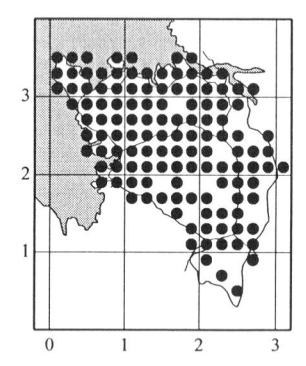

125 tetrads. Widespread and common, but a little less ubiquitous in limestone areas. Will grow in the open, although the most luxuriant specimens are in sheltered burn valleys and gorges. Dies back very quickly in the autumn. First recorded in 1767 near Inchnadamph by J.Robertson.

[*Athyrium distentifolium* Tausch ex Opiz
Raineach Moire Ailpeach
**Alpine Lady-fern**

There are vague records for this montane species from the 1950s and 1980s on recording cards for NC21, but either or both may refer to the eastern side of Breabag, which is in v.c.106.]

*Gymnocarpium dryopteris* (L.) Newman
(*Thelypteris dryopteris* (L.) Slosson)
Sgeamh Dharaich
**Oak Fern**

46 tetrads. Not a common species and usually only growing in small quantity where it does occur.

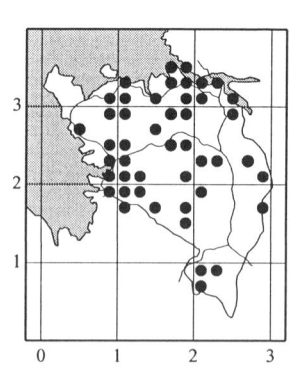

Found most often in rocky woodland where, in the shelter of the rocks, its distinctive bright green fronds are easily seen. We have found it to be very sparse in limestone areas. However, the species was first recorded in 1767 near Inchnadamph by J.Robertson, and in 1894 G.C.Druce recorded a form approaching *G. robertianum* from Knockan Crags.

*Gymnocarpium robertianum* (Hoffm.) Newman
(*Thelypteris robertiana* (Hoffm.) Slosson)
Raineach Cloich-aoil
**Limestone Fern**

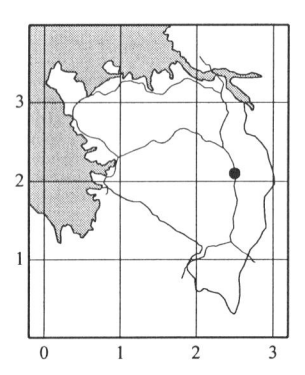

1 tetrad. Growing in some quantity on a ledge of the limestone cliffs of Creag Sròn Chrùbaidh. It was first recorded (from this site) in 1951 by A.McG.Stirling. This is the most northerly site in Britain for the species and its only station in West Sutherland.

*Cystopteris fragilis* (L.) Bernh.
Frith-raineach
**Brittle Bladder-fern**

45 tetrads. A calcicole, found widely in crevices and screes on

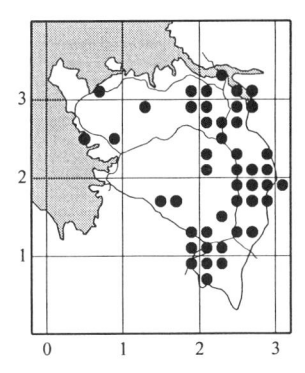

limestone and the Fucoid Beds, less often on other rock types and occasionally on mortared walls. Always in moist, sheltered, shady conditions. First recorded in 1767 near Inchnadamph by J.Robertson and in 1943 on walls in Lochinver by M.S.Campbell.

*Polystichum aculeatum* (L.) Roth
Ibhig Chruaidh
**Hard Shield-fern**

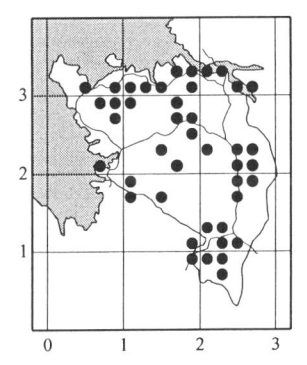

43 tetrads. Widespread, not only in limestone areas but also in the west, where its presence on the gneiss is presumably due to base-rich pockets. Found in gorges, on rock faces beside waterfalls and in boulder scree, but can be seen in the greatest profusion on the vertical walls of pots and cave entrances in the limestone. First recorded in 1890 at Inchnadamph by E.S.Marshall and F.J.Hanbury.

*Polystichum* x *illyricum* (Borbàs) Hahne
(*P. aculeatum* x *P. lonchitis*)
**Alpine Hybrid Shield-fern**

2 tetrads. The second record for the British Isles and the first for Scotland was made in Assynt in 1973 by A.McG.Stirling in limestone boulder scree at the foot of the cliffs of Creag Sròn Chrùbaidh, where it still grows with both parents. It is also found at Inchnadamph, in an exclosure on the limestone, where again both parents are present.

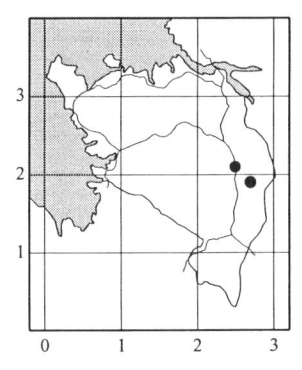

*Polystichum lonchitis* (L.) Roth
Raineach Chuilinn
**Holly Fern**

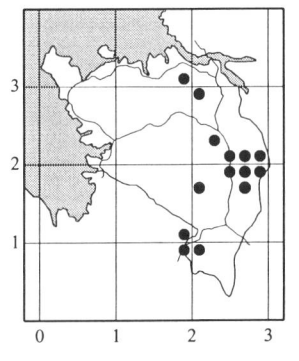

14 tetrads. Mainly on limestone or the Fucoid Beds. Notable exceptions are associated with calcareous bands in the Torridonian, one at 250m. on north-facing crags south of Lochan Fada, the other at 350m. on east-facing crags above Lochan Bealach Cornaidh; also, on similar bands in the gneiss, at the foot of Sàil Ghorm. First recorded in 1772 at '*Creg-a-chnocain*' by T.Pennant and J.Lightfoot.

DRYOPTERIDACEAE

*Dryopteris oreades* Fomin
(*D. abbreviata* (DC.) Newm.)
Marc-raineach an t-Slèibhne
**Mountain Male-fern**

Not seen during the present survey. There is a record from Suilven made by B.Flannigan in 1958. It is mapped in the first *Atlas* for NC02 and NC21; the former hectad seems to have little likely ground.

*Dryopteris filix-mas* (L.) Schott
Marc-raineach
**Male-fern**

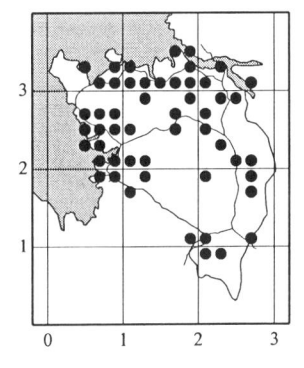

52 tetrads. In woodland and sheltered places amongst rocks. Not plentiful. First recorded in 1767 near Inchnadamph by J.Robertson.

*Dryopteris affinis* (Lowe) Fraser-Jenk.
(*D. borreri* Newm.)
Mearlag
**Golden-scaled Male-fern**

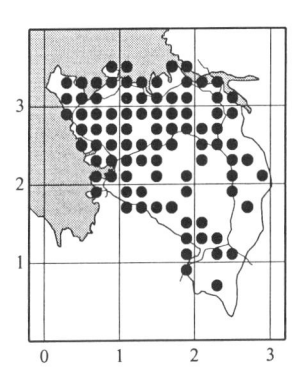

89 tetrads. Common and widespread, found in much more open situations than the preceding species. Stands in sunny boulder

scree make a fine sight early in the year, when it is at its most 'golden'. First recorded in 1890 at Lochinver by E.S.Marshall and F.J.Hanbury, as '*D. filix-mas* var. *paleacea*'. Subspecies *cambrensis* and *borreri* were recorded in the 1990s by C.E.K.Scouller.

*Dryopteris aemula* (Aiton) Kuntze
Raineach Phreasach
**Hay-scented Buckler-fern**

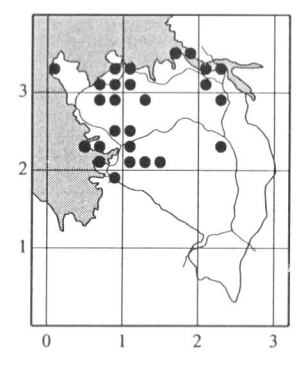

26 tetrads. We have found it to be a not infrequent component of the ground flora of rocky woodland, always in sheltered, shady situations. First recorded from NC03V in 1990.

*Dryopteris carthusiana* (Vill.) H.P.Fuchs
(*D. lanceolatacristata* (Hoffm.) Alston)
Raineach Chaol
**Narrow Buckler-fern**

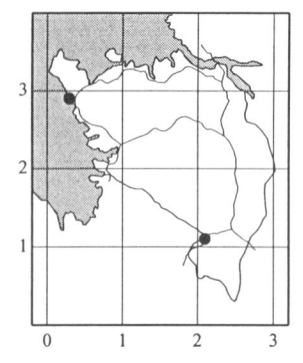

2 tetrads. Very rare. Found once at Knockan and once by wet rocks at Stoer. First recorded during the 1950s survey for the first *Atlas*. The *Fern Atlas* maps it for NC01, 02, 12, 22 and 23, suggesting that it may have been under-recorded during the present survey.

*Dryopteris dilatata* (Hoffm.) A.Gray
Raineach nan Radan
**Broad Buckler-fern**

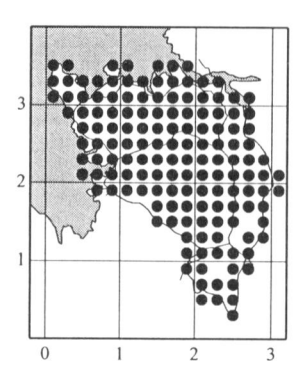

146 tetrads. An abundant fern of shady places in woodland or in the shelter of rocks. Absent from some tetrads in limestone areas. First recorded in 1894 at Knockan Crags by G.C.Druce.

*Dryopteris expansa* (C.Presl) Fraser-Jenk. & Jermy
(*D. assimilis* S.Walker)
Raineach nan Radan Thuatach
**Northern Buckler-fern**

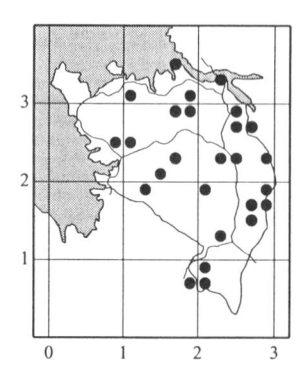

26 tetrads. Thinly distributed. Most often in scree or in well-sheltered positions between outcrop rocks. First recorded in 1955 at Achmelvich by J.Anthony.

BLECHNACEAE

*Blechnum spicant* (L.).Roth
Raineach Chruaidh
**Hard Fern**

154 tetrads. Almost ubiquitous. It would be difficult to name a habitat from which this species has not been recorded.

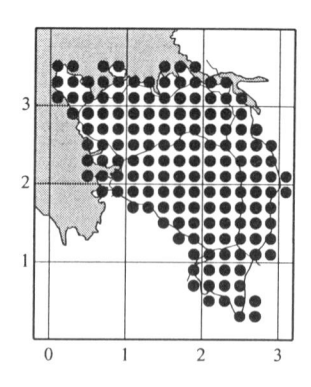

It seems impervious to exposure and manages to thrive almost anywhere, although only producing fertile fronds under favourable conditions. First recorded during the 1950s survey for the first *Atlas*.

# PINOPSIDA
## Conifers

PINACEAE

There are three main types of conifer plantings. Firstly the 'Inver Woods' and Culag Wood (the latter now a Communiy Woodland) which have a major conifer component, as well as some hardwoods. Secondly, a more extensive, commercial forestry plantation, planted in the mid 1980s at Ledmore, mostly in NC21. Thirdly, a number of small estate plantings, visible from the road between Inchnadamph and Ledmore, which were made about the middle of the 20th century to provide deer shelter. There is also an estate plantation just north-east of Glencanisp Lodge. Species in these three groups have been mapped because of their general ecological significance in the landscape.

Abies procera Rehder
**Noble Fir**

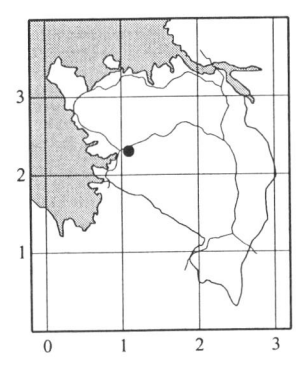

1 tetrad. There are occasional trees in the amenity woodlands surrounding Glencanisp Lodge and also a few on the edge of a nearby plantation.

*Pseudotsuga menziesii* (Mirb.) Franco
**Douglas Fir**

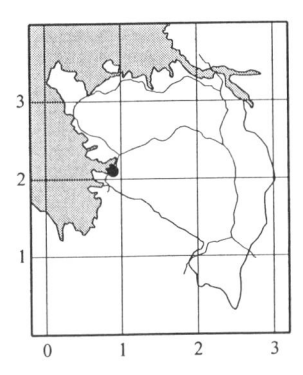

1 tetrad. A small number has been planted in the immediate vicinity of an old house on the edge of Culag Wood, but not in a commercial forestry context.

*Picea sitchensis* (Bong.) Carrière
Giuthas Siotga
**Sitka Spruce**

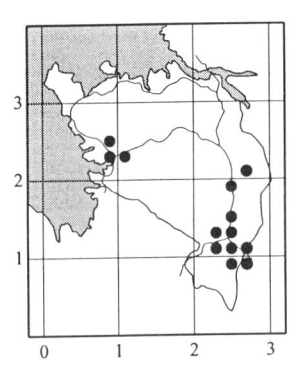

13 tetrads. Well suited to the high rainfall of the area and therefore widely used in commercial plantings, as well as in mixed amenity woodland. It is regenerating naturally in an extensive open area in the 'Inver Woods', from the seed source of mature trees nearby.

*Picea abies* (L.) H.Karst
Giuthas Lochlannach
**Norway Spruce**

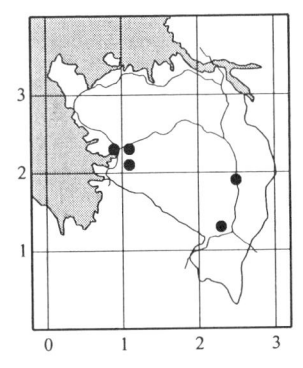

5 tetrads. Distribution much as for *Picea sitchensis,* although less widely planted. Similar also in its occurrence as self-sown specimens in the open area in 'Inver Woods' alongside the river.

*Larix decidua* Mill.
Learag
**European Larch**

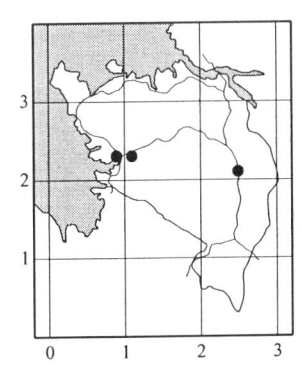

3 tetrads. Trees in the 'Inver Woods' and a small group of old trees alongside the road at Inchnadamph have been mapped. First recorded during the 1950s survey for the first *Atlas*.

Pinus sylvestris L.
Giuthas
**Scots Pine**

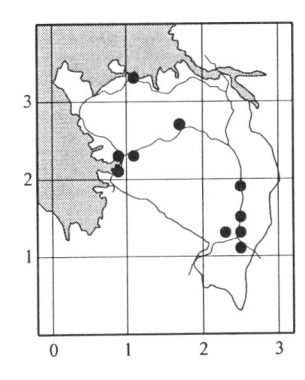

10 tetrads. The trees on the islands in Loch Assynt were once thought to be remnants of original pine forest, but it is now known that this is not the case. This species is the one most commonly planted in a non-commercial context and in many places does indeed look 'natural'. First recorded in 1767 by J.Robertson.

*Pinus contorta* Douglas ex Loudon
**Lodgepole Pine**

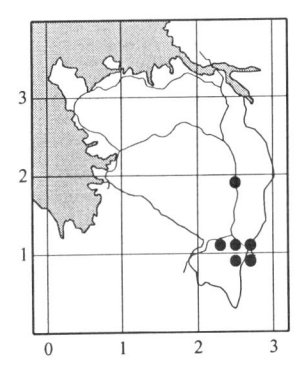

6 tetrads. As this is the pine most suited to Assynt's soil and climate, it has been extensively used in the large-scale commercial plantations at Ledmore and in amenity plantings elsewhere.

CUPRESSACEAE

*Juniperus communis* L.
Aiteann
**Juniper**

89 tetrads. A very widespread species, although there is great

variation in quantity in different areas, which may in some cases reflect the history of burning. The area around and south of Loch

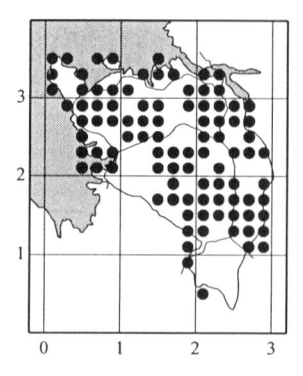

Poll Dhàidh is particularly rich in Juniper and there are some well grown bushes there, which must be of a considerable age. No really upright individuals have been seen, although a few old, spreading ones, growing up against a steep bank, may appear to be so. As intermediates between the two subspecies are known to occur elsewhere, we have only recorded the species, although it does seem that our bushes show strong affinities to ssp. *nana*. First recorded in 1767 by J.Robertson.

# MAGNOLIOPSIDA
# Flowering Plants

NYMPHACEAE

*Nymphaea alba* L.
Duilleag-bhàite Bhàn
**White Water-lily**

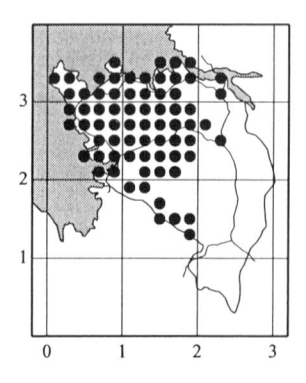

70 tetrads. Grows in sheltered waters, small lochs or bays of larger ones, where it is not exposed to wave action. With its rather specific depth requirement of 0.5–1.5m., the pattern of *Nymphaea* in a small loch may reflect the contours of the loch floor. The result is concentric rings of water-lilies, backed by emergent vegetation and fronted perhaps by pondweeds. It is most widespread in areas of small acid lochs and absent from the limestone lochs such as Mhaolach-coire, Urigill, Borralan and Awe. First recorded at Kylesku in Anderson (1834).

RANUNCULACEAE

*Caltha palustris* L.
Lus Buidhe Bealltainn
**Marsh-marigold**

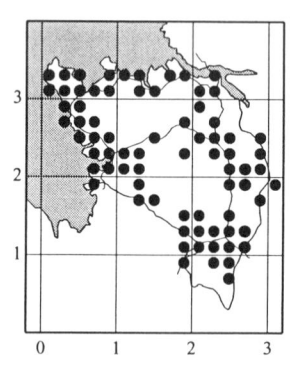

74 tetrads. Usually a plant of lowland marshes and the edge of rivers and lochs, but has also been recorded at a height of over 600m. on the headwaters of burns draining off Breabag. Absent from much of the interior, where small acid lochs and burns predominate. First recorded as '*C. radicans*' (now var. *radicans*) in 1907 'by a stream at Inchnadamph' by G.C.Druce.

*Trollius europaeus* L.
Leolaicheann
**Globeflower**

86 tetrads. A component of base-rich, usually wet, grassland and of the tall-herb vegetation on inland cliff ledges. From sea level to over 600m.

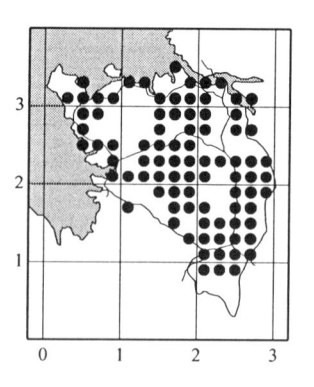

First recorded in 1886 at 'Achumore' by A.Gray.

*Anemone nemorosa* L.
Flùr na Gaoithe
**Wood Anemone**

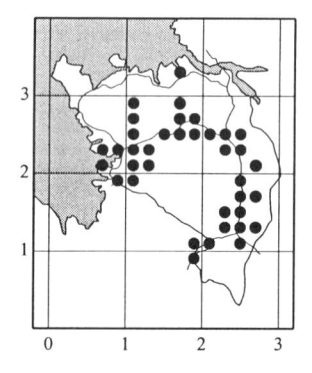

37 tetrads. Not common and often thinly distributed where it does occur. Only a small percentage of the records come from woodland; boulder scree, bracken and even heather can provide the necessary shelter and stony loch shores are also favoured. A high proportion of the plants do not flower. First recorded in 1767 near Inchnadamph by J.Robertson.

*Ranunculus acris* L.
Buidheag an t-Samhraidh
**Meadow Buttercup**

154 tetrads. Grows in most grassy places: fields, light woodland, roadsides, fanks, and patches along the edge of burns. The missing dot is in the most botanically poor of all the tetrads, NC21I, lying on quartzite on the lower slopes of Canisp.

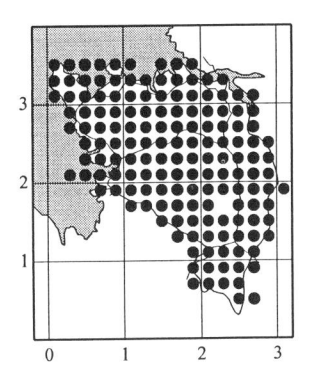

Flowering plants and ferns

First recorded in 1767 near Inchnadamph by J.Robertson.

*Ranunculus repens* L.
Buidheag
**Creeping Buttercup**

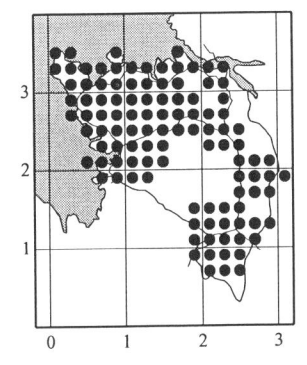

108 tetrads. Wet woodland, grassland, marshes, roadside verges and waste places. Absent from some of the higher and more remote ground. First recorded in 1767 near Inchnadamph by J.Robertson.

*Ranunculus bulbosus* L.
Fuile-thalmhainn
**Bulbous Buttercup**

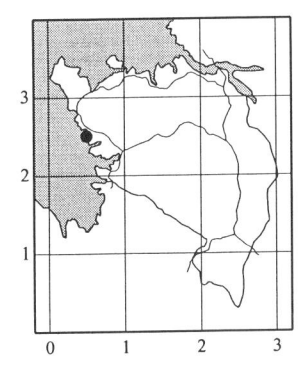

1 tetrad. The only record made during the present survey was in 2001 at Achmelvich. A small quantity was found in dune grassland alongside a fence. Its scarcity in what appears to be a suitable habitat may be due to blow-outs in, and subsequent restoration of, the grassland at the back of the dunes there. First recorded in 1984 by M.McC.Webster in the Achmelvich area.

[*Ranunculus arvensis* L.
**Corn Buttercup**

There is a dubious record made in 1972 from NC22.]

*Ranunculus auricomus* L.
Gruag Moire
**Goldilocks Buttercup**

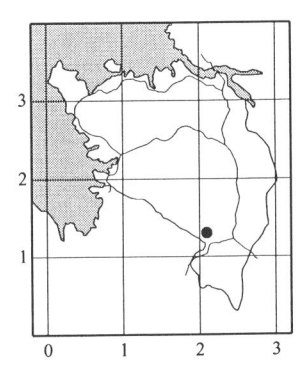

1 tetrad. The first and only record is from Eilean na Gartaig, known locally as 'Garlic Island', in Cam Loch, where a population was found in 1994 by T.Keatinge.

*Ranunculus flammula* L.
Glaisleun
**Lesser Spearwort**

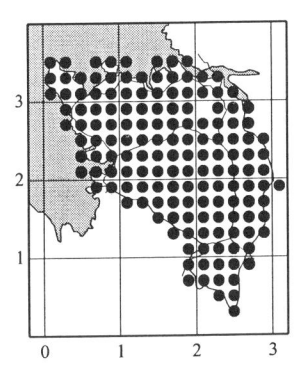

152 tetrads. Widespread and abundant in wet places: ditches, marshes, the margins of lochs, burns and rivers; sometimes growing actually in the water. First recorded in 1767 near Inchnadamph by J.Robertson. Subspecies *scoticus* was recorded in 1889 at Lochan Feòir by E.S.Marshall and F.J.Hanbury, and in 1976 by U.K.Duncan at the north-western end of Loch Borralan.

*Ranunculus ficaria* L.
Searragaich
**Lesser Celandine**

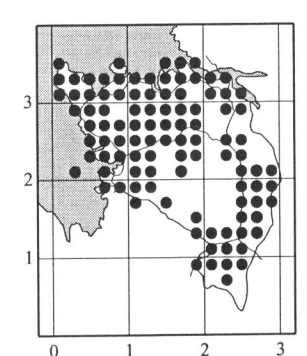

102 tetrads. Common in two rather different habitats; not only in damp sheltered ditches, marshes and wet woodland, but also as a component of the sward of sunny grassy slopes, perhaps with a light shading of bracken. First recorded in 1767 near Inchnadamph by J.Robertson.

[*Ranunculus hederaceus* L.
Fleann Uisge Eidheannach
**Ivy-leaved Water-crowfoot**

Not seen during the present survey. Although recorded for Assynt in *Anthony* (but with no further details), it was not mapped for any local hectads in the first *Atlas*.]

[*Ranunculus baudotii* Godr.
Fleann Uisge Shaillte
**Brackish Water-crowfoot**

Not seen during the present survey. A record of 'var.*marinus*' at Achmelvich is listed in A.G.Kenneth's botanical notebook for 1961, identified by 'Miss Muirhead'. This species does occur on the Uists, but a form of

77

*R. trichophyllus*, with some characters of *R. baudotii*, has been noted at Stoer.]

*Ranunculus trichophyllus* Chaix
Lìon na h-Aibhne
**Thread-leaved Water-crowfoot**

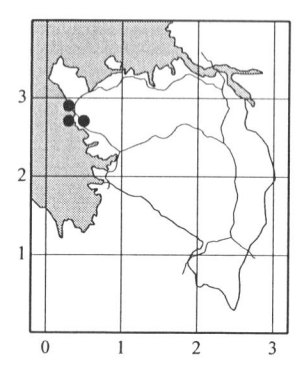

3 tetrads. The three records are from adjacent sites: firstly a small, rich burn running from Loch an Aigeil across dune grassland to the sea, secondly a broad, shallow, wet ditch, some yards from the edge of this loch and thirdly, a nearby drain leading to the shore. The burn has recently been dredged mechanically and it remains to be seen whether or not the species survives there. First reported in 1943 from 'lochan at Stoer' but 'without flower and needing confirmation' by A.J.Wilmott and M.S.Campbell. The modern record has provided that confirmation.

*Thalictrum minus* L.
Rù Beag
**Lesser Meadow-rue**

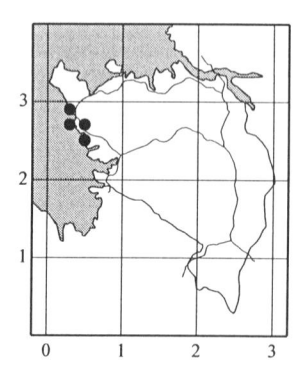

4 tetrads. At Achmelvich, Clachtoll and Stoer in sandy coastal grassland, a habitat of which there is little in Assynt.

First recorded in 1886 at Clachtoll by A.Gray. Recorded from NC13 in the 1950s survey for the first *Atlas*.

*Thalictrum alpinum* L.
Rù Ailpeach
**Alpine Meadow-rue**

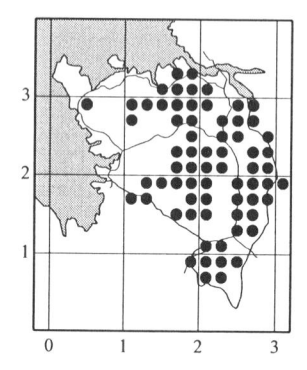

68 tetrads. Widespread in open hill country, where it can be found on the edge of wet stony flushes and in grassland along the margins of burns, ascending to over 600m. on Quinag. It is particularly frequent in limestone areas; elsewhere it indicates a degree of base-richness. First recorded in 1767 near Inchnadamph by J.Robertson.

PAPAVERACEAE

*Papaver somniferum* L.
Lus a'Chadail
**Opium Poppy**

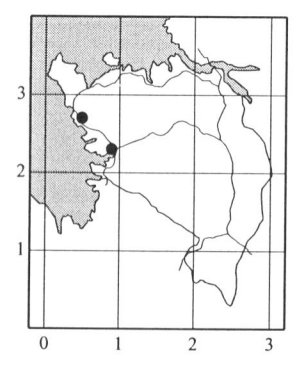

2 tetrads. A rare casual, turning up erratically in waste and disturbed ground, its distribution probably changing from year to year. First recorded in 1996 at Lochinver and later at Clachtoll.

*Papaver dubium* L.
Crom-lus Fad-cheannach
**Long-headed Poppy**

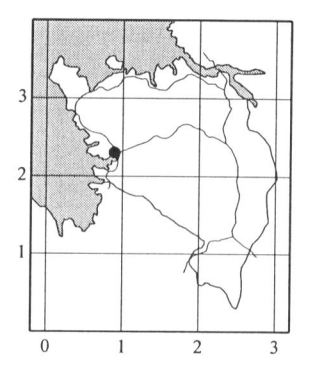

1 tetrad. A rare casual of disturbed ground, found at Lochinver in 1996. First recorded in *Anthony*, but with no further details.

*Meconopsis cambrica* (L.) Vig.
Crom-lus Cuimreach
**Welsh Poppy**

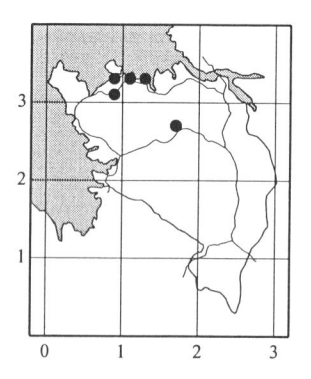

5 tetrads. Only found in the immediate vicinity of past or contemporary dwellings. Naturalised as a garden escape or relic of cultivation.

FUMARIACEAE

*Ceratocapnos claviculata* (L.) Liden
(*Corydalis claviculata* (L.) DC.
Fliodh an Tughaidh
**Climbing Corydalis**

3 tetrads. Rare and unpredictable. The records come from open rocky woodland at Duart, under bracken on an open hillside at Port Dhrombaig and in stone piles not far from the road at Nedd.

# Flowering plants and ferns

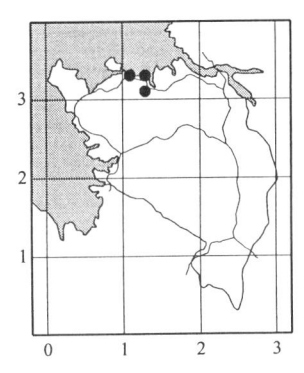

First recorded in 1886 at Elphin by A.Gray; also in 1953/4 on the main island in Loch Beannach by D.N.McVean.

*Fumaria bastardii* Boreau
Fuaim an t-Siorraimh Ard
**Tall Ramping-fumitory**

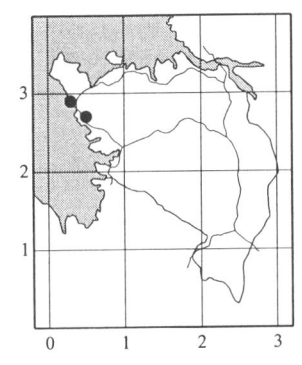

2 tetrads. The records come from two vegetable plots (lying in different tetrads), belonging to a converted croft house at Clachtoll. During the survey a probable *F. bastardii*, too young to identify with certainty, was seen at Achmelvich, but a return visit in 2000 found no trace of the plant. The first record was made in 1956 at Achmelvich by R.M.Graham and R.M.Harley and it was seen there again in 1959 by J.Anthony.

*Fumaria officinalis* L.
ssp. *officinalis*
Lus Deathach-thalmhainn
**Common Fumitory**

1 tetrad. The record comes from a potato patch in Clachtoll (not from the same place as *F. bastardii*).

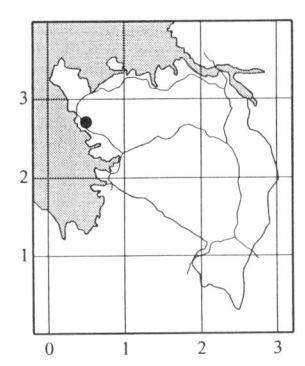

First recorded in 1943 as 'a weed in the garden of Culag Hotel, Lochinver' by A.J.Wilmott and M.S.Campbell; also in 1969/70 at Clachtoll by M.G.Coulson.

## ULMACEAE

*Ulmus glabra* Huds.
Leamhan
**Wych Elm**

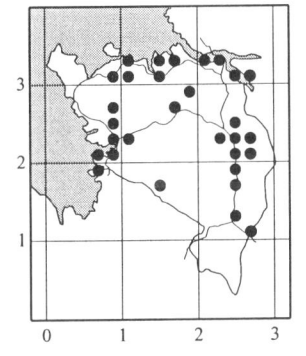

30 tetrads. Only occasionally found in woodland. With the exception of the Culag River gorge, where there are a number of young specimens, the majority of the records are of single trees. They may grow on the edge of a watercourse, in a gorge, or on an inland cliff, sometimes a long way from any other elm tree and posing an interesting question as to how they arrived in such inaccessible places. First recorded in 1767 near Inchnadamph by J.Robertson; his comment 'on precipices where there had never been plantations' (Balfour 1907) reflects concern as to whether the species was native, which continued for over a century.

## URTICACEAE

*Urtica dioica* L.
Deanntag
**Common Nettle**

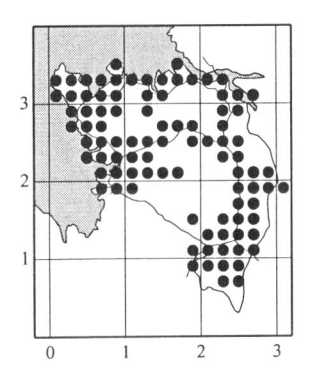

92 tetrads. Found in all the expected places, such as roadside verges and habitations, both contemporary and ruined. Nettles persist long after abandonment of the sheilings, fanks and other ruins in whose vicinity they were first established, and so may still be found in some very remote areas. They do not just indicate soil enrichment, but also an earlier presence of man and his stock. First recorded, as 'var. *horrida*', in 1943 at Stoer by A.J.Wilmott and M.S.Campbell.

*Urtica urens* L.
Deanntag Bhliadhnail
**Small Nettle**

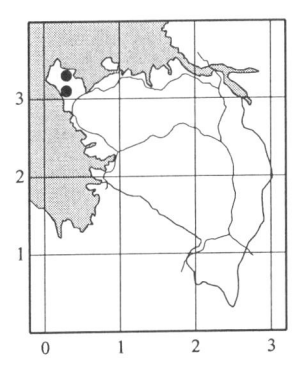

2 tetrads. A species that may have become less frequent with the reduction of cultivated land. It was found on a rubbish dump at Balchladich and in sandy grassland at the back of the shore, in the immediate vicinity of holiday chalets, at Culkein Stoer; a second

search at the end of the survey failed to re-find it at the latter site. First recorded in 1886 at Achmelvich by A.Gray, and in 1894 near Elphin and Ledbeg by G.C.Druce. There is also a 1950s record from NC02 and one in 1972 from NC22.

MYRICACEAE

*Myrica gale* L.
Roid
**Bog-myrtle**

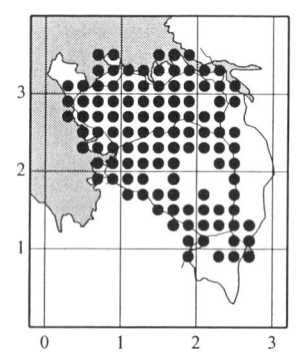

110 tetrads. Widespread and common over much of the acid hill country where conditions are suitably damp. Probably not over 500m. First recorded in 1767 near Inchnadamph by J.Robertson.

FAGACEAE

*Fagus sylvatica* L.
Faidhbhile
**Beech**

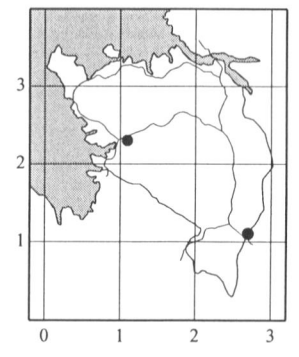

2 tetrads. Introduced. Occasional trees planted in gardens have not been mapped, only those in plantations or in a significant acreage of grounds as at Ledmore and Lochinver. The mixed plantings alongside the River Inver include an extensive open area; here beech is regenerating naturally from mature trees nearby. First recorded as 'beach' in 1774, at Tubeg, on the south side of Loch Assynt, by J.Home. The plantations near Lochinver were established by the time of A.Gray's visit in 1886.

*Quercus* spp.
**Oaks**

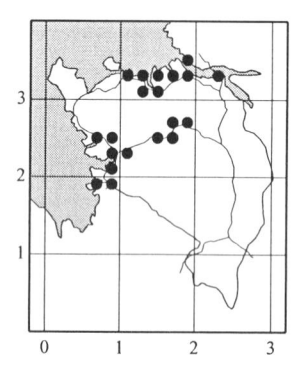

20 tetrads. Oaks are largely confined to the west and north coasts of Assynt, where they are usually found on south-facing rocky slopes, often in association with aspen and hazel. There are two main concentrations. The first is from Inverkirkaig north to Achmelvich, with particularly fine old trees on croft ground at Baddidarach. They occur again from Pollachapuill, where there are ancient coppice stools, through to the shores of Loch Nedd, with isolated outliers at Ardvar and Unapool. There is a further scatter of trees on the north shore of Loch Assynt and in adjacent areas.

Opinions on their precise identity differ. The first record made in 1767 near Inchnadamph by J.Robertson, is of *Q. robur*, as are subsequent records in 1886 at 'Loch Letteressie' by A.Gray, and in 1943 at Lochinver and Drumbeg by A.J.Wilmott and M.S.Campbell. However, D.N.McVean (1958) identified a single tree on the main island in Loch Beannach as *Q. petraea* and W.Yapp (1961) remarks on a 'few well grown specimens of *petraea*...by the roadside west of Loch Assynt Lodge', which is probably the same group as seen by Gray. *Anthony* hedges his bets, recording both species for Assynt.

We have attempted to identify a selection of trees from throughout the parish. Since oaks fruit only rarely this far north, we have had to rely almost entirely on leaf characters, using the excellent tables and illustrations in Wigston (1975). The heterogeneity of the local population is illustrated by a sample of twelve trees at the foot of Creag Dharaich on the east side of Loch Nedd. The only one bearing acorns was confidently assigned to *Q. petraea*; another was thought, because of the absence of stellate hairs and other characters, to be close to *Q. robur*. All the rest were intermediate, and are best perhaps regarded as *Q.x rosacea*? It is for this reason that we have not mapped the three taxa separately. We have detailed records of where oaks can be found in the parish, and would be pleased to help in their further investigation.

BETULACEAE

*Betula pendula* Roth
Beith Dhubhach
**Silver Birch**

There are records made during the 1950s survey for the first *Atlas* for NC02, 20 and 22 and another in 1981 for NC13. All are probably of planted trees, as is almost certainly the case with the small group which we noted on the south bank of the River Inver.

*Betula pubescens* Ehrh.
Beith Charraigeach
**Downy Birch**

121 tetrads. By far the commonest tree in the area, although not the one found at the greatest altitude; the Gaelic name

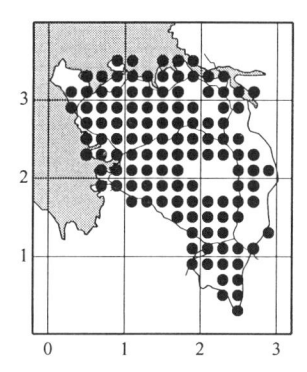

'Beith' often figures in the names of topographical features Woodlands which are almost entirely birch usually have a poor ground flora; those with an admixture of hazel, holly, rowan or aspen are often more interesting. The new crofter forestry schemes all have Downy Birch as one of the constituents and so the species will in the future occur in greater quantity than today. First recorded, at some thirty sites, in 1774 by J.Home. Subspecies *tortuosa* was recorded by E.S.Marshall and F.J.Hanbury in 1890 on 'an islet in Loch Awe'. A.J.Wilmott and M.S.Campbell (1946) express some doubt as to whether the northern species is in fact the same as the southern species known as *B. pubescens*. There is a further record of ssp. *tortuosa* from NC22 in 1969.

*Betula nana* L.
Beith Bheag
**Dwarf Birch**

This species was recorded 'in a single station higher up on Canisp', in association with *Vaccinium uliginosum* and *V. vitis-idaea,* by A.G.Kenneth in 1985. The southern half of NC2018 seems the most likely area, but it has not yet been refound.

*Alnus glutinosa* (L.) Gaertn.
Feàrna
**Alder**

28 tetrads. Grows occasionally on the edge of burns and lochs, flourishing on islands in the latter, where it is free from grazing. There are two Assynt lochs called Lochan Fearna; they no longer have alders on their margins. This is another species used frequently in crofter forestry plantings and will in future be more widespread. First recorded in 1953/54 on the 'main island' in Loch Beannach by D.N.McVean.

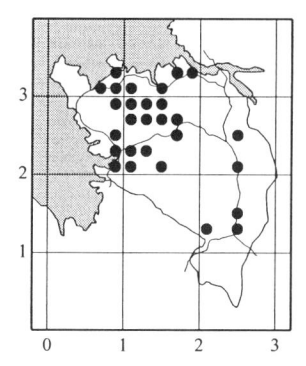

*Corylus avellana* L.
Calltuinn
**Hazel**

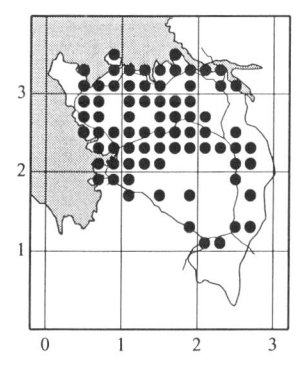

75 tetrads. Hazel woodlands can be seen at their best along the north coast of Assynt, where they clothe the hillsides coming down to the sea, producing some of our most attractive scenery. It does not tolerate exposure to strong wind and outside a woodland situation is most often found in the shelter of gorges and on the lee side of inland cliffs. In crofting areas, in the past, it was an important source of firewood and stock was kept away from newly developing coppice shoots. This is not the case today and hazel woodlands will inevitably deteriorate, as some wood is still cut but grazing is uncontrolled. A hopeful sign is the presence of hazel in the recently planted crofter forestry schemes, thus ensuring its continued presence. First recorded in 1767 near Inchnadamph by J.Robertson.

CHENOPODIACEAE

*Chenopodium rubrum* L.
**Red Goosefoot**

Not seen during the present survey. First recorded in 1886 at Lochinver by A.Gray.

*Chenopodium murale* L.
Praiseach a'Bhalla
**Nettle-leaved Goosefoot**

Not seen during the present survey. First recorded in 1969 on the shore at Clachtoll by M.G.Coulson.

*Chenopodium album* L.
Càl Slapach
**Fat-hen**

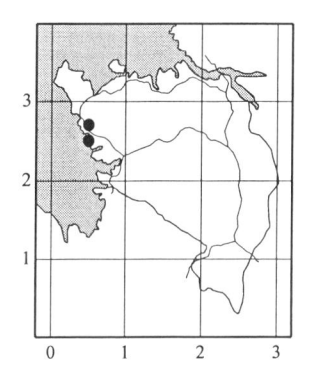

2 tetrads. A rare casual, recorded once from a garden at Clachtoll and once from a smallholding at Achmelvich. First recorded in 1886 at Inverkirkaig by A.J.Gray; subsequently at Lochinver in 1890 by E.S.Marshall and F.J.Hanbury and in 'cornfields' at 'Elphin and Knockain' in 1894 by G.C.Druce.

*Atriplex prostrata* Boucher ex DC.
(*A. hastata* L.)
Praiseach Mhìn Leathann
**Spear-leaved Orache**

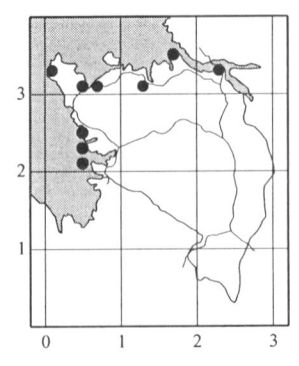

9 tetrads. Only one record was from cultivated ground. The remainder were coastal, growing above the high tide line on sandy or shingle beaches. First recorded for NC02 during the 1950s survey for the first *Atlas*.

*Atriplex glabriuscula* Edmonston
Praiseach Mhìn Cladaich
**Babington's Orache**

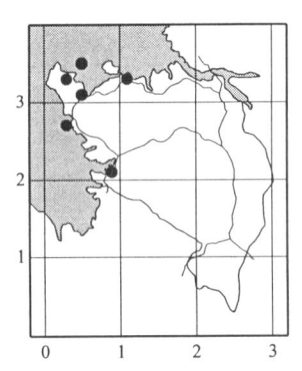

6 tetrads. All the records are from coastal habitats. It occurs above the high tide line on sand and shingle beaches, or on disturbed ground in the immediate vicinity of the shore. First recorded in 1890 at Lochinver by E.S.Marshall and F.J.Hanbury.

*Atriplex praecox* Hülph.
Praiseach Mhìn Thràth
**Early Orache**

Not seen during the present survey. A specimen collected at Lochinver in 1890 by E.S.Marshall and F.J.Hanbury has been confirmed as this species (Taschereau 1985). Another, found at Kylestrome (just outside the parish) in 1970, was confirmed by the same author; it may therefore occur more widely.

*Atriplex patula* L.
Praiseach Mhìn Chaol
**Common Orache**

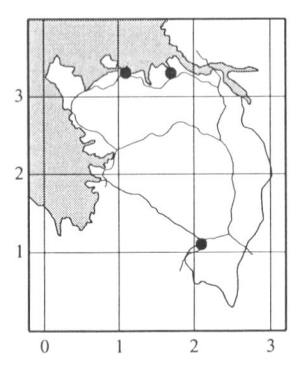

3 tetrads. Uncommon. There is one record from disturbed ground on a roadside verge at Elphin and two from coastal habitats at Loch Ardbhair and Drumbeg. First recorded during the 1950s survey for the first *Atlas* from NC02 and NC12. Taschereau (1985) maps one pre-1970s record from NC02.

*Atriplex* spp.
**Orache**

We have been notably unsuccessful in identifying this genus! Apart from the four species dealt with above, there remain a considerable number of plants, especially from shorelines, to which we were unable to put a name, sometimes because they were not at the right stage, but at other times because they did not appear to correspond adequately to any one species.

PORTULACACEAE

*Claytonia sibirica* L.
(*Montia sibirica* (L.) Howell)
Seachranaiche
**Pink Purslane**

Introduced. Not seen during the present survey. First recorded in 1962, from the side of a stream in Stoer by M.Reid.

*Montia fontana* L.
Fliodh Uisge
**Blinks**

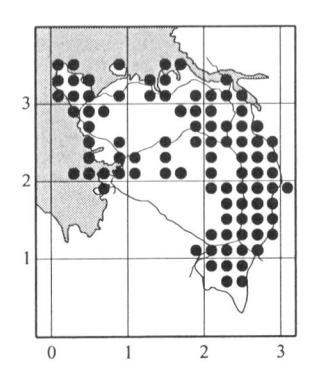

92 tetrads. Grows abundantly almost anywhere that is sheltered, wet and/or muddy. Ditches, small burns, muddy flushes, water-logged tracks and gardens. First recorded in 1767 near Inchnadamph by J.Robertson. Material collected since 1936 in NC02, 21 and 22 has all been determined as ssp. *fontana*.

CARYOPHYLLACEAE

*Arenaria serpyllifolia* L.
Lus nan Naoi Alt Tìomach
**Thyme-leaved Sandwort**

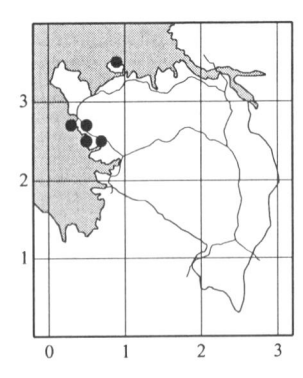

5 tetrads. An uncommon plant found near the coast, once in a garden, but otherwise in sandy grassland, at Achmelvich, Clachtoll and Oldany. Three of the five records were identified as ssp. *serpyllifolia*. First recorded during the 1950s survey for the first *Atlas*.

*Arenaria norvegica* Gunnerus ssp. *norvegica*
Lus nan Naoi Alt Lochlannach
**Arctic Sandwort**

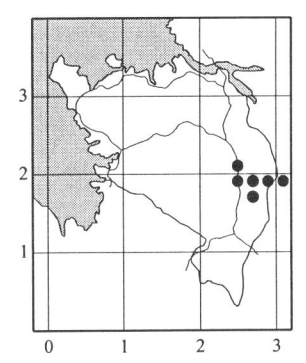

6 tetrads. This Red Data Book plant is a calcicole, confined to the Cambrian limestone at Inchnadamph and the gravels of the River Loanan which drains the limestone. In the Inchnadamph area it is at its most frequent on limestone at Cnoc Eilid Mhathain, but has also been recorded from Creag nan Uamh, Bealach Traligill and from rocks beside the River Traligill. The River Loanan, just south of where it is joined by the River Traligill, follows a very sinuous course and the margins on some of the stronger bends are heavily gravelled. When the water is high, the course of the river and the precise site of the gravels can change slightly and the populations of *Arenaria* become shifted. First recorded in 1886 at Inchnadamph by A.Gray.

*Honckenya peploides* (L.) Ehrh.
Lus a'Ghoill
**Sea Sandwort**

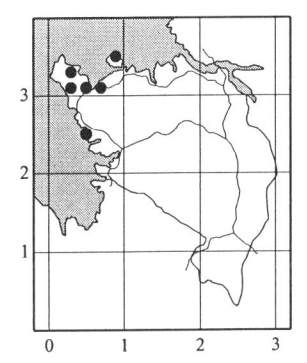

6 tetrads. Thinly distributed on shingle and stony stretches of shore, less often on sandy beaches. Found at Achmelvich, Balchladich, Clashnessie, Culkein Stoer, Imirfada and Oldany. First recorded in 1886 at Clachtoll by A.Gray and also, during the 1950s survey for the first *Atlas*, from NC13.

*Minuartia sedoides* (L.) Hiern
(*Cherleria sedoides* L.)
Lus an Tuim Chòinnich
**Cyphel**

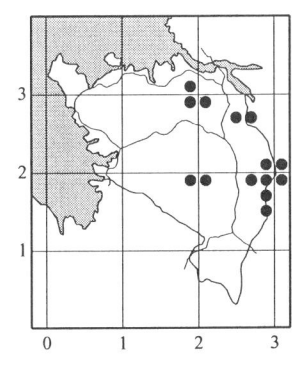

14 tetrads. Found regularly on the upper slopes of Canisp, the Conival/Breabag complex, Cnoc Eilid Mhathain, Glas Bheinn and Quinag. First recorded in 1907 by G.C.Druce as 'plentiful on Ben Mor, Assynt 108' (probably Conival).

*Stellaria media* (L.) Vill.
Fliodh
**Common Chickweed**

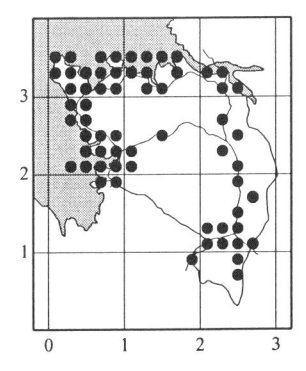

62 tetrads. An abundant plant, given the right habitat of open or cultivated ground. It is clear from the map that most records come from the vicinity of roads and habitations, but it is also common along the coast, where it may grow at the back of the shore in disturbed, weedy areas. First recorded during the 1950s survey for the first *Atlas*.

*Stellaria neglecta* Weihe
Fliodh Mhòr
**Greater Chickweed**

Not seen during the present survey. First recorded in 1890 on a 'ditchside, Inchnadamph' by E.S.Marshall and F.J.Hanbury. Reported also in 1969 from NC02 by M.G.Coulson.

*Stellaria holostea* L.
Tùrsach
**Greater Stitchwort**

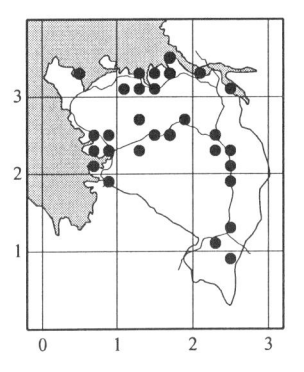

29 tetrads. Occurs only in the lowlands, where it grows beneath trees or scrub and occasionally in ditches. First recorded during the 1950s survey for the first *Atlas*.

*Stellaria graminea* L.
Tursarain
**Lesser Stitchwort**

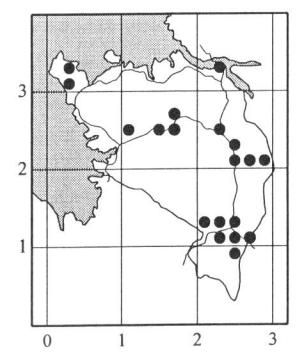

19 tetrads. Occasional, sprawling amongst the long grass of roadside verges or marshes. Less common than the preceding species. First

recorded during the 1950s survey for the first *Atlas*.

*Stellaria uliginosa* Murray
(*S. alsine* Grimm)
Flige
**Bog Stitchwort**

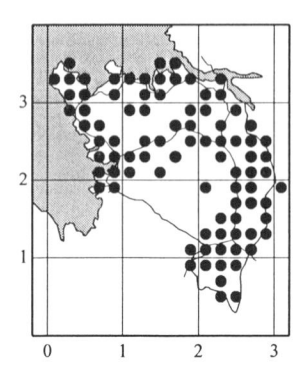

87 tetrads. Abundant in wet places; marshes and ditches, both shaded and in the open. First recorded during the 1950s survey for the first *Atlas*.

[*Cerastium arvense* L.
Cluas Luch Achaidh
**Field Mouse-ear**

A report in 1969/70 from Beinn nan Cnaimhseag may have been a large-flowered form of *C. fontanum*.]

*Cerastium arcticum* Lange
Cluas Luch Artach
**Arctic Mouse-ear**

Not seen during the present survey. First recorded in 1899 by C.E.Salmon from 'Canisp, at about 2500ft., scarce'. A specimen collected by him on 31st July of that year is labelled 'Conival' (which could be in v.c.108 or v.c.107). No v.c.108 records have appeared since then, but the species was recorded from the east side of Conival (v.c.107) by E.S.Marshall and W.A.Shoolbred in 1908, and refound there by D.A.Ratcliffe in 1959.

*Cerastium fontanum* Baumg.
(*C. holosteoides* Fr.)
Cluas Luch Choitcheann
**Common Mouse-ear**

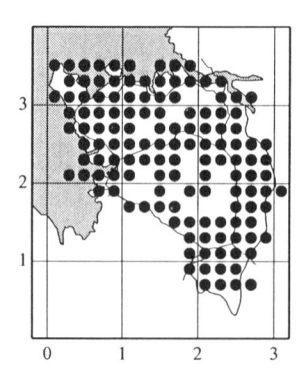

131 tetrads. Common and widespread in a variety of habitats: grassland, light woodland, waste and disturbed ground, and cultivated areas. A form with conspicuously large flowers has been noted at about 600m. beside burns on the ridge between Beinn Uidhe and Conival. First recorded during the 1950s survey for the first *Atlas*.

*Cerastium glomeratum* Thuill.
Cluas Luch Fhàireagach
**Sticky Mouse-ear**

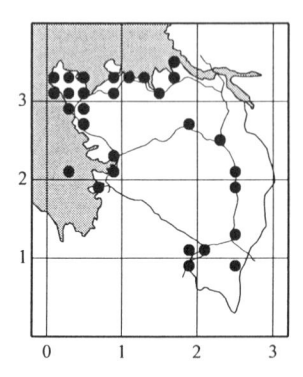

29 tetrads. Thinly scattered. Confined to roadside verges and disturbed ground. First recorded during the 1950s survey for the first *Atlas*.

*Cerastium diffusum* Pers.
(*C. atrovirens* Bab.)
Cluas Luch Mara
**Sea Mouse-ear**

12 tetrads. Occasional in sandy grassland near the sea.

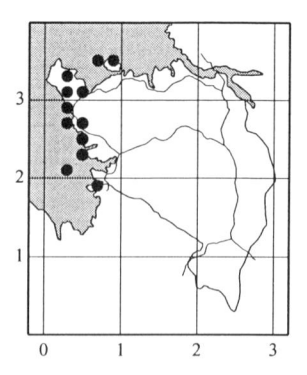

First recorded during the 1950s survey for the first *Atlas*.

*Cerastium semidecandrum* L.
Cluas Luch Bheag
**Little Mouse-ear**

Seen once, in NC02M, during the present survey. First recorded from NC02 during the 1950s survey for the first *Atlas*.

*Sagina nodosa* (L.) Fenzl
Mungan Snaimte
**Knotted Pearlwort**

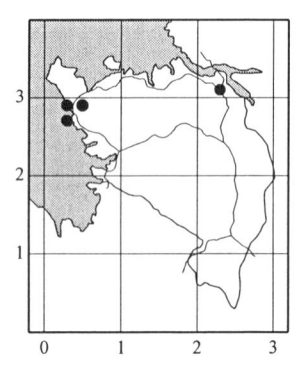

4 tetrads. Rare. Usually in damp or wet sandy grassland near the sea, but also south of Newton on a roadside verge. Found at three sites at Clachtoll, one of which was grassy fissures in the Torridonian seacliffs, and also at Stoer. First recorded in 1886 at Achmelvich by A.Gray and still to be found there in 1943.

*Sagina subulata* (Sw.) C.Presl
Mungan Mòintich
**Heath Pearlwort**

15 tetrads. In open sandy or gravelly ground and occasionally

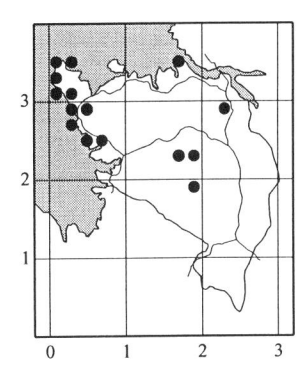

on rock. Var. *glabrata* was found once, growing on the bridge over the burn that runs out of Loch Dubh. First recorded in 1899 'near foot of *Coinnemheall*' by C.E.Salmon, and in 1907 at Inchnadamph by G.C.Druce.

[*Sagina saginoides* (L.) H.Karst.
Mungan Ailpeach
**Alpine Pearlwort**

Reported by J.Anthony from 'sea cliffs near Stoer' in 1959 (Anthony 1967). The herbarium specimen has since been re-determined as *S. subulata* var. *glabrata*.]

*Sagina procumbens* L.
Mungan Làir
**Procumbent Pearlwort**

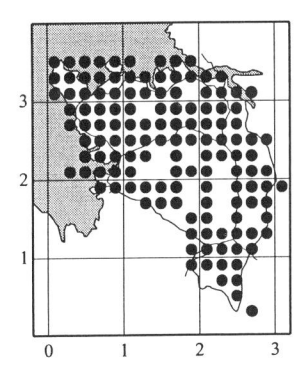

132 tetrads. Abundant in a wide variety of habitats: on rock, paths, open or disturbed ground at roadsides; in short grass, marsh and scree. First recorded in 1767 near Inchnadamph by J.Robertson.

*Sagina apetala* Ard.
Mungan
**Annual Pearlwort**

Not seen during the present survey. Recorded from NC22 in the 1950s survey for the first *Atlas* and, as 'ssp. *erecta*', in 1969 from Stoer Point and Allt nan Uamh.

*Sagina maritima* Don
Mungan Mara
**Sea Pearlwort**

Not seen during the present survey. First recorded in 1953 at Clachtoll/Stoer by M.S.Campbell and also, at about the same time, at Achmelvich.

*Spergula arvensis* L.
Cluain-lìn
**Corn Spurrey**

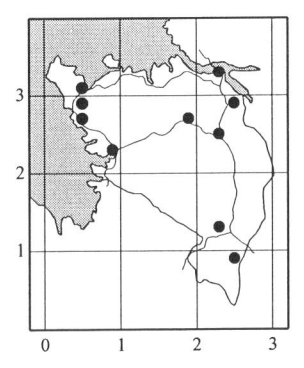

10 tetrads. Occurs very sparsely on cultivated and disturbed ground. Rather more surprisingly it was found on rocky parts of the shore of Cam Loch and Loch Urigill. First recorded in 1894 in 'cornfields at Elphin and Knockain' by G.C.Druce and again, in 1899, in a 'cultivated field, Inchnadamph' by C.E Salmon. Also recorded in 1981 in NC13.

*Spergularia media* (L.) C.Presl
Corran Mara Mòr
**Greater Sea-spurrey**

5 tetrads. The records come from saltmarsh at Achmelvich and Glenleraig, the strand line at Newton and Oldany and disturbed ground near the sea at Lochinver.

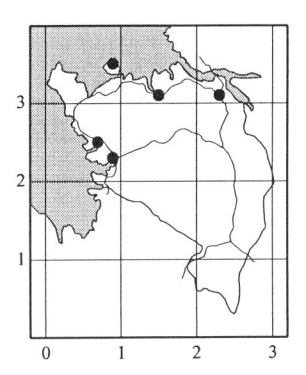

First recorded from NC02 during the 1950s survey for the first *Atlas*.

*Spergularia marina* (L.) Griseb.
Corran Mara Beag
**Lesser Sea-spurrey**

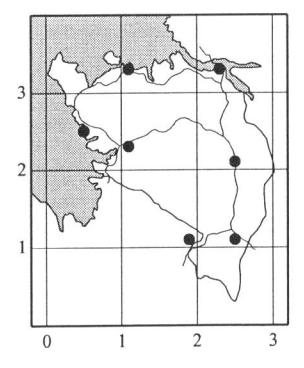

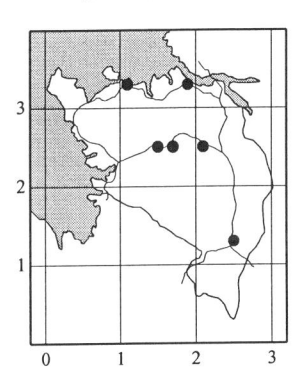

7 tetrads. Mostly found at roadsides, on the gravelly edge of the verge. One record comes from a saltmarsh and one from the deck of the long-abandoned Kylesku ferry. First recorded, as '*S. salina*', in 1943 in a 'saltmarsh by Oldany River, west of Drumbeg' by A.J.Wilmott and M.S.Campbell.

*Spergularia rubra* (L.) J.& C.Presl
Corran Gainmhich
**Sand-spurrey**

6 tetrads. All records are from roadsides, on the gravelly edge of the verge. First recorded in 1990 from the roadside at Gleann Ardbhair.

*Lychnis flos-cuculi* L.
Caorag Lèana
**Ragged-Robin**

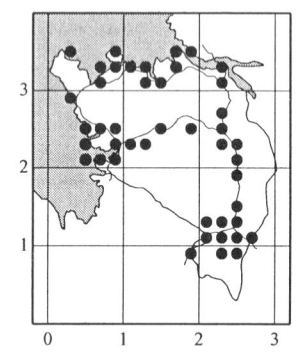

44 tetrads. A lowland plant of wet places, roadside ditches and riverside marshes. First recorded in 1767 near Inchnadamph by J.Robertson.

*Silene uniflora* Roth
(*S. maritima* With.)
Coirean na Mara
**Sea Campion**

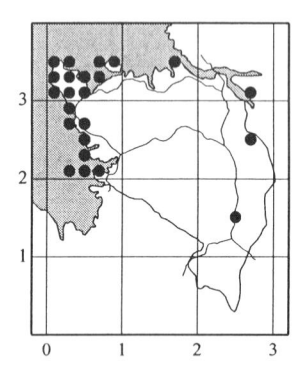

23 tetrads. The records are from sea cliffs, apart from two inland montane sites. One of these, at 700m. on the Ben Uidhe ridge, is almost pure scree and the other, at 450m. on Beinn an Fhuarain, is very sparse montane heath. First recorded during the 1950s survey for the first *Atlas*.

*Silene acaulis* (L.) Jacq.
Coirean Còinnich
**Moss Campion**

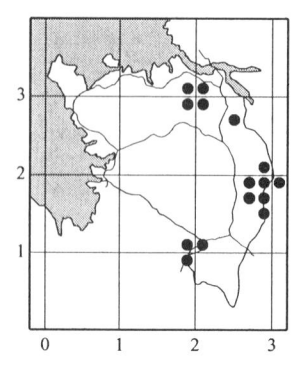

15 tetrads. Recorded from a number of the higher hills, often growing on limestone. Sites include Quinag, Glas Bheinn, Breabag and limestone rocks at Knockan and Allt nan Uamh. First recorded in 1886 at Knockan by A.Gray. Also, in 1899 by C.E.Salmon on the 'north slopes of Canisp'.

*Silene quadrifida* (L.) L.
**Alpine Campion**

A component of the Inchnadamph 'zoo'. Introduced and maintaining itself on a limestone outcrop at Glac Mhór. (See also records of *Campanula*, *Erinus*, *Gentiana* and *Phyteuma*). A single flower was identified by C.I.Pogson and A.J.Underhill, who photographed it at this site in 1997.

*Silene latifolia* Poir.
(*S. alba* (Mill.) E.H.L.Krause)
Coirean Bàn
**White Campion**

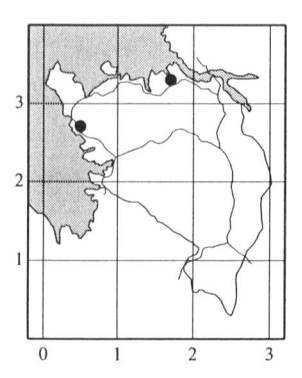

2 tetrads. Rarely seen. Our records are of garden weeds at Clachtoll and Ardvar. First recorded from NC02 and 03 during the 1950s survey for the first *Atlas*.

*Silene* x *hampeana* Meusel & K.Werner
*S. latifolia* x *S. dioica*

Not seen during the present survey. First recorded in 1955 from NC02 by P.H.Davis.

*Silene dioica* (L.) Clairv.
Cìrean Coilich
**Red Campion**

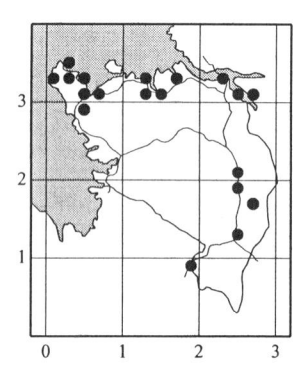

19 tetrads. At least a quarter of the records are from sea cliffs where, safe from grazing, the plants flower freely and the splash of crimson against the rock catches the eye. Other sites are woodland, scrub and occasionally in ditches. First recorded in 1767 near Inchnadamph by J.Robertson; also in 1953/4 on the 'main island in Loch Beannach' by D.N.McVean.

POLYGONACEAE

*Persicaria vivipara* (L.) Ronse Decr.
(*Polygonum viviparum* L.)
Biolur Ailpeach
**Alpine Bistort**

25 tetrads. Common in limestone grassland, as at Inchadamph, Elphin and Knockan; there is one record on gneiss, in a ravine on the headwaters of the Allt Poll an Droighinn.

Flowering plants and ferns

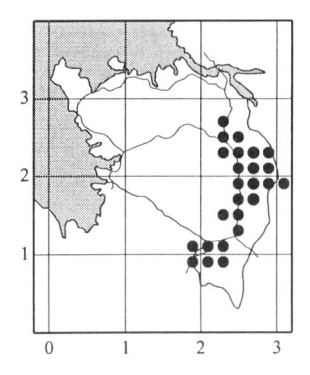

First recorded in 1767 near Inchnadamph by J.Robertson.

*Persicaria amphibia* (L.) Gray
(Polygonum amphibium L.)
Glùineach an Uisge
**Amphibious Bistort**

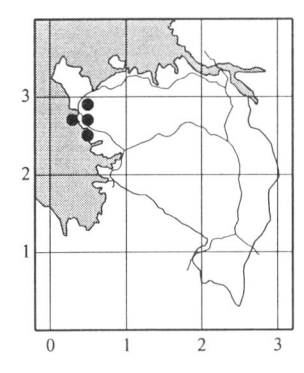

4 tetrads. Seen in aquatic form and at its best in Loch an Aigeil. The other three records are from marshy grassland at Achmelvich and Clachtoll. First recorded, in both aquatic and terrestrial forms, in 1886 at Clachtoll by A.Gray. Also recorded from NC13 during the 1950s survey for the first *Atlas*.

*Persicaria maculosa* Gray
(Polygonum persicaria L.)
Glùineach Dhearg
**Redshank**

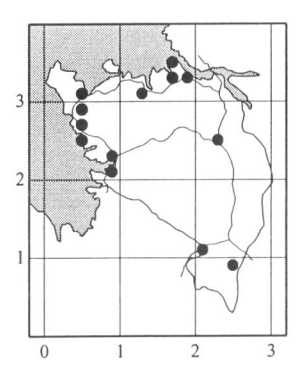

13 tetrads. A plant of open, cultivated or disturbed ground, its distribution following the line of roads and habitations. First recorded during the 1950s survey for the first *Atlas*.

*Persicaria lapathifolia* (L.)Gray
(Polygonum lapathifolium L.)

Not seen during the present survey. First recorded in 1886 at Clachtoll by A.Gray.

*Persicaria hydropiper* (L.) Spach
(Polygonum hydropiper L.)
Glùineach Theth
**Water-pepper**

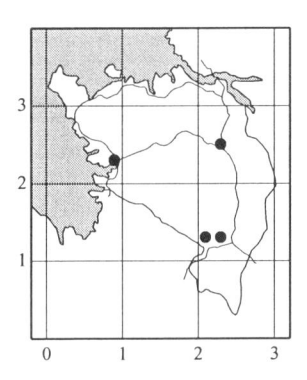

4 tetrads. Found in wet, open habitats. On muddy, stony margins of Loch Assynt and Cam Loch and at the back of saltmarsh at Baddidarach. First recorded in 1943 at Feadan by A.J.Wilmott and M.S.Campbell.

*Polygonum oxyspermum* ssp. *raii* (Bab.) D.A.Webb & Chater
Glùineach na Tràighe
**Ray's Knotgrass**

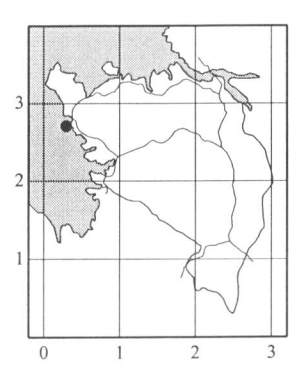

1 tetrad. The first and only record of this species was made in August 1991, in sandy beach shingle in Clachtoll Bay, by F.Rose.

*Polygonum arenastrum* Boreau
Glùineach Ghainmhich
**Equal-leaved Knotgrass**

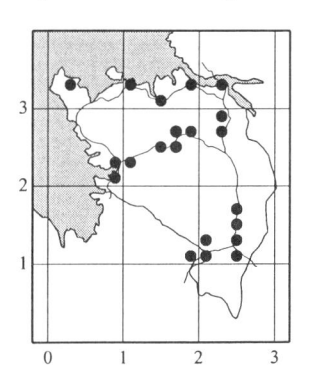

21 tetrads. Entirely confined to roadsides, where it grows on the gravelly edge of the verge. First recorded in 1992 just outside Lochinver.

*Polygonum aviculare* L. s.l.
Glùineach Bheag
**Knotgrass**

All records in *Anthony* were assigned to this taxon which was first recorded in the 1950s survey for the first *Atlas*. *P. aviculare* sensu stricto was not found during the present survey.

*Polygonum boreale* (Lange) Small
Glùineach Thuathach
**Northern Knotgrass**

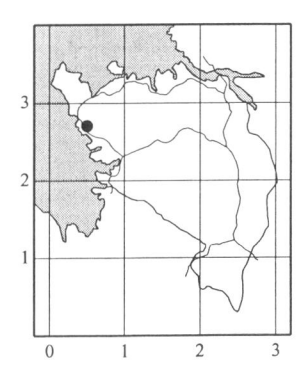

1 tetrad. The first and only record is from a garden in Clachtoll in 1998.

*Fallopia japonica* (Houtt.) Ronse Decr.
(*Polygonum cuspidatum* Siebold & Zucc.)
Glùineach Sheapanach
**Japanese Knotweed**

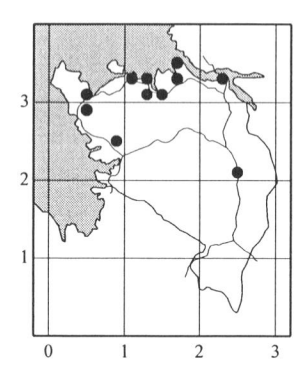

11 tetrads. Although relatively uncommon in Assynt, this introduced species shows the same persistence as in other parts of Britain. It survives in long-abandoned gardens and has spread to two adjacent roadside verges. First recorded in NC02 by A.J.Wilmott, probably in 1943/4.

*Fallopia baldschuanica* (Regel) Holub
Fìon-chrann Ruiseanach
**Russian-vine**

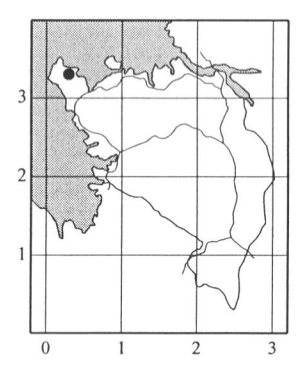

1 tetrad. Relic of cultivation or garden escape. The first and only record was made in 1996 at Culkein Stoer.

*Fallopia convolvulus* (L.) Á.Löve
(*Polygonum convolvulus* L.)
Glùineach Dhubh
**Black Bindweed**

1 tetrad. The only record comes from a garden in Unapool.

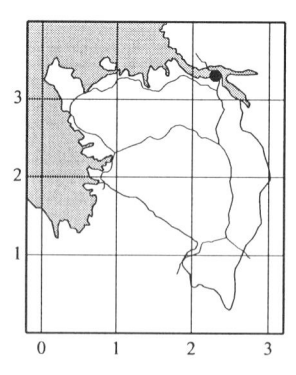

First recorded at Clachtoll during the 1950s survey for the first *Atlas*.

*Rumex acetosella* L.
Sealbhag nan Caorach
**Sheep's Sorrel**

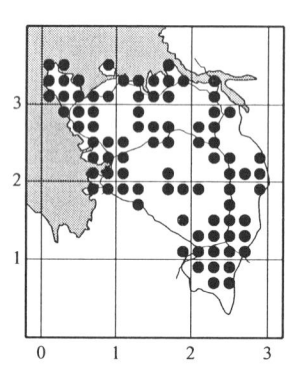

84 tetrads. Widespread in well-drained, heathy areas, acid grassland and around rocks. Regarded as a pernicious weed by local gardeners, who have applied to it the suitable name of 'rootus elasticus'! First recorded during the 1950s survey for the first *Atlas*.

*Rumex acetosa* L.
Samh
**Common Sorrel**

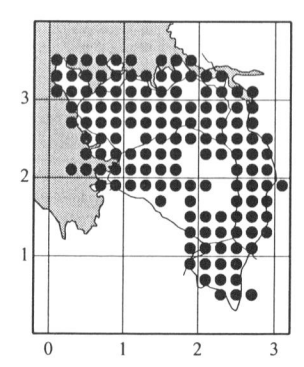

139 tetrads. Common in grassy places everywhere. First recorded in 1767 near Inchnadamph by J.Robertson.

*Rumex longifolius* DC.
Copag Thuathach
**Northern Dock**

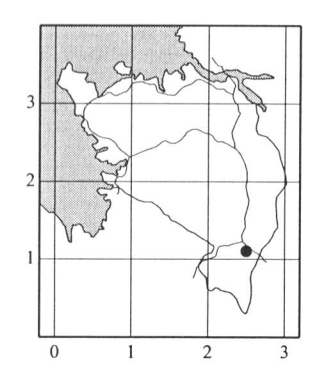

1 tetrad. The first and only record is of a single plant found growing beside the Ledmore River in 1997.

*Rumex crispus* L.
Copag Chamagach
**Curled Dock**

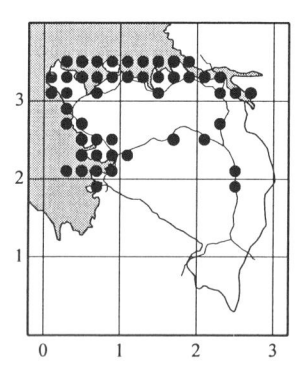

48 tetrads. The records are from the shore, or cultivated and disturbed ground, roadsides and areas of habitation, usually near the coast. In only three cases were plants growing on the shore sufficiently mature to be identified as ssp. *littoreus*; they came from Loch Ardbhair, Raffin and the island of A'Chleit. First recorded in 1886 at Elphin by A.Gray.

*Rumex obtusifolius* L.
Copag Leathann
**Broad-leaved Dock**

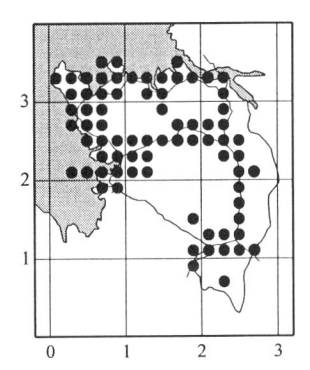

75 tetrads. Very common on roadsides, cultivated and waste ground. First recorded during the 1950s survey for the first *Atlas*.

*Oxyria digyna* (L.) Hill
Sealbhag nam Fiadh
**Mountain Sorrel**

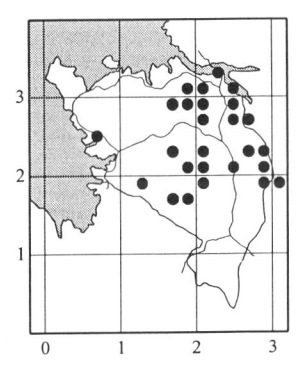

26 tetrads. Usually up in the hills in damp rocky places, particularly beside watercourses. Occasionally at much lower levels, where presumably it has rooted after being washed down from above. First recorded in 1899 at Inchnadamph and on the 'north slopes of Canisp' by C.E.Salmon.

PLUMBAGINACEAE

*Armeria maritima* Willd.
Neòinean Cladaich
**Thrift**

70 tetrads. Saltmarshes, sea cliffs and rocky shores, also inland on open, stony ground, well up on the hills.

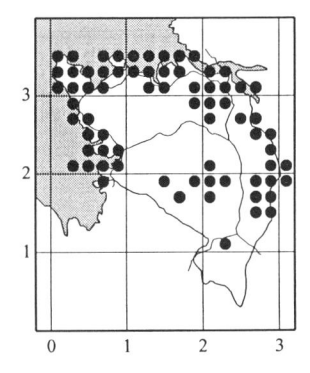

The only inland non-montane record is from the roadside by Loch na Gainmhich. First recorded in 1774 on 'the top of Braeback Hill' by J.Home.

CLUSIACEAE

*Hypericum androsaemum* L.
Meas an Tuirc Coille
**Tutsan**

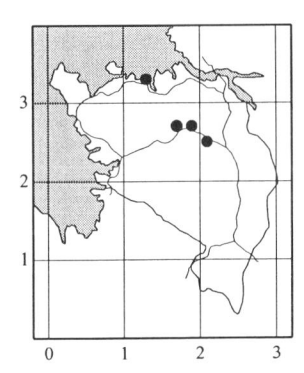

4 tetrads. The sites, small quarries and ditches along the road beside Loch Assynt, and the bank of a burn near houses in Nedd, suggest that the plants are of garden origin. They are maintaining themselves with varying degrees of success, often with other garden throwouts. First recorded in 1944 at Lochinver by A.J.Wilmott.

*Hypericum tetrapterum* Fr.
Beachnuadh Fireann
**Square-stalked St John's-wort**

1 tetrad. The first and only record, made in 1990, is from the verge of a relatively recently constructed road near Unapool.

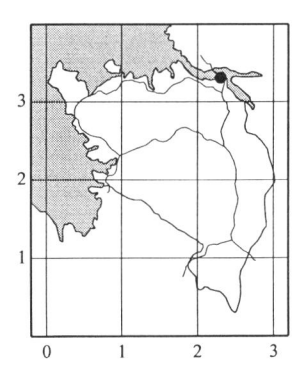

*Hypericum pulchrum* L.
Lus Chaluim Chille
**Slender St John's-wort**

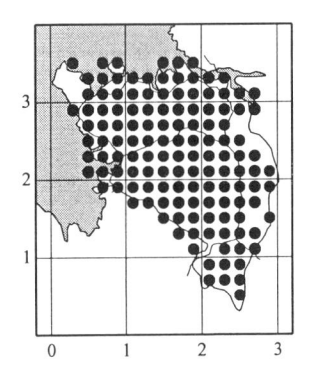

132 tetrads. Widespread and common in dry heath, stony grassland and on banks. The only St John's-wort in Assynt, apart from the garden escape and casual species mentioned above. First recorded in 1767 near Inchnadamph by J.Robertson.

DROSERACEAE

*Drosera rotundifolia* L.
Lus na Feàrnaich
**Round-leaved Sundew**

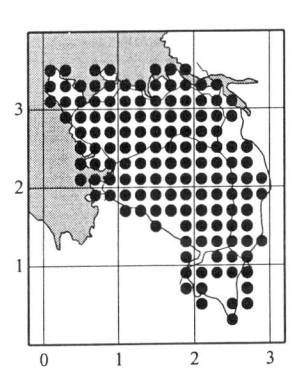

142 tetrads. Widespread and common in open, wet, peaty habitats; also in stony flushes, on *Sphagnum* tussocks, and on bare peat. First recorded in 1767 near Inchnadamph by J.Robertson.

*Drosera* x *obovata* Mert. & W.D.J.Koch
*D. rotundifolia* x *D. anglica*

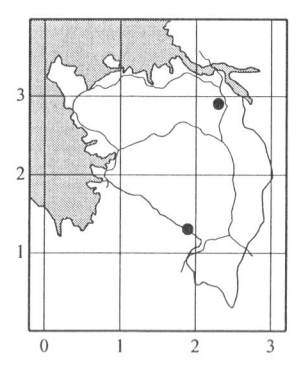

2 tetrads. Both records are from just above the water's edge, on the Unapool Burn and a small watercourse near Creagan Mór. In both cases the plants were observed to look 'odd' in the field and one population was clearly sterile. This hybrid may have been overlooked in other places. First recorded in 1890 'in a bog near Lochinver, with the parents, in plenty' by E.S.Marshall and F.J.Hanbury. Mapped for NC02 in the *Critical Supplement*. Anthony's card index also has a record for 'Skiag Bridge, 14.9.[18]87', the original source of which has not been traced.

*Drosera anglica* Huds.
Lus a'Ghadmainn
**Great Sundew**

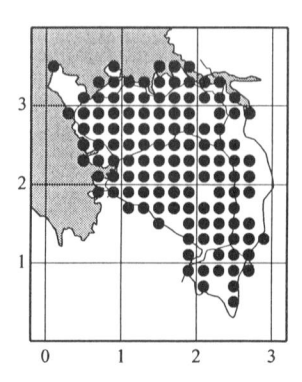

123 tetrads. Widespread and common in particularly wet, peaty places and stony flushes. Less tolerant of dryness than *D. rotundifolia*. First recorded in 1886 as 'more abundant in Assynt than *rotundifolia*' by A.Gray.

*Drosera intermedia* Hayne
Dealt Ruaidhe
**Oblong-leaved Sundew**

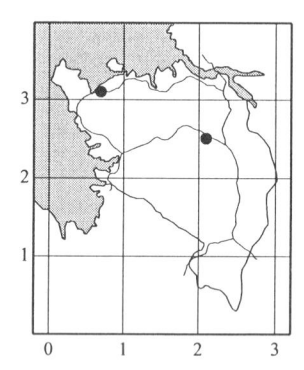

2 tetrads. Recorded from wet heath near Loch na Bruthaich, and from a *Schoenus* flush on the edge of Loch Assynt. We have probably under-recorded this species. First recorded in 1908 on the 'south shore of Loch Assynt' by E.S.Marshall and W.A.Shoolbred; subsequently from two bogs at the head of Loch Leathad nan Aighean and at Airigh Bheag in 1956 by A.O.Chater. A further record was made in 1967, west of Loch Leitir Easaidh, by P.Cobb.

VIOLACEAE

*Viola riviniana* Rchb.
Dail-chuach Coitcheann
**Common Dog-violet**

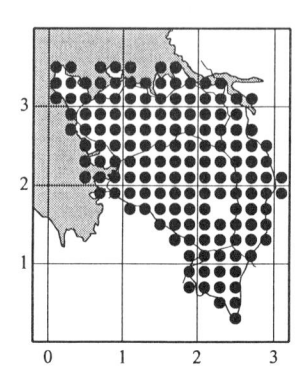

151 tetrads. Common and widespread throughout the area, in heathland, grassland and woodland. Not recorded from the famously barren tetrad, on quartzite on the lower slopes of Canisp. First recorded in 1943 'between Achadhantuir and Feadan' by A.J.Wilmott and M.S.Campbell.

*Viola canina* L.
Sàil-chuach Mointich
**Heath Dog-violet**

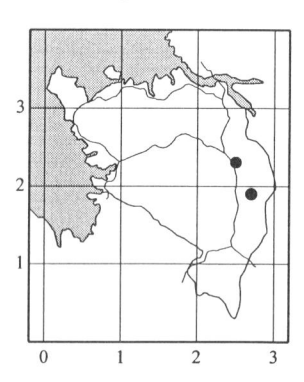

2 tetrads. Both records are from short grassland on the limestone, at Creagan Breaca and Eadar a'Chalda. As there seems to be other, equally suitable, ground, it is likely that this species has been under-recorded. The name is first used in 1767 by J.Robertson for a plant found near Inchnadamph, which may have been this species, or *V. riviniana*. In 1886 the species was recorded from Clachtoll by A.Gray, and there are more recent records from Achmelvich and Stoer which may be correct, since *Anthony* records the species as 'frequent in coastal areas in the west and north'.

*Viola palustris* L.
Dail-chuach Lèana
**Marsh Violet**

146 tetrads. Common and widespread in marshes, bogs and wet woodland.

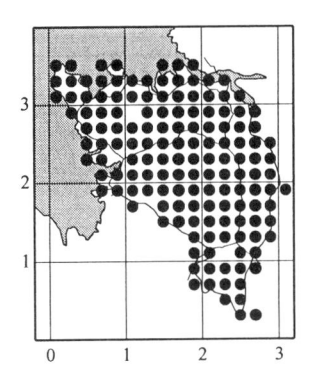

First recorded in 1767 near Inchnadamph by J.Robertson. Omitted, presumably accidentally, from *Anthony*.


*Viola lutea* Huds.
Sàil-chuach an t-Slèibhe
**Mountain Pansy**

Not seen during the present survey. First recorded in 1886 at Inchnadamph by A.Gray, and again in 1899 on 'cliffs immediately overlooking Inchnadamph' by C.E.Salmon, but not since.


*Viola tricolor* L.
Goirmean-searradh
**Wild Pansy**

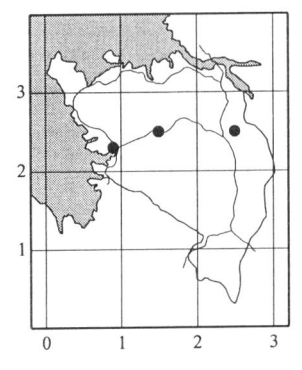

3 tetrads. An uncommon casual, found in waste or disturbed ground at Achmore, on the side of Loch Assynt and Lochinver. Two of the three records were identified as *V. tricolor* ssp. *tricolor*. *V. tricolor* ssp. *curtisii* has not been seen. First recorded in 1907 at Inchnadamph by G.C.Druce, and as ssp. *tricolor* in 1943 'in a potato patch near Inverkirkaig' by A.J.Wilmott and M.S.Campbell. There is also a 1950s record for NC03.

*Viola arvensis* Murray
Luibh Cridhe
**Field Pansy**

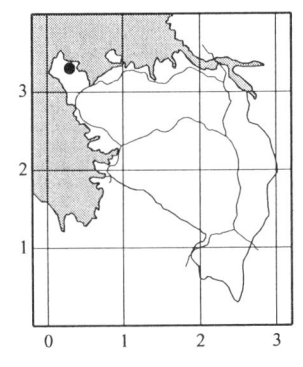

1 tetrad. Found only once, on disturbed ground near the sea at Culkein Stoer, growing with *Lycopsis arvensis*, another rare 'weed'. This could be the result of a seed bank of old cultivation having been uncovered, or of fresh seeds coming in with machinery. First recorded from NC02 during the 1950s survey for the first *Atlas*. Also recorded from Clashnessie in the 1960s by A.G.Kenneth.


SALICACEAE


*Populus tremula* L.
Critheann
**Aspen**

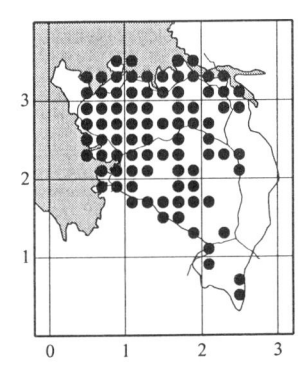

85 tetrads. Frequent in rocky woodland, gorges and on cliffs, where its suckers spread through the most unlikely substrates. Although it does not flower or fruit freely, it must sometimes set seed successfully, as the majority of its more isolated sites could not have been colonised in any other way. First recorded in 1767 near Inchnadamph by J.Robertson.

*Populus nigra* L.
Pobhlar Dubh
**Black-poplar**

The occasional planted trees which occur were not mapped. Recorded as 'introduced' in 1886 at Lochinver by A.Gray.


*Populus trichocarpa* Torr. & A.Gray ex Hook.
**Western Balsam-poplar**

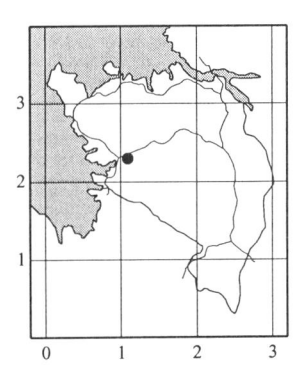

1 tetrad. Introduced. The first and only record, made in 1999, is of recently planted trees in the woods alongside the River Inver.


*Salix pentandra* L.
Seileach Labhrais
**Bay Willow**

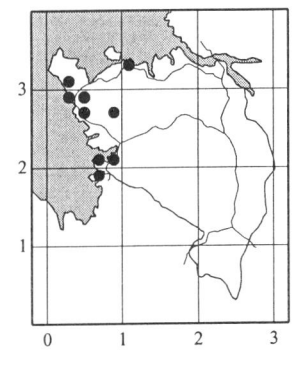

9 tetrads. Almost certainly introduced, as it nearly always occurs in the vicinity of dwellings, sometimes long-abandoned. It would have been planted for its attractive glossy leaves and handsome male catkins (all the trees whose sex we have noted have been male). Recorded as 'introduced?' in 1886 at Lochinver by A.Gray, and in 1890 at 'Lochinver, by a small woodland

swamp; evidently indigenous' by E.S.Marshall and F.J.Hanbury.

*Salix fragilis* L.
Seileach Brisg
**Crack Willow**

Introduced. Not seen, outside a garden situation, during the present survey. First recorded from NC03 in the 1950s survey for the first *Atlas,* and also by J.Anthony at Drumbeg in the same decade.

*Salix alba* L.
Seileach Bàn
**White Willow**

Introduced. Not seen, away from habitations, during the present survey. First recorded in 1886 from 'Achumore' by A.Gray.

*Salix purpurea* L.
Seileach Corcarach
**Purple Willow**

Not seen during the present survey. Recorded in 1886 on the River Kirkaig by A.Gray.

*Salix viminalis* L.
Seileach Uisge
**Osier**

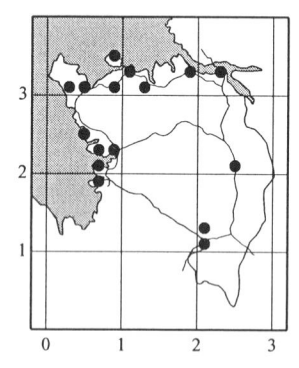

16 tetrads. Always associated with habitations, where presumably it was planted, although not in living memory, probably as a source of material for creels. First recorded from NC02 in the 1950s survey for the first *Atlas*.

*Salix* x *sericans* Tausch ex A.Kerner
*S. viminalis* x *S. caprea*
**Broad-leaved Osier**

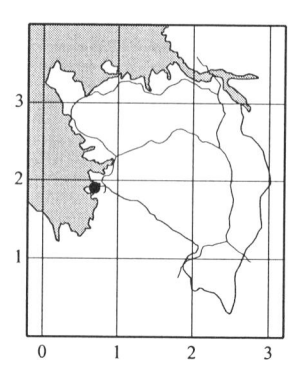

1 tetrad. Recorded in 1998 in a garden at Inverkirkaig. This small tree had grown from a stake pushed into the ground as a gatepost, some 60 years ago – a testimony to Assynt's moist climate! First recorded from material collected in 1908 on the Traligill River by E.S.Marshall and F.J.Hanbury and identified by E.F.Linton.

*Salix* x *smithiana* Willd.
*S. viminalis* x *S. cinerea*
**Silky-leaved Osier**

Not seen during the present survey. First recorded in 1943 at 'Lochinver, by stream going over to Lady Constance Bay' by A.J.Wilmott and M.S.Campbell.

*Salix caprea* L.
Geal-sheileach
**Goat Willow**

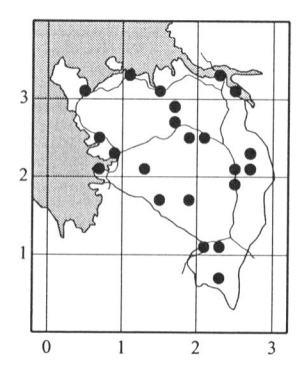

22 tetrads. Thinly scattered, usually as isolated trees in woodland or in gorges, some attaining an impressive size. First recorded in 1767 near Inchnadamph by J.Robertson. Specimens seen at Inchnadamph in 1887 were described by E.S.Marshall as an 'alpine exstipulate form' and identified by E.F.Linton as ssp. *sphacelata*; however this taxon is now only recognised from Scottish localities south of here and at altitudes over 300m.

*Salix cinerea* L.
ssp. *oleifolia* Macreight
Seileach Ruadh
**Rusty or Grey Willow**

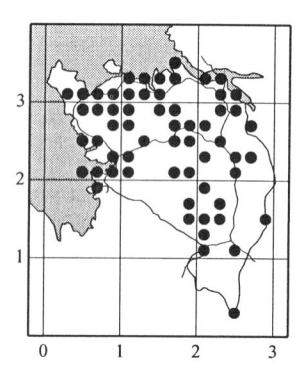

60 tetrads. The map gives a slightly false impression of the frequency of this species, which is actually spread thinly, occurring usually as isolated trees in woodland, gorges, along burns and roadsides. There are some exceptionally large specimens at Drumbeg and Nedd. First recorded in 1894 'near Elphin and Ledbeg' by G.C.Druce.

*Salix* x *multinervis* Doll
*S. cinerea* x *S. aurita*

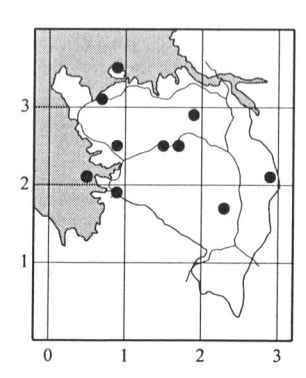

10 tetrads. Although these bushes caught the eye and were easily identified as the hybrid, it is likely

that there are many others, less obviously intermediate in character, which are of the same hybrid parentage. First recorded in 1890 'near Inchnadamph' by E.S.Marshall and F.J.Hanbury, and again in 1943 on a 'rocky hillside between Lochinver and Stoer' by A.J.Wilmott and M.S.Campbell.

*Salix* x *laurina* Sm.
*S. cinerea* x *S. phylicifolia* L.
**Laurel-leaved Willow**

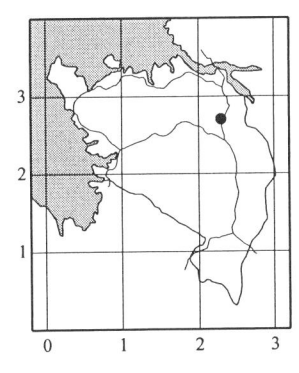

1 tetrad. The first and only record, made in 1992, is of one bush on the edge of Allt Sgiathaig, a small burn on the west side of the road which runs north from Skiag Bridge. *S. phylicifolia* has not been found in Assynt.

*Salix aurita* L.
Seileach Cluasach
**Eared Willow**

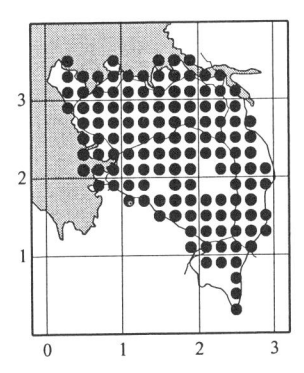

131 tetrads. By far the commonest willow, growing on moorland, burn banks, wet woodland and roadside verges. Sometimes forms low thickets on marshy ground; there are impressively extensive examples on Soyea Island. Hybridises freely. First recorded during the 1950s survey for the first *Atlas*.

*Salix* x *ludificans* F.B.White
*S. aurita* x *S. phylicifolia* L.

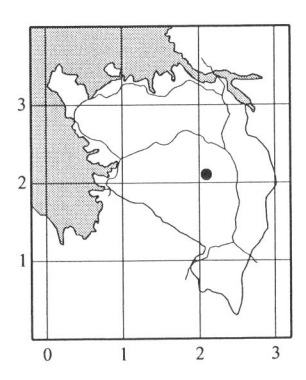

1 tetrad. The first and only record, made in 1997, is of a small bush on the bank of a burn, west of Beinn Reidh, in a very remote area. *S. phylicifolia* has not been recorded in Assynt, but is the parent of two different hybrids which have been found (a single bush in each case), about eight kilometres apart.

*Salix* x *ambigua* Ehrh.
*S. aurita* x *S. repens*

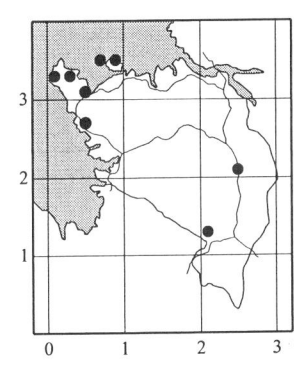

8 tetrads. Found on heathy ground, this is a fairly easily recognisable hybrid, intermediate between its parents. First recorded in 1890 'near Inchnadamph, in two forms' and 'coast, Lochinver' by E.S.Marshall and F.J.Hanbury.

*Salix repens* L.
Seileach Làir
**Creeping Willow**

105 tetrads. A very common willow, particularly abundant on coastal cliffs, but also on rocky heathland and in limestone areas. Variable in size, leaf shape and habit; no attempt has been made to assign varietal status to the records. First recorded in 1767 near Inchnadamph by J.Robertson.

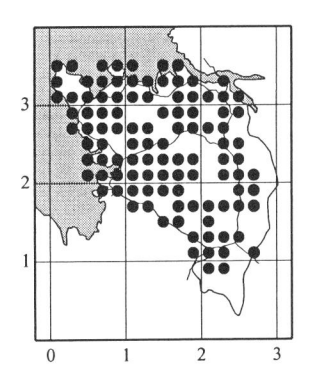

*Salix* x *cernua* E.F.Linton
*S. repens* x *S. herbacea*

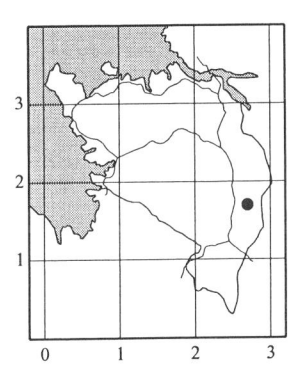

1 tetrad. The first and only record was made in 1999 on the northern slopes of Beinn an Fhuarain, in dry heath.

*Salix lapponum* L.
Seileach Clùimhteach
**Downy Willow**

Recorded in August 1956 by A.O.Chater beside a waterfall at the lower end of the Allt nan Uamh, and on the south side of Cnoc na Sròine. It has not been seen since and does not now occur at the Allt nan Uamh site.

93

*Salix myrsinites* L.
Seileach Miortail
**Whortle-leaved Willow**

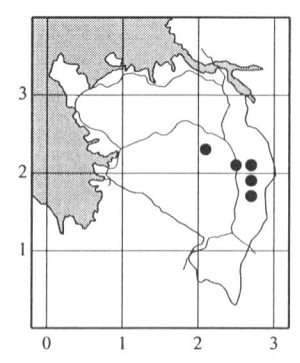

5 tetrads. In Assynt this species is largely confined to limestone areas. Four of the tetrads are on the limestones of the Cnoc Eilid Mhathain, River Traligill, Creag Sròn Chrùbaidh and Creag nan Uamh. The fifth is from Torridonian crags, at the foot of the north face of Beinn Gharbh, where it grows with the normally calcicolous *Asplenium viride*. First recorded in 1887 'by the Traligill' by E.S.Marshall. Two years later Marshall and F.J.Hanbury recorded it as 'very abundant on the limestone, between 300 and 700 feet...several hundreds of bushes...large bright-green patches in the heather formed by single shrubs often measuring several feet in diameter'. It has greatly diminished in quantity since then, presumably because the deer population has increased, and can now be seen at its most luxuriant in the Inchnadamph exclosures.

*Salix herbacea* L.
Seileach Ailpeach
**Dwarf Willow**

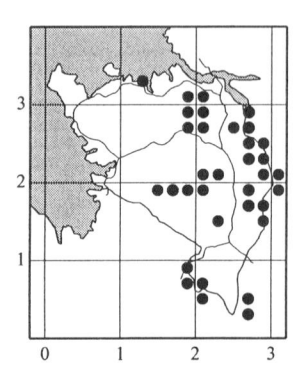

34 tetrads. Stony ground on hill tops, mostly above 350m. A striking exception is the site on the coast, at Rubh' Dhubhard near Drumbeg, where it grows at 15m. above sea level, on exposed peat on a rocky outcrop. First recorded in 1899, on 'Canisp, at 2500ft.', by C.E.Salmon. At first glance the record in the first *Atlas* for NC02 seems unlikely, but in the light of the Drumbeg record perhaps it is not so surprising.

BRASSICACEAE

*Sisymbrium altissimum* L.
**Tall Rocket**

Not seen during the present survey. First recorded in 1944 at Lochinver by A.J.Wilmott.

*Sisymbrium officinale* (L.) Scop.
Meilise
**Hedge Mustard**

Casual, not mapped. Recorded from NC02W and 13F. First recorded in 1886 by A.Gray for v.c.108 and attributed by J.Anthony to Assynt.

*Alliaria petiolata* (M.Bieb.) Cavara & Grande
Gàirleach Callaid
**Garlic Mustard**

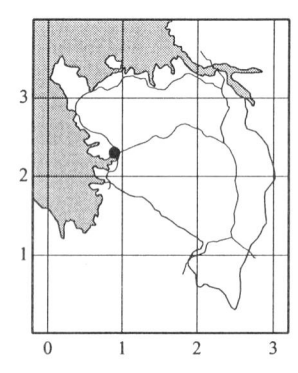

1 tetrad. A rare casual, absent from most of the north-west of Scotland. The first and only record was made in 2000, and was of a single plant, growing on the edge of a playground at Lochinver.

*Arabidopsis thaliana* (L.) Heynh.
Biolair Thàilianach
**Thale Cress**

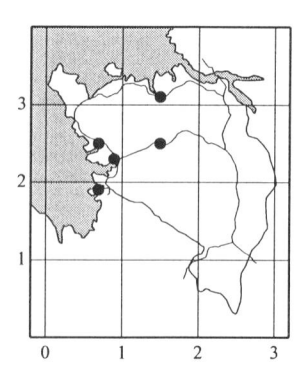

5 tetrads. Occurs rarely as a garden weed, and on roadside verges where imported soil has been spread. First recorded from NC02 during the 1950s survey for the first *Atlas*.

*Hesperis matronalis* L.
Feasgar-lus
**Dame's Violet**

Casual, not mapped. Recorded from one tetrad, NC02X. during the present survey. First recorded during the 1950s survey for the first *Atlas* from NC02.

*Rorippa nasturtium-aquaticum* (L.) Hayek
Biolair Uisge
**Water-cress**

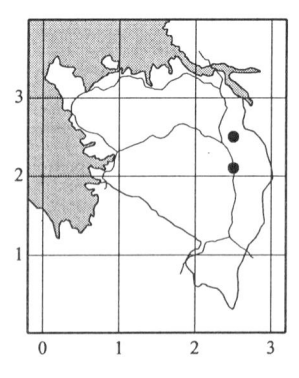

2 tetrads. Found only twice, at Inchnadamph and Achmore; seems to require slow-moving water, preferably affected by some enriching run-off from nearby houses! First recorded in 1993, at Achmore.

*Rorippa microphylla* (Boenn.)
N.Hylander ex Á & D.Löve
Mion-bhiolair
**Narrow-fruited Water-cress**

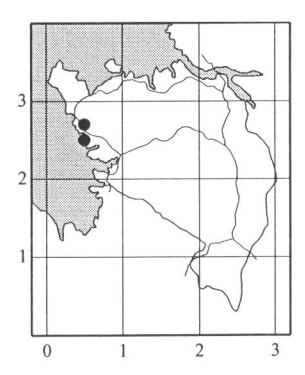

2 tetrads. Found at Clachtoll and Achmelvich. First recorded in 1998 from the second of these.

*Rorippa nasturtium-aquaticum* agg.
Biolair Uisge
**Water-cress**

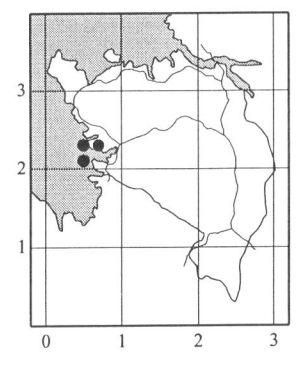

3 tetrads. Comprising records of plants which, in the absence of flowers or fruit, were not identifiable further. The most interesting was on the small, uninhabited island of Soyea where, on the top of the island, a damp area c.50m. across is completely covered with water-cress, with no sign of flower or fruit. The area is clearly used by gulls and/or geese as a roost. The tetrad line runs west-east through the site, thus accounting for two tetrads.

*Cardamine pratensis* L.
Flùr na Cuthaig
**Cuckooflower**

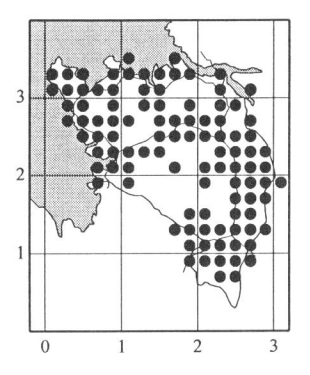

101 tetrads. Widespread and common in marshes, ditches and wet woodland. First recorded in 1767 near Inchnadamph by J.Robertson.

*Cardamine flexuosa* With.
Searbh-bhiolair Chasta
**Wavy Bitter-cress**

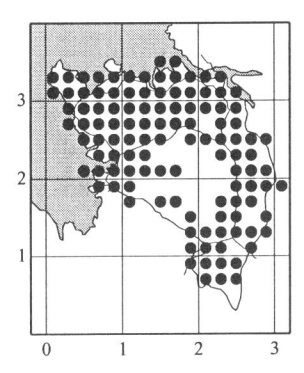

113 tetrads. Widespread and common in marshes, ditches, wet woodland and on river margins. Also found in wet disturbed ground, commonly as a garden weed. First recorded, in 1894 as '*C. sylvatica*', at Knockan Crags by G.C.Druce.

*Cardamine hirsuta* L.
Searbh-bhiolair Ghiobach
**Hairy Bitter-cress**

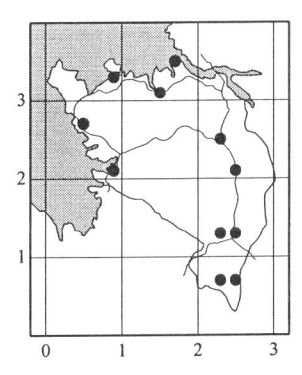

11 tetrads. Requires dry conditions. Found on the tops of old walls and occasionally as a garden weed. First recorded in 1767 near Inchnadamph by J.Robertson (although *C. flexuosa* and this species may not have been distinguished at that time). Found in 1894 'near Elphin and Ledbeg' by G.C.Druce.

*Arabis hirsuta* (L.) Scop.
Biolair na Creige Ghiobach
**Hairy Rock-cress**

Not seen during the present survey. First recorded from NC21 in the 1950s survey for the first *Atlas*, and from Ardvreck Castle by M.G.Coulson in 1969.

*Draba incana* L.
Biolradh Gruagain Liath
**Hoary Whitlowgrass**

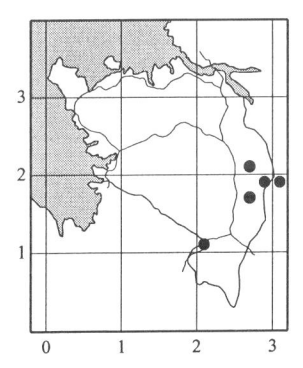

5 tetrads. A rare calcicole of crags, boulders, rock ledges and, occasionally, turf. Found in the limestone areas of Allt a'Bhealaich, Allt nan Uamh, Bealach Traligill, Elphin and Inchnadamph. First recorded in 1772 at '*Creag a-chnocaen*' by T.Pennant and J.Lightfoot, and again there in 1894 by G.C.Druce. Also, in 1949, from Cnoc an t-Sasunnaich (above Knockan crags) by A.A.P.Slack.

*Erophila verna* (L.) DC.
Biolradh Gruagain
**Common Whitlowgrass**

3 tetrads. In dry places. On a wall in Stoer village, outcrop rock at Oldany Farmhouse and

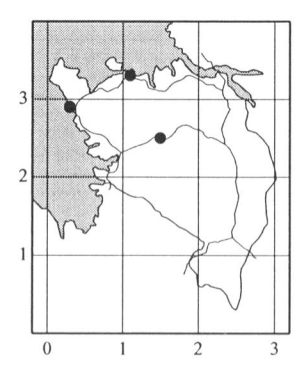

on the edge of an old road at Little Assynt. First recorded as *E. verna* s.l. from NC02 and NC22 during the 1950s survey for the first *Atlas*, and also from NC21 in 1976 by U.Duncan.

*Erophila glabrescens* Jord.
Biolradh Gruagain Mìn
**Glabrous Whitlowgrass**

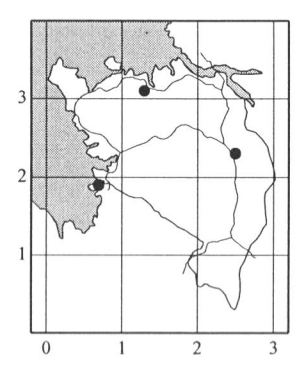

3 tetrads. Similar places to the above. On a concrete building base at Eadar a'Chalda, along the middle of the cemetery track at Nedd and on a roadside verge at Inverkirkaig. First recorded in 1989 from the last of these.

*Cochlearia pyrenaica* DC.
ssp. *alpina* (Bab.) Dalby
(*C. alpina* (Bab.) H.C.Wats.)
Carran Ailpeach
**Pyrenean Scurvygrass**

Not seen during the present survey. First recorded in 1907, somewhat ambiguously, as on 'Ben Mor Assynt, 108' by G.C.Druce (who was usually very precise about vice county boundaries). In Anthony's card index there is an entry for 'Coineval 2500ft.', and A.G.Kenneth recorded it from the 'N. face' of Canisp in 1985.

*Cochlearia officinalis* L.
Am Maraiche
**Scurvygrass**

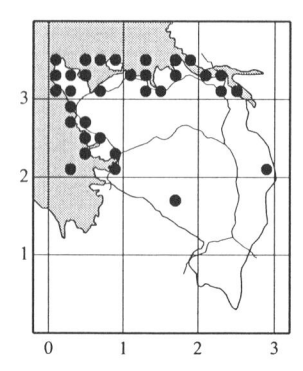

33 tetrads. Into this category have been placed all those plants which we did not or could not identify to subspecies. Very common around the coast, on rocky shores and salt marshes. Two records are from inland, montane sites, one on Suilven and one by a waterfall on the upper reaches of the River Traligill. First recorded during the 1950s survey for the first *Atlas*.

*Cochlearia officinalis* L.
ssp. *scotica* (Druce) P.S.Wyse Jacks.
(*C. scotica* Druce)
Carran Albannach
**Scottish Scurvygrass**

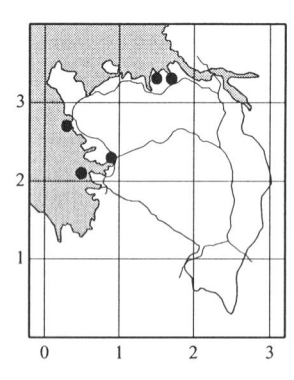

5 tetrads. Occurs on rocky shores and in saltmarsh at Baddidarach, Clachtoll, Loch Ardbhair, Loch Nedd and Soyea Island. First recorded in 1943 on 'sea rocks, Balchladich Bay' and again, in 1944, 'on a tiny extent of saltmarsh below the church, Lochinver' by A.J.Wilmott and M.S.Campbell.

*Capsella bursa-pastoris* (L.) Medik.
An Sporan
**Shepherd's-purse**

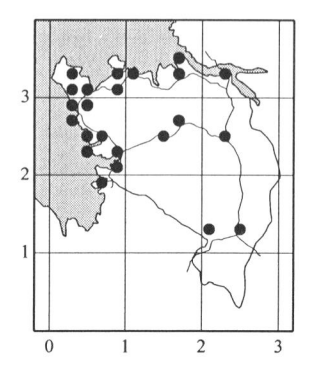

23 tetrads. All the records are from man-influenced habitats, cultivated and disturbed ground generally. First recorded during the 1950s survey for the first *Atlas*.

*Thaspi arvense* L.
Praiseach Fèidh
**Field Penny-cress**

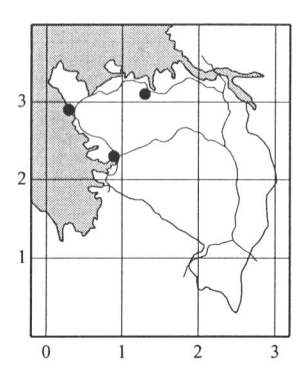

3 tetrads. A rare casual which has turned up twice in gardens, at Clachtoll and Nedd, and once on disturbed ground in Lochinver. First recorded in 1995 in the second of these sites.

*Subularia aquatica* L.
Lus a'Mhinidh
**Awlwort**

38 tetrads. Widespread in acid lochs where the substrate is sand or fine gravel. Usually found growing underwater with *Littorella* and *Lobelia*, but can also flourish

Flowering plants and ferns

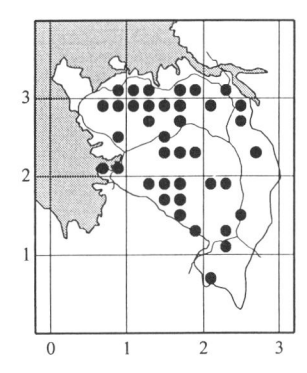

on 'dry land' when water levels are low. First recorded in 1886 in Loch Awe by A.Gray.

*Diplotaxis muralis* (L.) DC.
**Annual Wall-rocket**

Not seen during the present survey. First recorded in 1944 from Lochinver by A.J.Wilmott; also from NC02 in 1956 by J.Dickson.

*Brassica rapa* L.
Nèap Fiadhain
**Turnip**

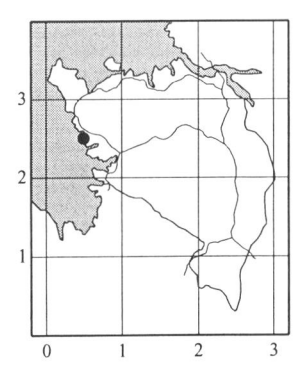

1 tetrad. When the record was made in 1995, it was a plentiful weed of cultivation on a patch of arable ground at Achmelvich, which has since been abandoned. First recorded from NC02, in the 1950s survey for the first *Atlas*.

*Cakile maritima* Scop.
Fearsaideag
**Sea Rocket**

2 tetrads. Found on the shore at Achmelvich and Stoer; a shortage of sandy beaches in Assynt may account for its rarity.

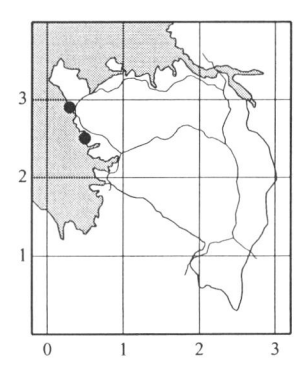

First recorded from Achmelvich during the 1950s survey for the first *Atlas*.

*Raphanus raphanistrum* L.
(R. raphanistrum var. aureum Wilmott)
Meacan Ruadh Fiadhain
**Wild Radish**

Not seen during the present survey. First recorded in 1894 from 'cornfields at Elphin and Knockan…the yellow flowered plant', by G.C.Druce; and in 1943 as var. *aureum*, at Drumbeg by A.J.Wilmott and M.S.Campbell.

EMPETRACEAE

*Empetrum nigrum* L.
Lus an Feannaig
**Crowberry**

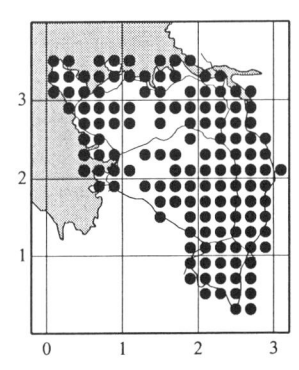

136 tetrads. Common and widespread on rocky moorland, hills and coastal cliffs. In the frequent absence of flowers or fruit, a large number of records were not identifiable to subspecies (but see next map). First recorded in 1767 near Inchnadamph by J.Robertson.

ssp. *hermaphroditum* (Hagerup) Böcher
Dearcag Fhithich
**Mountain Crowberry**

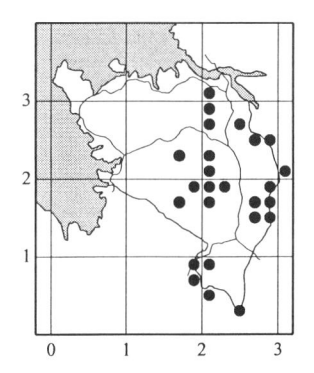

25 tetrads. The map comprises records from montane sites, where this subspecies was either definitely identified or was the more likely of the two.

ERICACEAE

*Loiseleuria procumbens* (L.) Desv.
Lusan Albannach
**Trailing Azalea**

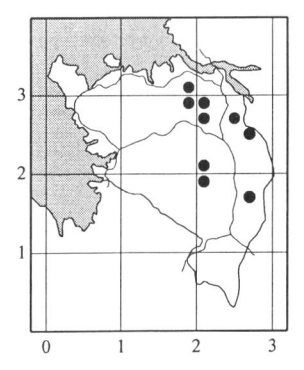

9 tetrads. An uncommon species of stony, acid hilltops. Recorded only from Beinn nan Cnaimhseag, Beinn Reidh, Beinn Uidhe, Canisp, Glas Bheinn and Quinag. First recorded in 1886 from 'Ben Reidhe-Quinag' by A.Gray.

*Arctostaphylos uva-ursi* (L.) Spreng.
Grainnseag
**Bearberry**

76 tetrads. Patchily distributed on rocky moorland and hillsides, often growing draped over outcrop rock. Can be particularly profuse

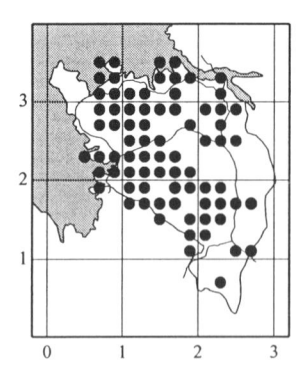

in sheltered gorges, where the growth may be luxuriant, but also occurs on broken quartzite pavement as on Canisp and Meallan Liath Mór. First recorded in 1886 from 'Quinag-shore at Kylesku' by A.Gray.

*Arctostaphylos alpinus* (L.) Spreng.
Cnaimhseag
**Arctic Bearberry**

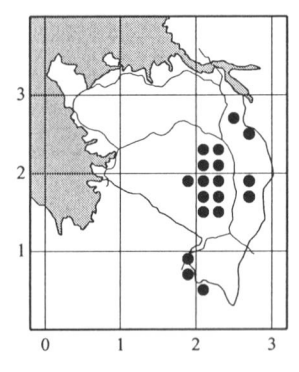

18 tetrads. On mostly high moorland amongst dwarf vegetation. It has occasionally been found at surprisingly low altitudes: at 250m. on Cnoc an Leathaid Bhuidhe and at 230m. on Druim nan Cnaimhseag. First recorded in 1886 at Beinn nan Cnaimhseag (hill of the bearberry) by A.Gray.

*Calluna vulgaris* (L.) Hull
Fraoch
**Heather**

159 tetrads. An almost complete map coverage. Although predominantly on dry moorland, rocky areas and open woodland, it also occurs commonly in wet heath, sometimes on *Sphagnum*.

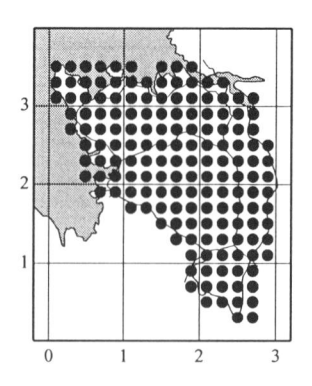

The white form occurs rarely. First recorded in 1767 near Inchnadamph by J.Robertson.

*Erica tetralix* L.
Fraoch Frangach
**Cross-leaved Heath**

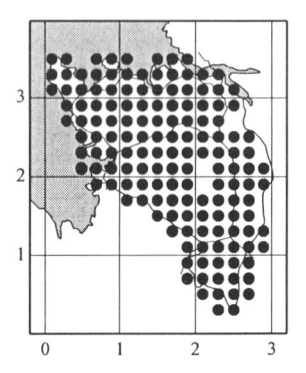

148 tetrads. Widespread in wet heath and bog. A white form occurs rather more commonly than in *Calluna*. First recorded in 1767 near Inchnadamph by J.Robertson.

*Erica cinerea* L.
Fraoch a'Bhadain
**Bell Heather**

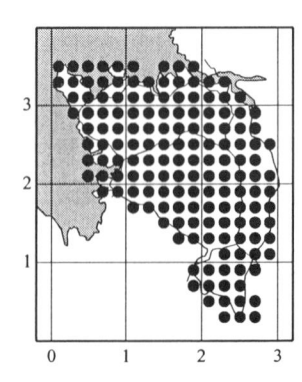

153 tetrads. Widespread and common on dry, heathy ground. We have seen a white form only twice. First recorded in 1767 near Inchnadamph by J.Robertson.

*Vaccinium vitis-idaea* L.
Lus nam Braoileag
**Cowberry**

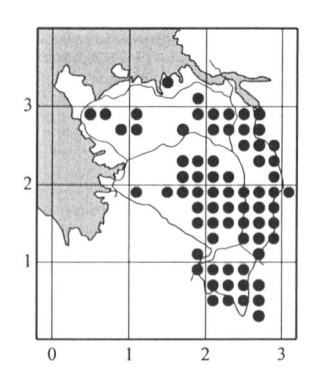

71 tetrads. Grows on peaty moorland and in open woodland, with a curious distribution which we do not really understand. It is apparently absent from much of the gneiss and is commonest at moderate altitudes, but could not be described as montane. First recorded in 1890 at Inchnadamph by E.S.Marshall and F.J.Hanbury.

*Vaccinium uliginosum* L.
Dearc Roide
**Bog Blaeberry**

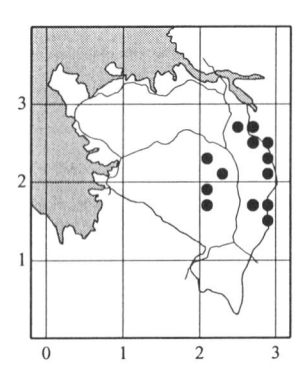

13 tetrads. Uncommon. Found on wet moorland at or above 500m.; the name 'Bog Bilberry' is a little misleading. First recorded in 1899 at Meallan Liath Mór by C.E.Salmon.

*Vaccinium myrtillus* L.
Caora-mhitheag
**Blaeberry** or **Bilberry**

141 tetrads. Common and widespread on moorland, in open woodland and on banks. Fruit not often seen except in places inaccessible to grazing animals, as

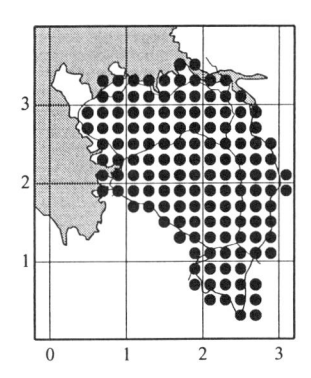

in forestry plantations. First recorded in 1767 near Inchnadamph by J.Robertson.

PYROLACEAE

*Pyrola media* Sw.
Glas-luibh Meadhanach
**Intermediate Wintergreen**

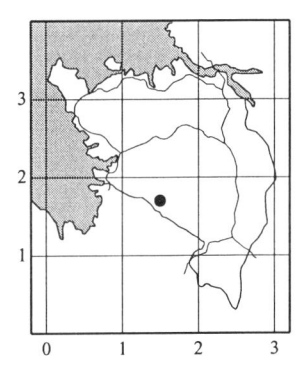

1 tetrad. The first and only record was made in 1999 by G.P.Rothero. The plants were found growing beneath *Calluna* in Gleannan a'Mhadaidh, at the point where a dark basic dyke forms part of the south-facing wall of this narrow gorge. None were flowering and the species was only certainly identified after one plant had flowered in cultivation. This is also one of the sites for *Asplenium septentrionale*.

*Orthilia secunda* (L.) House
Glas-luibh Fhiaclach
**Serrated Wintergreen**

1 tetrad. The first and only record was made in 1999 by G.P.Rothero.

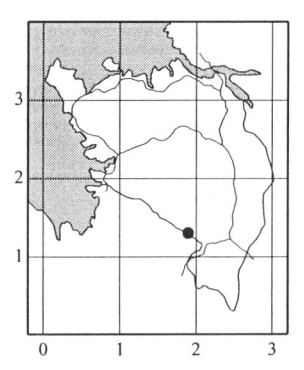

The plants were growing on a ledge of a steep Torridonian crag, on the north-facing side of a valley, east of Creagan Mór.

PRIMULACEAE

*Primula vulgaris* Huds.
Sòbhrach
**Primrose**

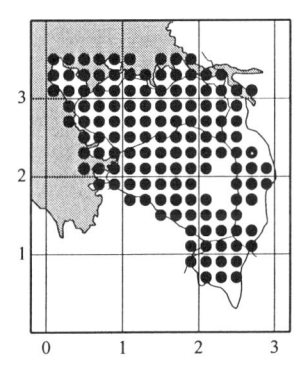

135 tetrads. Common and widespread where there is some shade or shelter, for example woodland, burn banks, boulder scree, gorges, and open grassland where bracken gives protection. First recorded in 1951 on the Traligill Burn by J.Raven.

*Primula veris* L.
Muisean
**Cowslip**

1 tetrad. Mentioned in *Anthony* as growing in sandy pastures by the sea. However, during the present survey this species was not recorded until 2000, when one non-flowering plant was seen at the back of the dunes at Achmelvich, and 2001, when two plants were noted in the same area.

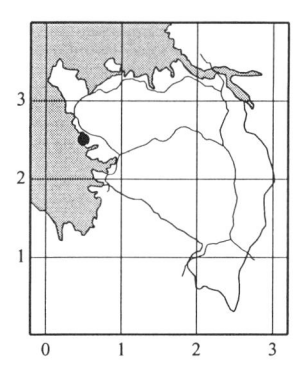

This area was extensively restored in the 1980s, following serious erosion. Recorded in 1767 near Inchnadamph by J.Robertson, but could this have been in error for the very much commoner *P. vulgaris*, which he did not mention? Recorded from NC02 and NC21 in the 1950s survey for the first *Atlas* and again from NC02 in 1969.

*Lysimachia nemorum* L.
Seamrag Moire
**Yellow Pimpernel**

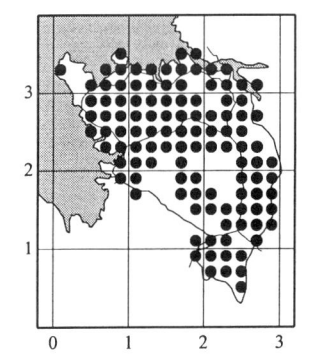

111 tetrads. Widespread and frequent, in woodland and on the edge of burns where it is sheltered by *Calluna* and other vegetation. First recorded in 1767 near Inchnadamph by J.Robertson.

*Trientalis europaea* L.
Reul na Coille
**Chickweed Wintergreen**

1 tetrad. Found once only, in 1987 at Duart, growing beneath *Calluna*. There are three other records from NC13. It was recorded in the 1950s survey for the first *Atlas*, there is a specimen in a collection of pressed flowers made in 1964 in the vicinity of

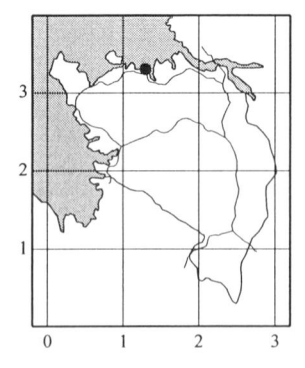

Drumbeg by S.MacLeod, and it was seen in the 1980s near Loch Drumbeg. First recorded in 1886 at 'Stronechrubie (very local, only this one locality known)' by A.Gray.

*Anagallis tenella* (L.) L.
Falcair Lèana
**Bog Pimpernel**

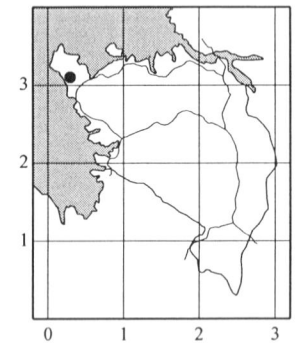

1 tetrad. The population extends over a small area of short, grazed, peaty vegetation, on a cliff top at Balchladich on the Stoer peninsula. First recorded in 1955 in a 'bog….Stoer' by J.V.Sutherland (which may be the present locality). Also recorded in 1981 near the mouth of the River Loanan by N.T.H.Holmes.

*Anagallis minima* (L.) E.H.L.Krause
Falcair Mìn
**Chaffweed**

15 tetrads. Nearly all the records are coastal and from sites within a few metres of high tide level. The plants occupy a very distinct habitat - small areas of short turf between outcropping rocks.

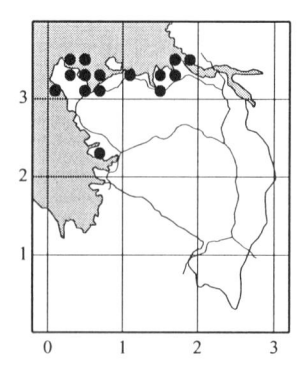

The only exceptions are a record from a saltmarsh, and a single plant found on rock splashed by water, a few hundred metres from the coast. First recorded in 1991 at Rubh'an Dùnain.

*Glaux maritima* L.
Lus na Saillteachd
**Sea-milkwort**

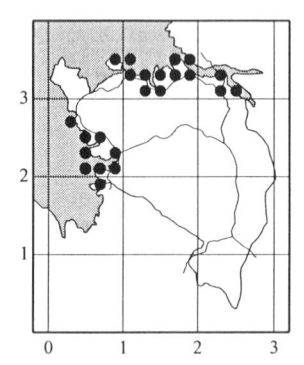

23 tetrads. Common on suitable sandy and rocky shores and in saltmarsh. First recorded in Anthony's card index as 'Lochinver, 16.7.[18]90 E.S.M[arshall].' Also recorded in 1943 at 'Lochinver, in saltmarsh on the north side of the bay' by A.J.Wilmott and M.S.Campbell.

GROSSULARIACEAE

*Ribes rubrum* L.
(*R. sylvestre* (Lam.) Mert. & Koch)
Dearc Dhearg
**Red Currant**

2 tetrads. Introduced or garden escapes, at Clashmore and Oldany, in each case growing on a wall.

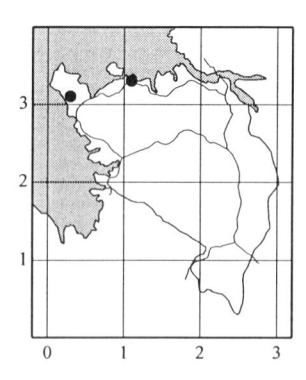

First recorded in 1767 by J.Robertson , 'in a small Island within the crooked Loch there is abundance of *Ribes rubrum*'. This is interpreted by Henderson and Dickson (1994) as Loch Lurgainn in Wester Ross, but the 1768 catalogue of J.Hope's *Hortus Siccus,* which contained Robertson's specimen, has 'Crooked Loch Assynt' and Eilean na Gartaig, the wooded island nearest to the road on Cam Loch ('cam' means 'crooked' in Gaelic), is a much more likely locality, especially as the species was recorded in 1959 from there by A.A.Slack. Also recorded in 1886 from Lochinver and in the 1950s survey for the first *Atlas*.

*Ribes nigrum* L.
Dearc Dhubh
**Black Currant**

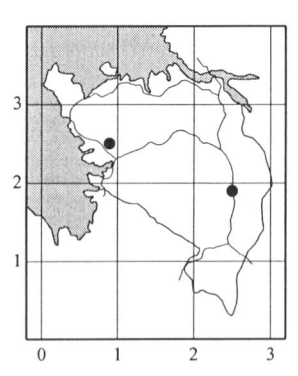

2 tetrads. Introduced. Two apparently bird-sown bushes, on the side of ditches at Rhicarn and Stronechrubie. Presumably of garden origin. First recorded from Rhicarn in 1996.

*Ribes uva-crispa* L.
Gròiseid
**Gooseberry**

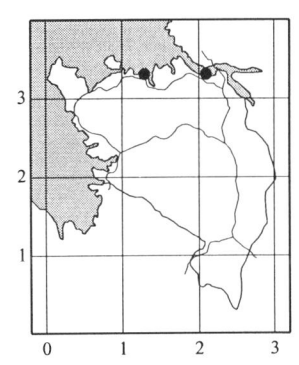

2 tetrads. Introduced. Found on the wall of an old ruin at Duart and a roadside ditch near Torgawn. Probably bird-sown and of garden origin. First recorded in 1907 'growing on the limestone cliffs of Blar nam Fiadhag…perhaps bird-sown' by G.C.Druce. Also recorded in 1943 'in boulder scree of gorge above Port Alltan na Bradhan' by A.J.Wilmott and M.S.Campbell.

CRASSULACEAE

*Sedum rosea* (L.) Scop.
Lus nan Laoch
**Roseroot**

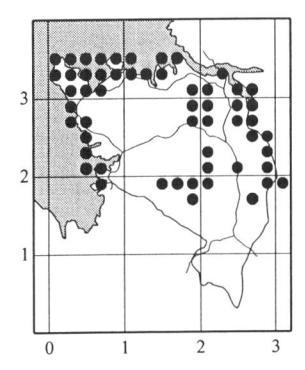

55 tetrads. Widespread along the coast where it grows luxuriantly on the cliffs. Also occurs inland, on cliffs and ledges on the higher hills. First recorded in 1952 at Achmelvich by J.Anthony.

*Sedum acre* L.
Gràbhan nan Clach
**Biting Stonecrop**

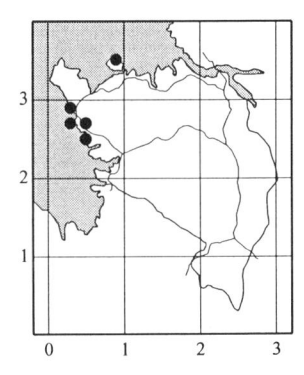

5 tetrads. Rare. Found only on the coast, on sandy or rocky shores or in dune grassland. First recorded in 1886 at Achmelvich Bay by A.Gray.

*Sedum album* L.
Gràbhan nan Clach Bàn
**White Stonecrop**

Not seen during the present survey. First recorded from NC02 and NC03 during the 1950s survey for the first *Atlas*.

*Sedum anglicum* Huds.
Biadh an t-Sionnaidh
**English Stonecrop**

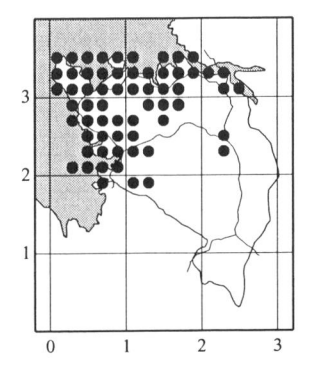

62 tetrads. Common on rock around the coast and extending some distance inland. Not found high in the hills or in limestone areas. First recorded in 1886 'Kylesku-Clachtoll' by A.Gray.

SAXIFRAGACEAE

*Saxifraga nivalis* L.
Clach-bhriseach an t-Sneachda
**Alpine Saxifrage**

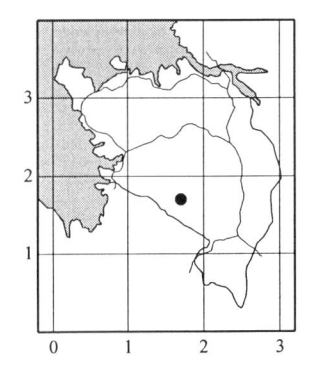

1 tetrad. The first and only record was made in 1999 by G.P.Rothero. Three rosettes were noted at 600m., on a vertical wall in the gulley between Meall Meadhonach and Meall Beag, at the eastern end of Suilven.

*Saxifraga stellaris* L.
Clach-bhriseach Reultach
**Starry Saxifrage**

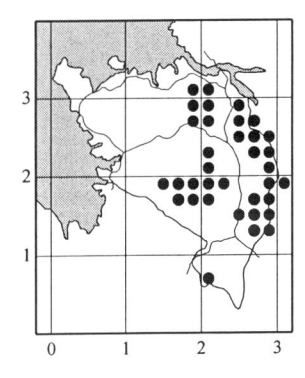

35 tetrads. Frequent in wet rocky places in the hills, growing on burn sides and at waterfalls, either on bare rock or on moss. First recorded during the 1950s survey for the first *Atlas*.

*Saxifraga* x *urbium* D.A.Webb
*S. umbrosa* L. x *S. spathularis* Brot.
**Londonpride**

1 tetrad. The first and only record was made in 1995 at Nedd.

Flora of Assynt

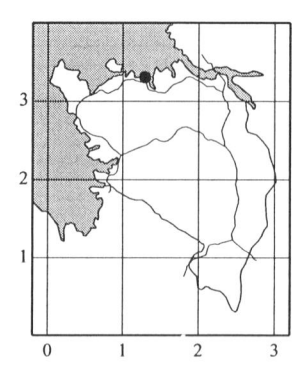

The plant flourishes with other garden escapes on a damp, roadside cliff.

*Saxifraga oppositifolia* L.
Clach-bhriseach Phurpaidh
**Purple Saxifrage**

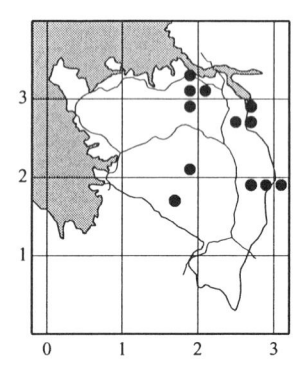

12 tetrads. Rocks and screes on the higher hills. Although most of the sites are montane and above 300m., one is at only 90m. on wet rocks beside a burn in Gleann Ardbhair, presumably washed down from the higher ground. First recorded in 1886 as 'common on mountain tops in Assynt' by A.Gray. A record for NC02 in the first *Atlas* may be an error.

*Saxifraga aizoides* L.
Clach-bhriseach Bhuidhe
**Yellow Saxifrage**

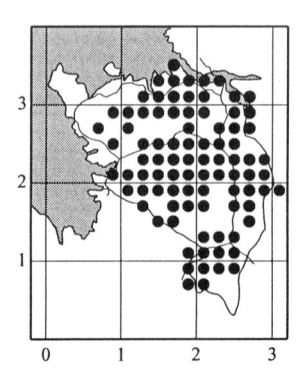

88 tetrads. Ranges from quite high altitudes down almost to sea level, on wet rocks, screes and burn sides. Particularly characteristic of stony basic flushes. First recorded in 1767, as '*S. autumnalis*', near Inchnadamph by J.Robertson.

*Saxifraga hypnoides* L.
Clach-bhriseach Còinnich
**Mossy Saxifrage**

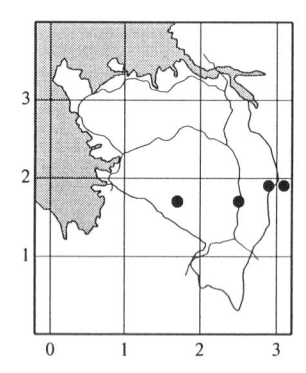

4 tetrads. Rare, on damp rocks and burn sides. Found at Allt nan Uamh, Bealach Traligill and on Suilven. First recorded in 1767 near Inchnadamph by J.Robertson; again in 1890 'on limestone about Inchnadamph, descending to 600ft' by E.S.Marshall and F.J.Hanbury, and in 1907 'on Quinag' by G.C.Druce.

*Tolmiea menziesii* (Pursh) Torr. & A.Gray
**Pick-a-back-plant**

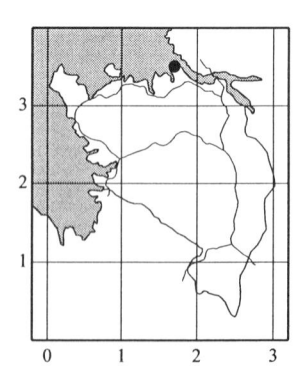

1 tetrad. Naturalised and thriving on a damp, shady, cliff face, in close proximity to an old house. First recorded in 1996 from Kerrachar.

*Chrysosplenium oppositifolium* L.
Lus nan Laogh
**Opposite-leaved Golden-saxifrage**

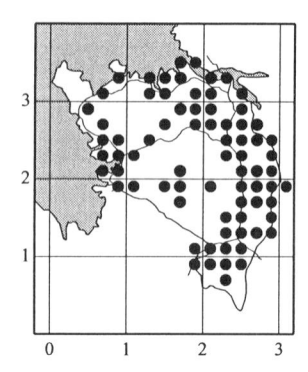

74 tetrads. Frequent in wet stony flushes, ditches, wet woodland and occasionally marshes. First recorded during the 1950s survey for the first *Atlas*.

*Parnassia palustris* L.
Fionnan Gael
**Grass-of-Parnassus**

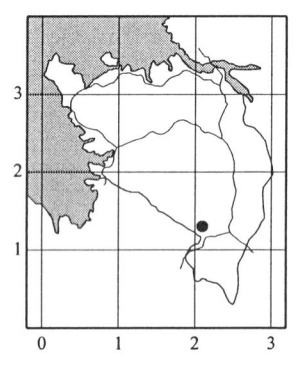

1 tetrad. Considering the frequency with which it occurs north of Assynt, it is surprising that the present survey has only produced one site for this species, in wet grassland close to the head of Loch Veyatie. Increased grazing pressure may be responsible for the disappearance of plants from sites known in earlier centuries. First recorded in 1767 near Inchnadamph by J.Robertson. Also recorded in 1886 at Loch Awe by A.Gray, and at Achmelvich during the 1950s survey for the first *Atlas*.

102

## ROSACEAE

*Filipendula ulmaria* (L.) Maxim.
Cneas Chù Chulainn
**Meadowsweet**

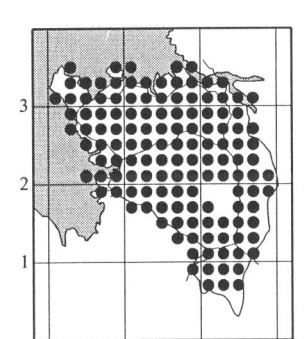

134 tetrads. Widespread and common on damp or wet ground where there is coarse vegetation. In marshes, wet woodland, burn sides, ditches and roadside verges. First recorded during the 1950s survey for the first *Atlas*.

*Rubus chamaemorus* L.
Lus nan Oighreag
**Cloudberry**

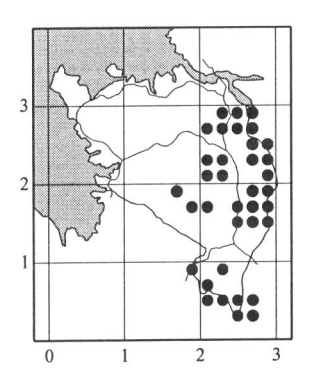

36 tetrads. On high, damp, peaty moorland, often growing with Dwarf Cornel. Occurs at the unusually low altitude of 100m. at the eastern end of An Coimhleum. First recorded, from NC21 and NC22, during the 1950s survey for the first *Atlas*.

*Rubus saxatilis* L.
Caor Bad Miann
**Stone Bramble**

75 tetrads. Typically trailing from ledges, on inland or coastal cliffs, over outcrop rock or on scree.

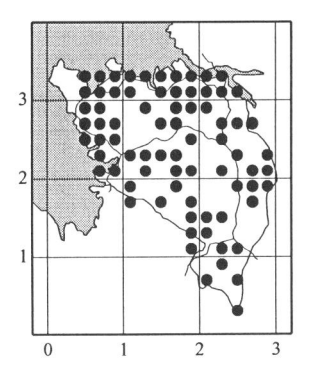

First recorded in 1894 on Knockan Crags 'in flower, with an odour of hawthorn' by G.C.Druce.

*Rubus idaeus* L.
Sùbh-craoibh
**Raspberry**

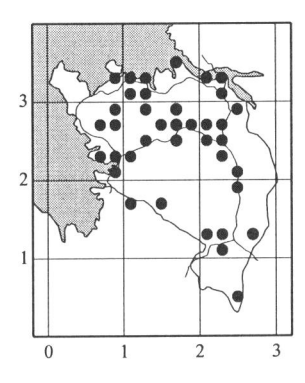

38 tetrads. In woodland and scrub, in rough vegetation along the base of walls, on verges and in ditches. First recorded during the 1950s survey for the first *Atlas*.

*Rubus fruticosus* agg.
Dris
**Bramble**

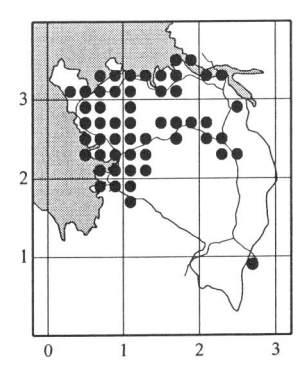

53 tetrads. This far north it only succeeds under the optimum conditions of a sunny, sheltered site, with light grazing pressure. The majority of records are of non-flowering plants and even those that bloom do not necessarily fruit. First recorded in 1890 as '*R. villicaulis* Kochl.....plentiful at Lochinver' by E.S.Marshall and F.J.Hanbury. The only records that accord with the modern concepts of the group are of *R. septentrionalis* W.C.R.Watson, collected in 1958 at Achadhantuir and Achmelvich by E.S.Edees and since determined by him.

*Potentilla palustris* (L.) Scop.
Còig-bhileach Uisge
**Marsh Cinquefoil**

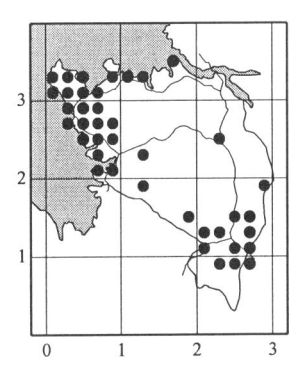

40 tetrads. The most frequent habitats are loch shores, where these are flat and merge into marshy ground. With the rise and fall in loch levels, it must be able to tolerate short periods of partial submersion. It also grows in isolated bogs and marshes. First recorded in 1767, as '*Comarum palustre*', near Inchnadamph by J.Robertson.

*Potentilla anserina* L.
Brisgean
**Silverweed**

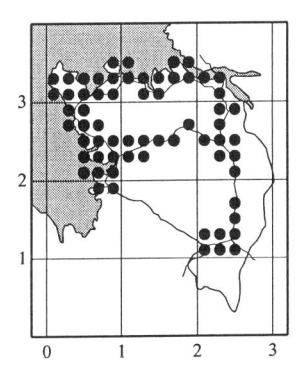

64 tetrads. It is clear from the map that this species is most

frequent along roadside verges, but it is also found in a variety of other habitats such as coastal grassland, river gravels, stony shores of the larger lochs and above high water mark on shingle and sandy beaches. First recorded during the 1950s survey for the first *Atlas*.

*Potentilla crantzii* (Crantz) Beck ex Fritsch
Leamhnach Ailpeach
**Alpine Cinquefoil**

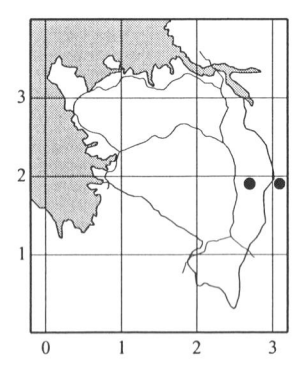

2 tetrads. Rare, on the limestone rocks of Cnoc Eilid Mhathain and in Bealach Traligill. First recorded in 1953 at the first of these sites by J.E.Raven and S.M.Walters.

*Potentilla erecta* (L.) Räusch.
Cairt Làir
**Tormentil**

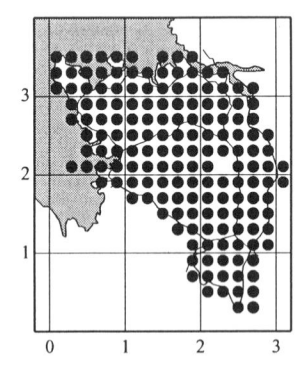

161 tetrads. Widespread and common. Tolerant of almost any conditions: dry or wet grassland, woodland, heath and rocky hillsides. First recorded in 1767 near Inchnadamph by J.Robertson.

*Potentilla reptans* L.
Còig-bhileach
**Creeping Cinquefoil**

Not seen during present survey. First recorded from NC22 in the 1950s survey for the first *Atlas*.

*Potentilla sterilis* (L.) Garcke
Sùbh-làir Brèige
**Barren Strawberry**

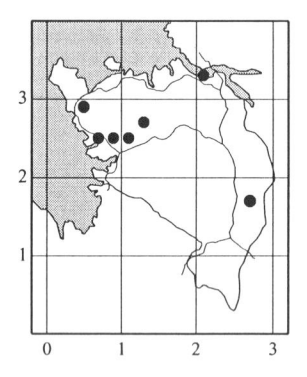

7 tetrads. An uncommon species of woodland and grassland on both basic and acidic soils. First recorded in 1992 from Tòrr a'Ghamhna.

*Sibbaldia procumbens* L.
Siobaldag
**Sibbaldia**

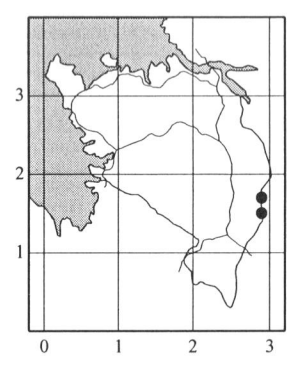

2 tetrads. Both the recent records are from Breabag: at 640m. on the edge of a small burn which drains an area of late snow-lie west of Meall Diamhain, and in an area of flushed stony grassland at 660m. on Fuarain Ghlasa. First recorded in 1946 by E.C.Wallace from NC21, and from NC22 during the 1950s survey for the first *Atlas*. It is interesting that the record in McVean and Ratcliffe (1962) is from the same grid reference as the first of the recent records mentioned above.

*Fragaria vesca* L.
Sùbh-làir Fiadhain
**Wild Strawberry**

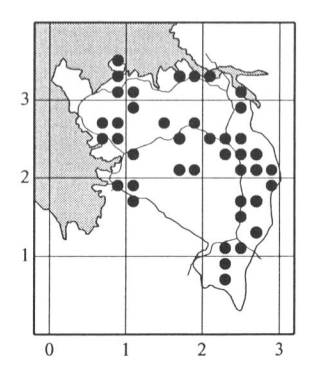

41 tetrads. Scattered in woodland, scrub, grassland and boulder scree. First recorded in 1943 on a 'roadside bank, Achadhantuir' by A.J.Wilmott and M.S.Campbell.

*Geum rivale* L.
Machall Uisge
**Water Avens**

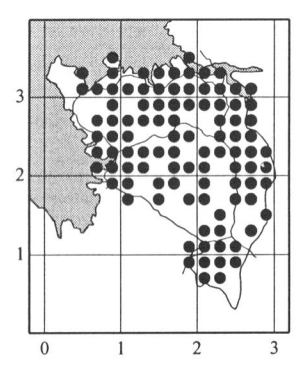

98 tetrads. Common in wet woodland and alongside burns, in all but the most barren areas. First recorded in 1894 as 'not infrequent' at Knockan Crags by G.C.Druce.

*Geum urbanum* L.
Machall Coille
**Wood Avens**

2 tetrads. Rare in the north-west of Scotland as a whole. One record is from a shady roadside verge at Nedd and the other from woodland at Rhicarn.

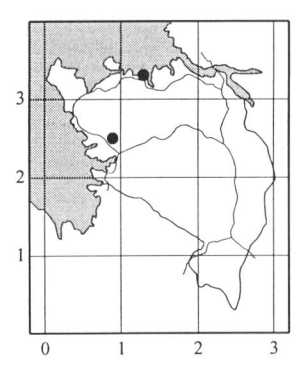

First recorded in 1993 at the second of these sites.

*Dryas octopetala* L.
Machall Monaidh
**Mountain Avens**

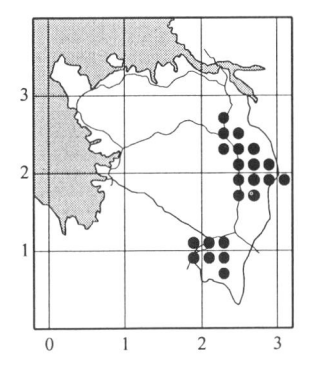

22 tetrads. Calcicole. Where there is outcropping limestone, Mountain Avens can grow abundantly on ledges and in the short turf around the rocks. It occurs down to about 100m., decorating rocks quite close to the road at the south-eastern end of Loch Assynt. First recorded in 1772 by T.Pennant and J.Lightfoot as 'still more abundantly for two miles together upon a vast limestone rock called *Creg-achnocaen*'. Mapped, presumably in error, for NC02 and NC12 in the first *Atlas*.

*Sanguisorba minor* Scop.
A'Bhileach Losgainn
**Salad Burnet**

1 tetrad. The first and only record for Assynt was made in 1997 at Ardvar. A plant was noted growing on the edge of a small industrial area and a second one on the verge of a private road nearby.

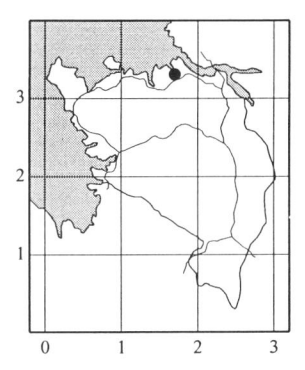

The former site had eight plants in 1998. Presumably the plants were brought in on machinery; it will be interesting to see if they establish.

*Alchemilla alpina* L.
Trusgan
**Alpine Lady's-mantle**

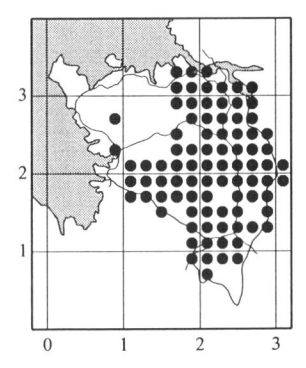

87 tetrads. In dry rocky places, screes, mountain ledges and well-drained grassland. Although typically a plant of higher ground, there are a few records at sea level. First recorded in 1767 near Inchnadamph by J.Robertson.

*Alchemilla vulgaris* agg.
Fallaing Moire
**Lady's-mantle**

First recorded in 1894 at Knockan Crags 'in the hairy and glabrous forms', by G.C.Druce.

*Alchemilla glaucescens* Wallr.
Fallaing Moire
**Lady's-mantle**

1 tetrad. This Red Data Book species was first recorded in 1953 by S.M.Walters and referred to in *Mountain Flowers* (Raven and Walters 1956).

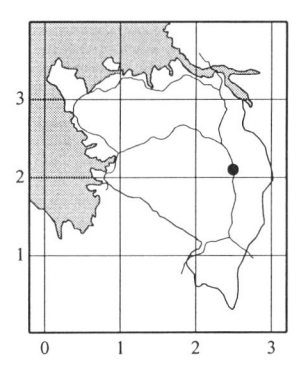

The site, just north of the Inchnadamph Hotel, is described there as the 'rabbit-grazed south-facing slopes of a shallow valley'. Subsequently, F.J.Roberts located it there in 1973, 1988 and 1998, commenting on its close association with a long outcrop of dark grey limestone running down the valley side and contrasting with paler country rock.

*Alchemilla filicaulis* Buser
Fallaing Moire Chaol
**Lady's-mantle**

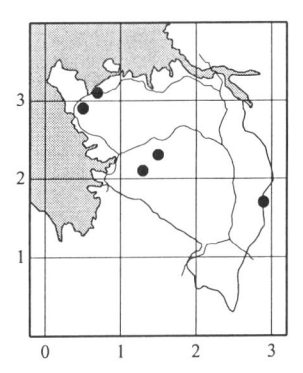

5 tetrads. This map shows records of plants that were not flowering and therefore not identifiable to subspecies. First recorded in 1907, as '*A. vulgaris* var. *filicaulis*' on Canisp by G.C.Druce.

The two maps following represent the results of our best efforts at identification to subspecies; we make no great claims for their accuracy, as it is accepted (Stace 1997) that there are intermediates between the two subspecies in Scotland. A.J.Wilmott and M.S.Campbell were the first, in 1943/44, to record what are now ssp. *filicaulis* and ssp. *vestita*, not far apart on the Achmelvich road, but queried their distinctiveness in the Highlands.

ssp. *filicaulis*

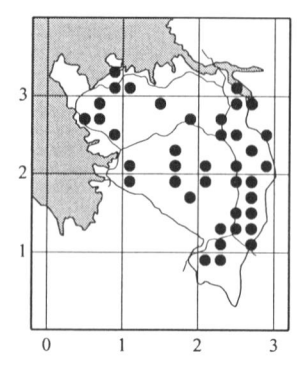

39 tetrads. Damp grassland in sun or light shade. No discernible difference in habitat requirements between this and ssp. *vestita*.

ssp. *vestita* (Buser) M.E.Bradshaw

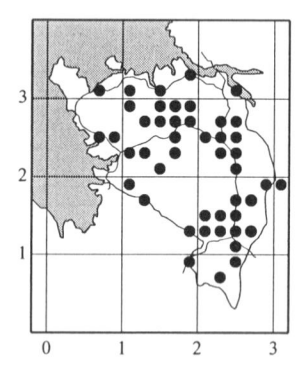

46 tetrads. Damp grassland in sun or light shade. No discernible difference in habitat requirements between this and ssp. *filicaulis*.

*Alchemilla glomerulans* Buser
Fallaing Moire
**Lady's-mantle**

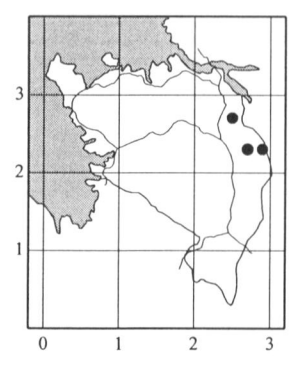

3 tetrads. Two records are from the Allt Poll an Droighinn, one high on rocks near its source and the other on shingle on its lower reaches. It has also been found amongst gneiss boulders by the outflow of Loch a'Choire Dheirg on Glas Bheinn. First recorded by G.P.Rothero in 1995 at the last named of these three sites.

*Alchemilla wichurae* (Buser) Stefánsson
Fallaing Moire
**Lady's-mantle**

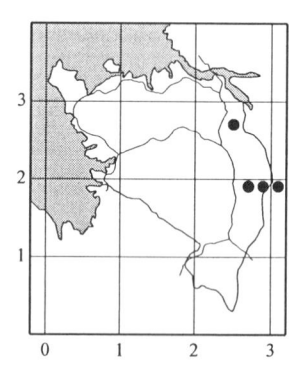

4 tetrads. One record is from water-side rocks in a base-rich area of Glas Bheinn; two are from Bealach Traligill, one in basic grassland and one on an outcrop of the Fucoid Beds, and the fourth above Loch Mhaolach-coire. First recorded in 1953 by S.M.Walters, at an altitude of 90m. near Inchnadamph, and at 305m. above Loch Mhaolach-coire.

*Alchemilla glabra* Neygenf.
Fallaing Moire Mhìn
**Lady's-mantle**

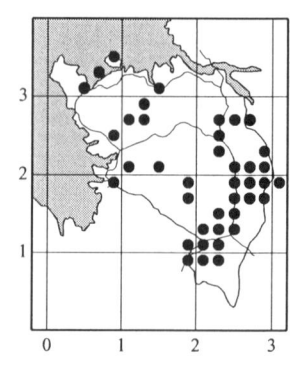

40 tetrads. On rocky grassland, beside burns and waterfalls, where it tolerates constant drenching. Scattered throughout the parish, but most frequent on the higher ground and in limestone areas. First recorded in 1899 at Inchnadamph as '*A. vulgaris* L. var. *alpestris*' by C.E.Salmon.

*Alchemilla mollis* (Buser) Rothm.
**Lady's-mantle**

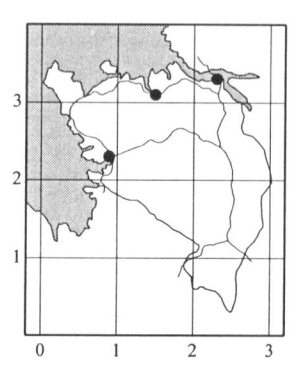

3 tetrads. A garden escape, self-seeding readily and able to establish itself given reasonably open conditions. Recorded from Lochinver, Nedd and Unapool. First recorded in 1998 from the second of these sites.

*Aphanes arvensis* L.
Spìonan Moire
**Parsley-piert**

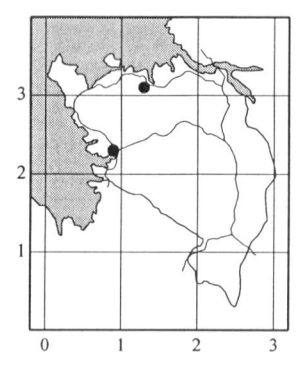

2 tetrads. On tracks and disturbed ground, at Nedd and Lochinver. There is in addition, in tetrad NC13W, one record of *A. arvensis* s.l., which was too immature to identify further. The aggregate was first recorded from NC12 during the 1950s survey for the first *Atlas*.

*Aphanes australis* Rydb.
(*A. microcarpa* (Boiss. & Reut.) Rothm.
Spìonan Moire Caol
**Slender Parsley-piert**

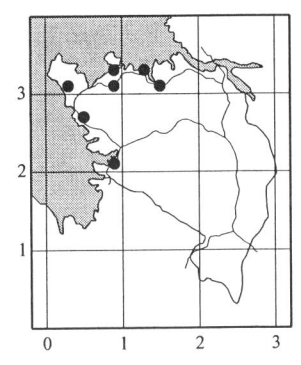

7 tetrads. Rare, in small disused quarries, on tracks and bare disturbed ground generally. Mapped from NC12 in the *Critical Supplement*.

*Rosa* spp.
**Roses**

A very high proportion of the roses in Assynt, especially those in exposed situations in the hills, neither flower nor fruit and therefore could not be identified; their occurrence has not been mapped. In the early years of the survey, all roses were determined by A.L.Primavesi or G.G.Graham. Latterly, with this experience as a guide, they have been determined by P.A.Evans using Graham & Primavesi (1993). Hedges do not exist in Assynt and habitats for all species, apart from *R. pimpinellifolia*, seem to be a random mix of light woodland, scrub, walls, burn courses, cliffs and gorges. In most cases therefore, a habitat description for a particular species has not been attempted.

The first records of roses in Assynt were made in 1886 by A.Gray, who reported '*R. canina, R. mollis* and *R. tomentosa*'. E.S.Marshall and colleagues, on their visits from 1887 onwards, collected material of a bewildering mix of species, hybrids, varieties and forms. A.J.Wilmott and M.S.Campbell, on their visits in 1943 and 1944, also collected material which was 'named with the assistance of Dr.Melville'. A summary of these and other records made up to 1960 or so may be found in *Anthony*. As part of a recent reassessment of the British rose taxa, which is summarised in Graham and Primavesi (1993), a large number of specimens collected by earlier botanists, especially Marshall, have been re-examined and re-determined, by these authors and others. The first records given below are based on their work.

*Rosa pimpinellifolia* L.
Ròs Beag Bàn na h-Alba
**Burnet Rose**

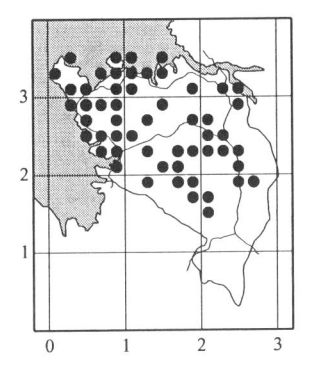

54 tetrads. Most frequent on coastal cliffs and crags but also spreading across almost to the eastern boundary of the parish, on outcrop rocks, along burn courses and in gorges. Only rarely seen out on the open hillside. First recorded in 1890 at Inchnadamph.

*Rosa canina* L.
Ròs nan Con
**Dog-rose**

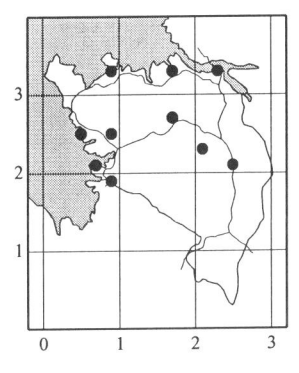

10 tetrads. It is now accepted that *R. canina*, once thought to be virtually absent from Scotland, does occur occasionally. First collected in 1955 at Oldany by J.Anthony (det. A.L.Primavesi). Groups represented are: Dumales (1996, Achadhantuir, R.Maskew), Pubescentes (1995, NC13, P.A.Evans; 1996, Achmelvich, R.Maskew) and Transitoriae (1955, Oldany, J.Anthony; 1994, Ardroe, P.A.Evans; 1996, Achmelvich, R.Maskew).

*Rosa* x *dumalis* Bechst.
*R. canina* x *R. caesia*

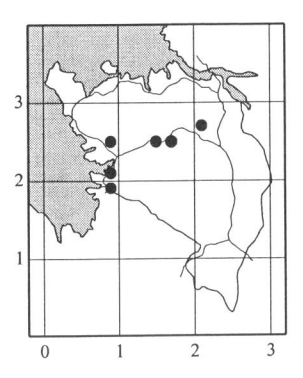

6 tetrads. Found on the river bank at Inverkirkaig, in grassland by Loch Culag, on a roadside verge at Little Assynt and in two places on the stony shores of Loch Assynt. First recorded in 1996 at Achadhantuir by R.Maskew.

*Rosa caesia* Sm.

ssp. *caesia*
**Hairy Northern Dog-rose**

Not recorded during the present survey. First collected in 1890 at Lochinver and Inchnadamph by E.S.Marshall.

ssp. *glauca* (Nyman) G.G.Graham & Primavesi
Ròs nan Con Tuathach
**Glaucous Northern Dog-rose**

5 tetrads. Found beside a burn north of Suilven, in a wooded gorge near Creagan Mór, on the shore of Cam Loch and on a roadside verge at Unapool.

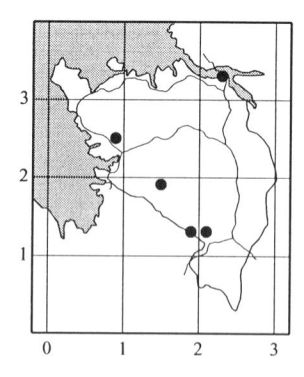

First collected in 1890 at Kylesku by E.S.Marshall, and also by him in 1908 at Inchnadamph. In 1955 it was found at Stoer by J.Anthony and in 1996 at Achadhantuir by R.Maskew.

*Rosa caesia* x *R. sherardii*

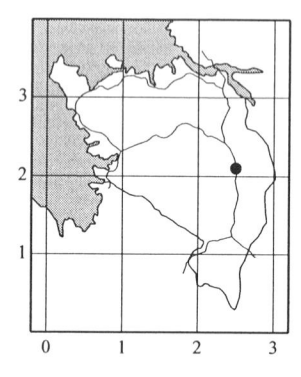

1 tetrad. Recorded in 1994 by the River Traligill. First collected in 1908 at Inchnadamph and Kylesku by E.S.Marshall.

*Rosa sherardii* Davies
Ròs Shioraird
**Sherard's Downy-rose**

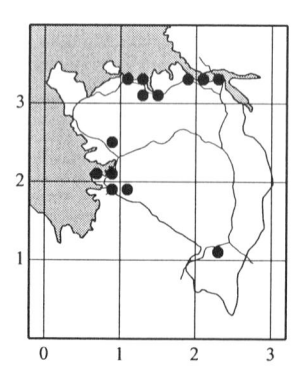

13 tetrads. This is the species most often identified and the map therefore gives the best idea of the most favourable areas for roses.

First collected in 1889 at Lochinver by G.C.Druce. Later records are from Kylesku by F.J.Hanbury in 1890 and by E.S.Marshall in 1908, and at Achmelvich and Oldany in 1955 by J.Anthony.

*Rosa* x *involuta* Sm.
*R. sherardii* x *R. pimpinellifolia*

Not seen during the present survey. First collected in 1908 at Inchnadamph by E.S.Marshall.

*Rosa* x *rothschildii* Druce
*R. sherardii* x *R. canina*

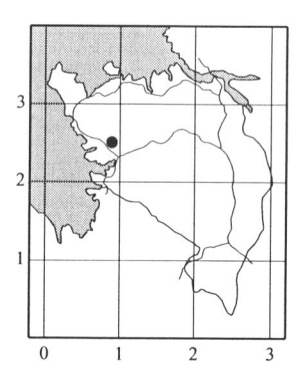

1 tetrad. The first and only record was made in 1996 at Achadhantuir by R.Maskew.

*Rosa* x *suberecta* (Woods) Ley
*R. sherardii* x *R. rubiginosa*

Not seen during the present survey. First collected in 1908 at Kylesku by E.S.Marshall.

*Rosa mollis* Sm.
Ròs Bog
**Soft Downy-rose**

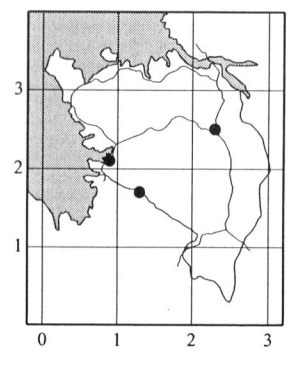

3 tetrads. Found on the shores of Loch Assynt and Fionn Loch and also in Culag Wood. First collected in 1890 at Inchnadamph and Kylesku by E.S.Marshall.

*Rosa* x *sabinii* Woods
*R. mollis* x *R. pimpinellifolia*

Not seen during the present survey. First collected in 1890 near Loch Assynt by F.J.Hanbury.

*Rosa rubiginosa* L.
Dris Chùbhraidh
**Sweet-briar**

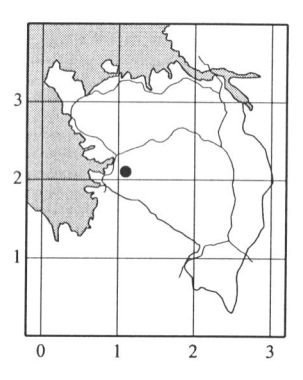

1 tetrad. An isolated bush found in an isolated place, on the path to Cnocnaneach in 1995; an unlikely record, but has been confirmed. First recorded in 1989 at Inchnadamph by R.Maskew and also by him, in 1996, at Lochinver.

*Rosa* x *suberecta* (Woods) Ley
*R. rubiginosa* x *R. sherardii*

Not seen during the present survey. First collected in 1908 at Inchnadamph by E.S.Marshall.

*Prunus spinosa* L.
Preas nan Airneag
**Blackthorn**

9 tetrads. Blackthorn is rare in the north-west of Scotland and in Assynt it occupies two distinct habitats: small, suckering patches of roadside or riverside scrub, and very occasional scattered populations in heathland or rough grassland.

Flowering plants and ferns

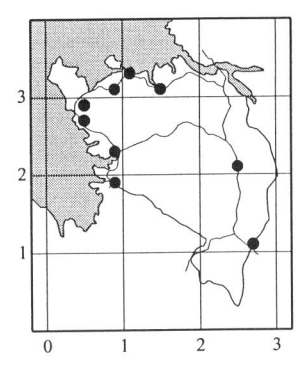

The bushes are usually quite densely covered with lichen and often carry flowers, but little fruit. One exception was a single, young bush at Glenleraig, in a sheltered, sunny position, which had no lichen cover but was bearing a heavy crop of sloes. First recorded in 1886 'Torbreck-Inchnadamph' by A.Gray. Although many are associated with areas of past habitation, the most south-easterly record, in NC21Q, consists of a few well-grazed bushes in scree on the south-facing slopes of Cnoc na Sròine, in an area shown as woodland by Home in 1774 and far from any habitation.

*Prunus avium* (L.)
Geanais
**Wild Cherry**

During the present survey, planted trees were noted at Glencanisp Lodge in NC12B and Rientraid in NC13W. First recorded in NC21 during the 1950s survey for the first *Atlas*.

*Prunus padus* L.
Fiodhag
**Bird Cherry**

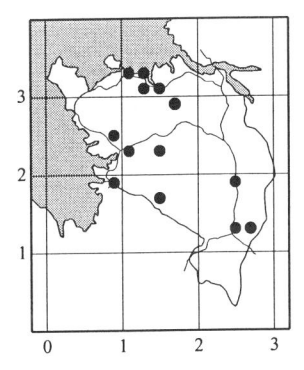

13 tetrads. Isolated trees, often making fine specimens, on cliffs or outcrop rock beside burns. Also in woodland especially at Nedd. First recorded in 1886 'Calda Burn –Ledbeg' by A.Gray

*Sorbus aucuparia* L.
Caorann
**Rowan**

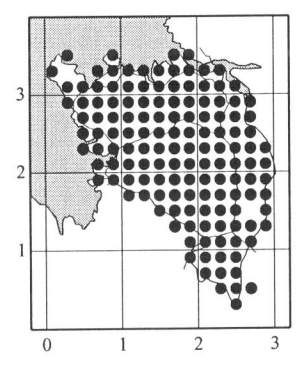

140 tetrads. The most widespread, although not the commonest, tree in Assynt, ascending higher up the hills than birch. The fact that the berries are favoured food of the thrush family is responsible for its appearance in some unlikely places. At 500m. in the southernmost tip of the parish, small, grazed 'buns' of rowan can be found at the foot of the boundary marker stones which provide tempting perches for birds in this most barren landscape. First recorded in 1767 near Inchnadamph by J.Robertson.

*Sorbus rupicola* (Syme) Hedl.
Gall-uinnsean na Creige
**Rock Whitebeam**

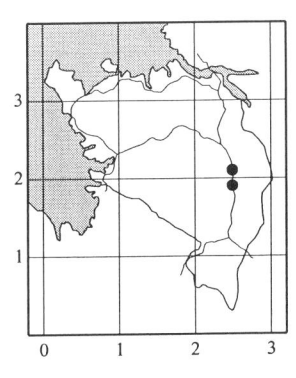

2 tetrads. A very rare calcicole, confined to two separate stretches of the Creag Sròn Chrùbaidh cliffs.

These west-facing limestone cliffs rise from a very steep, scree-littered grassy slope and the trees grow a few metres up from the base of the cliff, where they are safe from grazing animals. A southern group of 13 trees is scattered over a length of 300m. and nearly a kilometre to the north there are two further specimens. Both groups are at an altitude of about 150m. First recorded as '*Pyrus aria*' in 1827 on 'limestone rocks, Assynt' by R.Graham. The locality is given as Inchnadamph in Anderson (1834), and further refined in 1907 to 'growing out of the limestone precipices of Blàr nam Fiadhag' by G.C.Druce.

*Cotoneaster simonsii* Baker
Cotaineastar Hiomàilianach
**Himalayan Cotoneaster**

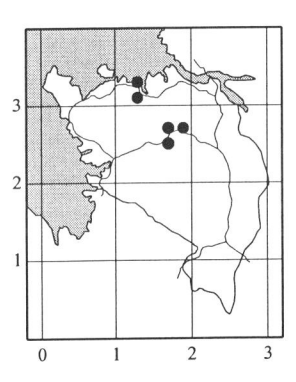

5 tetrads. All presumably bird-sown, with a recognisable garden seed source not far away. They have colonised, and thrive on, cliffs and rocky loch shores. First recorded in 1994 on the banks of a burn at Tumore.

*Crataegus monogyna* Jacq.
Sgìtheach
**Hawthorn**

11 tetrads. The status of hawthorn in the north of Scotland is difficult to determine. In some places, such as Lochinver, it has clearly been planted, and has given rise to bird-sown offspring in the vicinity. Some well-established and probably very old trees grow on limestone crags from the Ardvreck peninsula south to Stronechrubie.

109

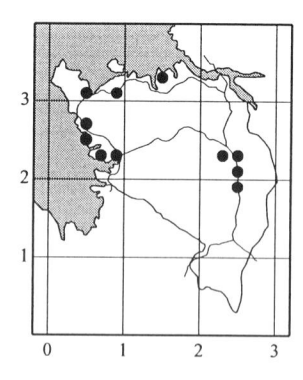

Hawthorn is not a woodland tree in Assynt. First recorded in 1833 as 'one bush...on a rock at Loch Assynt' by R.Graham.


FABACEAE


*Astragalus danicus* Retz.
Bliochd-pheasair Chorcra
**Purple Milk-vetch**

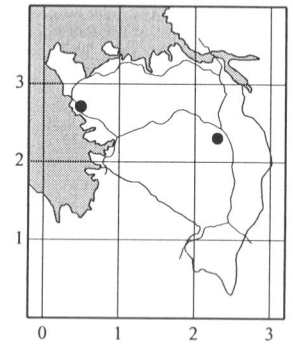

2 tetrads. An unknown lady botanist mentioned to T.H.Fowler, a fellow guest at the Inchnadamph Hotel, that she had seen a small population of this plant in limestone grassland on the Ardvreck Castle peninsula. He saw it there first in 1987 and it still flourishes in the same restricted area. The other Assynt site is a patch of sandy grassland beside the track leading to Clachtoll car-park. It was found there in 1993 by A.McG.Stirling and A.A.P.Slack and here also it still flourishes. This is essentially an east coast plant and these sites represent a great extension of its range, but it should be borne in mind that the Clachtoll dunes have been extensively restored and planted with introduced Marram. First recorded by 'the unknown lady botanist' at Ardvreck, but in the absence of more precise information the credit must go to T.H.Fowler for his record on this site in 1987.


*Anthyllis vulneraria* L.
Cas an Uain
**Kidney Vetch**

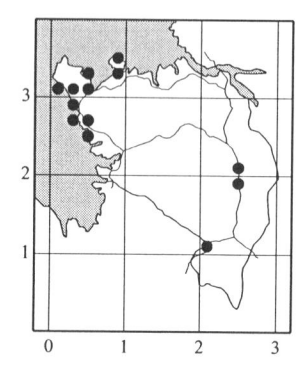

13 tetrads. Found most frequently on coastal cliffs and in sandy grassland, but there are inland sites on the limestone at Elphin and Inchnadamph. First recorded during the 1950s survey for the first *Atlas*.


*Lotus corniculatus* L.
Barra-mhìslean
**Common Bird's-foot-trefoil**

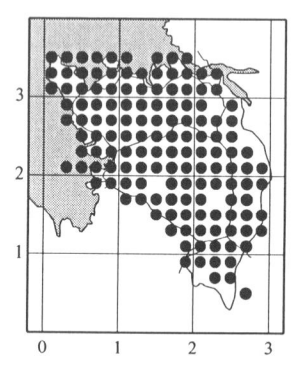

139 tetrads. Widespread, flourishing in even small patches of grassland. On burn banks, at the base of cliffs and outcrop rock and on the occasional patch of richer turf up in the hills. Only absent from high and rocky terrain. A component of commercial seed mixes, var. *sativus*, which at first glance resembles *L. pedunculatus*, has been found once on a roadside verge at Nedd. The species was first recorded in 1890 near Inchnadamph by E.S.Marshall and F.J.Hanbury 'with flowers twice as large as usual'.


*Lotus pedunculatus* Cav.
(*L. uliginosus* (Schkuhr))
Barra-mhìslean Lèana
**Greater Bird's-foot-trefoil**

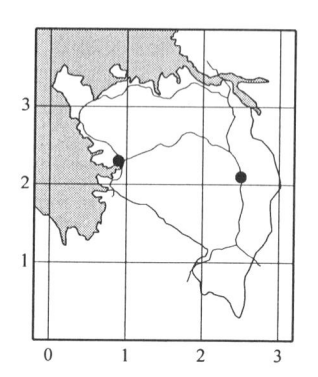

2 tetrads. In rough wet grassland by the road to Glencanisp Lodge and alongside a ditch bordering some species-rich grassland at Inchnadamph. First recorded in 1981 by N.T.H.Holmes at two sites on the River Inver, near Brackloch and Little Assynt.


*Vicia orobus* DC.
Peasair Shearbh
**Wood Bitter-vetch**

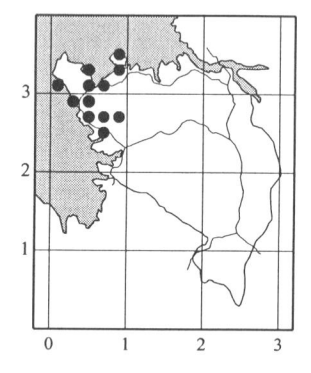

12 tetrads. Confined to cliffs and crags too steep to graze, within three km. of the sea. First recorded in 1943 between Achmelvich and Stoer by A.J.Wilmott and M.S.Campbell.


*Vicia cracca* L.
Peasair nan Luch
**Tufted Vetch**

49 tetrads. The main distribution follows the coast and roads, where

Flowering plants and ferns

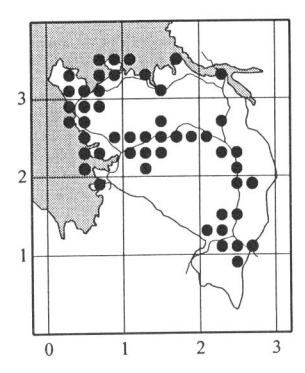

it is found on clifftops, roadside verges, in ditches and other rough, lowland grasslands. First recorded in 1943 on coastal rocks at Imirfada by A.J.Wilmott and M.S.Campbell.

*Vicia sylvatica* L.
Peasair Coille
**Wood Vetch**

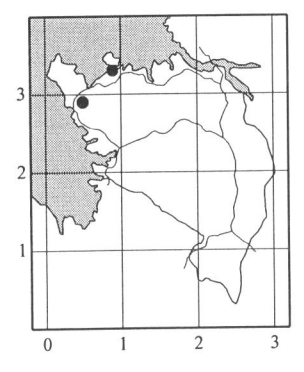

2 tetrads. A rare species found only on cliffs, perhaps because of grazing pressure. Noted once from a sea cliff at Port a'Ghleannain and once from a cliff at Stoer, about one km. inland. First recorded in 1886 at Clachtoll by A.Gray. Subsequent records came from the Stoer/Clachtoll area.

*Vicia sepium* L.
Peasair nam Preas
**Bush Vetch**

62 tetrads. Widespread and common in rough grassy places, along roadside verges and in light woodland.

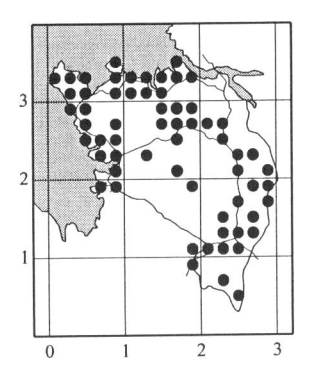

First recorded in 1907 near Inchnadamph by G.C.Druce.

*Vicia sativa* L.
Peasair Chapaill
**Common Vetch**

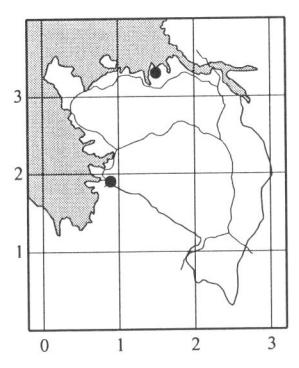

2 tetrads. These are two, not very meaningful records; the one in the south, is a plant in rough grassland alongside a road at Inverkirkaig and the one in the north, which was identified as ssp. *nigra*, in heath vegetation on the edge of Loch Nedd. First recorded in the 1950s survey for the first *Atlas* from three hectads, in none of which it has been seen during the present survey. This may reflect the reduction in the amount of cultivation in the area.

*Lathyrus linifolius* (Reichard) Bässler
(*L. montanus* Bernh.)
Cairt Leamhna
**Bitter-vetch**

119 tetrads. Common and widespread in grassland, scrub, boulder scree and amongst heather on the hill.

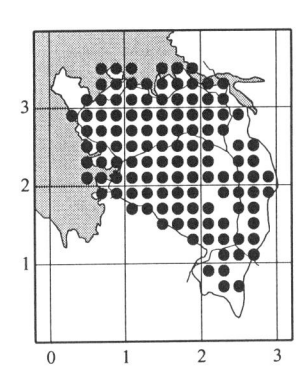

First recorded in 1767 near Inchnadamph by J.Robertson as '*Orobus tuberosus*'. Lightfoot (1777) remarks that 'The Highlanders have great esteem for the tubercles of the roots of this plant'.

*Lathyrus pratensis* L.
Peasair Bhuidhe
**Meadow Vetchling**

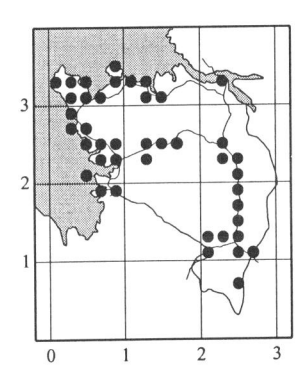

42 tetrads. As the map shows, the rough lowland grassland which this species prefers is best provided by roadside verges and coastal areas. First recorded in 1943, at the edge of a marsh near Drumbeg, by A.J.Wilmott and M.S.Campbell.

*Trifolium repens* L.
Seamrag Bhàn
**White Clover**

133 tetrads. Widespread and common, growing almost everywhere there is grass; not found on bare rocky hills.

111

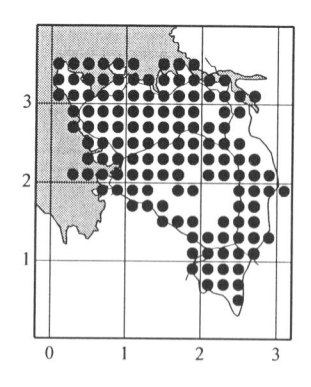

First recorded during the 1950s survey for the first *Atlas*.

*Trifolium campestre* Schreb.
Seamrag Bhuidhe
**Hop Trefoil**

Not seen during the present survey. First recorded in 1886 at Inchnadamph by A.Gray.

*Trifolium dubium* Sibth.
Seangan
**Lesser Trefoil**

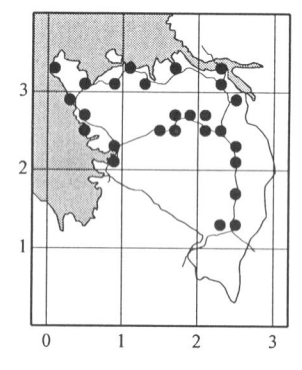

26 tetrads. Confined to dry, thinly vegetated places, such as roadsides, quarries and waste ground. First recorded in 1894 'near Elphin and Ledbeg', by G.C.Druce as '*T. minus*'.

*Trifolium pratense* L.
Seamrag Dhearg
**Red Clover**

68 tetrads. Roadside verges, coastal grassland and in patches of richer vegetation along burn and river valleys.

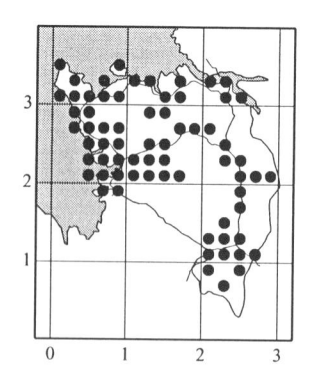

First recorded in 1774 at Little Assynt by J.Home, in 'the first specimen of sown Grass that ever was made in the Country…such stalks of red Clover the surveyor never saw, each stalk being incredibly great, resembling the thickness of a large bean when cut in the Month of July.' Perhaps similar to today's agricultural varieties?

*Cytisus scoparius* (L.) Link
(Sarothamnus scoparius (L.) Wimmer ex Koch)
Bealaidh
**Broom**

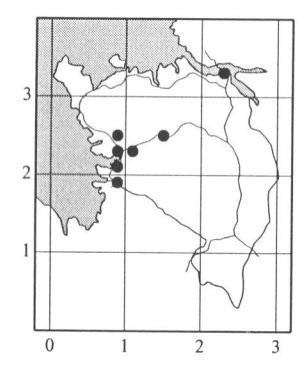

7 tetrads. May well be an introduction to this area. Its few occurrences are confined to roadsides, walls and the vicinity of houses. First recorded during the 1950s survey for the first *Atlas*.

*Ulex europaeus* L.
Conasg
**Gorse**

58 tetrads. Patchy in its distribution, but often plentiful where it does occur. It covers some areas of hillside, usually south-facing, is scattered along roadsides and occurs in patches in old shieling areas, where it may have been used for browse or to shelter stock. It has also colonised gravels on the River Loanan. First recorded in 1886 'Inchnadamph-Loch Inver' by A.Gray.

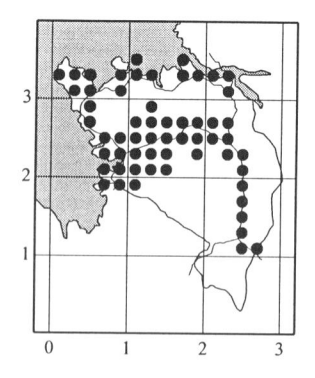

*Ulex gallii* Planch.
Conasg Siarach
**Western Gorse**

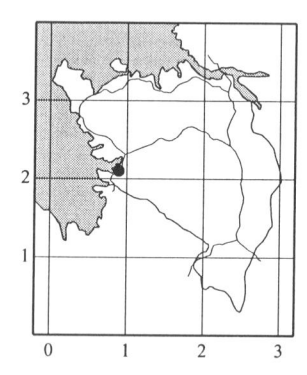

1 tetrad. First recorded in 1943 by A.J.Wilmott and M.S.Campbell 'on a hillside south of Lochinver, towards Lady Constance Bay'. This record must surely be represented by the small population, holding its own but not increasing, on one area of open hillside in Culag Wood.

HALORAGACEAE

*Myriophyllum spicatum* L.
Snàthainn Spìceach
**Spiked Water-milfoil**

Not seen during the present survey. First recorded in 1886 from 'Elphin-Inchnadamph' by A.Gray.

*Myriophyllum alterniflorum* DC.
Snàthainn Bhàthaidh
**Alternate Water-milfoil**

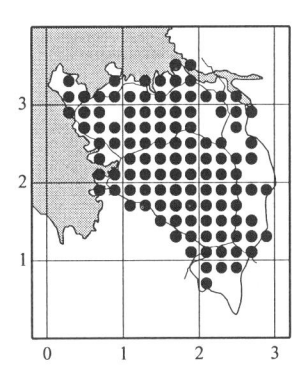

113 tetrads. Widespread and common in burns and the shallow part of lochs. First recorded in 1886 in the River Loanan at Inchnadamph by A.Gray.

LYTHRACEAE

*Lythrum salicaria* L.
Lus na Sìochaint
**Purple Loosestrife**

Not seen during the present survey. A probable garden escape, recorded by A.J.Wilmott and M.S.Campbell in 1943, on the 'brick wall of the garden of the Culag Hotel, Lochinver' and in 1944 from a rocky valley at Achmelvich. Found again at Achmelvich in 1972 and at Loch Ruighean an Aitinn in 1988 (a garden runs down to the shore of this loch).

ONAGRACEAE

*Epilobium hirsutum* L.
Seileachan Mòr
**Great Willowherb**

1 tetrad. Not previously recorded for Assynt, it was found during the present survey on a roadside verge just outside Lochinver in 1999 and was still there in 2001. Apart from this, it appears to be absent from the west coast of mainland Scotland north of the Great Glen; it will be interesting to see if it survives.

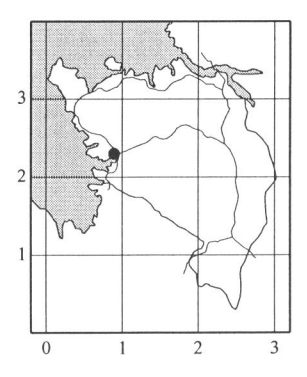

*Epilobium parviflorum* Schreb.
Seileachan Liath
**Hoary Willowherb**

Not seen during the present survey. Recorded for NC02 during the 1950s survey for the first *Atlas*.

*Epilobium montanum* L.
Seileachan Coitcheann
**Broad-leaved Willowherb**

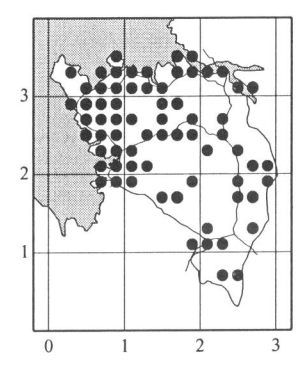

69 tetrads. Flourishes in a variety of habitats, both dry and damp. Widespread in open, ruderal situations, cracks in rock faces and in marshy ground. First recorded in 1767 near Inchnadamph by J.Robertson.

*Epilobium tetragonum* L.
(E. adnatum Griseb.)
Seileachan Gas Ceithir-cheàrnach
**Square-stalked Willowherb**

1 tetrad. Not previously recorded for Assynt, this species was noted during the present survey in 1998 as a garden weed at Ardvar' Judging by its national distribution it has almost certainly been introduced.

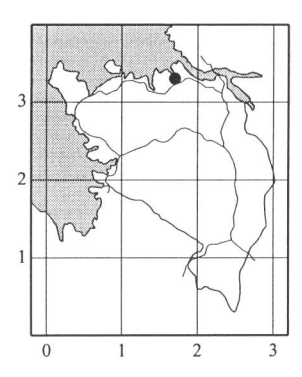

*Epilobium obscurum* Schreb.
Seileachan Faireagach
**Short-fruited Willowherb**

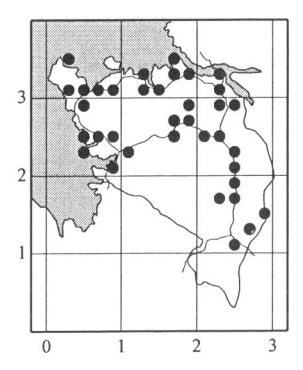

36 tetrads. Occasional on roadside verges, in ditches and marshy ground. First recorded in 1887 'near Skaig [sic] Bridge, Loch Assynt' by E.S.Marshall, and in 1944, by A.J.Wilmott and M.S.Campbell as a 'garden weed, Culag Hotel, Lochinver.' It was only recorded during the 1950s survey for the first *Atlas* in NC02 and 22, since when it seems to have spread considerably.

*Epilobium* x *marshallianum* Hausskn.
*E. obscurum* x *E. anagallidifolium*

'A very curious plant' was collected in 1887 by E.S.Marshall 'by a rill above the path from Inchnadamph to Ben Mor of Assynt at about 1600ft.' It was identified by him as a hybrid new to science and named in his honour by Haussknecht. Not recorded since in West Sutherland, and only

113

known elsewhere in the British Isles from Stirlingshire.

*Epilobium* x *rivulicola* Hausskn.
*E. obscurum* x *E. alsinifolium*

Not seen during the present survey. First recorded in 1899 'by the Traligill Burn above *Inchnadamph* by C.E.Salmon and also by him, according to *Anthony*, in 1900 in the Allt nan Uamh valley.

*Epilobium ciliatum* Raf.
Seileachan Aimeireaganach
**American Willowherb**

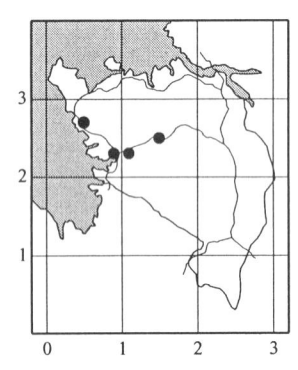

4 tetrads. Introduced into Great Britain in 1891, it has appeared in Assynt only rarely: as a garden weed at Clachtoll, in a small roadside quarry at Little Assynt, on the sea wall at Lochinver and beside the track leading to the Star Pool on the River Inver. First recorded in 1992 from the last of these sites.

*Epilobium palustre* L.
Seileachan Lèana
**Marsh Willowherb**

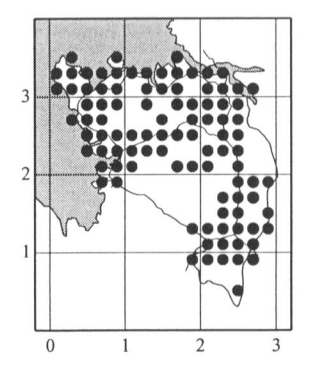

97 tetrads. The commonest willowherb in Assynt, growing in generally, marshes, bogs and ditches. First recorded in 1943 in a 'ditch north of Lochinver' by A.J.Wilmott and M.S.Campbell.

*Epilobium anagallidifolium* Lam.
Seileachan Ailpeach
**Alpine Willowherb**

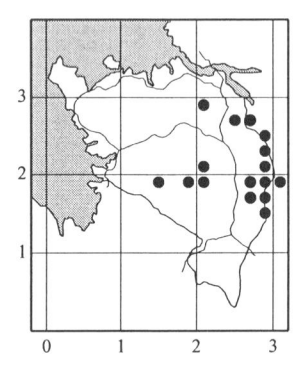

16 tetrads. In bryophyte flushes, stony flushes and on burn sides, usually at over 500m. First recorded in 1899 on Glas Bheinn and the north slopes of Canisp by C.E.Salmon.

*Epilobium* x *boissieri* Hausskn.
*E. anagallidifolium* x *E. alsinifolium*

Not seen during the present survey. First recorded in 1899 on Glas Bheinn by C.E.Salmon.

*Epilobium alsinifolium* Vill.
Seileachan Fliodhach
**Chickweed Willowherb**

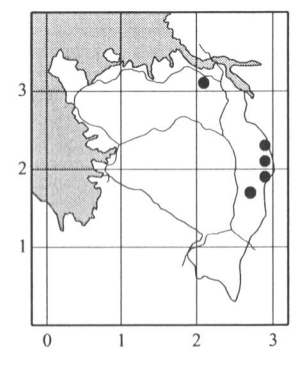

5 tetrads. Considerably rarer than the preceding species, although found in similar habitats. Recorded from Airigh na Beinne, Allt nan Uamh valley, Bealach Traligill, Beinn an Fhuarain and Imir Fada. First recorded in 1899 by C.E.Salmon 'by the Traligill Burn above Inchnadamph' and in the Allt nan Uamh valley.

*Epilobium brunnescens*
(Cockayne) P.H.Raven & Engelhorn
(E. nerterioides Cunn.)
Seileachan Làir
**New Zealand Willowherb**

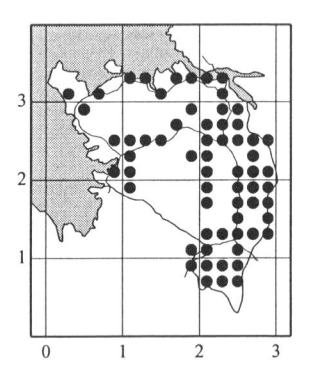

63 tetrads. A notable success story, even for a willowherb. First recorded in Britain in 1908, it was another 60 years before it was seen in Assynt, although judging by its present distribution, it has probably been here longer than that. Unlike many introductions, which rely solely on man-made habitats and disturbed ground for their spread, this one has succeeded by colonising river shingle, waterside boulders and shady cliffs, as well as wall-tops and stonework generally. First recorded in 1969 on the banks of the River Loanan by A.W.Punter.

*Chamerion angustifolium* (L.) Holub
(Chamaenerion angustifolium (L.) Scop.)
Seileachan Frangach
**Rosebay Willowherb**

22 tetrads. Apart from a few occurrences in small roadside quarries, this species is scattered thinly on cliffs, both coastal and inland.

## Flowering plants and ferns

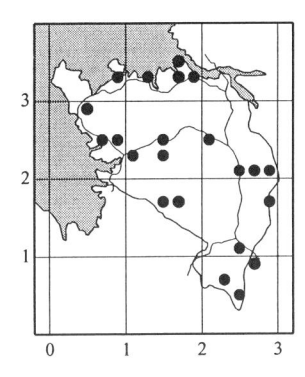

This is the natural habitat of rosebay willowherb in the north-west and it is seen here at its best, in small groups rising elegantly from a rock crevice – not in gaudy bands along a roadside!

*Circaea* spp.

There has, in the past, been considerable confusion in the recording of this genus. In the light of the comments on *C. alpina* in Perring (1968), it seems likely that all the historic records of this species should be referred to *C.* x *intermedia*.

*Circaea lutetiana* L.
Fuinseagach
**Enchanter's-nightshade**

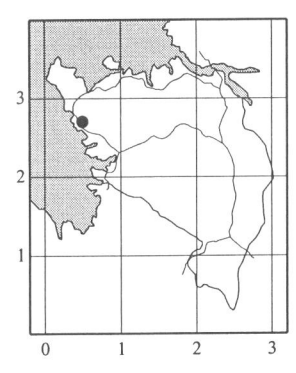

1 tetrad. Only found once during the present survey (and identified with some care!), in 1998, from scree at Clachtoll. *Anthony* cites this species from Lochinver and Achmelvich. The Lochinver record appears, from Anthony's card index, to be that of '*C. alpina*', made in 1944 by A.J.Wilmott and M.S.Campbell 'in the crevices of a roadside stone wall about a mile north of Lochinver'. In view of the comments in Perring (1968) quoted above, this may well refer to *C.* x *intermedia*. There are no further details of the Achmelvich record.

*Circaea* x *intermedia* Ehrh.
*C. lutetiana* x C. *alpina*
Lus na h-Oighe
**Upland Enchanter's-nightshade**

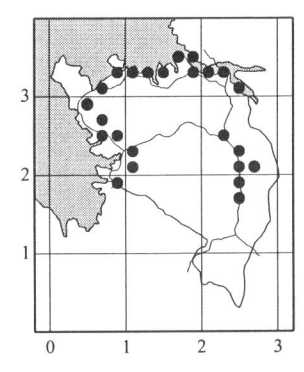

24 tetrads. Woodland is one of its main habitats and this accounts for the coastal distribution of the records. Scree and shaded rocks are other, less common sites. First recorded as '*C. alpina*' at Inchnadamph in 1886 by A.Gray and in 1890 by E.S.Marshall and F.J.Hanbury; probably!

[*Circaea alpina* L.
Lus na h-Oighe Ailpeach
**Alpine Enchanter's-nightshade**

Not seen during the present survey and unlikely to occur. See preceding notes on the genus.]

### CORNACEAE

*Cornus suecica* L.
(Chamaepericlymenum suecicum (L.) Aschers. & Graebn.)
Lus a'Chraois
**Dwarf Cornel**

19 tetrads. Amongst heather on high, heathy grassland.

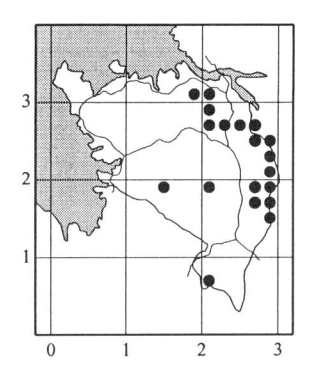

First recorded in 1767 near Inchnadamph by J.Robertson.

### AQUIFOLIACEAE

*Ilex aquifolium* L.
Cuileann
**Holly**

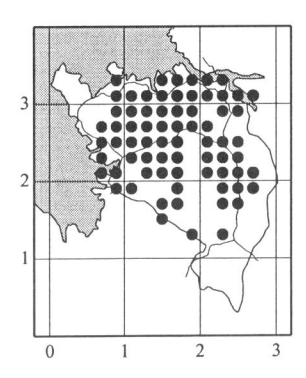

71 tetrads. Widespread in woodland and also on crags, out of the way of browsing animals. Essentially a lowland tree, but given the shelter of a gorge or a burn valley it can extend some distance up into the hills. It does not tolerate exposure well and trees which grow up out of their shelter soon become stunted. Despite its occurrence throughout much of the parish, it does not fruit readily, perhaps suggesting that it is not entirely suited by the climate. First recorded in 1774 in the 'Wood of Tumore' by J.Home.

### EUPHORBIACEAE

*Euphorbia helioscopia* L.
Lus nam Foinneachan
**Sun Spurge**

5 tetrads. All the records are of

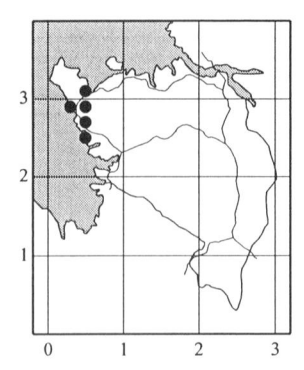

garden weeds, often in potato patches. Found at Achmelvich, Clachtoll, Clashnessie and Stoer. It may have been commoner years ago, when there was more extensive cultivation. First recorded, from NC02, in the 1950s survey for the first *Atlas*.

*Euphorbia peplus* L.
Lus Leighis
**Petty Spurge**

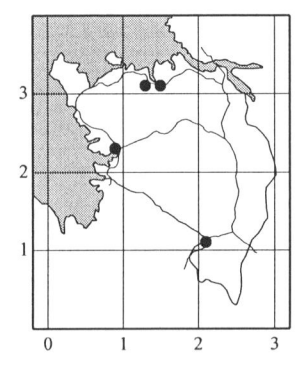

4 tetrads. On newly made roadside verges at Elphin, Lochinver and Nedd and once in a garden, also at Nedd. This suggests that it has been brought into the area during roadworks, unlike the preceding species, which is surviving in cultivated ground. First recorded in 1890 as 'a weed at Lochinver' by E.S.Marshall and F.J.Hanbury.

LINACEAE

*Linum catharticum* L.
Lìon nam Ban-sìdh
**Fairy Flax**

124 tetrads. Widespread and common in coastal and inland grassland. Also on the hills, where there are patches of sweeter grass amongst the heather, but absent from some of the very high ground. First recorded, throughout the parish, in the 1950s survey for the first *Atlas*.

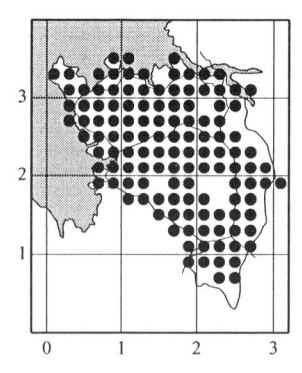

*Linum usitatissimum* L.
**Flax**

Found in two places in a garden in Nedd in 1995 but it did not appear the following year. Recorded, in 2001, on a verge at Baddidarach.

POLYGALACEAE

*Polygala vulgaris* L.
Lus a'Bhainne
**Common Milkwort**

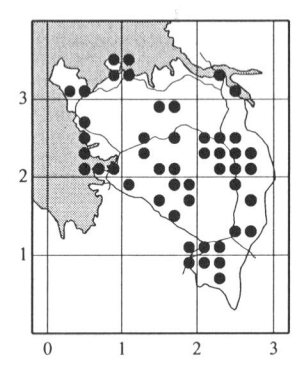

48 tetrads. Frequent in grassland in the limestone areas and in sandy coastal grassland. Scattered records from the gneiss, often in the vicinity of ultrabasic dykes. First recorded (possibly) in 1767, near Inchnadamph by J.Robertson (*P. serpyllifolia* was not differentiated at that time). Recorded unambiguously in 1894, at Knockan by G.C.Druce.

*Polygala serpyllifolia* Hosé
Siabann nam Bansìdh
**Heath Milkwort**

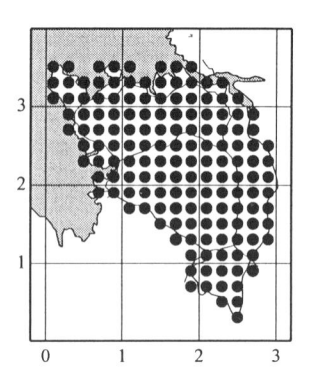

151 tetrads. Widespread and common in grassland and amongst heather. Apparently tolerant of a range of soils. First recorded in 1886, without a specific locality, by A.Gray, as '*P. depressa*'.

HIPPOCASTANACEAE

*Aesculus hippocastanum* L.
Craobh Geanm-chnò Fhiadhaich
**Horse-chestnut**

Occasional trees planted in gardens and grounds.

ACERACEAE

*Acer pseudoplatanus* L.
Craobh Pleantrainn
**Sycamore**

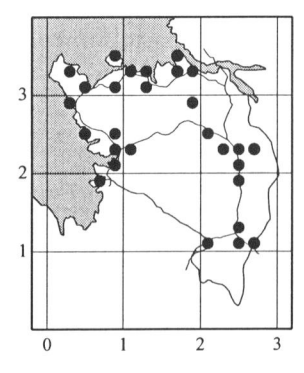

28 tetrads. Introduced, but established along roadsides, near houses, on rocky loch shores and in screes. Reproduces successfully, judging by the appearance of seedlings in a number of places. There are some

fine specimens of Sycamore in the vicinity of Lochinver Harbour. First recorded in 1886 at Inchnadamph by A.Gray with the caveat 'probably not native'.

OXALIDACEAE

*Oxalis acetosella* L.
Feada Coille
**Wood-sorrel**

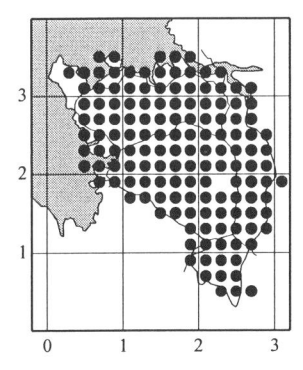

142 tetrads. Common and widespread wherever there is good drainage and a little shade and shelter provided by vegetation or rock. It grows in woodland, on moorland and on open hillsides under bracken. For a plant of such apparent fragility it flourishes in surprisingly open conditions, perhaps because of the high atmospheric moisture. First recorded in 1767 near Inchnadamph by J.Robertson.

GERANIACEAE

*Geranium dissectum* L.
Crobh Preachain Geàrrte
**Cut-leaved Crane's-bill**

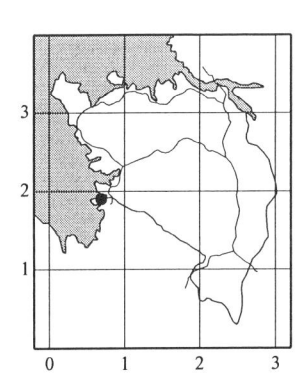

1 tetrad. Found once during the present survey, at Inverkirkaig. First recorded in 1944 as a 'garden weed, Culag Hotel, Lochinver' by M.S.Campbell.

*Geranium molle* L.
Crobh Preachain Mìn
**Dove's-foot Crane's-bill**

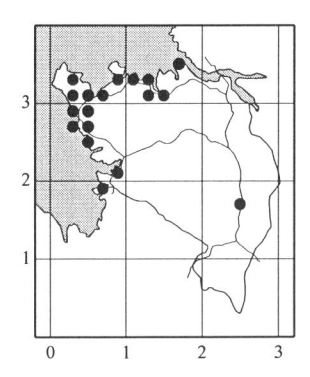

18 tetrads. All but one of the records are from sandy coastal grassland or roadside verges near the sea. The only exception is from the banks of the River Loanan. First recorded during the 1950s survey for the first *Atlas*.

*Geranium lucidum* L.
Crobh Preachain Deàlrach
**Shining Crane's-bill**

Imported with garden material to Nedd in 1991 and has survived and spread within the garden for 10 years. Recorded, rather surprisingly, in 1767 near Inchnadamph by J.Robertson.

*Geranium robertianum* L.
Lus an Ròis
**Herb-Robert**

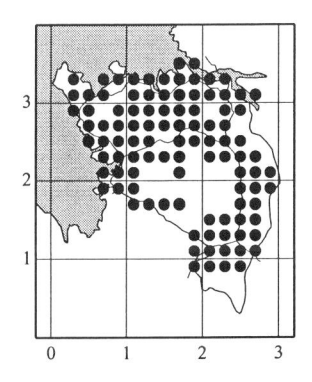

101 tetrads. Widespread and common except on the highest ground. Will tolerate shade and is found in woodland, in gorges, around outcrop rock and is particularly characteristic of boulder scree. First recorded in 1767 near Inchnadamph by J.Robertson.

*Erodium cicutarium* (L.) L'Hér.
Gob Corra
**Common Stork's-bill**

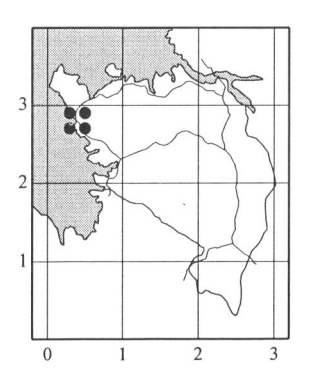

4 tetrads. Sandy grassland near the sea at Clachtoll and Stoer. First recorded, from NC02, in the 1950s survey for the first *Atlas*.

TROPAEOLACEAE

*Tropaeolum speciosum* Poepp. & Endl.
**Flame Nasturtium**

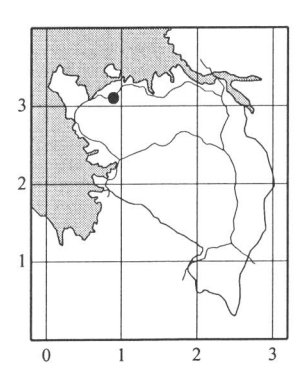

1 tetrad. Introduced. Probably an escape from a nearby garden, but long established and flourishing in a patch of Blackthorn scrub by the roadside at Strathcroy (NC03).

117

## BALSAMINACEAE

*Impatiens glandulifera* Royle
Lus a'Chlogaid
**Indian Balsam**

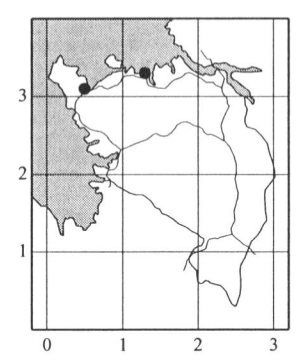

2 tetrads. A recent, and very invasive, garden escape, it has established itself on the banks of burns close to two gardens, one at Clashnessie and one at Nedd. If left alone there is every reason to suppose that it will continue to flourish. First recorded in 2000 at both sites.

## ARALIACEAE

*Hedera helix* L.
Eidheann
**Ivy**

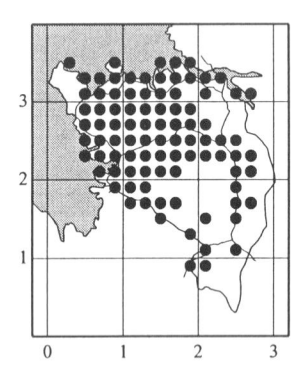

91 tetrads. It is characteristically associated with rock in its many forms: in scree, on both inland and coastal cliffs, on outcrops and in gorges. It is quite uncommon to see ivy clothing a tree trunk or sprawling on the ground in woodland. First recorded in 1767 near Inchnadamph by J.Robertson.

## APIACEAE

*Hydrocotyle vulgaris* L.
Lus na Peighinn
**Marsh Pennywort**

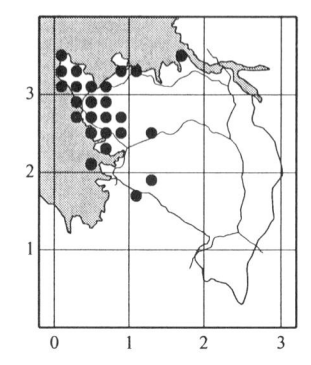

25 tetrads. In wet places; marshes, river margins and loch shores, but with a strong coastal tendency. It is able to tolerate seasonal changes in loch levels and appears sometimes to be growing in water, with floating leaves. First recorded in 1767 near Inchnadamph by J.Robertson.

*Sanicula europaea* L.
Bodan Coille
**Sanicle**

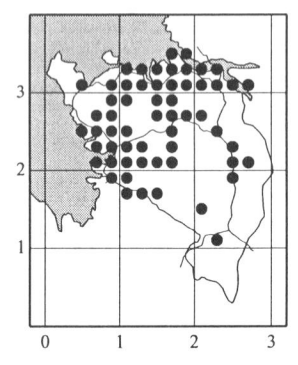

58 tetrads. Frequent, as might be expected, in woodland, but also in other habitats which provide similar sheltered and moist conditions such as boulder scree, gorges and rocky loch shores. First recorded in 1942 on a 'wooded rocky slope between Achadhantuir and Feadan' by A.J.Wilmott and M.S.Campbell.

*Anthriscus sylvestris* (L.) Hoffm.
Costag Fhiadhain
**Cow Parsley**

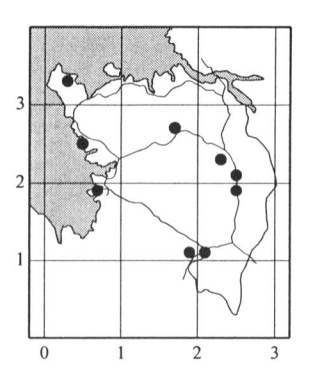

9 tetrads. In lowland, grassy places, almost always those affected in some way by man or his activities. Ditchsides, disturbed ground near houses and a small field where cattle are fed, have been recorded. Apart from one at Elphin, roadside populations are rare. First recorded in the 1950s survey for the first *Atlas*.

*Myrrhis odorata* (L.) Scop.
Mirr
**Sweet Cicely**

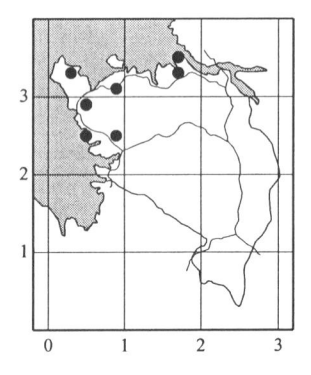

7 tetrads. Introduced and naturalised in a few places, surviving in abandoned gardens, and on roadside verges, with other garden throw-outs. First recorded during the 1950s survey for the first *Atlas*.

*Conopodium majus* (Gouan) Loret
Cnò-thalmhainn
**Pignut**

93 tetrads. Common and widespread, except in areas where most of the ground lies over 500m.

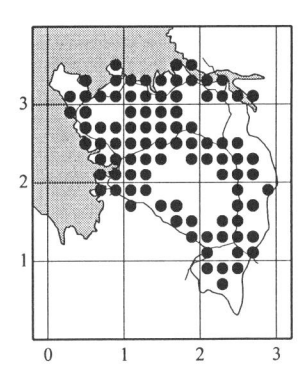

Found in grassy heath where there is some nutrient enrichment, open woodland and grassland. First recorded as '*Bunium bulbocastanum*' (see Henderson and Dickson 1994) in 1767 near Inchnadamph by J.Robertson.

*Aegopodium podagraria* L.
Lus an Easbaig
**Ground-elder**

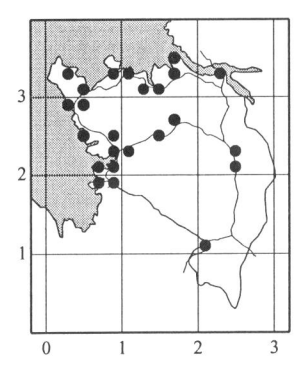

24 tetrads. In gardens, disturbed ground near houses and on roadside verges. First recorded in 1886 on an 'island in Loch Assynt (garden escape)' by A.Gray.

*Oenanthe crocata* L.
Dàtha Bàn Iteodha
**Hemlock Water-dropwort**

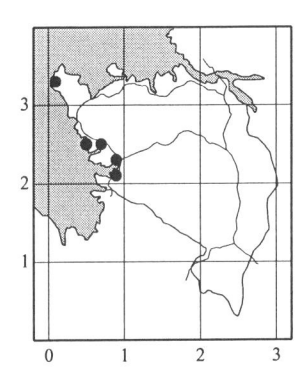

5 tetrads. All sites are coastal.

There are good populations around Bàgh an t-Strathain and on the banks of the River Culag by the bridge in Lochinver. It occurs also at Achmelvich and one plant was found growing at the back of the shingle in a small inlet just south of Stoer Lighthouse. First recorded in 1943 in a 'drainage channel by sea in bay south of Strathan' by A.J.Wilmott and M.S.Campbell.

*Ligusticum scoticum* L.
Sunais
**Scots Lovage**

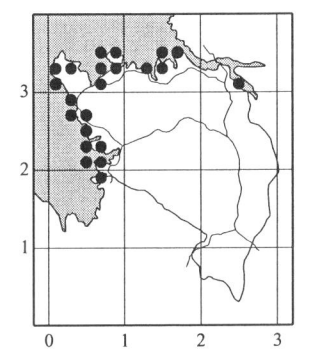

22 tetrads. Thinly scattered on maritime cliffs and rocks, usually where grazing sheep cannot reach. First recorded in 1890 by E.S.Marshall and F.J.Hanbury.

*Angelica sylvestris* L.
Lus nam Buadh
**Wild Angelica**

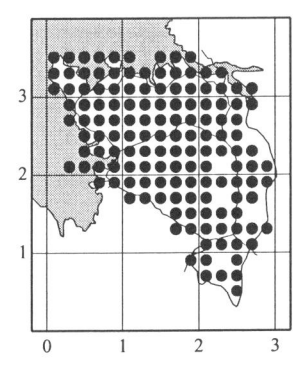

139 tetrads. Common and widespread in a great variety of habitats. Open woodland, maritime cliffs and wet places such as marshes, burn and river margins. First recorded in 1767 near Inchnadamph by J.Robertson.

*Peucedanum ostruthium* (L.) W.D.J.Koch
Mòr-fhliodh
**Masterwort**

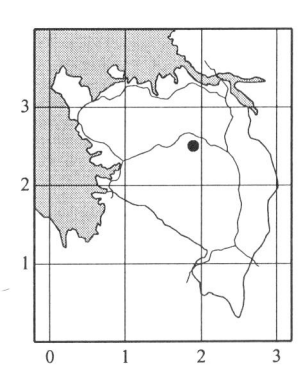

1 tetrad. The first and only record was made by A.Scott in 1994 on Eilean Assynt, an island in Loch Assynt, where it flourishes. There was a 'fortalice'on this site in the 14th century and other remains indicate some mediaeval presence there. The date of the introduction of Masterwort is quite unknown.

*Heracleum sphondylium* L.
Odharan
**Hogweed**

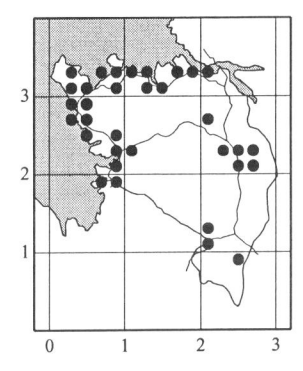

33 tetrads. Occasional on rough, ungrazed grassland and roadside verges. First recorded in 1767 near Inchnadamph by J.Robertson.

*Daucus carota* L.
Curran Fiadhain
**Wild Carrot**

10 tetrads. Dune grassland and roadside verges near the sea.

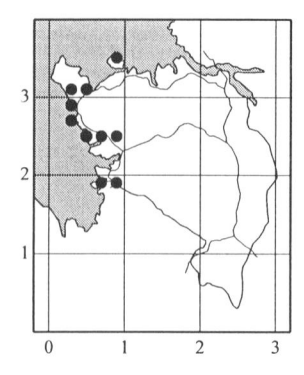

First recorded in 1943 at Achadhantuir (where it is still common by the roadside) by A.J.Wilmott and M.S.Campbell.

## GENTIANACEAE

*Gentianella campestris* (L.) Börner
Lus a'Chrùbain
**Field Gentian**

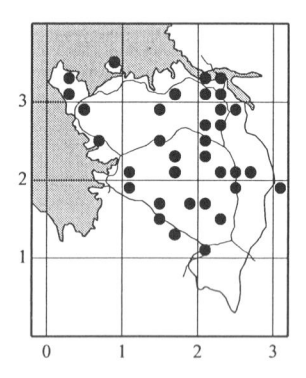

34 tetrads. Frequent in acid grassland, occasionally on roadside verges. The maps for this species and the following are to some extent complementary, reflecting the different soils on which they occur.

*Gentianella amarella* (L.) Börner ssp. *septentrionalis* (Druce) N.M.Pritch.
Lus a'Chrùbain Tuathach
**Autumn Gentian**

19 tetrads. Scattered on basic grassland, dune grassland and roadside verges.

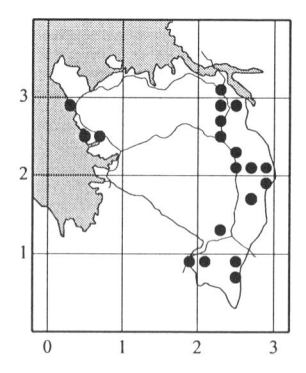

First recorded in 1886 at Inchnadamph by A.Gray.

*Gentiana verna* L.
**Spring Gentian**

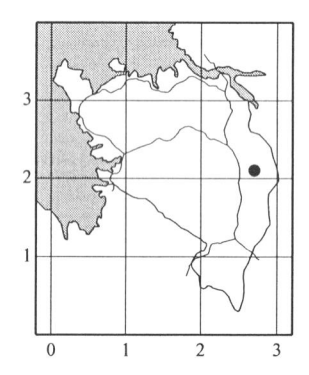

1 tetrad. The rarest of the plants comprising the Inchnadamph 'zoo', a group of species introduced and maintaining themselves on a limestone outcrop at Glac Mhòr. First recorded in 1997. (See also records of *Campanula*, *Erinus*, *Silene* and *Phyteuma*).

[*Gentiana nivalis* L.
Lus a'Chrùbain Sneachada
**Alpine Gentian**

Reported (in error for *Gentianella amarella* ssp. *septentrionalis*?) in 1896 'on rocks by Ardvreck Castle' by Lowe (1899); referred to by Druce (1908).]

## SOLANACEAE

*Solanum dulcamara* L.
Searbhag Mhilis
**Bittersweet**

Not seen during the present survey.

First recorded as 'Lochinver 1944' in *Anthony* and in 1958 in NC03 by J.G.Urquhart and R.B.Knox.

*Datura stramonium* L.
**Thorn-apple**

A plant was found in 1992 in a long-unused, cold greenhouse at Glenleraig. Provenance unknown.

## CONVOLVULACEAE

*Convolvulus arvensis* L.
Iadh-lus
**Field Bindweed**

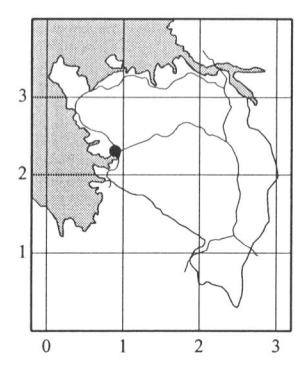

1 tetrad. The first and only record was made by C.Warwick in 1999, in an overgrown part of her garden in Lochinver.

*Calystegia sepium* (L.) R.Br.
Dùil Mhial
**Hedge Bindweed**

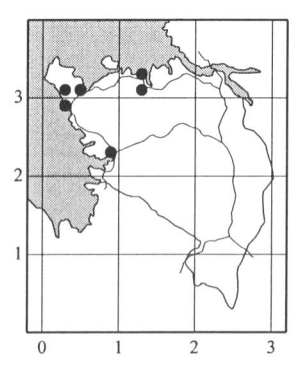

6 tetrads. All the records are from gardens, where it persists, scrambling over hedges and fences. Found at Balchladich, Clashnessie, Drumbeg, Lochinver, Nedd and Stoer. First recorded, in

*Calystegia pulchra* Brummit & Heywood
Dùil Mhial Ghiobach
**Hairy Bindweed**

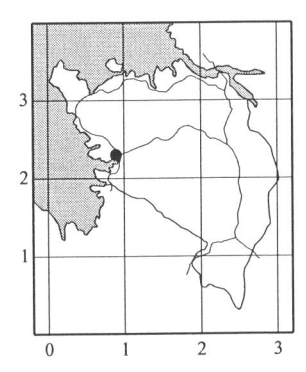

1 tetrad. One record only, from Lochinver, where it grows over a garden hedge. First recorded from Inchnadamph in *Anthony*, but with no further details. Mapped for NC22 in the *Critical Supplement*.

*Calystegia silvatica* (Kit.) Griseb.
Dùil Mhial Mhòr
**Large Bindweed**

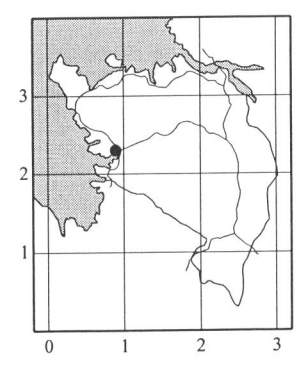

1 tetrad. The first and only record was made in 1994, in a Lochinver garden.

## MENYANTHACEAE

*Menyanthes trifoliata* L.
Trì-bhileach
**Bogbean**

120 tetrads. Common and widespread in marshes, bog pools and shallow water in lochs. Spreading by rhizomes, it usually forms extensive patches and flowers most freely when growing actually in water. First recorded in 1767 near Inchnadamph by J.Robertson.

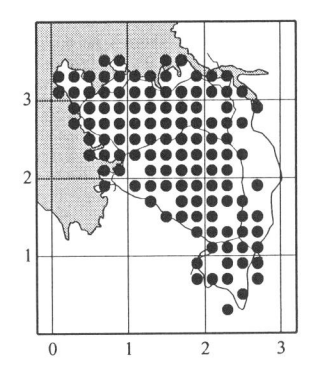

## HYDROPHYLLACEAE

*Phacelia tanacetifolia* Benth.
**Phacelia**

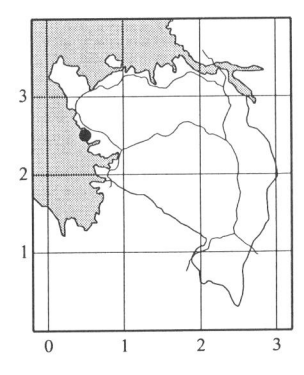

1 tetrad. The first and only record was made in 1995, on a piece of arable land at Achmelvich. The land is no longer cultivated, but the species was surviving there in 2000.

## BORAGINACEAE

*Echium vulgare* L.
Lus-na-nathrach
**Viper's-bugloss**

1 tetrad. The first and only record was made in 1991 at Clachtoll, in sandy grassland near the sea.

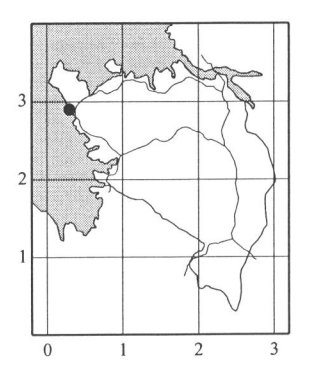

*Symphytum officinale* L.
Meacan Dubh
**Common Comfrey**

Not seen during the present survey. First recorded, as 'var. *purpureum*', in 1943 'by stream below cottage, Inverkirkaig' by A.J.Wilmott and M.S.Campbell; and, as the species, during the 1950s survey for the first *Atlas*.

*Symphytum* x *uplandicum* Nyman
*S. officinale* x *S. asperum* Lepech.
Meacan Dubh Ruiseanach
**Russian Comfrey**

Not seen during the present survey. First recorded in 1987 'at Drumbeg, south side of the road by the hotel' by A.Showler. The site was probably destroyed by the expansion of the hotel carpark.

*Anchusa arvensis* (L.) M.Bieb.
(*Lycopsis arvensis* L.)
Lus Teanga an Daimh
**Bugloss**

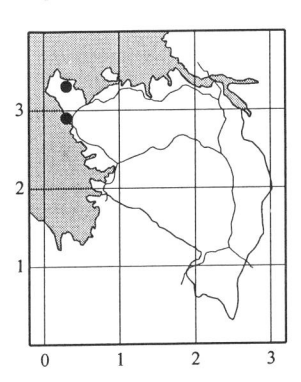

2 tetrads. A rare weed of cultivation, whose scarcity

probably reflects the changes in farming patterns. The records come from sandy grassland by the road at Culkein Stoer, where it was growing with *Viola arvensis*, suggesting that there is a seed bank of weeds beneath the grass; also in a blow-out in dune grassland at Clachtoll. First recorded in 1886 at Clachtoll by A.Gray.

*Pentaglottis sempervirens* (L.) Tausch ex L.H.Bailey
Bog-lus
**Green Alkanet**

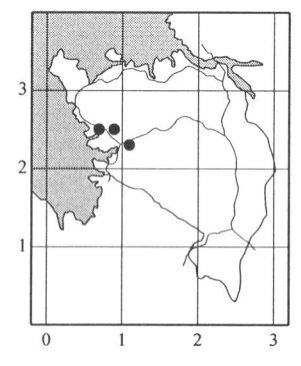

3 tetrads. Two records are from roadside verges at Torbreck and Achmelvich, where it grows with garden throw-outs and builder's rubble. The third comes from more 'natural-looking' surroundings, in the wooded grounds of Glencanisp Lodge. First recorded in 1886 at Elphin by A.Gray and subsequently from Stoer in 1958 by J.Anthony.

*Mertensia maritima* (L.) Gray
Tìodlach na Mara
**Oysterplant**

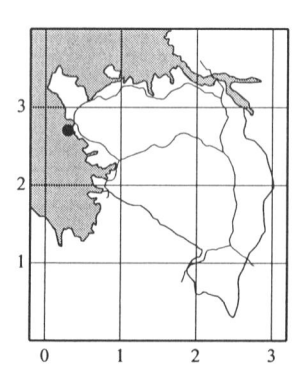

1 tetrad. Surviving, but only just, in its one remaining site on a storm beach at Clachtoll well above high water mark. Protected from possible damage by grazing and by the feet of wandering sheep, it is just managing to hold its own. First recorded in 1886 at Inverkirkaig, where it has never since been seen, although the site may have disappeared under the road embankment. First recorded at its present site at Clachtoll in 1949 by M.McC.Webster. It was recorded at Stoer in 1955 and from Clachtoll *and* Stoer in 1965, suggesting that there may have been a second site a little north of the present one, the precise location of which is unknown.

*Myosotis scorpioides* L.
Cotharach
**Water Forget-me-not**

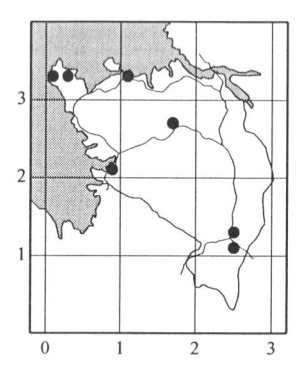

7 tetrads. Uncommon in marshes and ditches. First recorded in 1907 as '*M. palustris* var. *strigulosa*' at Inchnadamph by G.C.Druce.

*Myosotis secunda* Al.Murray
Lus Midhe Ealaidheach
**Creeping Forget-me-not**

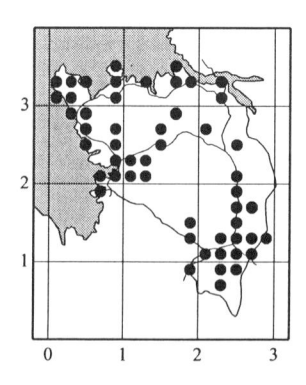

52 tetrads. Found in marshes, ditches and wet places generally. The commonest forget-me-not, probably because it is more tolerant of of acid conditions than the other two water forget-me-nots. First recorded in 1907, as '*M. repens*', at Inchnadamph by G.C.Druce.

*Myosotis laxa* Lehm.
(M. caespitosa Schultz)
Lus Midhe Dosach
**Tufted Forget-me-not**

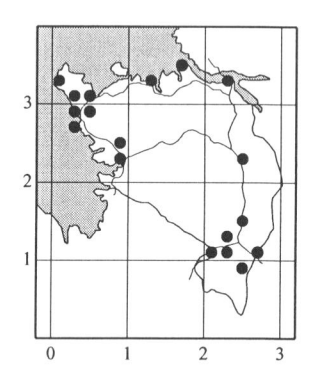

18 tetrads. Occasional in ditches and marshes. First recorded in 1943, as '*M. caespitosa*', at the 'north end of Balchladich Bay' by A.J.Wilmott and M.S.Campbell.

*Myosotis arvensis* (L.) Hill
Lus Midhe Aitich
**Field Forget-me-not**

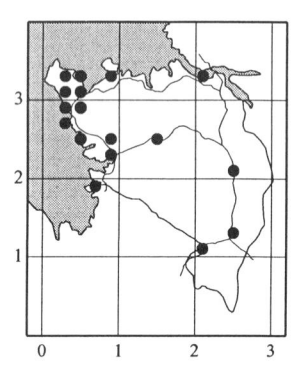

17 tetrads. Occasional in dry places on roadsides, in gardens and disturbed ground. First recorded in 1943 as a 'garden weed, Culag Hotel, Lochinver' by A.J.Wilmott and M.S.Campbell.

*Myosotis discolor* Pers.
Lus Midhe Caochlaideach
**Changing Forget-me-not**

20 tetrads. In a variety of habitats: verges, disturbed ground,

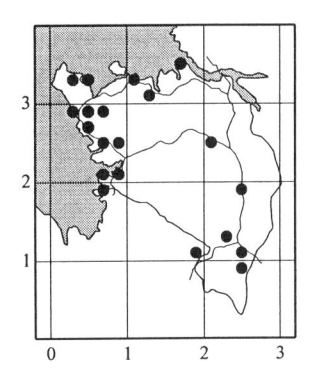

gardens and marshes. First recorded in 1894, as '*M. versicolor*', 'near Elphin and Ledbeg' by G.C.Druce.

LAMIACEAE

*Stachys sylvatica* L.
Lus nan Sgor
**Hedge Woundwort**

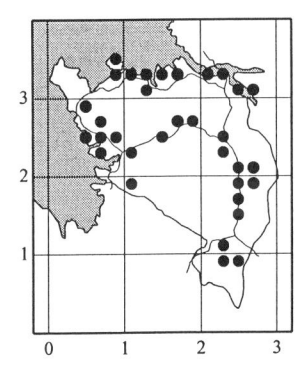

33 tetrads. Occasional in woodland, particularly alongside burns, on roadside verges and in boulder scree. First recorded during the 1950s survey for the first *Atlas*.

*Stachys* x *ambigua* Sm.
*S. sylvatica* x *S. palustris*
**Hybrid Woundwort**

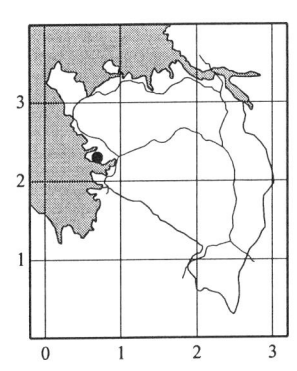

1 tetrad. The only record is of a stand growing beside a small burn at Ardroe. First recorded in NC02, during the 1950s survey for the first *Atlas*.

*Stachys palustris* L.
Brisgean nan Caorach
**Marsh Woundwort**

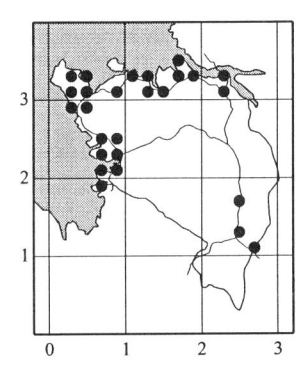

26 tetrads. Occasional in marshes, ditches, and alongside rivers, usually in larger stands than Hedge Woundwort. First recorded in 1894 in 'cornfields...at Elphin and Knockain' by G.C.Druce. There are also records in the 1950s from NC12 and in 1977 from NC22.

*Lamium album* L.
Teanga Mhìn
**White Dead-nettle**

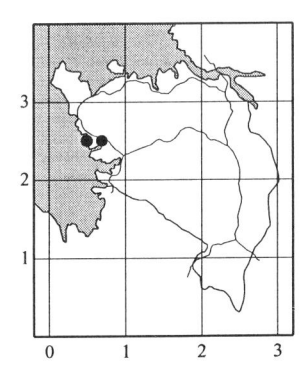

2 tetrads. Both records were in the Achmelvich area; one growing as a garden weed and the other amongst builder's rubble on a verge. Recorded from NC01 in the 1950s survey for the first *Atlas* (but much of this hectad is in West Ross).

*Lamium purpureum* L.
Caoch-dheanntag Dhearg
**Red Dead-nettle**

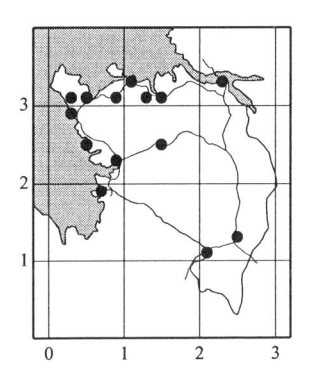

14 tetrads. A weed of cultivation, found mostly in gardens, but occasionally on disturbed ground along roadsides. First recorded during the 1950s survey for the first *Atlas*; also, in 1966, from Inchnadamph by A.P.Conolly.

*Lamium hybridum* Vill.
Caoch-dheanntag Gheàrrte
**Cut-leaved Dead-nettle**

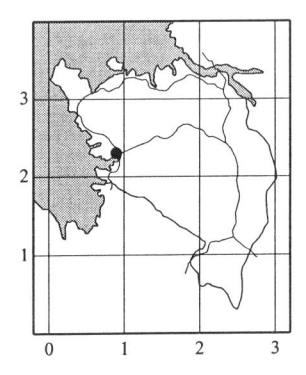

1 tetrad. A rare casual, found on a newly constructed soil bank in Lochinver. First recorded in 1943 as a 'garden weed, Culag Hotel, Lochinver' by A.J.Wilmott and M.S.Campbell.

*Lamium confertum* Fr.
(*L. molucellifolium* Fr.)
Caoch-dheanntag Thuathach
**Northern Dead-nettle**

1 tetrad. The only record was from a garden in Clachtoll. This weed of cultivation was probably much commoner in the past when arable land was more plentiful.

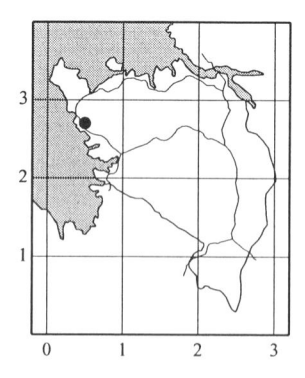

First recorded in 1890, as '*L. intermedium*', in 'cultivated ground, Lochinver' by E.S.Marshall and F.J.Hanbury; also in 1894, in 'cornfields…at Elphin and Knockain' by G.C.Druce. It was recorded in NC02 during the 1950s survey for the first *Atlas*.

*Lamium amplexicaule* L.
Caoch-dheanntag Chearc
**Henbit Dead-nettle**

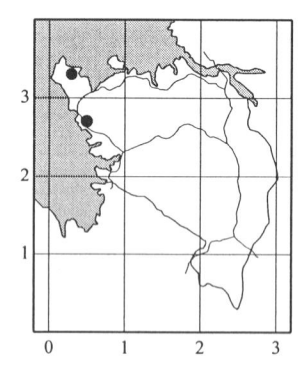

2 tetrads. In gardens at Clachtoll and Culkein Stoer. This is another example of a plant which would have been more frequent years ago when there was a greater area of cultivation. First recorded in NC02 during the 1950s survey for the first *Atlas*.

*Galeopsis tetrahit* L.
Deanntag Lìn
**Common Hemp-nettle**

10 tetrads. An uncommon weed of gardens and disturbed ground.

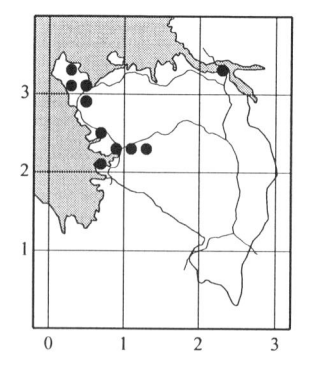

First recorded, *s.l.*, during the 1950s survey for the first *Atlas*; also in 1962 from NC22 by R.A.R.Clarke.

*Galeopsis bifida* Boenn.
**Bifid Hemp-nettle**

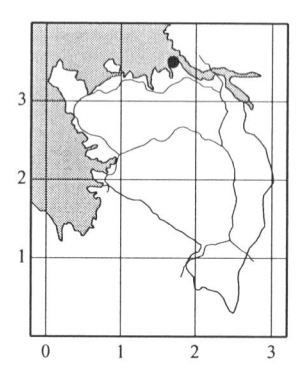

1 tetrad. The first and only record was made in 1995, from a garden at Kerrachar.

*Scutellaria galericulata* L.
Cochall
**Skullcap**

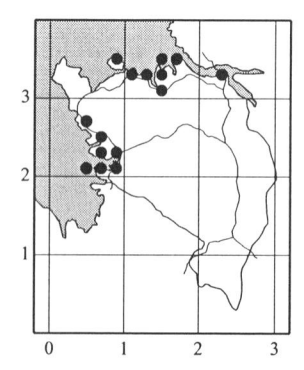

15 tetrads. The distribution is tied very closely to the coast. It may grow in the shingle at the back of the shore or in marshy ground amongst *Iris pseudacorus* immediately inland. First recorded in 1886 on a 'shingle beach, Inverkirkaig' by A.Gray; also in 1943 'amongst stones at the top of the beach, Clashnessie Bay, where the water comes down from a small marsh above', by A.J.Wilmott and M.S.Campbell.

*Teucrium scorodonia* L.
Sàisde Coille
**Wood Sage**

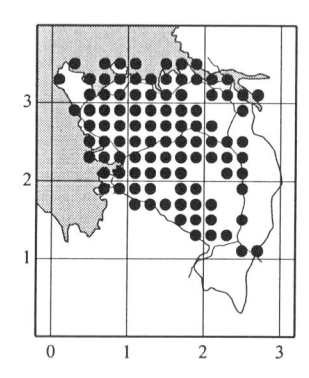

100 tetrads. Common and widespread in well-drained situations; most frequently on cliffs, outcrop rock and in rocky woodland. First recorded in 1943 on 'rocks south of Lochinver' and 'in a gully near Inverkirkaig' by A.J.Wilmott and M.S.Campbell.

*Ajuga reptans* L.
Glasair Choille
**Bugle**

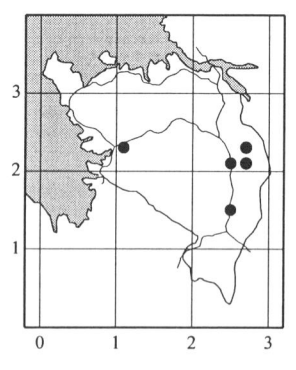

5 tetrads. Rare, on shady banks beside water. Found on the stony banks of the River Traligill just above the Inchnadamph Hotel and further upstream in wooded limestone gorges of both the main river, by G.P.Rothero, and its tributary Allt Poll an Droighinn, by A.E.White. The largest population was recorded on a wooded island in Loch Awe by R.E.C.Ferreira

Flowering plants and ferns

and A.Scott. The record from the wooded grounds of Glencanisp Lodge is probably an introduction. First recorded in 1767 near Inchnadamph by J.Robertson and seen there again in 1890 by E.S.Marshall.

*Ajuga pyramidalis* L.
Glasair Bheannach
**Pyramidal Bugle**

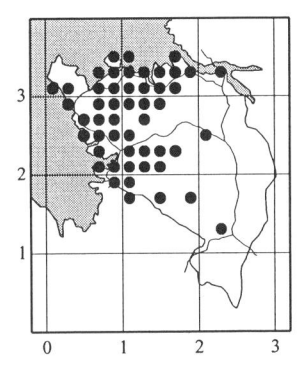

50 tetrads. Widespread at low altitudes in the western half of the parish. Once we were familiar with its vegetative characters, we found it to be surprisingly common on the gneiss; within the 50 tetrads, 142 sites were recorded. It grows on well-drained, sunny slopes, sometimes under a light covering of vegetation, usually *Calluna*, and in rock crevices. Hill ground which has been burned a year or two previously often carries a large number of rosettes. Only a small percentage of rosettes bear flowering spikes and the apparent anomaly of a small seed source and a large number of plants can be explained by the presence of fine roots, up to 20 cm. long, which run horizontally a few inches below the surface of the soil, producing new plantlets at their ends. First recorded in 1886 at Torbreck by A.Gray.

*Glechoma hederacea* L.
Eidheann Thalmhainn
**Ground-ivy**

3 tetrads. Found beneath a garden hedge at Glenleraig, on the edge of an old field at Rientraid and in the garden of a long-abandoned croft at Rhicarn. At all three places it was probably introduced at some time in the past, possibly as a substitute for hops! First recorded in 1988 at the last of these three sites.

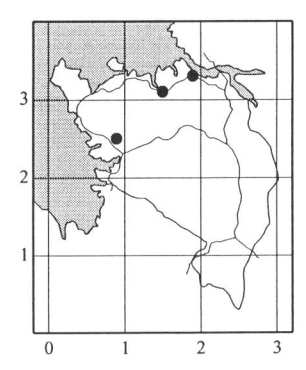

*Prunella vulgaris* L.
Dubhan Ceann-chòsach
**Selfheal**

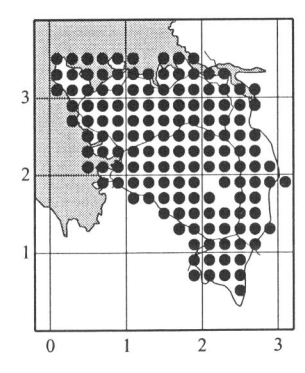

145 tetrads. Widespread and very common in grassland, woodland, heathland and on roadsides. Apart from some high level squares, only apparently absent from the notoriously barren tetrad, NC21I, on the Canisp quartzite. Both pink and white-flowered forms have been seen, as well as the familiar blue-violet. First recorded during the 1950s survey for the first *Atlas*.

*Thymus polytrichus* A. Kern. ex Borbás
(T. drucei Ronn.)
Lus an Rìgh
**Wild Thyme**

146 tetrads. Common and widespread in dry siuations. Grows on both coastal and inland grassland and on rock outcrops.

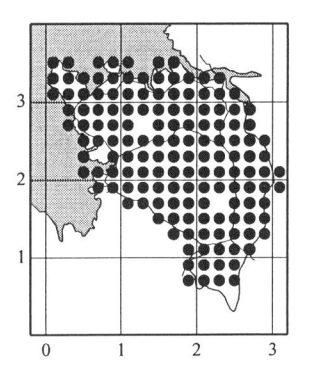

First recorded in 1767, as '*T. serpyllum*', near Inchnadamph by J.Robertson.

[*Thymus pulegioides* L.
Lus an Rìgh Mòr
**Large Thyme**

Reported in 1907 on Quinag by G.C.Druce; current knowledge of its distribution in Scotland makes this seem an unlikely site.]

*Lycopus europaeus* L.
Feòran Curraidh
**Gypsywort**

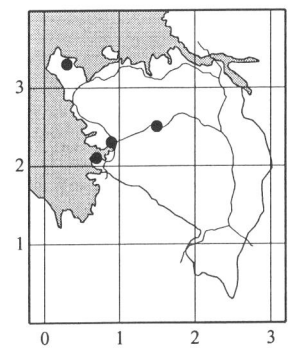

4 tetrads. Rare, usually coastal. In a garden at Little Assynt, in saltmarsh at Baddidarach and Badnaban and in a ditch leading to the sea at Culkein Stoer. First recorded in 1992 at the last of these four sites.

*Mentha arvensis* L.
Meannt an Arbhair
**Corn Mint**

2 tetrads. Rare. Found beside Loch Culag and at Culkein Stoer; in both instances in marshy grassland.

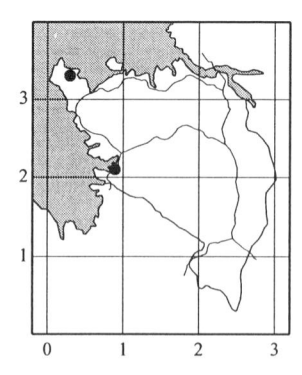

First recorded in NC21 during the 1950s survey for the first *Atlas*, and in the same hectad in 1974 by U.K.Duncan.

*Mentha* x *gracilis* Sole
*M. arvensis* x M. *spicata*
(Mentha x gentilis auct. non L.)
Meannt Tomach
**Bushy Mint**

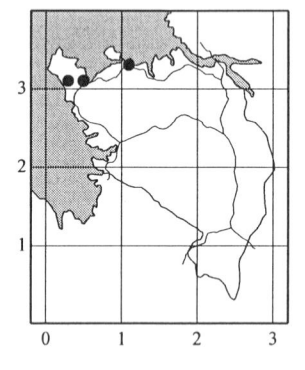

3 tetrads. All the records are garden escapes of the variegated form. There is a well established population in the reedswamp of Loch na Claise, one in the vicinity of an old house at Clashnessie and one in disturbed ground at Culkein Drumbeg. First recorded in 1943, as '*M. gentilis* L. var. *variegata* (Sole) Sm.', 'escaped among houses at Stoer, and cultivated in gardens at Balchladich', by A.J.Wilmott and M.S.Campbell.

*Mentha aquatica* L.
Meannt an Uisge
**Water Mint**

7 tetrads. Found in marshy grassland and on river margins. Considering the amount of wet ground that exists in the area, this species is surprisingly uncommon.

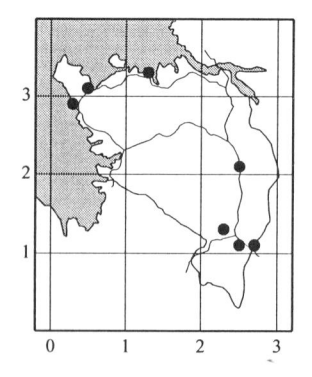

First recorded in 1886 as '*M. hirsuta*', 'Loch Assynt-Loch Awe' by A.Gray.

*Mentha* x *piperita* L.
*M. aquatica* x *M. spicata*
Meannt a'Phiobair
**Peppermint**

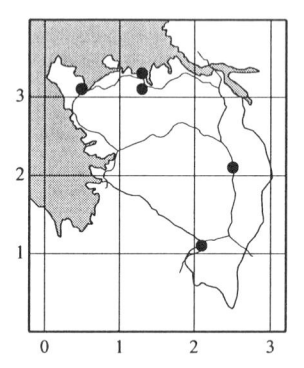

5 tetrads. Found in ditches at Nedd and Inchnadamph, river margins at Clashnessie and Elphin and a loch at Drumbeg. First recorded in 1943 by A.J.Wilmott and M.S.Campbell, who noted that it was 'given us at Balchladich as the culinary mint used there'. Further records were made in NC02 in the 1950s and 1960s.

*Mentha spicata* L.
Meannt Gàrraidh
**Spear Mint**

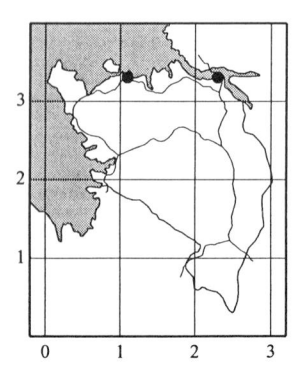

2 tetrads. Garden escapes, in the vicinity of houses at Culkein Drumbeg and on a verge at Unapool. First recorded in 1886 as '*M. viridis*' at Lochinver by A.Gray.

*Mentha* x *villosonervata* Opiz
*M. spicata* x *M. longifolia* (L.) Huds
**Sharp-toothed Mint**

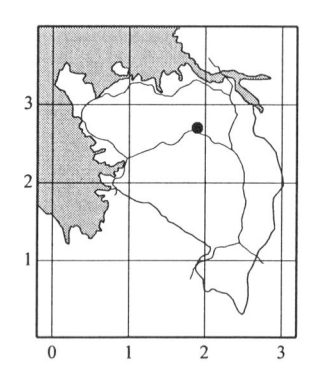

1 tetrad. The first and only record was made in 1992, from the edge of Loch Assynt near Tumore.

*Mentha* x *villosa* Huds.
*M. spicata* x *M. suaveolens* Ehrh.
(Mentha x cordifolia Opiz)
Meanntas
**Apple-mint**

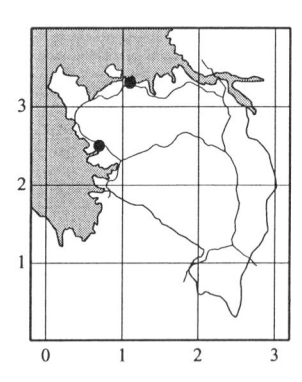

2 tetrads. Two records of garden escapes, in disturbed ground at Culkein Drumbeg and Achmelvich. First recorded in 1995 from the second of these sites.

*Mentha* x *rotundifolia* (L.) Huds.
*M. longifolia* (L.) Huds. x *M. suaveolens* Ehrh.
(*Mentha* x *niliaca* Juss.ex Jacq.)
**False Apple-mint**

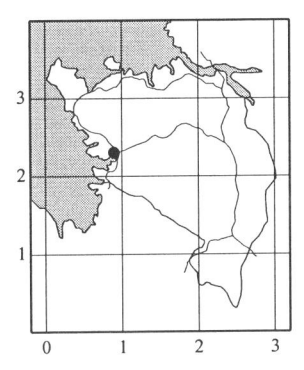

1 tetrad. Found on waste ground at Lochinver in 1994. The record in *Anthony* of this taxon, as '*M.* x *niliaca*' Juss.ex Jacq., states 'Lochinver, 1944, A.J.W'., but this does not appear to relate to any of the entries under *Mentha* in Wilmott and Campbell (1946). The record for NC02 under *M.* x *niliaca* var. *webberi* in the *Critical Supplement* may derive from this.

HIPPURIDACEAE

*Hippuris vulgaris* L.
Earball Capaill
**Mare's-tail**

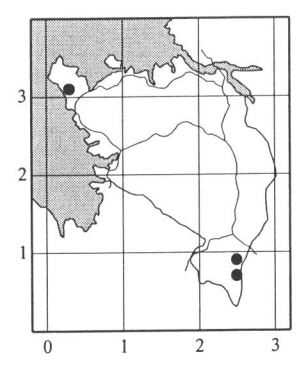

3 tetrads. A rare aquatic. It is in or near three of the burns running into the southern end of Loch Urigill: along the Allt nam Meur in a number of places, on an unnamed burn to the west of that, and in a pool beside the lower reaches of the Allt an Achaidh. It also grows in a small, sheltered bay at that end of the loch. Elsewhere, only known from Loch na Claise, at the mouth of a small burn flowing into its northern shore at the western end. First recorded in 1977 at Loch Urigill by D.H.N.Spence and E.D.Allen.

CALLITRICHACEAE

*Callitriche stagnalis* Scop.
Biolair Ioc
**Common Water-starwort**

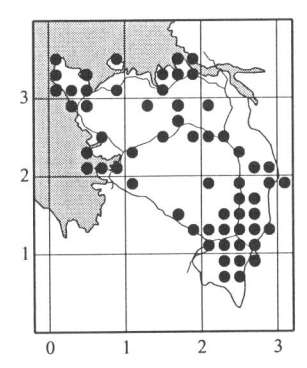

59 tetrads. A common plant in its preferred habitat of slow-moving, shallow water and wet mud. The first record was made in 1990 from Achnacarnin, although in 1767 J.Robertson identified *C. stagnalis* s.l. near Inchnadamph.

*Callitriche platycarpa* Kütz.
Biolair Ioc Leathann
**Various-leaved Water-starwort**

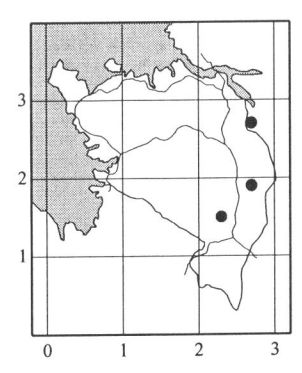

3 tetrads. Found in the burn running into Loch Mhaolach-coire from the south, in the exit burn from Lochan a'Choire Ghuirm and in Loch a'Chroisg. First recorded in 1894 in the 'river by falls...Ledbeg' by G.C.Druce.

*Callitriche hamulata* Kütz. ex W.D.J.Koch
Biolair Ioc Meadhanach
**Intermediate Water-starwort**

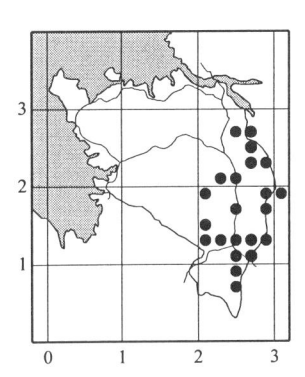

22 tetrads. The distribution of this normally calcifuge species is puzzling in Assynt. It appears to be absent from the gneiss, and is concentrated in the east, where at least some of the sites are base-rich. It is characteristic of high-level lochs and the burns that feed them and can grow most luxuriantly in deep cushions in these situations, which are most frequent in the east. In view of the difficulty of separating this species from *C. brutia* Petagna, it is wise to treat our records as representing *C. hamulata* s.l. (see Preston and Croft 1997). First recorded in 1908 in the River Loanan at Inchnadamph by E.S.Marshall.

PLANTAGINACEAE

*Plantago coronopus* L.
Adharc-fèidh
**Buck's-horn Plantain**

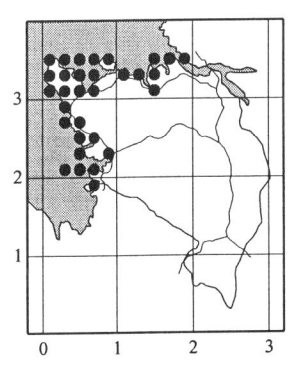

31 tetrads. Found only on the coast; on rocks, saltmarsh and

other very short turf near the sea. Recorded in 1767 by J.Robertson, supposedly near Inchnadamph (see chapter on history of recording), but in the light of its distribution today this seems unlikely. If so, the first record is 1943, at Lochinver and Balchladich, by A.J.Wilmott and M.S.Campbell.

*Plantago maritima* L.
Slàn-lus na Mara
**Sea Plantain**

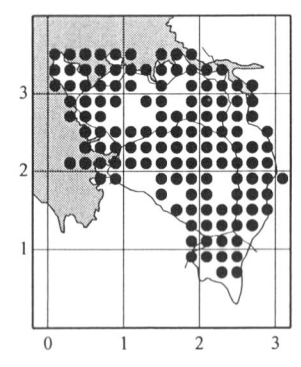

131 tetrads. A misleading name in Assynt, as the map shows. As well as rock, saltmarsh and short turf by the sea, this plantain is widespread inland, on roadsides and in stony and rocky places in the hills. First recorded in 1767 near Inchnadamph by J.Robertson.

*Plantago major* L.
Cuach Phàdraig
**Greater Plantain**

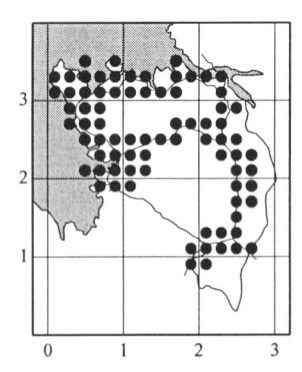

78 tetrads. Confined to roadsides and waste places near buildings. It may grow for a short distance along tracks running into the hills, presumably carried on the feet of people or animals. First recorded during the 1950s survey for the first *Atlas*.

*Plantago lanceolata* L.
Slàn-lus
**Ribwort Plantain**

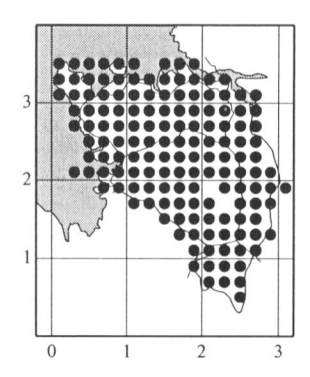

147 tetrads. Widespread and very common in grassland of all types. First recorded in 1767 near Inchnadamph by J.Robertson.

*Littorella uniflora* (L.) Asch.
Lus Bòrd an Locha
**Shoreweed**

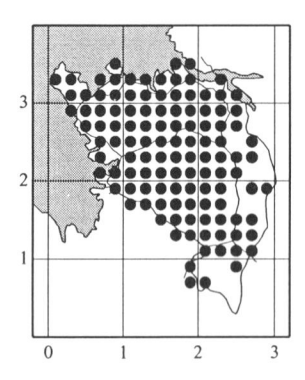

116 tetrads. Common and widespread in all lochs which have suitably shallow water around the margins. Also found in slow-moving rivers and burns. First recorded during the 1950s survey for the first *Atlas*.

OLEACEAE

*Fraxinus excelsior* L.
Uinnseann
**Ash**

19 tetrads. The status of Ash trees in Assynt is often difficult to determine. In some places they are clearly planted, well established and reproducing successfully from seed.

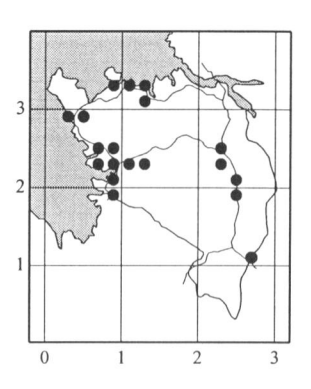

In others the trees may be descendants of native stock, or of trees introduced so long ago that no trace of them now remains. First recorded in 1774 at '*Knockneach*' by J.Home.

SCROPHULARIACEAE

*Verbascum thapsus* L.
Coinneah Moire
**Great Mullein**

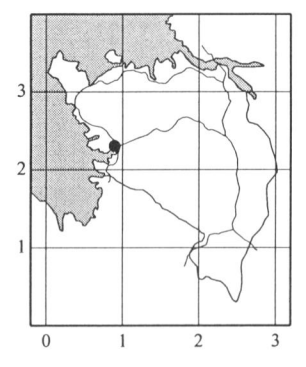

1 tetrad. A rare casual. The first record was in 1996 from waste ground at Lochinver; also on the imported soil of a new roadside verge at Baddidarach in 2001.

*Scrophularia nodosa* L.
Lus nan Cnapan
**Common Figwort**

18 tetrads. Occasional in light woodland, scrub, ditches and boulder scree; also a frequent weed around Lochinver.

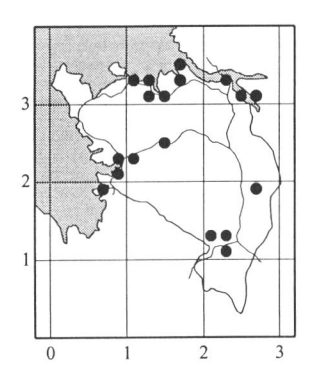

First recorded in 1886 as 'Loch Assynt – Loch Inver' by A.Gray.

*Mimulus guttatus* DC.
Meilleag an Uillt
**Monkeyflower**

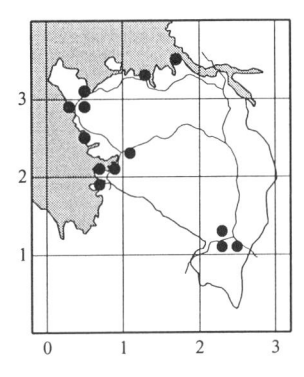

13 tetrads. Introduced but well established on a number of burns, ditches and river margins. First recorded in 1988 from Clashnessie. This may be the species referred to by E.S.Marshall in 1887 as '*M. luteus* L....well established at Inchnadamph'.

*Mimulus* x *robertsii* Silverside
*M. guttatus* x *M. luteus* L.
**Hybrid Monkeyflower**

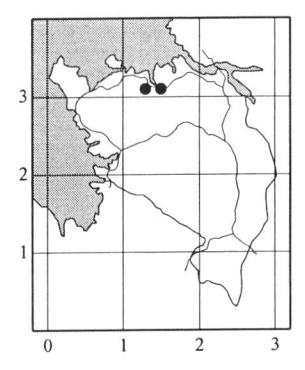

2 tetrads. Established in burns and ditches at Nedd and Glenleraig. First collected in 1961 at Lochinver by V.Gordon, whose material was determined by R.H.Roberts.

*Mimulus* x *burnetii* S.Arn.
*M. guttatus* x *M. cupreus* Dombrain
**Coppery Monkeyflower**

3 tetrads. In ditches and burns at Badnaban, Culkein Stoer and Inverkirkaig, near gardens from which it has presumably escaped. First collected in 1958 at Clashnessie by M.M.George and determined as this hybrid. Material from Inverkirkaig collected by V.Gordon was thought by R.H.Roberts to be of the same parentage.

*Mimulus* x *maculosus* T.Moore
*M. luteus* L. x *M. cupreus* Dombrain
**Scottish Monkeyflower**

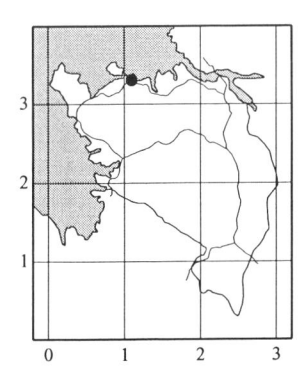

1 tetrad. The first and only record was made in 1999 at Culkein Stoer, where the plant was growing in a ditch.

*Digitalis purpurea* L.
Lus nam Ban-sìdh
**Foxglove**

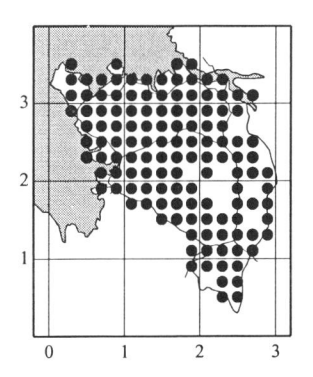

128 tetrads. Widespread and common in well-drained, sheltered, but not shaded, places. Banks of watercourses, scree, heaths, very light woodland, disturbed ground and roadsides. First recorded during the 1950s survey for the first *Atlas*.

*Erinus alpinus* L.
Meuran Sìdh
**Fairy Foxglove**

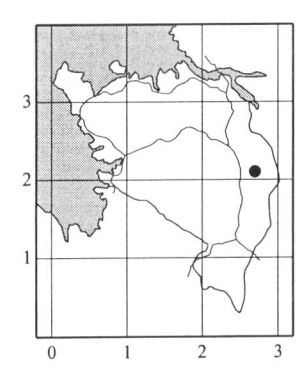

1 tetrad. Introduced and maintaining itself in the Inchnadamph 'zoo', on a limestone outcrop at Glac Mhór. (See also *Campanula, Gentiana, Phyteuma, Silene*). First recorded in 1992 at this site.

*Veronica serpyllifolia* L.
Lus-crè Talmhainn
**Thyme-leaved Speedwell**

ssp. *serpyllifolia*

79 tetrads. Common in the hills in grassland, as a garden weed and in waste ground.

129

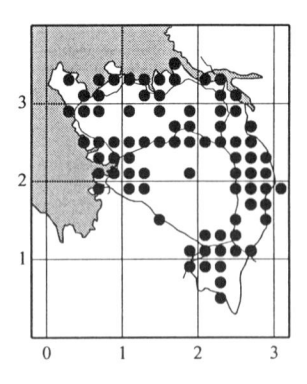

First recorded during the 1950s survey for the first *Atlas*.

ssp. *humifusa* (Dicks.) Syme

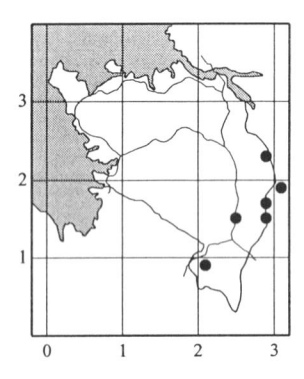

6 tetrads. Rare. In wet places in the hills, particularly in bryophyte-dominated flushes. First recorded in 1886 on Conival by A.Gray; also in 1907 on Canisp by G.C.Druce.

*Veronica officinalis* L.
Lus-crè Monaidh
**Heath Speedwell**

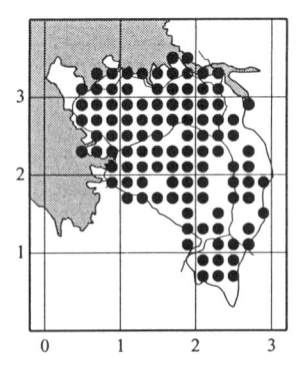

105 tetrads. Widespread in dry heath, well-drained grassland and rock crevices. First recorded in 1943 'in a gully near Inverkirkaig' by A.J.Wilmott and M.S.Campbell.

*Veronica chamaedrys* L.
Nuallach
**Germander Speedwell**

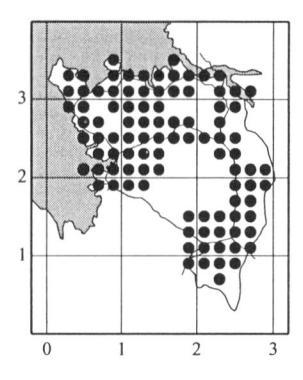

96 tetrads. Common in light woodland and lowland grassland, both basic and acidic, often growing beneath bracken. First recorded in 1767 near Inchnadamph by J.Robertson.

*Veronica scutellata* L.
Lus-crè Lèana
**Marsh Speedwell**

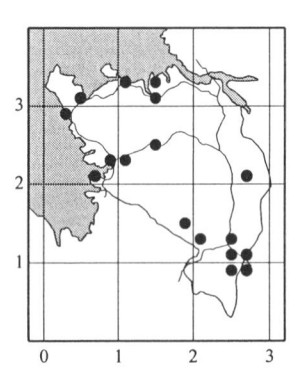

17 tetrads. Occasional in marshes, ditches and on river margins. First recorded in 1949 in a 'bog by Loch Culag' by M.McC.Webster.

*Veronica beccabunga* L.
Lochal Mothair
**Brooklime**

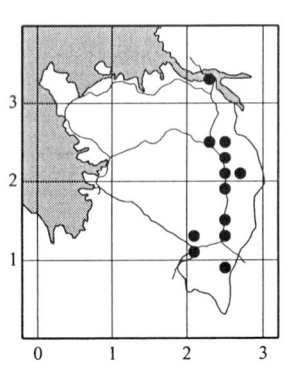

12 tetrads. Uncommon. In ditches, or rarely on the edge of burns, mainly in limestone areas. Shows a definite preference for the vicinity of houses, where adjacent ditches and watercourses may benefit from some 'enrichment'! First recorded during the 1950s survey for the first *Atlas* from NC21, 22 and 23.

*Veronica anagallis-aquatica* L.
Fualachdar
**Blue Water-speedwell**

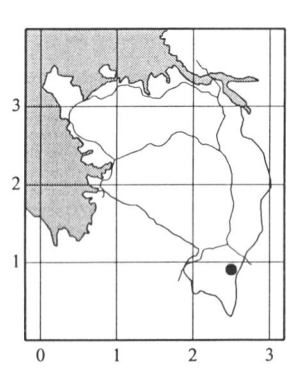

1 tetrad. In a deep peaty ditch running into the south-eastern end of Loch Urigill. First recorded in 1886 as '*V. anagallis*' in the 'Cromalt Hills' by A.Gray. Record queried by Anthony and the species was not shown in the Assynt area in the first *Atlas*, but Gray's record has now been vindicated.

*Veronica arvensis* L.
Lus-crè Balla
**Wall Speedwell**

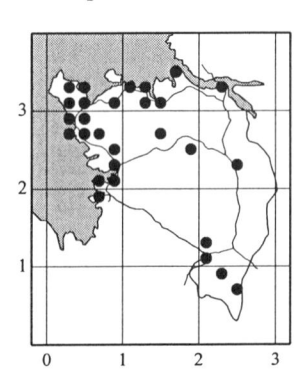

28 tetrads. Occasional in open, dry habitats such as walls, disturbed and cultivated ground and occasionally on rocks. First

recorded during the 1950s survey for the first *Atlas*.

*Veronica agrestis* L.
Lus-crè Arbhair
**Green Field-speedwell**

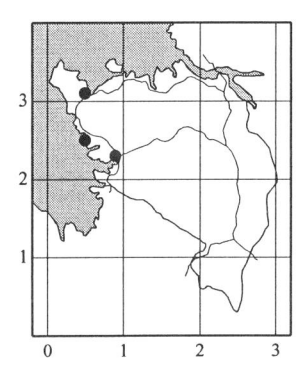

3 tetrads. Disturbed ground at Achmelvich, Clashnessie and Lochinver. First recorded in 1894 in 'cornfields at Elphin and Knockain' by G.C.Druce; also in 1966 at Inchnadamph by A.P.Conolly.

*Veronica persica* Poir.
Lus-crè Gàrraidh
**Common Field-Speedwell**

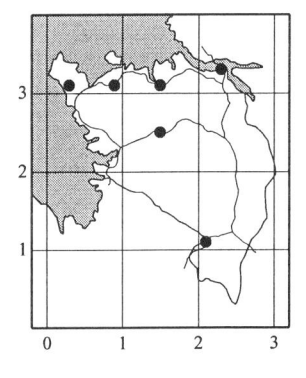

6 tetrads. A weed of cultivation and on disturbed ground along roadsides. First recorded in 1943 in 'Culag Hotel garden, Lochinver' by A.J.Wilmott and M.S. Campbell.

*Veronica filiformis* Sm.
Lus-crè Claidh
**Slender Speedwell**

9 tetrads. Introduced but spreading, in short grass associated with gardens and other mown areas.

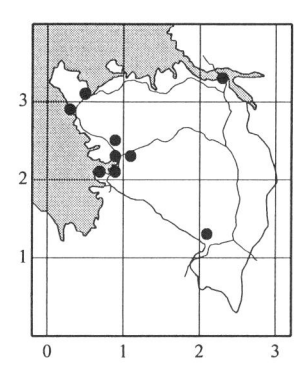

First recorded in 1957 on a 'grassy roadside, Lochinver' by R.A.Graham and R.M.Harley.

*Veronica hederifolia* L.
Lus-crè Eidheannach
**Ivy-leaved Speedwell**

Not seen during the present survey. First recorded in 1886 at Elphin by A.Gray.

*Melampyrum pratense* L.
Càraid Bhuidhe
**Common Cow-wheat**

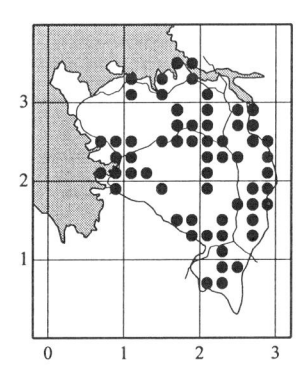

60 tetrads. Scattered on dry heaths and in open woodland. Occasional plants were noted with very striking dark red-purple marks on the corolla. First recorded in 1887 on Loch Assynt by E.S.Marshall.

*Euphrasia officinalis* agg.
Lus nan Leac
**Eyebright**

We have recorded *Euphrasia* spp. from 115 of Assynt's 164 tetrads, so the genus is clearly widespread, although the number of species is not great. With initial help from A.J.Silverside, nine species have been recorded and mapped, but the hybrids, which are legion, were beyond our capabilities!

*Euphrasia arctica* Lange ex Rostrup
ssp. *borealis* (F.Towns.) Yeo

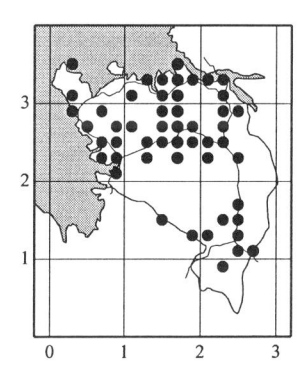

51 tetrads. The commonest eyebright in grassy places and particularly common along roadsides. First recorded as '*E. brevipila*' in 1899 in a meadow at Inchnadamph by C.E.Salmon, and under its current name by Anthony in the 1950s.

*Euphrasia nemorosa* (Pers.) Wallr.

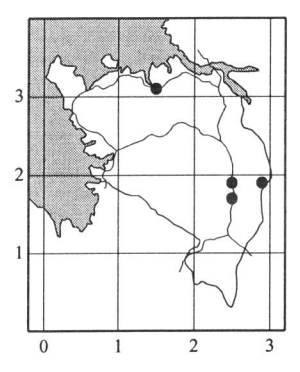

4 tetrads. One coastal record at Glenleraig and three from basic grassland at Stronechrubie, banks of the River Loanan and the Allt a'Bhealaich. First recorded in 1998 from the last of these sites. No historic record has been found which can safely be attributed to this species.

*Euphrasia confusa* Pugsley

11 tetrads. Grassland, often basic, in the hills. First recorded in 1943 on a 'roadside bank north of Stoer'

131

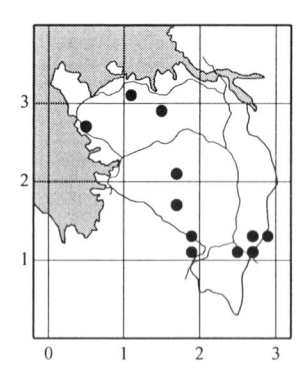

by A.J.Wilmott and M.S.Campbell (det. H.W.Pugsley). Other validated records were made from the Inchnadamph area and at Clachtoll during the 1950s survey for the first *Atlas*.

*Euphrasia frigida* Pugsley

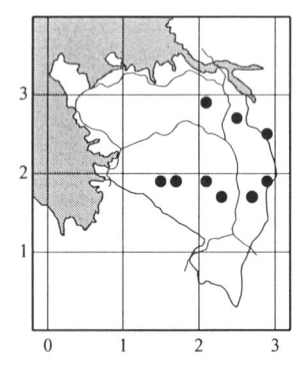

9 tetrads. A hill species, from Bealach Traligill, Beinn an Fhuarain, Canisp, Corrag Ghorm, Quinag, Suilven and Glas Bheinn (where it was recorded on the summit at 746m.). First recorded in 1956, from two localities at the southern end of Breabag by D.N.McVean.

*Euphrasia foulaensis* F.Towns. ex Wettst.

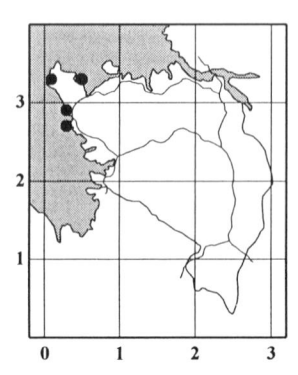

4 tetrads. Short coastal turf at Clachtoll, Culkein Stoer, Stoer Lighthouse and Stoer village. First recorded in 1953 at Clachtoll by M.McC.Webster.

*Euphrasia ostenfeldii* (Pugsley) Yeo

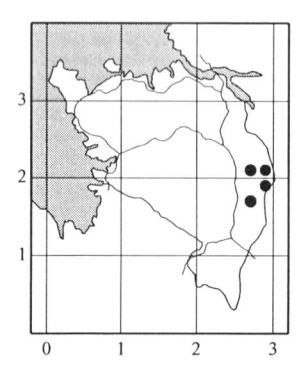

4 tetrads. On open, gravelly soils on the limestone, alongside the River Traligill and the Allt nan Uamh, and in the Bealach Traligill. First recorded in 1908 as 'a dwarfed form of '*E. curta*...on limestone in the valley of the Allt nan Uamh' by E.S.Marshall and F.J.Hanbury and subsequently, in 1956, by G.Halliday in NC21.

*Euphrasia marshallii* Pugsley

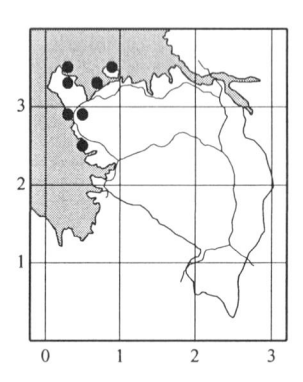

7 tetrads. A Red Data Book species, growing in short turf by the sea. First recorded in 1953 on 'basic rocks' at Stoer by U.K.Duncan.

*Euphrasia micrantha* Rchb.

76 tetrads. The most widespread eyebright in dry habitats in the hills, usually associated with *Calluna*.

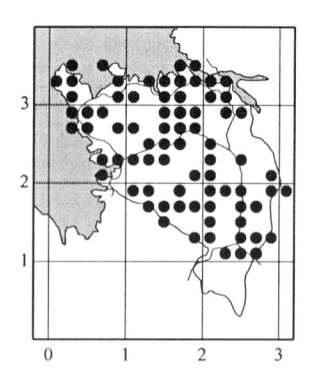

First recorded in 1899 at Inchnadamph by C.E.Salmon; also in 1943 'on a slope south of Lochinver' by A.J.Wilmott and M.S.Campbell.

*Euphrasia scottica* Wettst.

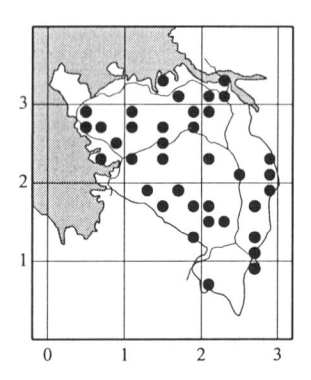

37 tetrads. Frequent in damp flushes. First recorded in 1899 at Inchnadamph by C.E.Salmon.

An as yet unpublished taxon, provisionally called *E. fharaidensis* by A.J.Silverside, was recorded in 1998 from the back of the saltmarsh at Loch Ardbhair, NC13R, (conf. A.J.S.).

Although no consistent effort was made to collect material of suspected hybrids, specimens of the following were identified by A.J.Silverside:
*Euphrasia arctica borealis* x *micrantha* from NC13B.
*Euphrasia arctica borealis* x *scottica* from NC11N and 22L.
*Euphrasia foulaensis* x *micrantha* from NC13B.

The records of hybrids from Assynt recorded in *Anthony* are difficult to evaluate in terms of current nomenclature.

*Odontites vernus* (Bellardi) Dumort.
Modhalan Coitcheann
**Red Bartsia**

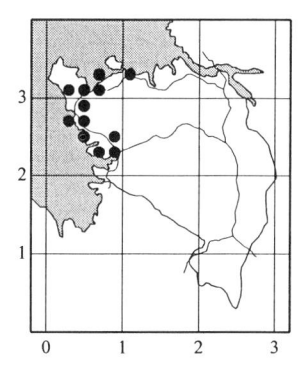

12 tetrads. Grassland, disturbed ground and roadsides, in coastal areas. First recorded in 1890 at Inchnadamph by E.S.Marshall and F.J.Hanbury; from NC22 by K.M.Goodway et al. in 1953 and by B.Burrow in 1977.

*Rhinanthus minor* L.
Modhalan Buidhe
**Yellow-rattle**

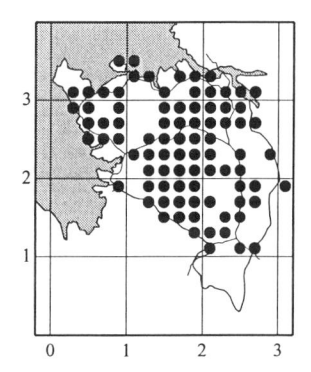

85 tetrads. Widespread in grassy places, but in small quantities. Not seen growing abundantly, as in southern haymeadows. In a very large number of cases it was not possible to identify to subspecies because the top of the main stem had been bitten off by sheep or deer. Five subspecies were recorded and mapped, but we have not emerged with any very clear idea of their habitat preferences. First recorded as '*R. crista-galli*' in 1767 near Inchnadamph by J.Robertson. A variety of montane and lowland forms were reported from 1899 onwards, and sspp. *stenophyllus*, *monticola* and *borealis* are recorded for Assynt in *Anthony*.

ssp. *minor*

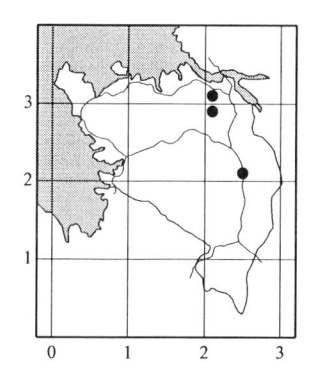

3 tetrads. In two places on Quinag and at Stronechrubie. Also recorded in NC02, 03 and 12 in the *Critical Supplement*.

ssp. *stenophyllus* (Schur) O.Schwartz

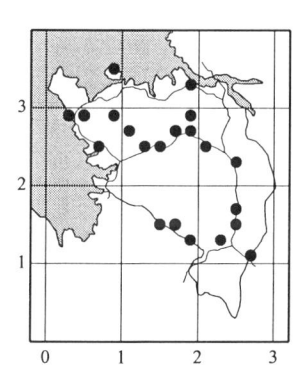

21 tetrads. The most widespread of the subspecies identified and the most consistently lowland.

ssp. *monticola* (Sterneck) O.Schwartz

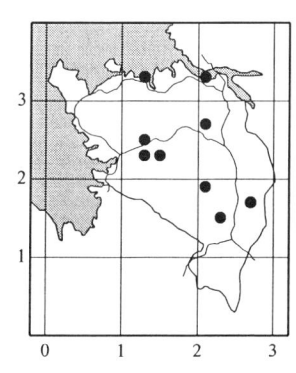

9 tetrads. In both acidic and basic grassland, including roadsides. Also recorded in NC02 and 03 in the *Critical Supplement*.

ssp. *lintonii* (Wilmott) P.D.Sell

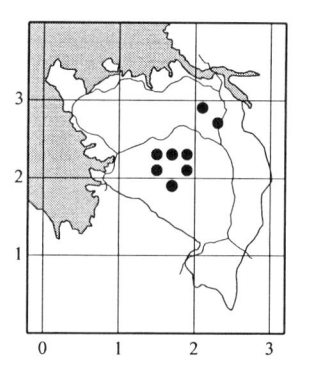

8 tetrads. In grassy places quite high in the hills.

ssp. *borealis* (Sterneck) P.D.Sell

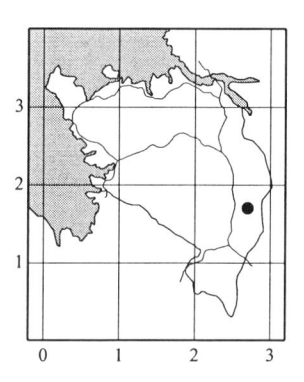

1 tetrad. In basic grassland by the Allt nan Uamh. Also recorded in NC11, 12 and 22 in *Critical Supplement*.

*Pedicularis palustris* L.
Lus Riabhach
**Marsh Lousewort**

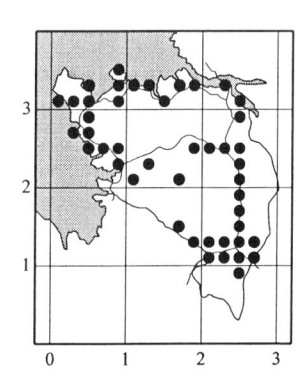

45 tetrads. Frequent in lowland wet places. The dots on the map illustrate well the course of the river valleys, and therefore the position of the marshes. First recorded in 1767 near Inchnadamph by J. Robertson.

133

*Pedicularis sylvatica* L.
Lus Riabhach Monaidh
**Lousewort**

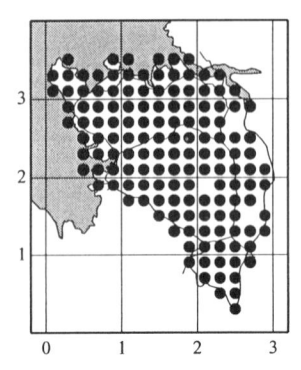

142 tetrads. Widespread and common on heaths and acid grassland. Usually in drier places than the preceding species. First recorded in 1767 near Inchnadamph by J.Robertson.

LENTIBULARIACEAE

*Pinguicula lusitanica* L.
Mòthan Beag Bàn
**Pale Butterwort**

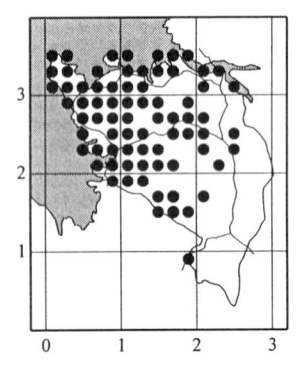

74 tetrads. Common in bogs and wet heaths on the western side of the parish. Sometimes in quite bare stony flushes and demanding wetter conditions than *P. vulgaris*. First recorded in 1886 'Lochinver – Beinn Garbh' by A.Gray.

[*Pinguicula alpina* L.
**Alpine Butterwort**

E.S.Marshall (1885) notes that 'In the late summer of 1884 Mr.W.J.Ball of Harrow gathered *P. alpina* near the sea at no great elevation in the Lochinver district of Sutherland (a new record).' This record does not seem to have been queried later, but could perhaps have referred to an aberrant form of *P. lusitanica*.]

*Pinguicula vulgaris* L.
Mòthan
**Common Butterwort**

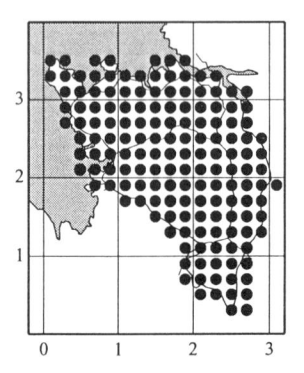

156 tetrads. Widespread and common on wet heaths. First recorded in 1767 near Inchnadamph by J.Robertson.

*Utricularia* spp.
**Bladderworts**

Flowering material occurs only very rarely in this part of Scotland and all species have therefore been determined by microscopic examination of the quadrifid hairs. Bladderworts in Assynt fall into three 'groups', which are vegetatively distinct. *U. vulgaris* and *australis* are splits of the old *U. vulgaris,* but in the absence of flowers they are indistinguishable from one another; on geographical grounds therefore, they have all been mapped as *U. australis,* as also have those recorded as '*U. vulgaris* s.l.' in the Scottish Loch Survey of 1988. What was formerly known as *U. intermedia* is now considered to consist of three species: *U. intermedia* s.s., *U. stygia* and *U. ochroleuca. U. intermedia* s.l. was first recorded in 1886 in Loch Awe by A.Gray. *U. intermedia* s.s. has never been recorded from Assynt. The third 'group' is *Utricularia minor* only (thank goodness).

*Utricularia australis* R.Br.
Lus nam Balgan
**Bladderwort**

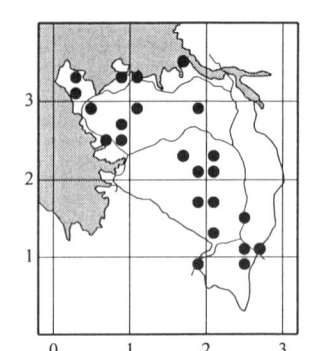

23 tetrads. Occasional in shallow peaty pools and small, sheltered, loch inlets. We have never been seen it in flower. First recorded in 1938, as *U. neglecta*, at Little Assynt by P.M.Hall.

*Utricularia stygia* G.Thor
Lus nam Balgan Lochlannach
**Nordic Bladderwort**

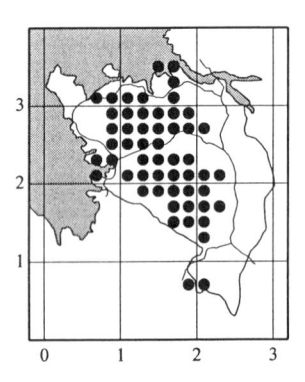

54 tetrads. Widespread, particularly on the gneiss, in shallow, still water, able to survive on wet mud during periods of temporary low water. We have never seen it in flower. First recorded in 1992 from Camas nam Bad.

*Utricularia ochroleuca* R.W.Hartm.
Lus nam Balgan Bàn
**Pale Bladderwort**

3 tetrads. In habitats similar to the other species, but rarely seen.

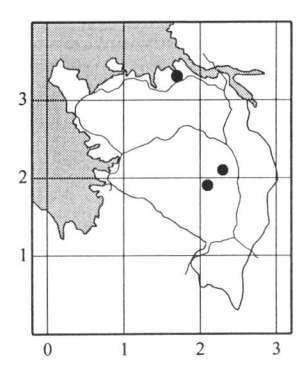

Flowers, as usual, absent. First recorded in 1993 at Leathad Lianach.

*Utricularia minor* L.
Lus nam Balgan Beag
**Lesser Bladderwort**

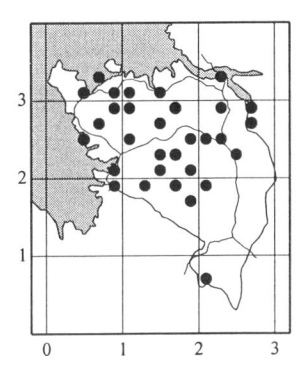

31 tetrads. Scattered apparently randomly throughout the parish, mostly, although not entirely, away from the limestone areas. We have once seen this species in flower. Recorded in 1883 by R.Graham, 'found in flower only once in a small pool near the base of 'Speckanconich, Assynt, by Mr Parnell'. We have been unable to trace a locality of this name in Assynt, but there is one on Ben Mór Coigach in Wester Ross. The first definite record is from 'Achumore' in 1886 by A.Gray.


CAMPANULACEAE

[*Campanula latifolia* L.
Guc Mòr
**Giant Bellflower**

Not seen during the present survey. Recorded as an introduction in 1953 by V.Gordon from NC22.]

*Campanula rotundifolia* L.
Currac Cuthaige
**Harebell**

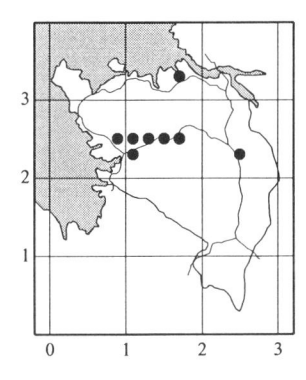

8 tetrads. The 'bluebell of Scotland' is conspicuous by its absence from most of the parish. Most of the records are from the verges of new sections of the road between Skiag Bridge and Lochinver, which underwent major re-construction and re-alignment in the 1970s. Exceptions are at a point near Calda House, where harebells are growing away from the road on limestone grassland, and at Inveruplan, on a disused section of the road, cut off during the roadworks. The most interesting case, which shows that the species is not really suited by conditions here, is a long-disused, small, roadside quarry at Ardvar, where one or two plants have appeared and flowered every year for about 20 years without increasing. First recorded in 1767 at Inchnadamph by J.Robertson. Recorded during the 1950s survey for the first *Atlas* from NC03 and 21.

*Campanula cochleariifolia* Lam.
**Fairy's-thimble**

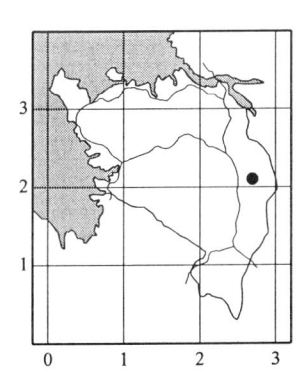

1 tetrad. Introduced and maintaining itself in the Inchnadamph 'zoo', on a limestone outcrop at Glac Mhór. (See also *Erinus*, *Gentiana*, *Phyteuma* and *Silene*). First recorded in 1992 at this site.

*Phyteuma scheuchzeri* All.
**Oxford Rampion**

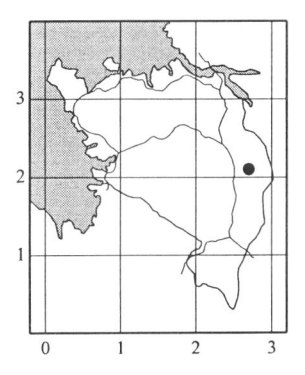

1 tetrad. A highly successful member of the Inchnadamph 'zoo', where it was introduced by persons unknown at some time in the past. It thrives in fissures of a limestone outcrop at Glac Mhór. (See also *Campanula*, *Erinus*, *Gentiana* and *Silene*). As it is no longer found on the wall of the Oxford college, this is now probably its only station in the British Isles. First recorded by G.M.Richards and A.E.White in 1992.

*Lobelia dortmanna* L.
Flùr an Lochain
**Water Lobelia**

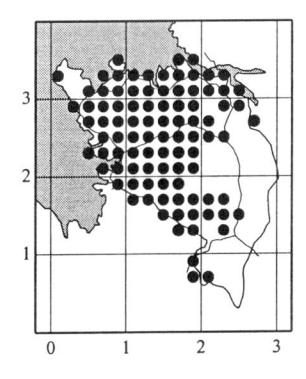

93 tetrads. A calcifuge, widespread and common except in the east of the parish where most of the lochs are base-rich. Grows around the margins of lochs where

the substrate is peat, sand or fine stones. The second most common, after *Littorella*, of the rosette aquatics. Characteristic of our most nutrient-poor lochs. First recorded during the 1950s survey for the first *Atlas*.

RUBIACEAE

*Sherardia arvensis* L.
Màdar na Machrach
**Field Madder**

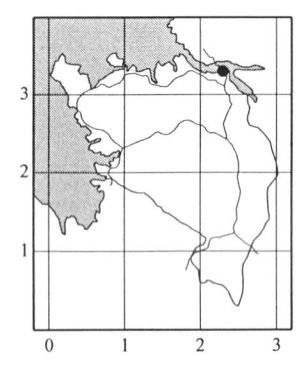

1 tetrad. One plant on the edge of a small carpark at Unapool. First recorded in the 1950s survey for the first *Atlas* from NC03.

*Galium boreale* L.
Màdar Cruaidh
**Northern Bedstraw**

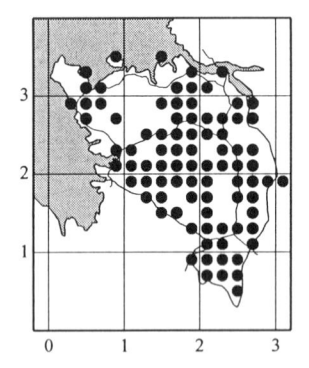

86 tetrads. Frequent on rocky stream banks, dry heath and cliff ledges. First recorded in 1943 from 'rocks in the River Inver, Lochinver' by A.J.Wilmott and M.S.Campbell.

*Galium odoratum* (L.) Scop.
Lus a'Chaitheimh
**Woodruff**

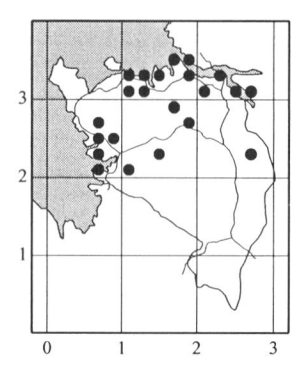

22 tetrads. Occasional in woodlands and boulder scree; sometimes growing in very dry conditions on crags. First recorded in 1886 at '*An Coilean*' by A.Gray.

[*Galium uliginosum* L.
Màdar Uaine
**Fen Bedstraw**

Unlikely, but reported in 1886 by A.Gray at Lochinver.]

*Galium palustre* L.
Mèdar Lèana
**Common Marsh Bedstraw**

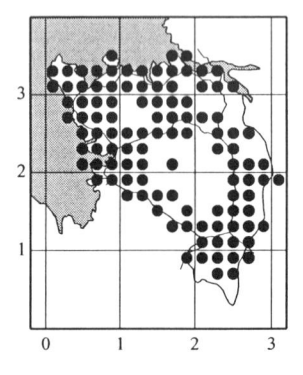

108 tetrads. Widespread and common in marshes, ditches, wet woodland and on river banks. First definitely recorded during the 1950s survey for the first *Atlas*. Subspecies *elongatum* was recorded in 1969 from NC02.

*Galium verum* L.
Lus an Leisaich
**Lady's Bedstraw**

29 tetrads. Occasional in basic grassland, coastal grassland and on roadside verges. First recorded in 1767 near Inchnadamph by J.Robertson.

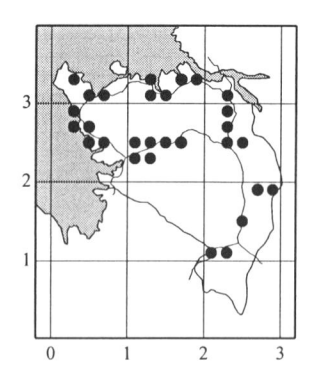

*Galium sterneri* Ehrend.
Màdar Cloich-aoil
**Limestone Bedstraw**

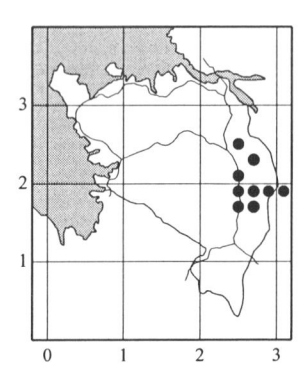

9 tetrads. Confined to limestone areas, where it grows in grassland and on rocks. First recorded in 1890 'on limestone, about Inchnadamph' by E.S.Marshall and F.J.Hanbury and then, in 1925, at Knockan by G.C.Druce.

*Galium saxatile* L.
Màdar Fraoich
**Heath Bedstraw**

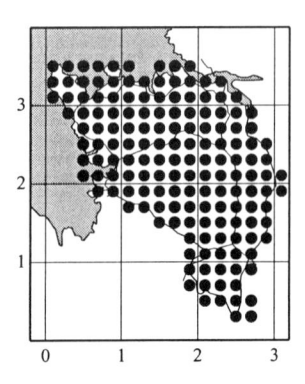

156 tetrads. Common and widespread in woodland, dry

heath, rocks, boulder scree and all types of grassland. First recorded, as '*G. montanum*', in 1767 near Inchnadamph by J.Robertson.

*Galium aparine* L.
Garbh-lus
**Cleavers**

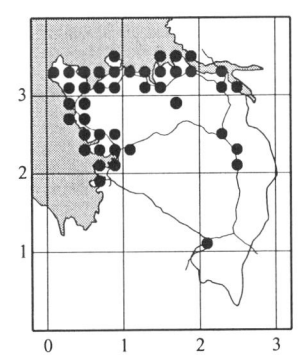

42 tetrads. Occasional on disturbed ground and roadsides; more often amongst the stones at the back of a beach. First recorded in 1943 'on the beach at Clashnessie' by A.J.Wilmott and M.S.Campbell.

[*Cruciata laevipes* Opiz
(Galium cruciata (L.) Scop.)
Lus na Croise
**Crosswort**

Reported, as '*Valantia cruciata*', in 1767 near Inchnadamph by J.Robertson.]

CAPRIFOLIACEAE

*Sambucus nigra* L.
Droman
**Elder**

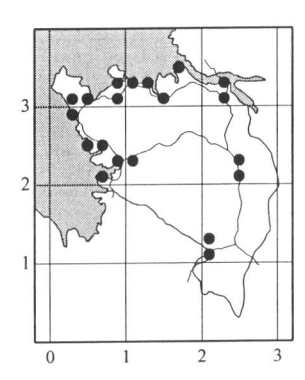

20 tetrads. Apart from one, presumably bird-sown, tree on limestone scree at Stronechrubie, its distribution shows a clear link with houses; what is less clear is the reason for this. The bushes only flower occasionally and fruit rarely, so it would be unrewarding as a crop. However, in the past, the plant was valued as a defence against witches' charms and the leaves used for treating wounds. First recorded in 1767 near Inchnadamph by J.Robertson.

*Viburnum opulus* L.
Caor-chon
**Guelder-rose**

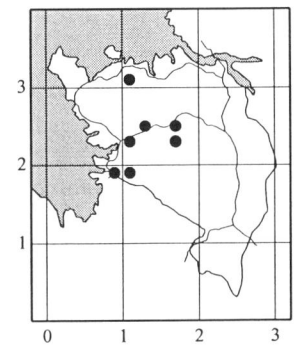

7 tetrads. A rare shrub, usually growing singly on or near the margin of a watercourse. The Rivers Inver and Kirkaig provide good examples. First recorded in 1886 at Inchnadamph by A.Gray.

*Symphoricarpos albus* (L.) S.F.Blake
Sùbhag Sneachda
**Snowberry**

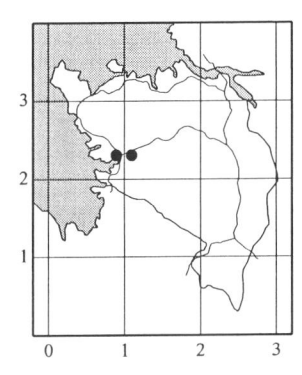

2 tetrads. Naturalised in wooded grounds in the Lochinver area. First recorded in 1992 in the grounds of Glencanisp Lodge.

*Lonicera periclymenum* L.
Lus na Meala
**Honeysuckle**

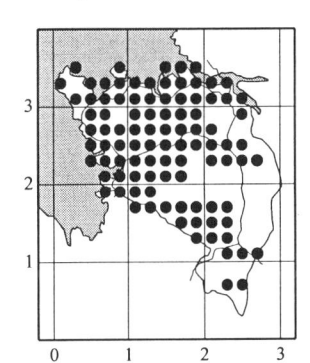

96 tetrads. Widespread except on high ground. Sometimes in open woodland, but most often on rock faces in gorges and on outcrops along sheltered burn and river courses. First recorded in 1767 near Inchnadamph by J.Robertson.

ADOXACEAE

*Adoxa moschatellina* L.
Mosgadal
**Moschatel**

Not seen during the present survey. The first and only record was made by G.C.Druce in 1894 at Knockan Crags 'sparingly at the base of an overhanging rock...most northerly station yet recorded in Britain'.

VALERIANACEAE

*Valerianella locusta* (L.) Laterr.
Leiteis an Uain
**Common Cornsalad**

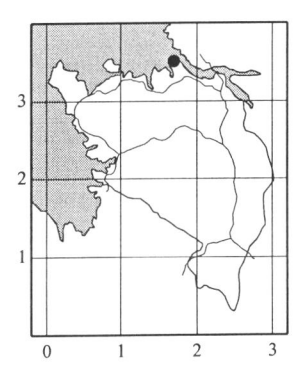

1 tetrad. A casual, growing in a

137

garden at Kerrachar. First recorded, as '*V. olitoria*', in 1886 at Achmelvich Bay by A.Gray.

*Valerianella carinata* Loisel.
**Keel-fruited Cornsalad**

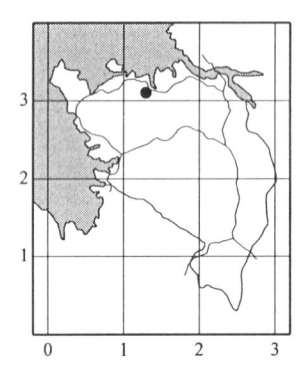

1 tetrad. A chance introduction into a garden, where it has seeded itself for a number of years. First recorded in 1997 at Nedd.

*Valeriana officinalis* L.
Carthan Curaidh
**Common Valerian**

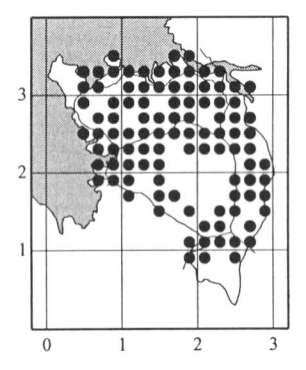

100 tetrads. Widespread and frequent in marshes, wet roadside ditches and alongside burns and rivers. First recorded in 1767 near Inchnadamph by J.Robertson.

DIPSACACEAE

*Succisa pratensis* Moench
Ura-bhallach
**Devil's-bit Scabious**

156 tetrads. A ubiquitous species seemingly unaffected by soil type or aspect, growing from sea level to the hilltops. In the latter it survives and flowers in stony montane heath, only growing to a few centimetres in height. Usually blue, but white and pink forms have also been seen. First recorded in 1767 near Inchnadamph by J.Robertson.

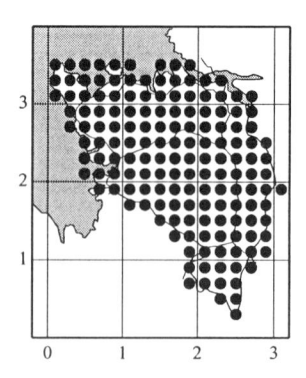

ASTERACEAE

*Arctium minus* s.l.
Leadan Liosda
**Burdock**

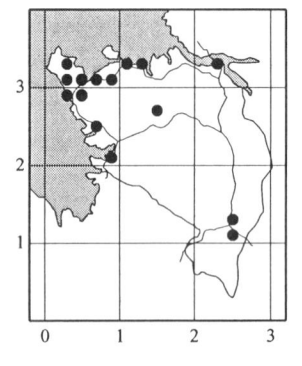

15 tetrads. Regrettably not enough attention was paid to more precise determinations and we are unable to say whether records should be ascribed to *A. minus*, s.s. or *A. nemorosum*. Grows in waste places, disturbed ground and gardens. Occasionally also up in the hills, where fleece-borne burrs have introduced plants to places where sheep lie up. First recorded in 1943 as '*A. vulgare*', 'in a midden at Feadan', by A.J.Wilmott and M.S.Campbell.

*Saussurea alpina* (L.) DC.
Sàbh-lus Ailpeach
**Alpine Saw-wort**

18 tetrads. In the hills, on cliffs, rocks and outcrops by burns. First recorded in 1886 as 'Ben More, Assynt – Cama Loch (descends to 400 feet)' by A.Gray. The 1950s record for NC02, mapped in the first *Atlas*, seems unlikely.

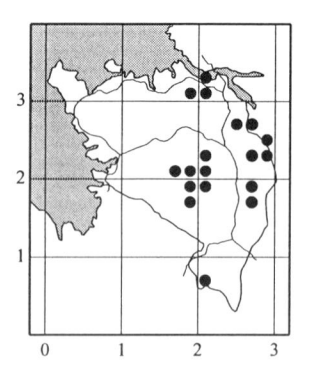

*Carduus nutans* L.
Fòthannan Crom
**Musk Thistle**

A casual, not seen during the present survey. First and only record is of 'one plant in a waste field opposite the Youth Hostel at Achmelvich', found in 1958 by M.McC.Webster.

*Cirsium vulgare* (Savi) Ten.
Cluaran Deilgneach
**Spear Thistle**

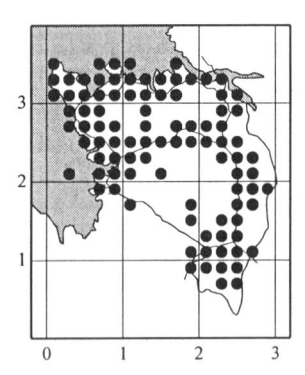

94 tetrads. Common in lowland areas, predominantly on roadsides, disturbed ground and in gardens. Windblown thistledown has given rise to occasional plants by ruins and sheep folds in the hills. First recorded during the 1950s survey for the first *Atlas*.

*Cirsium heterophyllum* (L.) Hill
Cluas an Fhèidh
**Melancholy Thistle**

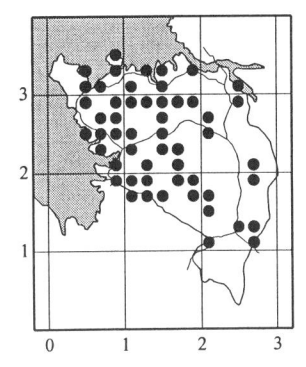

53 tetrads. In scree, wet heath and beside burns, particularly in gorges where there is some shelter. Occasionally in woodland, where, because of the shade, it rarely flowers. First recorded in 1767 as '*C. helenoides*' near Inchnadamph by J.Robertson.

*Cirsium palustre* (L.) Scop.
Cluaran Lèana
**Marsh Thistle**

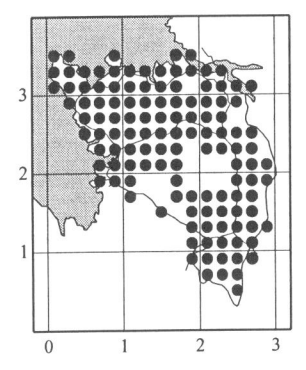

123 tetrads. Widespread and common, not only in ditches, marshes, wet woodland and burn sides, but also in drier grassland and roadside verges. White flowered plants occur, sometimes in quite extensive patches. First recorded in 1767 near Inchnadamph by J.Robertson.

*Cirsium* x *celakovskianum* Knaf
*C. arvense* x *C. palustre*

1 tetrad. A single specimen of this hybrid was confirmed from a small quarry beside the road between Skiag Bridge and Kylesku in 1998. This is the first and only record.

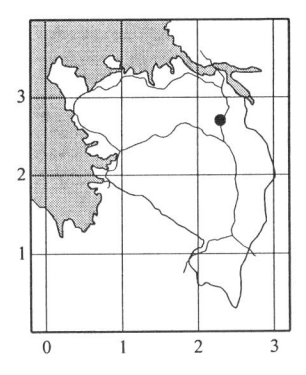

*Cirsium arvense* (L.) Scop.
Fòthannan Achaidh
**Creeping Thistle**

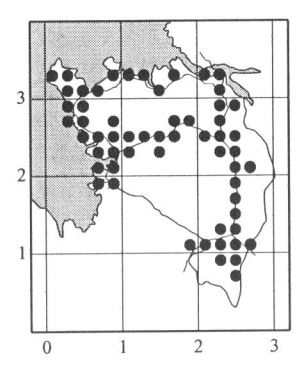

58 tetrads. Almost entirely confined to waste ground and verges, where it was probably brought in during roadworks. Rarely seen as a component of grassland. First recorded during the 1950s survey for the first *Atlas*.

*Centaurea cyanus* L.
Gorman
**Cornflower**

Not seen during the present survey. The first and only record was made in 1899 from 'a cultivated field, Inchnadamph' by C.E.Salmon.

*Centaurea nigra* L.
Cnapan Dubh
**Common Knapweed**

70 tetrads. Most frequent on roadsides and in coastal grassland. Also in the hills, where there is some better soil in the patches of grassland alongside burns. First recorded in 1944, in a plethora of forms, at Achmelvich by A.J.Wilmott and M.S.Campbell.

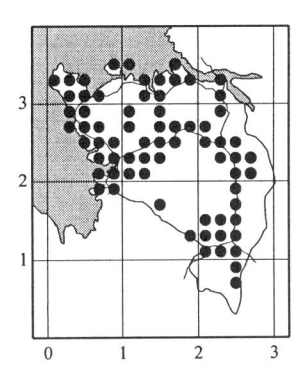

*Lapsana communis* L.
Duilleag Bhràghad
**Nipplewort**

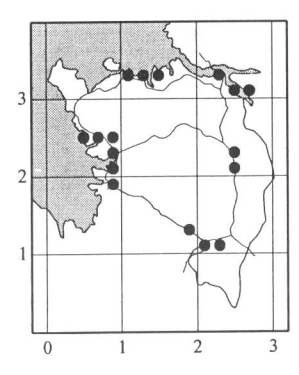

17 tetrads. Occasional in woodland, waste places and along roadsides. First recorded in 1890 'on limestone rocks near Inchnadamph' by E.S.Marshall and F.J.Hanbury, with the comment 'at least a mile from house or garden…can hardly be otherwise than indigenous here'.

*Hypochaeris radicata* L.
Cluas Cait
**Cat's-ear**

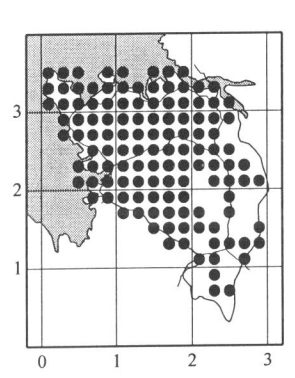

123 tetrads. Widespread and common in unshaded habitats; all types of grassland and in rock crevices. First recorded during the 1950s survey for the first *Atlas*.

*Leontodon autumnalis* L.
Caisearbhan Coitcheann
**Autumn Hawkbit**

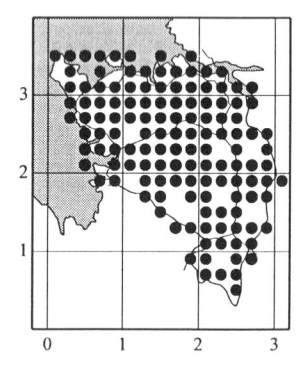

132 tetrads. Widespread and common on roadsides, waste ground and grassy places. First recorded in 1887 near Inchnadamph by E.S.Marshall.

*Tragopogon pratensis* L.
Feusag a'Ghobhair
**Goat's-beard**

Not seen during the present survey. The first and only record was made in the early 1960s at Clashnessie by A.G.Kenneth.

*Sonchus arvensis* L.
Bliochd Fochainn
**Perennial Sow-thistle**

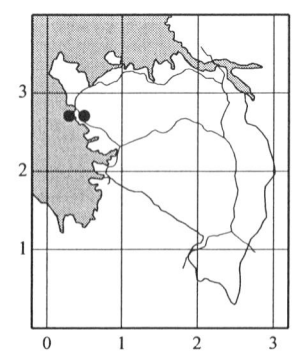

2 tetrads. At Clachtoll, where it occurs rarely on shingle and boulder beaches and in one place at the foot of a roadside wall. A species whose frequency reflects the lessening of cultivation, as the first record, made in 1894 'in cornfields...at Elphin and Knockain' by G.C.Druce, suggests.

*Sonchus oleraceus* L.
Bainne Muice
**Smooth Sow-thistle**

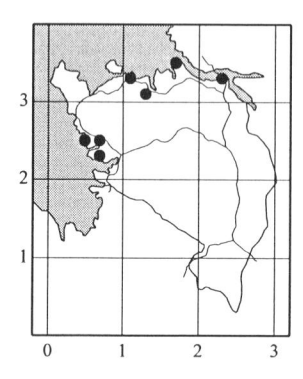

7 tetrads. A rare weed of cultivation and disturbed ground. First recorded during the 1950s survey for the first *Atlas*.

*Sonchus asper* (L.) Hill
Searbhan Muice
**Prickly Sow-thistle**

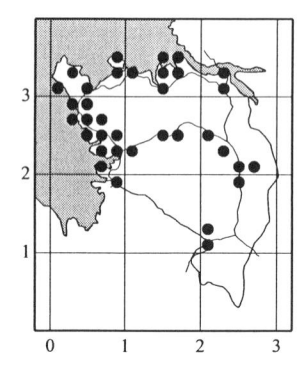

35 tetrads. Occasional along roadsides, in waste ground and rock fissures, in coastal and, less frequently, inland areas. First recorded during the 1950s survey for the first *Atlas*.

*Taraxacum* spp.
Beàrnan Brìde
**Dandelions**

151 tetrads. We have made no attempt to have the dandelions investigated and the map below shows how widespread the genus is. Found in all types of grassland and in woodland, as well as on cultivated and disturbed ground. First recorded, as '*T. spectabile*', in 1908 on the River Traligill by E.S.Marshall and W.A.Shoolbred. *Anthony* records four 'species' from Assynt: *officinale* (Section Ruderalia); *palustre* (Section Palustria); *spectabile* (Section Spectabilia) and *laevigatum* (Section Erythrosperma).The two maps that follow are the results of the efforts of visiting botanists.

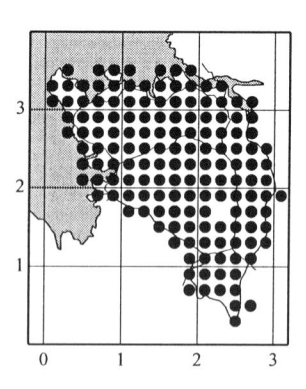

*Taraxacum faeroense* (Dahlst.) Dahlst.

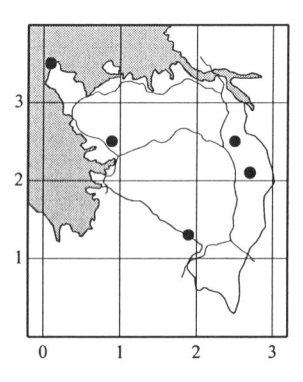

5 tetrads. First recorded in 1988 from NC11W during the BSBI Monitoring Survey.

*Taraxacum unguilobum* Dahlst.

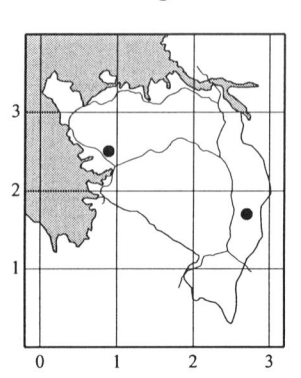

2 tetrads. First recorded in 1993 by A.McG.Stirling at Manse Loch.

*Crepis paludosa* (L.) Moench
Lus Curain Lèana
**Marsh Hawk's-beard**

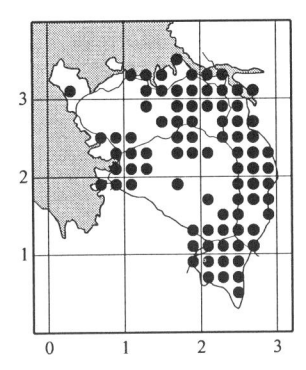

83 tetrads. An uneven distribution for which there seems no obvious reason. Commonest along burn courses and in wet woodland. First recorded in 1894 'near Elphin and Ledbeg' by G.C.Druce.

*Crepis capillaris* (L.) Wallr.
Lus Curain Mìn
**Smooth Hawk's-beard**

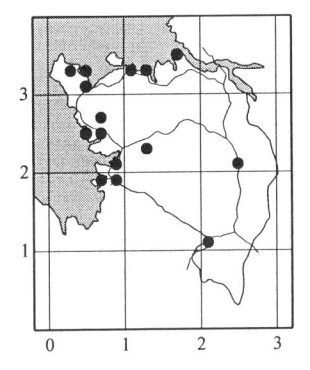

15 tetrads. Occasional beside roads and tracks and in coastal grassland. Two records, from Achnacarnin and Inverkirkaig, are of var. *glandulosa*. First recorded in 1894, as '*C. virens*', 'in cornfields [at] Elphin and Knockain' by G.C.Druce.

*Pilosella officinarum* F.W.Schultz & Sch.Bip.
(*Hieracium pilosella* L.)
Srubhan na Muice
**Mouse-ear Hawkweed**

60 tetrads. In dry well-drained places. Outcrop rocks, walls and rocky grassland on light soils. First recorded in 1767 near Inchnadamph by J.Robertson.

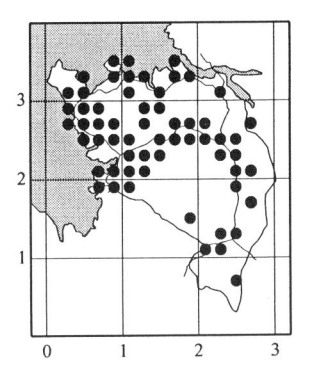

*Hieracium* spp.
**Hawkweeds**

Hawkweeds are one of the largest and most difficult groups in the British flora, and about a quarter of the 261 British species are found in the north-western Highlands, some being restricted to it. A few distinctive species may be named in the field, but identification generally depends on the examination of carefully collected voucher specimens by one of a handful of specialists in this group. Changes, still ongoing, in nomenclature and species concepts, require the re-examination of herbarium material and can render difficult the interpretation of older records.

Happily, much of the historical material collected in Assynt, for instance by E.S.Marshall, has been checked comparatively recently. We have therefore been able to use most of the records summarised in *Anthony*. Where they do not appear to accord with the distribution maps in the *Critical Supplement*, or species concepts have changed, we have sought expert opinion.

The earliest record of a hawkweed from Assynt is of '*H. murinum*', by J.Robertson in 1767, but this name does not equate to any modern species. Over a century later, in 1886, A.Gray recorded the montane species *H. holosericeum* on Beinn nan Cnaimhseag, where it still occurs.

Serious recording of local hawkweeds began with the visit of E.S. Marshall and F.J.Hanbury in 1890 (the latter being one of the acknowledged experts in the group), with substantial additions in 1908, when Marshall returned with W.A.Shoolbred. G.C.Druce had recorded a few species in 1894 and 1907, C.E.Salmon some in 1899. After this flurry of interest, especially in the species of the limestone around Inchnadamph, there was only sporadic recording in the following fifty years, by R.H.Williamson in the 1920s, A.J.Wilmott in the 1940s, and D.C.McClintock, R.A.Graham and J.Anthony in the 1950s.

Between 1961 and 1986, A.G.Kenneth made a major contribution to our knowledge of the hawkweeds of West Sutherland, especially its hills, his material being verified by C.West and P.D.Sell. *H. kennethii*, so named 'in recognition of his fine work over many years', was collected by him on Cnoc na Creige north of Glas Bheinn. His real enthusiasm was reserved for areas to the north of Assynt, especially Foinaven, but he did visit most of the hills in the parish. He was fairly scathing about the paucity of the hawkweeds on some of them, for example Canisp, visited in 1985, recording in his notebook 'good day, wretched hill...not a single hawkweed seen in ground covered by me'! R.J.Pankhurst contributed records from the limestone in 1972 and T.Edmondson some from further afield in 1974.

This brings us the present day. In 1996 one of us (I.M.E.) was persuaded by the enthusiasm of Vincent Jones, on a field course at Kindrogan, to start collecting material in the course of general tetrad recording. Vincent paid two visits to Assynt in 1997 and 1998, when we concentrated on the limestone, and he also helped to weed out, from the material collected, large numbers of specimens of the commonest species, *H. vulgatum*. All other material collected was kindly determined by David McCosh and it is on these records and Vincent's that the tetrad maps are based. Since geographical coverage was inevitably uneven we have not used the phrase 'Not seen during the present survey'. We are most grateful to David for his critical reading of a draft of this account and his very helpful comments.

Section Foliosa

*Hieracium reticulatum* Lindeb.

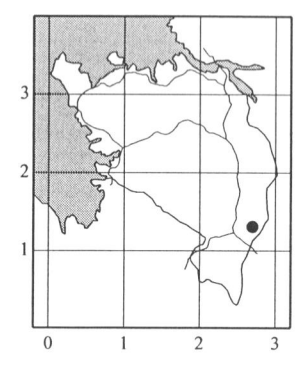

1 tetrad. Falls on the Ledbeg River, 1999. Not previously recorded.

*Hieracium strictiforme* (Zahn) Roffey

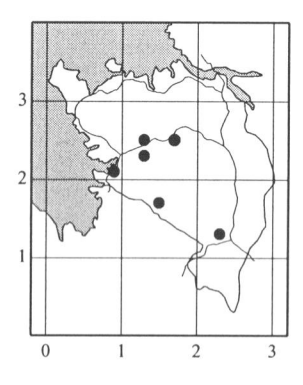

6 tetrads. Often close to water, as on the banks of Loch Assynt and the River Inver and the shore of Loch Gleannan a'Mhadaidh. Elsewhere on crags south-east of Badnaban, at Brackloch and on the north shore of Cam Loch. Recorded in 1943 in a 'rocky gulley n.e. of Inverkirkaig' by A.J.Wilmott and M.S.Campbell (probably the Badnaban locality), and in 1953 at Inchnadamph by C.West.

*Hieracium subcrocatum* (E.F.Linton) Roffey

Recorded in *Anthony* as collected in 1944 at Inverkirkaig and Achmelvich by A.J.Willmott and M.S.Campbell.

*Hieracium latobrigorum* (Zahn) Roffey

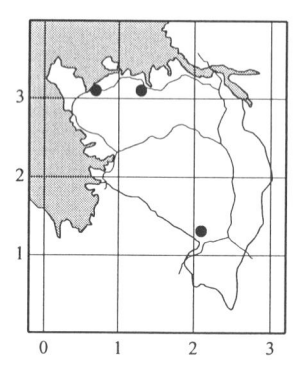

3 tetrads. In the gorge between Cam Loch and Loch Veyatie, on gneiss at Nedd and Clashnessie. Recorded in 1904 at 'Inver' by F.R.Tennant.

Section Tridentata

*Hieracium sparsifolium* Lindeb.

Recorded in 1923 at Inchnadamph by R.H.Williamson and in 1977 at Ledmore.

Section Vulgata

*Hieracium vulgatum* Fries.

22 tetrads. The commonest hawkweed by far in Assynt, on crags, seacliffs and in rocky gorges, on a variety of rock types, usually in shady situations.

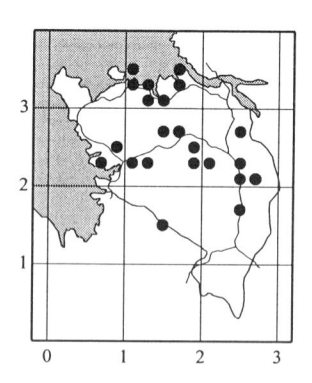

Recorded in 1887 'about Inchnadamph' by E.S.Marshall and many times since.

*Hieracium rubiginosum* F.Hanb.

Recorded in 1908 on Beinn Gharbh by A.Ley, in 1959 on the Allt Poll an Droighinn, and in 1974 at Nedd by T.Edmondson. This species is currently under revision.

*Hieracium orcadense* W.R.Linton

Recorded in 1908 on the Allt Poll an Droighinn by W.A.Shoolbred, and in 1956 at Stoer by R.A.Graham.

*Hieracium caesiomurorum* Lindeb.

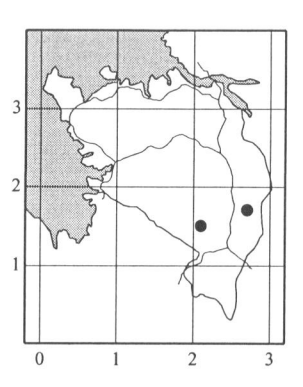

2 tetrads. In a wet cleft in a Torridonian crag north of Cam Loch, 1997, and in the Allt nan Uamh valley 1998 (V.Jones). Recorded in 1899 on Glas Bheinn by C.E.Salmon, in 1952 at Clachtoll by D.C.McClintock, and in 1974 at Achmelvich by T.Edmondson.

*Hieracium subtenue* (W.R.Linton) Roffey

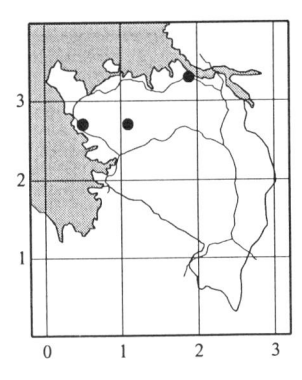

2 tetrads. On a crag south of Loch Crocach and a picrite dyke at Alltana'bradhan. Recorded in 1899 on Canisp by C.E.Salmon, in 1908 on the Allt Poll an Droighinn and 'Chalda Burn' by E.S.Marshall and W.A. Shoolbred, in 1956 at Stoer by R.A.Graham, and between 1966 and 1985 on Quinag, 'Conival/ Breabag' and Canisp by A.G.Kenneth.

*Hieracium rivale* F.Hanb.

Recorded in 1890 on the Traligill and Allt a'Chalda Mor by F.J.Hanbury and E.S.Marshall, and in 1908 on the Allt Poll an Droighinn by the latter.

*Hieracium pollinarioides* Pugsley

Recorded in 1890 at Lochinver by E.S. Marshall, in 1944 at Inverkirkaig and Lochinver by A.J.Willmott, in 1951 at Stoer by J.E.Raven, and in 1956 at Stoer and Clashnessie by R.A.Graham.

*Hieracium pictorum* E.F.Linton

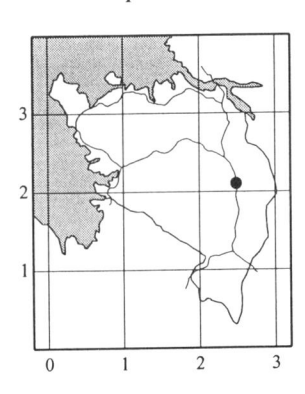

1 tetrad. Traligill valley, 1997 (V.Jones). Recorded in 1966 on Quinag by A.G.Kenneth.

*Hieracium duriceps* F.Hanb.

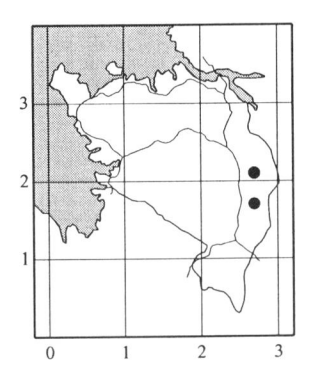

2 tetrads. Traligill valley and near Creag nan Uamh (V.Jones). Recorded in 1890 at Inchnadamph by E.S. Marshall and in 1974 at Strathan by T.Edmondson.

*Hieracium camptopetalum* (F.Hanb.) Sell and C.West

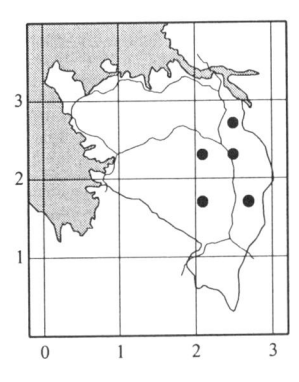

5 tetrads. On a wet gneiss crag east of Loch a'Choire Dheirg (Glas Bheinn), on Creag na h-Iolaire (Beinn nan Cnaimhseag), on Torridonian crags below Beinn Gharbh and south of Lochan Fada, and on limestone at Creag a'Choimleum and in the Allt nan Uamh valley (both V.Jones). Recorded in 1908 at Allt nan Uamh, Inchnadamph and Kylesku by E.S. Marshall and in 1951 at Inchnadamph by C.West. An endemic species, with few localities in northern Scotland.

Section Oreadea

*Hieracium orimeles* F.Hanb. ex W.R.Linton.

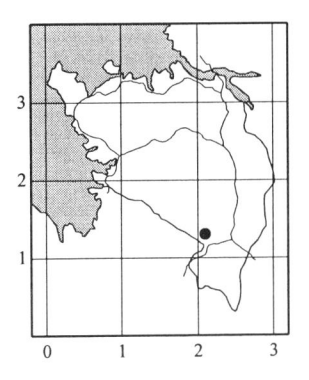

1 tetrad. On a north-facing Torridonian crag south of Creagan Beag, 1996. Only once found elsewhere in West Sutherland, in 1966 at Durness by A.G.Kenneth.

*Hieracium caledonicum* F. Hanb.

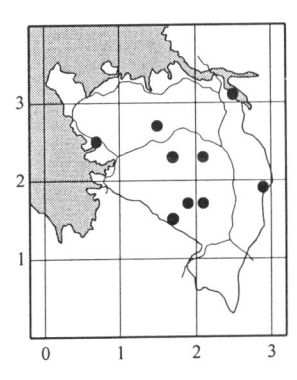

9 tetrads. On crags of a variety of rock types, including gneiss, Torridonian and the Fucoid Beds, and also on the rocky shore of Loch Veyatie. Recorded in 1887 and 1908 at Inchnadamph by E.S.Marshall and W.A.Shoolbred, in 1890 at Lochinver and on Canisp by E.S.Marshall, in 1951 at Inchnadamph by C.West and in 1974 'between Conival and Breabag' [Bealach Traligill?] by A.G. Kenneth. This species is currently under revision.

*Hieracium argenteum* Fries.

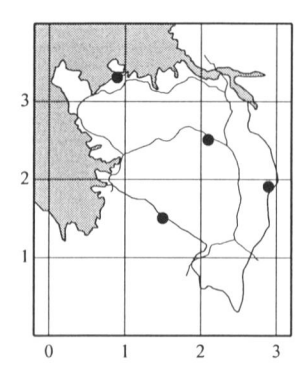

4 tetrads. In a quarry beside the Allt na Doire Cuilinn, on gneiss above the Allt nan Damph (Loch Veyatie) and on Cnoc Phollain Beithe, and on quartzite to the east of Cuil Dhubh. Apparently catholic in its choice of substrate. Recorded in 1894 at Knockan by G.C.Druce, in 1897 at Inchnadamph by E.S. Marshall and in 1962 on Quinag by A.G. Kenneth.

*Hieracium nitidum* Backh.

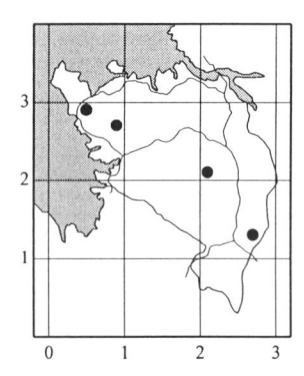

4 tetrads. On gneiss crags at Ruigh Chnoc, near Loch Meall a'Chuna Beag and Loch an Aigeil, and on granite on the north side of Cnoc na Sròine. Recorded in 1908 at Inchnadamph by E.S.Marshall, and in 1964 at Culkein Drumbeg by A.G.Kenneth. Mapped in the *Critical Supplement* for NC03, 13, 21, 22 and 23 under the title '*H. jovimontis*'.

*Hieracium eucallum* Sell and C.West

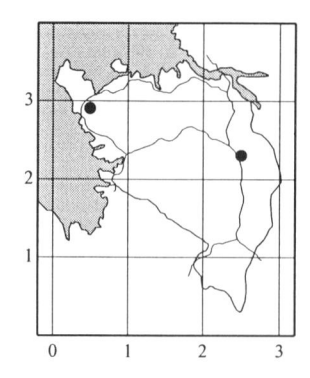

2 tetrads. On crags beside Loch an Aigeil, 1997, and on Creag a'Choimhleum, 1998 (both V. Jones). Not previously recorded.

*Hieracium saxorum* (F.Hanb.) Sell and C.West

Recorded in 1951 from the upper part of the Traligill valley by J.E.Raven.

[*Hieracium jovimontis* (Zahn) Roffey

This species was mapped in error for *H. nitidum* (q.v.) in the *Critical Supplement*.]

*Hieracium dicella* Sell and C. West

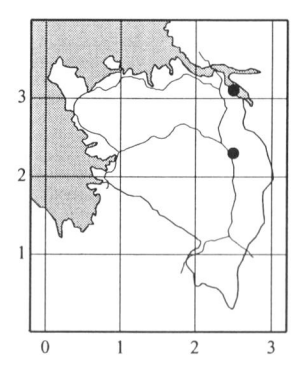

2 tetrads. On the Fucoid Beds at Liath Bhad, 1990, and on Creag a'Choimhleum, 1998. Not previously recorded.

*Hieracium sarcophylloides* Dahlst.

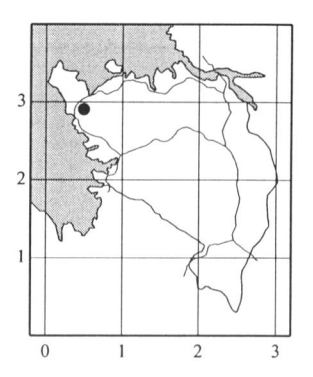

1 tetrad. On crags beside Loch an Aigeil, 1998 (V.Jones). Not previously recorded.

Section Cerinthoidea.
A group of hawkweeds that are very well represented in Assynt (seven of the ten native species), especially on the limestone.

*Hieracium iricum* Fries

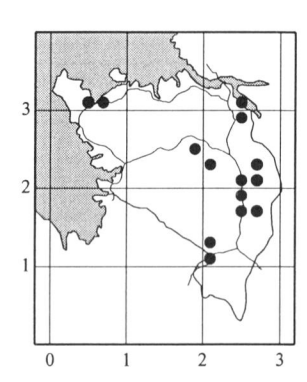

14 tetrads. A handsome plant, widely distributed both on and off the limestone. Recorded in 1890 by E.S.Marshall at 'Inchnadamph, ascending to 1200 ft.' and 'very fine by the river side, above Lochinver', and on a number of occasions since.

*Hieracium anglicum* Fries.

13 tetrads. Frequent on the limestone and occasional elsewhere.

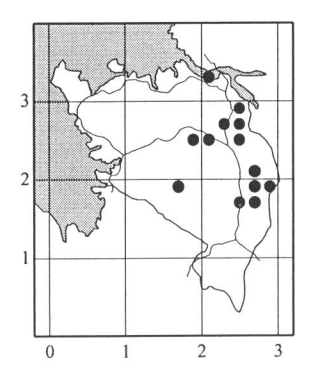

Recorded in 1908 at Knockan and Inchnadamph by E.S.Marshall, and in 1966 on Quinag by A.G.Kenneth. This species is currently under revision.

*Hieracium langwellense* F. Hanb.

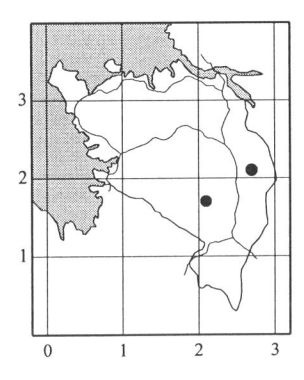

2 tetrads. On limestone in the Traligill valley and beside burns on Torridonian crags at two places south of Lochan Fada. Recorded in 1908 on the Allt a'Chalda Mor and Beinn Gharbh by E.S.Marshall.

*Hieracium ampliatum* (W.R. Linton) Ley

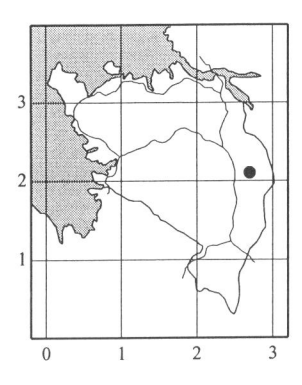

1 tetrad. On limestone in the Traligill valley, 1997 (V.Jones).

Recorded in 1890 at Lochinver and in 1908 around Inchnadamph by E.S.Marshall, in 1971 near Ledmore by A.G.Kenneth, and in 1974 at Strathan by T.Edmondson. This species is currently under revision.

*Hieracium hebridense* Pugsley

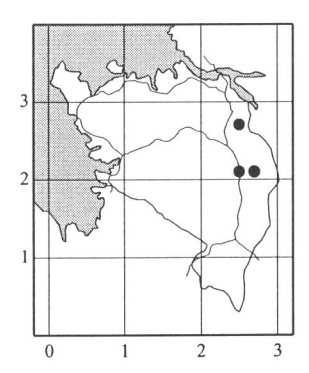

3 tetrads. In the Traligill valley, and on the north side of Glas Bheinn (V.Jones). Recorded in 1899 on Glas Bheinn by C.E.Salmon, in 1908 at Loch Mhaolach-Coire by E.S.Marshall and in 1958 in the Traligill valley by C.West.

*Hieracium flocculosum* Backh.

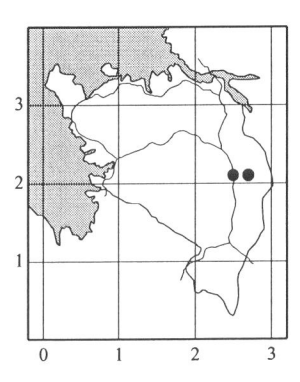

2 tetrads. In the Traligill valley, on an outcrop on Glac Mhór, and on roadside rocks at Sròn Chrùbaidh (V.Jones). Recorded in 1908 in the Traligill valley by E.S.Marshall. This species is currently under revision.

*Hieracium shoolbredii* E.Marshall

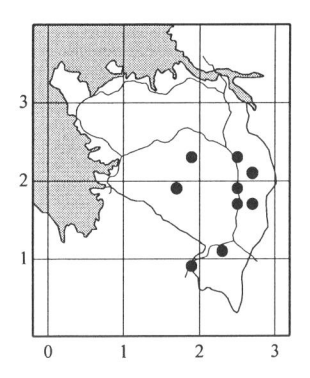

9 tetrads. Mainly on the limestone, but also on the gneiss at Coire Dubh (north of Suilven) and in a wooded gorge south of Tubeg. Recorded in 1899 on Beinn an Fhuarain by C.E.Salmon and on a number of occasions since, most recently in 1974 at Achmelvich by T.Edmondson.

Section Subalpina.

*Hieracium dasythrix* (E.F.Linton) Pugsley

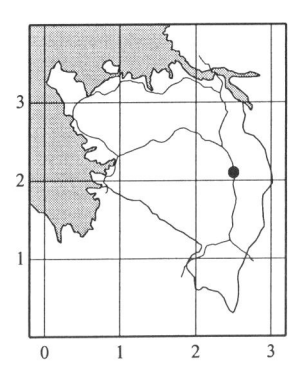

1 tetrad. Traligill valley, 1997 V.Jones. Mapped for NC21 in the *Critical Supplement*.

*Hieracium hyparcticoides* Pugsley

Recorded in 1890 and 1908 around Inchnadamph by E.S.Marshall and W.A.Shoolbred, and also off the limestone on the Allt Poll an Droighinn and on Beinn Gharbh; also in 1957 at Inchnadamph by C.West. In the British Isles this species is confined to West Sutherland, apart from one locality in East Ross.

*Hieracium glandulidens* Sell and C.West

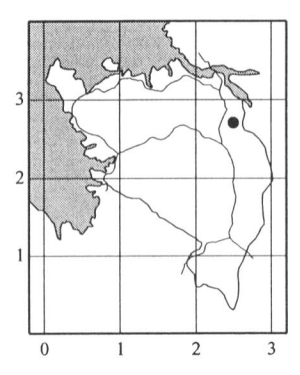

1 tetrad. Glas Bheinn, 1997 (V.Jones). Described new to science from material collected in 1908 at Inchnadamph by E.S.Marshall; also found in 1964 on Quinag by A.G. Kenneth.

*Hieracium lingulatum* Backh. ex. Hook. & Arn.

Recorded in 1908 at Inchnadamph, on Beinn Gharbh and Canisp by E.S.Marshall, in 1957 at Inchnadamph by R.C.Pankhurst, and in 1984 on 'the approach to Beinn Uidhe' by A.G.Kenneth.

Section Alpina.
Since some of these species are not in Kent (1992), they are arranged alphabetically.

*Hieracium alpinum* L.

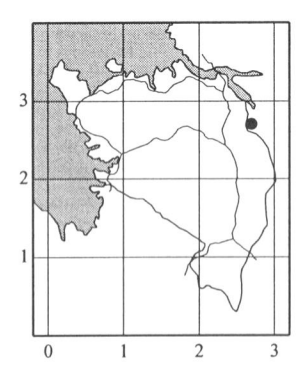

1 tetrad. Bealach na h-Uidhe, at c.600m., 1999. Not previously recorded.

*Hieracium holosericeum* Backh.

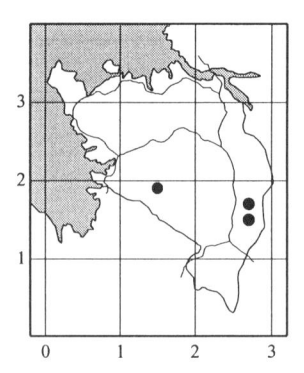

3 tetrads. A lovely plant, growing in stony ground on the summit ridges of Beinn an Fhuarain and Suilven, and also on the south side of Beinn na Cnaimhseag. Recorded in 1900 at Inchnadamph by T.J.Foggitt, in 1960 on Glas Bheinn and in 1986 on Canisp by A.G.Kenneth.

*Hieracium kennethii* Sell and Tennant

Collected in 1987 by A.G.Kenneth at c. 450-550m. on Cnoc na Creige.

*Hieracium marginatum* Sell and C.West

Recorded in 1986 in the vicinity of Canisp by A.G.Kenneth.

*Hieracium perscitum* Sell and C.West

Recorded in 1908 on Canisp by W.A.Shoolbred.

*Hieracium subglobosum* Pugsley

Recorded in 1890 on Canisp by E.S. Marshall, and in 1966 on Quinag by A.G. Kenneth. Formerly referred to as the dark-styled form of '*H. globosiflorum*'.

*Antennaria dioica* (L.) Gaertn.
Spòg Cait
**Mountain Everlasting**

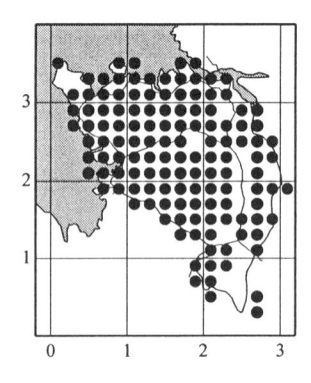

128 tetrads. Widespread in dry situations. Heaths, rocks and short grassland, both coastal and inland. First recorded in 1767, as '*Gnaphalium dioicum*', near Inchnadamph by J.Robertson.

*Gnaphalium sylvaticum* L.
Cnàmh-lus Mòintich
**Heath Cudweed**

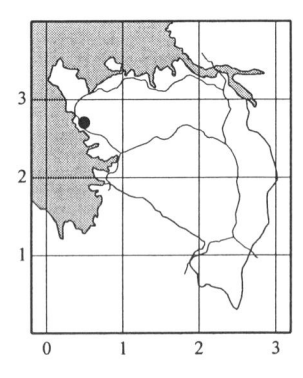

1 tetrad. Rather surprisingly, just the one record, from heath grassland beside the Clachtoll peat track. First recorded in 1767 near Inchnadamph by J.Robertson. Recorded during the 1950s survey for the first *Atlas* from NC02, 03, 12, 13 and 21, since when it has apparently declined.

*Gnaphalium supinum* L.
Cnàmh-lus Beag
**Dwarf Cudweed**

8 tetrads. Rare on hilltops in bare, stony or gravelly places: Beinn an Fhurain, Beinn Uidhe, Breabag, Glas Bheinn, Quinag and Suilven.

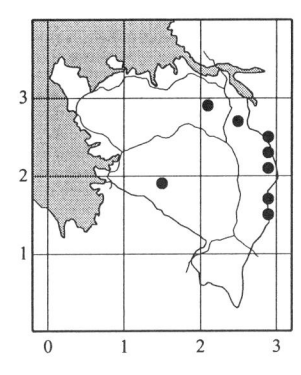

First recorded in 1886 on '*Coinne mheall, Assynt*' by A.Gray; also in 1899 by C.E.Salmon at 2,300ft. on Canisp, where we have not seen it.

*Gnaphalium uliginosum* L.
Cnàmh-lus Lèana
**Marsh Cudweed**

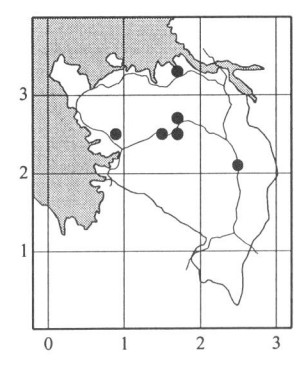

6 tetrads. Typically a plant of roadsides, along the gravelly strip where the road meets the grass. Seen once as a garden weed. First recorded during the 1950s survey for the first *Atlas* from NC02.

*Solidago virgaurea* L.
Slat Oir
**Goldenrod**

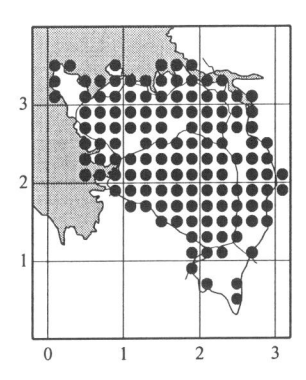

132 tetrads. Widespread and common in well-drained, often rocky, places. In dry heath, grassland, light woodland and also on hilltops, where it survives and flowers on stony ground, only growing to a few cm. in height. There are some limestone areas in which we have not found it. First recorded in 1767 near Inchnadamph by J.Robertson.

*Aster tripolium* L.
Neòinean Sàilein
**Sea Aster**

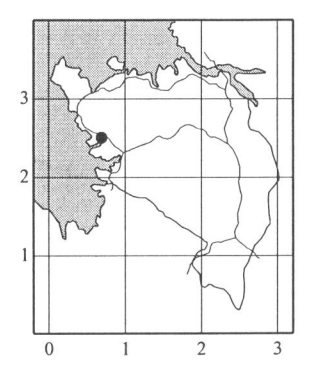

1 tetrad. Only found at Loch Roe and, although saltmarsh is not a common habitat in Assynt, there seems no obvious reason for it not to be elsewhere. First recorded during the 1950s survey for the first *Atlas* from NC02; possibly the current locality.

*Bellis perennis* L.
Neòinean
**Daisy**

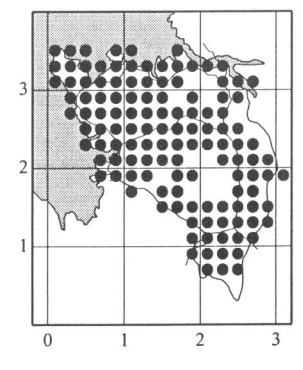

125 tetrads. Abundant in grassland, beside tracks and as a weed of cultivation and disturbed ground. First recorded in 1767 from near Inchnadamph by J.Robertson.

*Tanacetum parthenium* (L.) Sch.Bip.
(*Chrysanthemum parthenium* (L.) Bernh.)
Meadh Duach
**Feverfew**

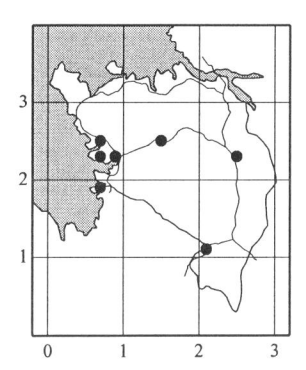

7 tetrads. Introduced in the past, probably for medicinal purposes, and maintaining itself near ruined houses, on roadsides and on the sea wall in Lochinver. First recorded in 1943 at Lochinver by A.J.Wilmott and M.S.Campbell, with the comment 'garden outcast on the shore above high water mark, but subject to much spray'. Recorded in 1981 from NC13.

*Tanacetum vulgare* L.
(*Chrysanthemum vulgare* (L.) Bernh.)
Lus na Frainge
**Tansy**

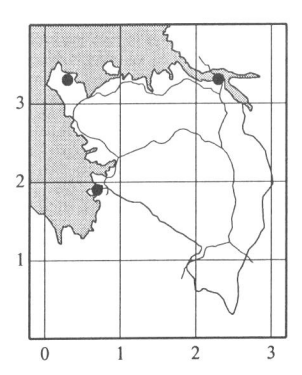

3 tetrads. Rare. Surviving in or near abandoned gardens or the disused parts of present ones. Found in Culkein Stoer, Inverkirkaig and Unapool. First recorded in 1886 at 'Loch Inver' by A.Gray. There are post-1950 records from NC02 and 13.

*Artemisia vulgaris* L.
Liath-lus
**Mugwort**

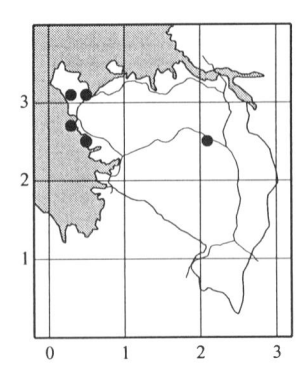

5 tetrads. In gardens at Achmelvich, on a wall at Clashmore, in a sandy gully near the sea at Clachtoll, in sandy grassland at Clashnessie and in a roadside quarry west of Skiag bridge First recorded during the 1950s survey for the first *Atlas* from NC13 and 21, in addition to the current hectads.

*Artemisia absinthium* L.
Burmaid
**Wormwood**

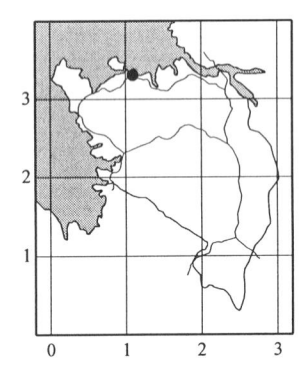

1 tetrad. Found only once, beside an old, uninhabited house at Culkein Drumbeg. First recorded in 1944 by A.J.Wilmott in a 'neglected old garden.... Achmelvich'.

*Achillea ptarmica* L.
Craidh-lus
**Sneezewort**

105 tetrads. Widespread and common in marshes, damp grassland and beside watercourses.

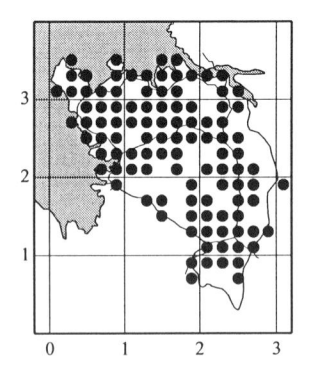

First recorded during the 1950s survey for the first *Atlas*.

*Achillea millefolium* L.
Eàrr-thalmainn
**Yarrow**

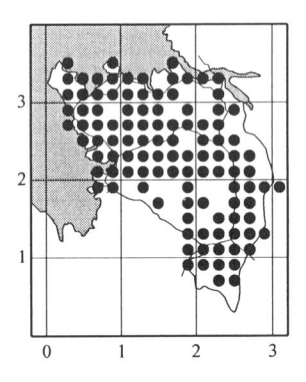

108 tetrads. Widespread and very common on roadsides and in short grassland. First recorded during the 1950s survey for the first *Atlas*.

*Chrysanthemum segetum* L.
Bile Bhuidhe
**Corn Marigold**

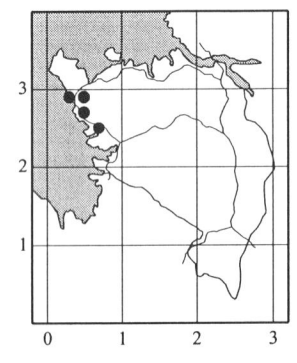

4 tetrads. A good example of how lack of present-day arable has affected a plant's distribution. The species re-appeared, for the first time in 20 years, in a patch of former cultivation at Stoer when it was newly planted with potatoes in

1997. It was seen once in Clachtoll in 1980, but has not been seen since. It was noted in a garden at Achmelvich from 1994 to 1996, but not since. First recorded during the 1950s survey for the first *Atlas,* from NC02, 03, 12 and 13.

*Leucanthemum vulgare* Lam.
(Chrysanthemum leucanthemum L.)
Nèoinean Mòr
**Ox-eye Daisy**

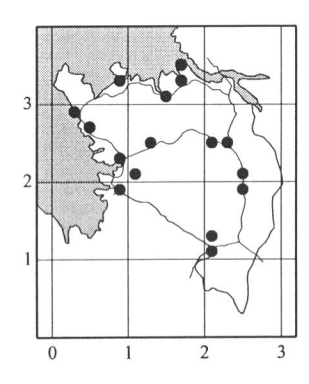

16 tetrads. Only occasional. The records are mostly from roadside verges and waste ground. It is probably under-recorded in the intensively grazed areas.of grassland near the coast First recorded during the 1950s survey for the first *Atlas*.

*Matricaria recutita* L.
Buidheag an Arbhair Chùbhraidh
**Scented Mayweed**

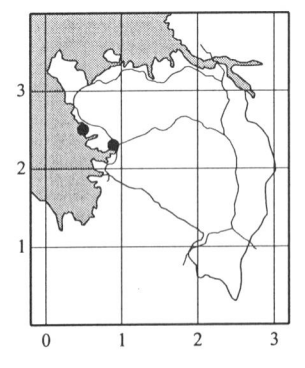

2 tetrads. Rare casual, presumably brought in with machinery. First recorded (new to the vice-county) in 1994 at Lochinver, and in the following year from Achmelvich.

*Matricaria discoidea* DC.
Lus Anainn
**Pineappleweed**

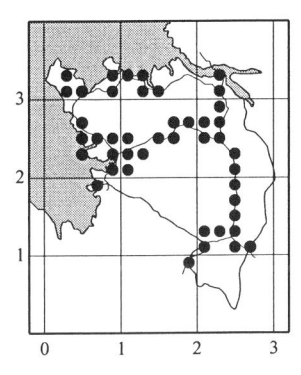

44 tetrads. Frequent along the gravelly strip between the road and the grass verge. First recorded during the 1950s survey for the first *Atlas*.

*Tripleurospermum maritimum* (L.) W.D.J.Koch
Buidheag na Mara
**Sea Mayweed**

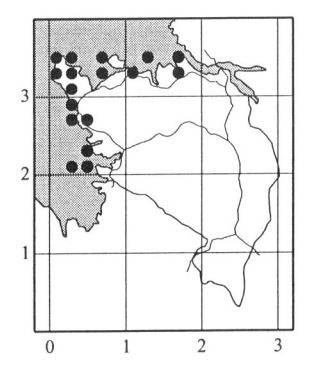

17 tetrads. Frequent on sea cliffs, rocky shores and shingle beaches. First recorded in 1886, as '*Matricaria inodora* b. *salina*', at Clachtoll by A.Gray.

*Tripleurospermum inodorum* (L.) Sch.Bip.
(T. maritimum ssp. inodorum (L.) Hyl. ex Vaar.)
Buidheag an Arbhair
**Scentless Mayweed**

10 tetrads. Very occasional in waste places and coastal grassland. First recorded in 1990 from Stoer village.

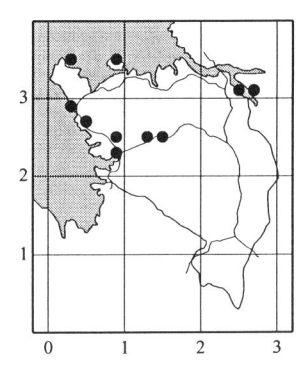

There are no earlier records, presumably because of the relatively recent change in status.

*Senecio jacobaea* L.
Buaghallan
**Common Ragwort**

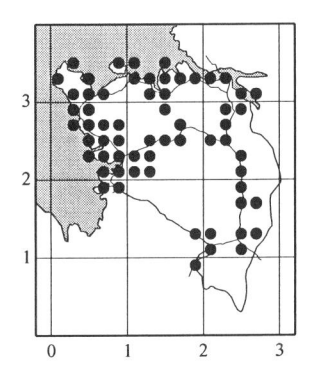

62 tetrads. Common in rough grassland, scrub and along the sides of roads and tracks. First recorded during the 1950s survey for the first *Atlas*.

*Senecio aquaticus* Hill
Caoibhreachan
**Marsh Ragwort**

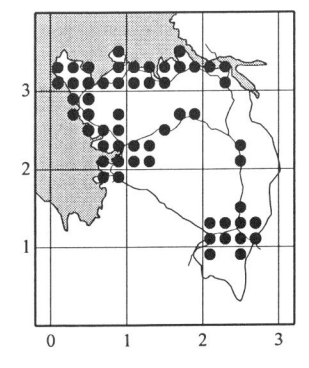

56 tetrads. Common in lowland wet places: ditches, marshes, river sides and wet grassland. First recorded during the 1950s survey for the first *Atlas*.

*Senecio vulgaris* L.
Grunnasg
**Groundsel**

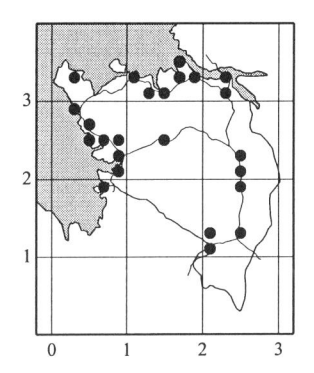

24 tetrads. Surprisingly uncommon. Noted occasionally as a garden weed and on disturbed areas of roadside verges. First recorded during the 1950s survey for the first *Atlas*.

*Senecio sylvaticus* L.
Grunnasg Monaidh
**Heath Groundsel**

Not seen during the present survey. First recorded from NC02 during the 1950s survey for the first *Atlas*.

*Senecio viscosus* L.
Grunnasg Leantalach
**Sticky Groundsel**

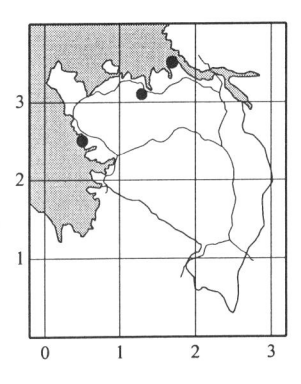

3 tetrads. Rare garden weed seen at Achmelvich, Kerrachar and Nedd. First recorded in 1995 from the last of these localities.

*Tussilago farfara* L.
Cluas Liath
**Colt's-foot**

30 tetrads. Predominantly a plant of river gravels, but also

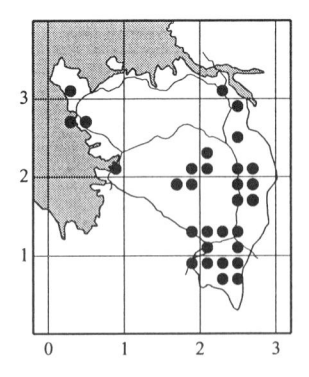

occasionally on scree and roadsides. First recorded during the 1950s survey for the first *Atlas*.

*Eupatorium cannabinum* L.
Cainb-uisge
**Hemp-agrimony**

Not seen during the present survey. Recorded in 1767 near Inchnadamph by J.Robertson.

JUNCAGINACEAE

*Triglochin palustre* L.
Barr a'Mhilltich Lèana
**Marsh Arrowgrass**

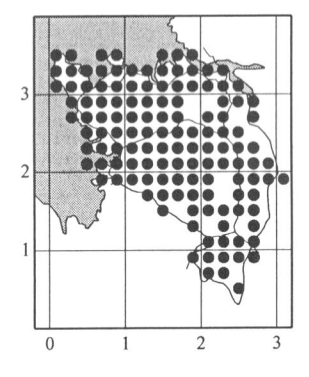

130 tetrads. Widespread and abundant in marshes, flushes, on loch shores and sometimes at the back of salt marshes, where it may grow alongside *T. maritimum*. First recorded in 1887 at Inchnadamph by E.S.Marshall.

*Triglochin maritimum* L.
Barr a'Mhilltich Mara
**Sea Arrowgrass**

19 tetrads. In saltmarshes and wet coastal turf amongst rocks. First recorded during the 1950s survey for the first *Atlas*.

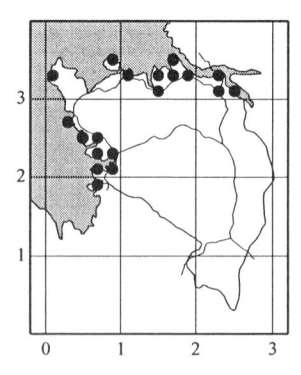

POTAMOGETONACEAE

*Potamogeton natans* L.
Duileasg na h-Aibhne
**Broad-leaved Pondweed**

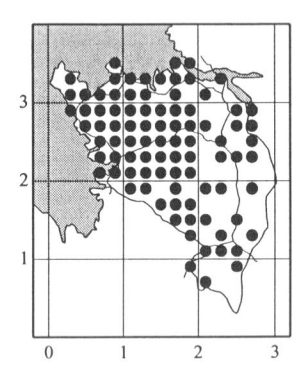

89 tetrads. Widespread in lochs and slow-moving stretches of rivers. First recorded in 1886 at Elphin by A.Gray.

*Potamogeton* x *sparganiifolius* Laest. ex Fr.
*P. natans* x *P. gramineus*
Lìobhag Stiallach
**Ribbon-leaved Pondweed**

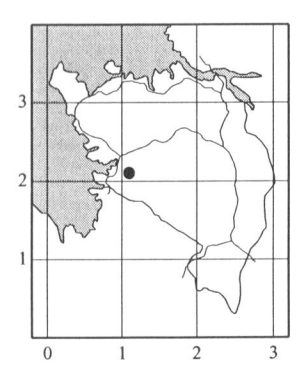

1 tetrad. Recorded in 1998 in Loch Druim Suardalain during the Scottish Loch Survey (det. C.D. Preston).

*Potamogeton polygonifolius* Pourr.
Lìobhag Bogaich
**Bog Pondweed**

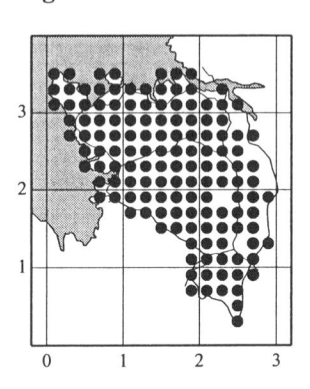

137 tetrads. Widespread and common in ditches, bogs and wet mud, as well as in open water in bog pools, lochs and river backwaters. First recorded in 1887 in Lochan Feòir by E.S.Marshall.

[*Potamogeton lucens* L.
Lìobhag Loinnreach
**Shining Pondweed**

A record for Assynt in 1886 is ascribed in *Anthony* to A.Gray, but Gray and Hinxman (1888) has no specific locality. Preston and Croft (1997) show no recent record north of Skye.]

*Potamogeton gramineus* L.
Lìobhag Fheurach
**Various-leaved Pondweed**

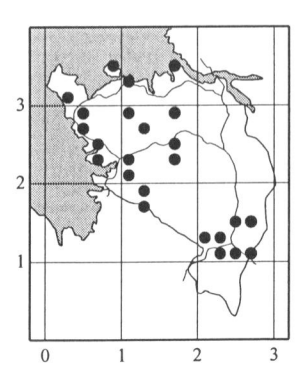

24 tetrads. Occasional in the shallow water of lochs. First

recorded in 1886 in Loch Awe by A.Gray.

*Potamogeton* x *nitens* Weber
*P. gramineus* x *P. perfoliatus*
Lìobhag Shoilleir
**Bright-leaved Pondweed**

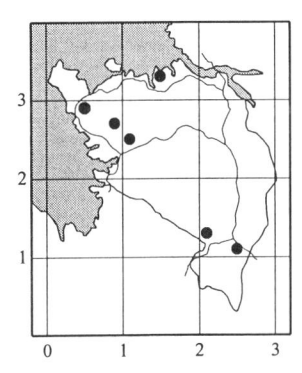

6 tetrads. Recorded from only a few lochs and not always ones in which the parents have been found. First recorded in 1886 at Alltana'bradhan by A.Gray and subsequently, in 1943, at Loch an Aigeil by A.J.Wilmott and M.S.Campbell.

*Potamogeton alpinus* Balb.
Lìobhag Dhearg
**Red Pondweed**

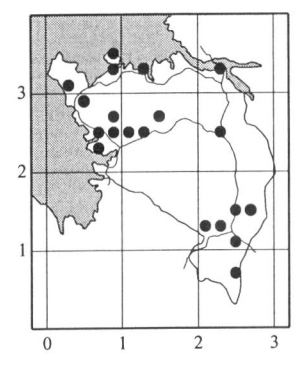

20 tetrads. Occasional in lochs. First recorded in 1943 in Loch an Aigeil by A.J.Wilmott and M.S.Campbell.

*Potamogeton praelongus* Wulfen
Lìobhag Fhada
**Long-stalked Pondweed**

11 tetrads. Occasional in the larger, base-rich, lochs. As it grows at depths inaccessible from the shore, most records were from washed-up material. First recorded in 1890 in the 'Gillaroo Loch' by E.S.Marshall and F.J.Hanbury.

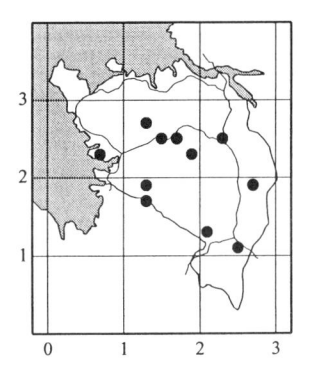

*Potamogeton perfoliatus* L.
Dreimire Uisge
**Perfoliate Pondweed**

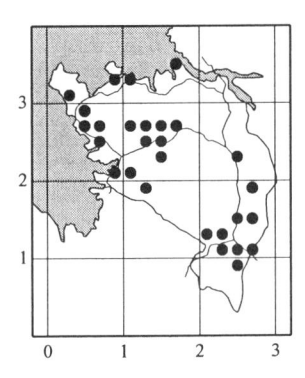

28 tetrads. Occasional in lochs and rarely in rivers. As with the preceding species, it was usually identified from washed-up material since it grows beyond wading depth. First recorded in 1886 from Loch Borralan by A.Gray.

*Potamogeton rutilus* Wolfg.
Lìobhag Ruadh
**Shetland Pondweed**

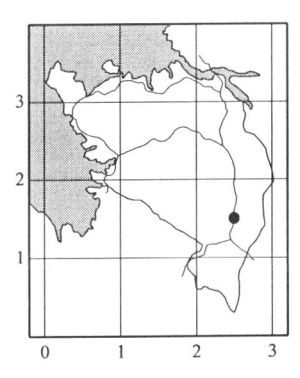

1 tetrad. Red Data Book species. A very rare pondweed known locally only from Loch Awe, where it was discovered in 1988 by E.Charter during the Scottish Loch Survey.

*Potamogeton berchtoldii* Fieber
Lìobhag Bheag
**Small Pondweed**

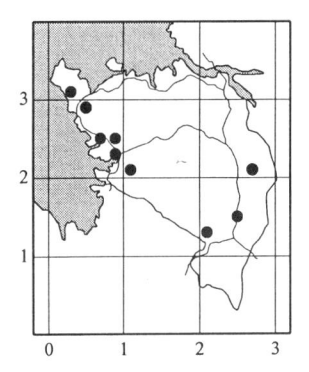

9 tetrads. Occasional in shallow water in lochs. First recorded in 1943 in Loch an Aigeil by A.J.Wilmott and M.S.Campbell.

*Potamogeton filiformis* Pers.
Lìobhag Chaol
**Slender-leaved Pondweed**

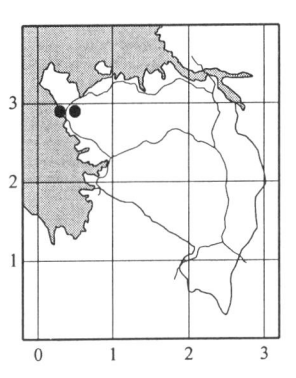

2 tetrads. Occurs in Loch an Aigeil, a rich loch at Clachtoll, and in the burn which links it to the sea. This burn is subject to occasional cleaning out, which certainly affects the flora temporarily, if not in the longer term. First recorded in 1886 in Loch Urigill by A.Gray, and in Loch an Aigeil in 1943 by A.J.Wilmott and M.S.Campbell.

[*Potamogeton pectinatus* L.
Lìobhag Fhineil
**Fennel Pondweed**

Not seen during the present survey. Recorded in 1886 in Loch Urigill by A.Gray, but not included in *Anthony*.]

RUPPIACEAE

*Ruppia maritima* L.
Snàth-lus Mara
**Beaked Tasselweed**

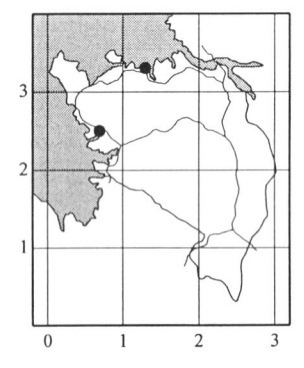

2 tetrads. Records were made during the Scottish Loch Survey in 1988, from Loch Fasg an t-Seana Chlaidh and the adjoining Lochan Sàile, and in 2001 from the brackish loch at Duart.

ZOSTERACEAE

*Zostera marina* L.
Bilearach
**Eelgrass**

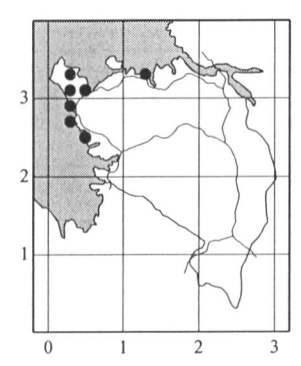

7 tetrads. A bed of the growing plant was located in Port Dhrombaig, NC13G, by T.Lockie in 2001. Prior to that we mapped the six beaches where washed-up material had been found, in the hope that this would provide some clue as to the location of *Zostera* beds. While the two northern beaches, at Clashnessie and Culkein Stoer, could conceivably have received the plant on currents from Port Dhrombaig or from Oldany (which was the site of the first record), it seems unlikely that this could be the case at Achmelvich, Balchladich, Clachtoll and Stoer Bay. There must be other *Zostera* beds out there, still to be discovered by someone who makes a systematic search in a boat at Low Spring Tides. First recorded from Oldany; the greatest detail is to be found in J.Anthony's card index, which has an entry 'sea bed…gravelly…Oldany Is….7/1958…29/03'. *Anthony* has 'Oldany, 1955, B.F[lannigan].'

LEMNACEAE

*Lemna minor* L.
Mac gun Athair
**Common Duckweed**

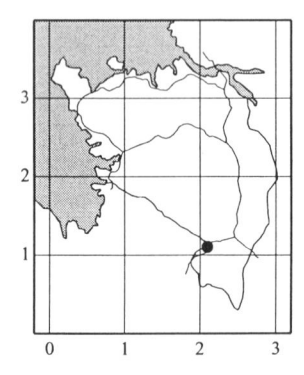

1 tetrad. The first and only record was made in 1993 at Elphin. The plants were growing in a recently re-excavated ditch by Assynt Primary School, but had gone the next year.

JUNCACEAE

*Juncus squarrosus* L.
Brù-chorcan
**Heath Rush**

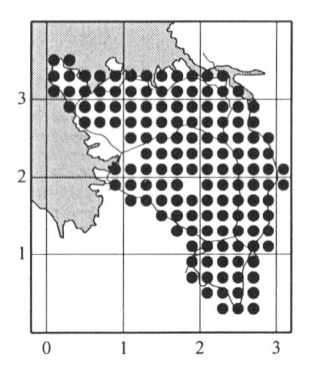

142 tetrads. Widespread and common on wet and dry heaths. It is tolerant of trampling and so is often particularly noticeable on or beside tracks. First recorded during the 1950s survey for the first *Atlas*.

*Juncus tenuis* Willd.
Luachair Chaol
**Slender Rush**

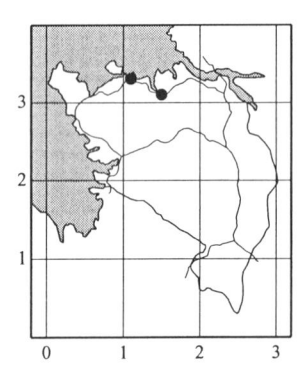

2 tetrads. Introduced. Growing on the track running from the road to the sea at Oldany and on a small disused roadside quarry, which acts as a layby, about 1km. east of Glenleraig. First recorded in 1943 as '*J.macer*', in a 'roadside runnel north of Clashnessie' by A.J.Wilmott and M.S.Campbell. *Anthony* cites 'Eddrachillis (Kylesku)', but Kylesku is not in Eddrachillis parish. Kylestrome, on the other hand, is, and as the rush is still to be found there, this may just be a confusion of names.

*Juncus gerardii* Loisel.
Luachair Rèisg Ghoirt
**Saltmarsh Rush**

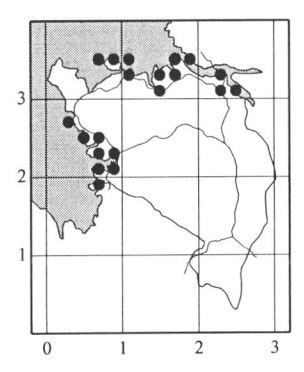

20 tetrads. Frequent in saltmarshes and patches of damp coastal turf. First recorded in 1886 at Inverkirkaig by A.Gray.

*Juncus trifidus* L.
Luachair Thrì-bhileach
**Three-leaved Rush**

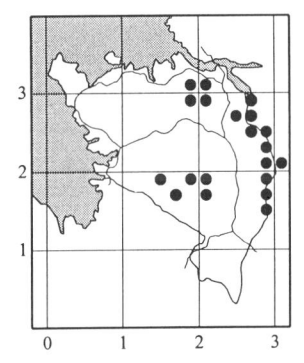

20 tetrads. Occasional, but may be common where it does occur, in bare places and rock crevices on the higher hills. First recorded in 1899 on 'Canisp, north side' by C.E.Salmon.

*Juncus bufonius* L.
Buabh-luachair
**Toad Rush**

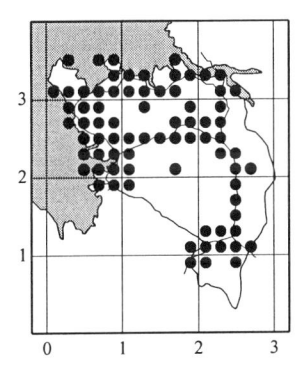

76 tetrads. Frequent along roadsides, tracksides and on disturbed ground, where conditions are suitably damp. First recorded in 1897 at Inchnadamph by E.S.Marshall.

*Juncus ambiguus* Guss.
Buabh-luachair Bheag
**Frog Rush**

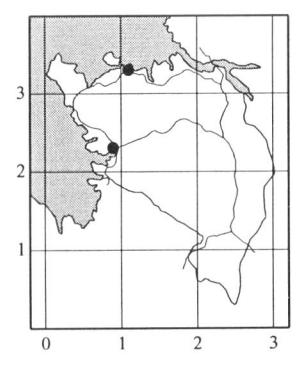

2 tetrads. Prefers brackish conditions, where a marsh meets a saltmarsh, or a burn runs through a saltmarsh to the sea. Only found at Baddidarach and Culkein Drumbeg, but almost certainly under-recorded. First recorded in 1988 in 'damp, sandy ground at head of beach, Clachtoll' by E.Norman.

*Juncus alpinoarticulatus* Chaix
Luachair Ailpeach
**Alpine Rush**

Not seen during the present survey. Found in 1887 by E.S.Marshall 'near Inchnadamph..growing with '*J. lamprocarpus*' [*J. articulatus*] at about 400ft.' There is a voucher specimen at Edinburgh, with another labelled as 'Loch Assynt…1908'. There have been no other records.

*Juncus articulatus* L.
Lachan nan Damh
**Jointed Rush**

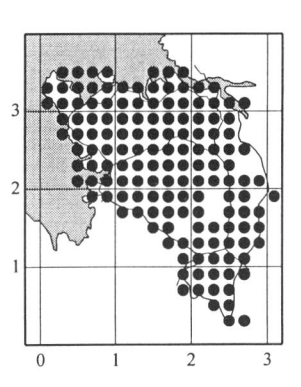

145 tetrads. Extremely common in wet places, particularly around loch margins and often actually in the water. There can be few lochs in Assynt from which this rush has not been recorded. First recorded in 1887 at Inchnadamph by E.S.Marshall.

*Juncus* x *surrejanus* Druce ex Stace & Lambinon
*J. articulatus* x *J. acutiflorus*

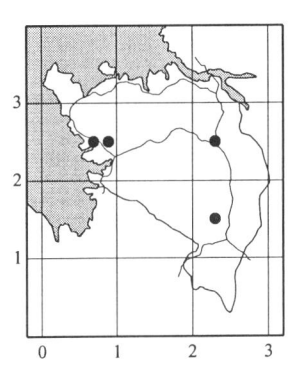

4 tetrads. Rare. Recorded from a marsh at Achadhantuir, a verge by Loch an Ordain, the edge of Loch Assynt and flushed grassland at Cnoc an Leathaid Bhuidhe. First recorded in 1992 from the last of these sites.

*Juncus acutiflorus* Ehrh. ex Hoffm.
Luachair a'Bhlàth Ghèir
**Sharp-flowered Rush**

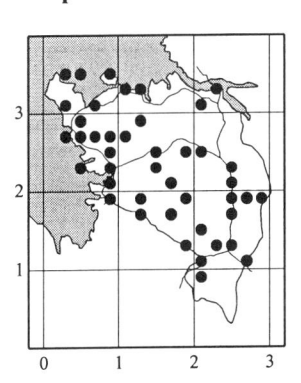

43 tetrads. Scattered in suitably wet habitats: ditches, marshes, loch and river margins. First recorded

during the 1950s survey for the first *Atlas*.

*Juncus bulbosus* L.
Luachair Bhalgach
**Bulbous Rush**

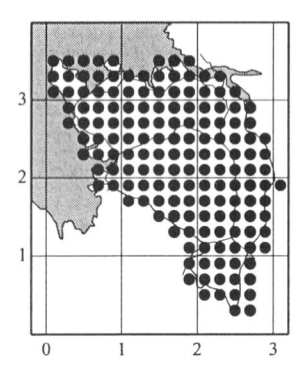

157 tetrads. Recorded in one of its forms from almost every tetrad. The terrestrial form is ubiquitous in marshes, bogs, on loch shores and disturbed ground. In shallow, still water it looks similar, but in deeper or in running water makes long, streaming plants which are usually flowerless. First recorded in 1887 by E.S.Marshall as form '*fluitans*' in a 'slow stream below Quinag' and as form '*uliginosus*' in Lochan Feòir.

*Juncus triglumis* L.
Luachair Thrì-lusan
**Three-flowered Rush**

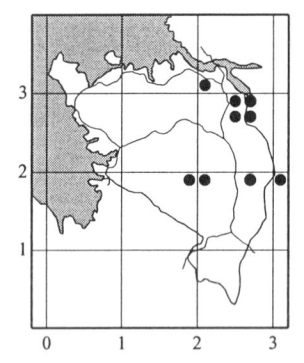

9 tetrads. In basic flushes in the hills. Found in Bealach Traligill, on the north side of Canisp, Cnoc Eilid Mhathaid, Cnoc na Creige Glas Bheinn and Sàil Gharbh. First recorded in 1886 on '*Coinne mheal* Assynt' by A.Gray. E.S.Marshall in 1889 commented 'descends to 700ft. near Inchnadamph'. This is rather lower than we have seen it.

*Juncus effusus* L.
Luachair Bhog
**Soft Rush**

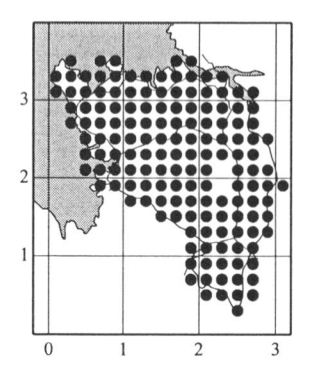

150 tetrads. Widespread and abundant throughout the parish, in bogs, marshes, wet grassland, woodland and on river margins. Var. *spiralis* is not uncommon on unmetalled tracks and occasionally elsewhere. First recorded in 1944 on a 'roadside at Inver (Lochinver)' by A.J.Wilmott and M.S.Campbell, with the comment 'uncommon (rare?)'.

*Juncus conglomeratus* L.
Bròdh Bràighe
**Compact Rush**

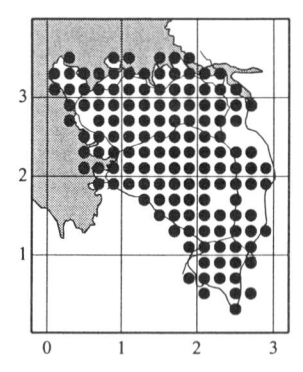

139 tetrads. Widespread, but not as abundant as the preceding species. In a similar range of habitats, but a little more prevalent on drier ground. First recorded in 1943 as 'common everywhere' by A.J.Wilmott and M.S.Campbell, with the comment 'collected at Inver (Lochinver) with what appeared to be hybrids with the next species [*J.effusus*] 1944'.

*Luzula pilosa* (L.) Willd.
Learman Fionnach
**Hairy Woodrush**

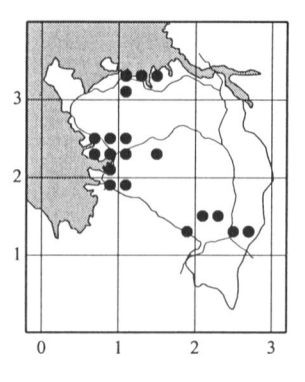

19 tetrads. Occasional in open woodland. First recorded during the 1950s survey for the first *Atlas*.

*Luzula sylvatica* (Huds.) Gaudin
Luachair Coille
**Great Woodrush**

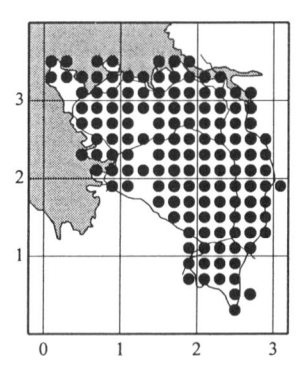

138 tetrads. Common and widespread in the shelter of woodland and rocky burn banks and a frequent component of the tall herb vegetation on north-facing crags in the hills. More surprising is its occurrence on sea cliffs, open moorland and occasionally on summit ridges, where the exposure appears to be great. First recorded in 1894 at 'Knockan Crags' by G.C.Druce.

*Luzula campestris* (L.) DC.
Learman Raoin
**Field Woodrush**

51 tetrads. In short grassland, generally at low altitudes. First recorded in 1907 'on Ben More 108' by G.C.Druce (who had

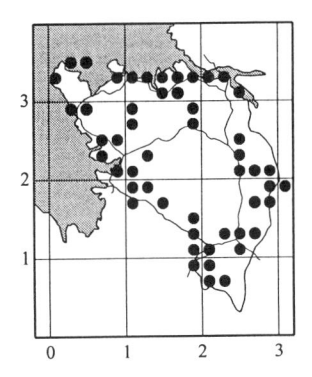

a keen perception of vice-county boundaries).

*Luzula multiflora* (Ehrh.) Lej.
Learman Monaidh
**Heath Woodrush**

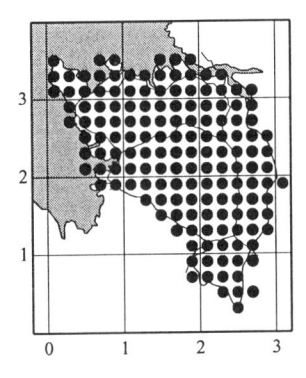

152 tetrads. Ubiquitous in dry situations in heath, acid grassland and woodland, from the coast to high in the hills. We did not make a consistent attempt to identify it to subspecies, but ssp. *congesta* was noted five times and ssp. *multiflora* three. First recorded in 1894 as '*L. congesta*....near Elphin and Ledbeg' by G.C.Druce.

*Luzula arcuata* Sw.
Learman Crom
**Curved Woodrush**

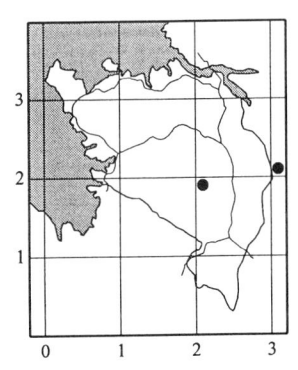

2 tetrads. A Red Data Book species found on bare, unstable hill tops. A previously unknown population of 20+ plants in fine 'soily' scree, perhaps derived from quartzite, was discovered by G.P.Rothero in 1999, at c. 800m. to the east of the summit of Canisp. A colony at 940m., bestriding the north ridge of Conival, was found by him in 1998. This ridge, along which runs the v.c.107/108 boundary, is the usual route up Ben More Assynt from Inchnadamph. The Conival site may well be where the species was first found in 1833 by A.Graham, 'on the ridge leading to the top of Ben More Assynt from Inchnadamf', and subsequently by C.E.Salmon in 1899, with the comment 'still plentiful in one compact patch on the way to the summit where it was recorded many years ago.'

*Luzula spicata* (L.) DC.
Learman Ailpeach
**Spiked Woodrush**

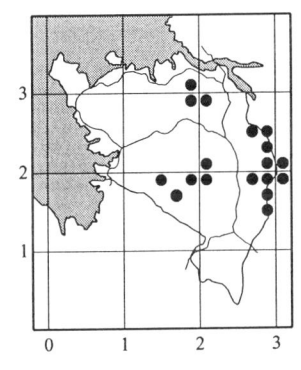

18 tetrads. On open stony ground on the hills, at lower altitudes than the preceding. First recorded in 1907 on 'Ben More, Assynt' by G.C.Druce (see above under *L. campestris*).

CYPERACEAE

*Eriophorum angustifolium* Honck.
Canach
**Common Cottongrass**

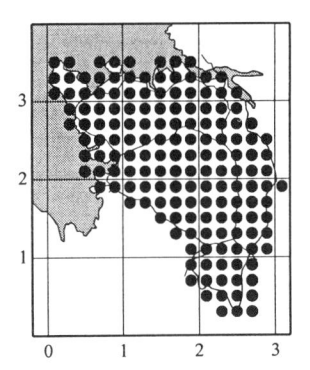

158 tetrads. Widespread and abundant in wet bogs and shallow pools. First recorded in 1767, as '*E. polystachion*', near Inchnadamph by J.Robertson.

*Eriophorum latifolium* Hoppe
Canach an t-Slèibh
**Broad-leaved Cottongrass**

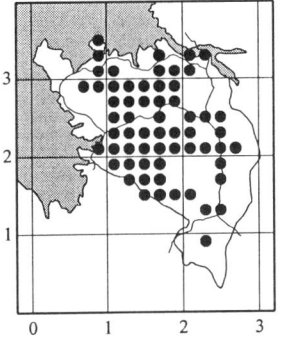

61 tetrads. In small populations scattered across the gneiss in base-rich flushes; also occasionally on the limestone. First recorded in 1886 at Achmore by A.Gray.

*Eriophorum vaginatum* L.
Sìoda Monaidh
**Hare's-tail Cottongrass**

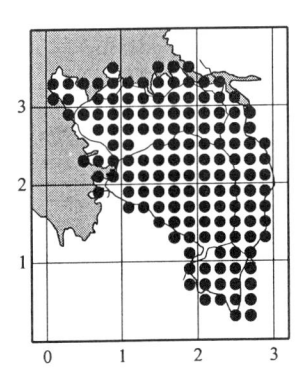

143 tetrads. Although equally widespread, it is nothing like as abundant as *E. angustifolium*. It grows in damp, peaty places, often in small tussocks, and its early

155

flowers are a welcome first sign of life on the hill. First recorded in 1767 near Inchnadamph by J.Robertson.

*Trichophorum cespitosum* (L.) Hartm.
(Scirpus cespitosus L.)
Cìob
**Deergrass**

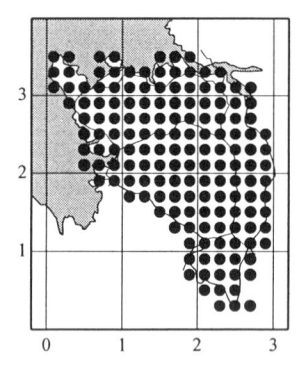

155 tetrads. Ubiquitous in bogs, wet and dry heaths and acid grasslands. Subspecies *cespitosum*, although recorded elsewhere in West Sutherland, has not yet been seen in Assynt, although it seems likely that it will be found, because of the sheer quantity of suitable habitat. First recorded in 1767 near Inchnadamph by J.Robertson.

*Eleocharis palustris* (L.) Roem. & Schult.
Bioran Coitcheann
**Common Spike-rush**

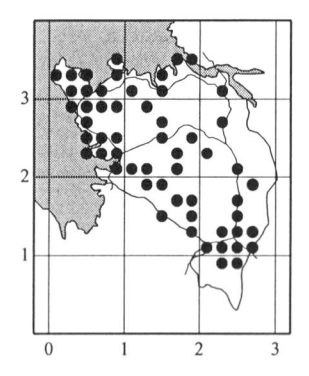

56 tetrads. Locally frequent as emergent vegetation along the margin of slow-moving rivers and in the more fertile lochs. Not usually seen in the acid hill lochs. Recorded in 1767 near Inchnadamph by J.Robertson.

*Eleocharis uniglumis* (Link) Schult.
Bioran Caol
**Slender Spike-rush**

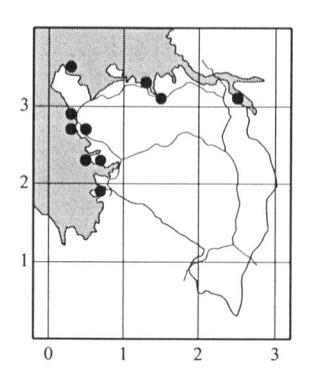

10 tetrads. Uncommon, in saltmarshes and freshwater marshes near the sea. First recorded during the 1950s survey for the first *Atlas*.

*Eleocharis multicaulis* (Sm.) Desv.
Bioran Badanach
**Many-stalked Spike-rush**

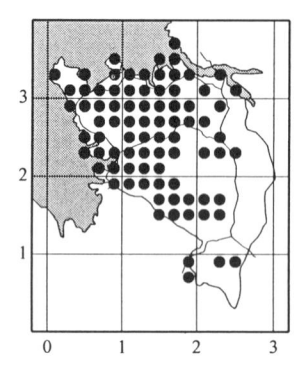

82 tetrads. Although found in stony flushes, it is most common in the shallow water on the edge of lochs. Rare in limestone lochs and those at very high levels. First recorded in 1886 at Tumore by A.Gray.

*Eleocharis quinqueflora* (Hartmann) O.Schwartz
Bioran nan Lusan Gann
**Few-flowered Spike-rush**

105 tetrads. Common and widespread in open communities, in wet stony flushes or on wet peat.

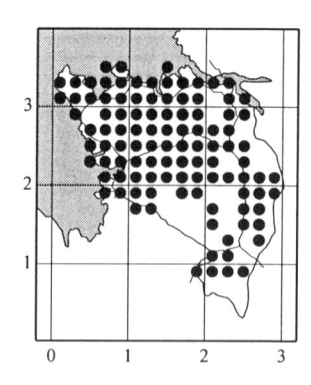

Absent from the higher hills. First recorded in 1907 at Inchnadamph by G.C.Druce.

[*Eleocharis acicularis* (L.) Roem. & Schult.
Bioran Dealgach
**Needle Spike-rush**

Recorded, presumably in error, in 1886 at Loch Awe by A.Gray.]

*Schoenoplectus lacustris* (L.) Palla
(Scirpus lacustris L.)
Luachair Ghòbhlach
**Common Club-rush**

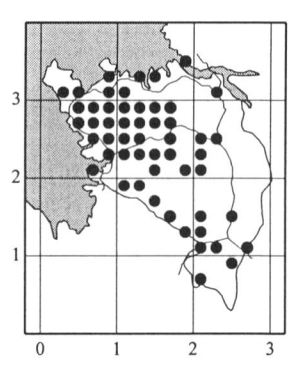

53 tetrads. With two exceptions, all the records are from lochs; rivers in Assynt are mostly too fast and shallow to suit this species. First recorded in 1886 in Lochan Feòir by A.Gray.

*Isolepis setacea* (L.) R.Br.
(Scirpus setaceus L.)
Curcais Chalgach
**Bristle Club-rush**

33 tetrads. Occasional in damp open ground or amongst very short vegetation. In acid flushes, wet grassland and along tracks.

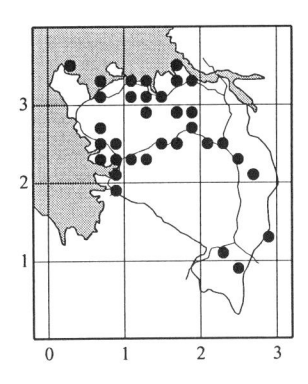

First recorded in 1943 at Lochinver by A.J.Wilmott and M.S.Campbell.

*Eleogiton fluitans* (L.) Link
(Scirpus fluitans L.)
Curcais air Bhog
**Floating Club-rush**

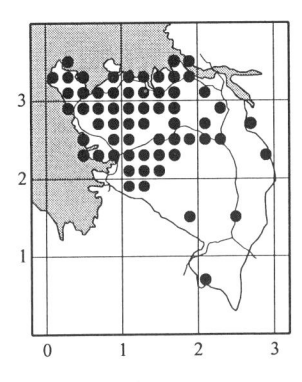

61 tetrads. Common in lochs on the gneiss and on mud which is only seasonally flooded. Flowers freely, both terrestrially and as an aquatic. It is an easily recognised bright green, subtly different from the colour of any other filiform aquatic found in this area. First recorded in 1890 at Lochan Feòir by E.S.Marshall.

*Blysmus rufus* (Huds.) Link
Seisg Rèisg Ghoirt
**Saltmarsh Flat-sedge**

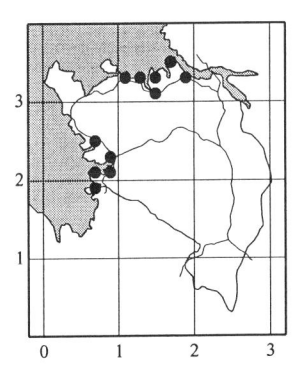

11 tetrads. Uncommon, in saltmarshes and wet stony places at the back of the shore. First recorded during the 1950s survey for the first *Atlas*.

*Schoenus nigricans* L.
Sèimhean Dubh
**Black Bog-rush**

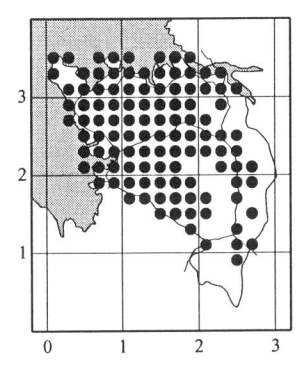

110 tetrads. May be dominant in an extensive mire or form a band along the margins of a small watercourse, nearly always where there is base-rich flushing. See Ferreira (1995) for a detailed account of its ecology. Although large stands of *Schoenus* are generally species-poor, it is interesting that *Dactylorhiza lapponica* is associated with it in Assynt and that *Platanthera bifolia* is often found in its vicinity. First recorded in 1767 near Inchnadamph by J.Robertson.

*Rhynchospora alba* (L.) Vahl
Gob-sheisg
**White Beak-sedge**

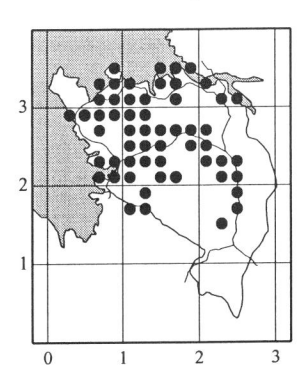

56 tetrads. Locally frequent in very wet places, often fringing bog pools; mainly on the gneiss. Usually in extensive patches, the pale spikes conspicuous at flowering time. First recorded in 1887 near Inchnadamph by E.S.Marshall.

*Cladium mariscus* (L.) Pohl
Sàbh-sheisg
**Great Fen-sedge**

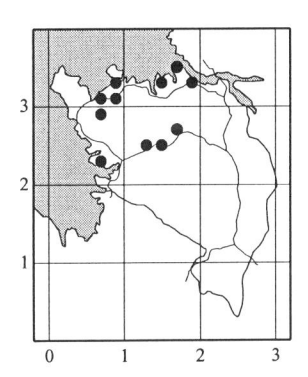

11 tetrads. Sparsely distributed in, or adjacent to, a few of the more base-rich lochs, where it makes good stands in sheltered bays. First recorded during the 1950s survey for the first *Atlas*.

*Carex paniculata* L.
Seisg Bhadanach Mhòr
**Greater Tussock-sedge**

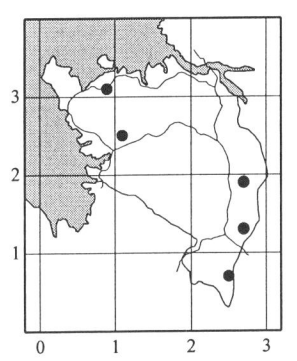

5 tetrads. A rare sedge, with a strangely disjunct distribution. There is a single tussock in a bog at Achadh Mór, and one 7 km. away in a bog at Bad a'Bhainne. It is then 17 km. to six plants in miry loops at Loch Mhaolach-coire, where the burn runs in from the south, and another 6 km. to the Ledbeg River, south-east of Feur Loch, where there is a respectable group of 20 plants! Three tussocks in a flush by the Allt nam Meur at Cromalt, complete the picture and leave the botanist wondering how it makes such leaps. First recorded

in 1983 at Loch Mhaolach-coire by A.S.MacLennan.

*Carex otrubae* Podp.
Seisg Gharbh Uaine
**False Fox-sedge**

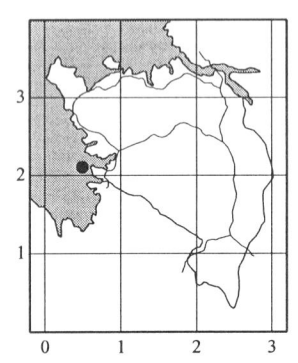

1 tetrad. Not previously recorded for Assynt, its last minute discovery on Soyea Island was a complete fluke. While scrambling over rocks to return to the boat, we spotted the plants on the edge of a brackish pool just above high-water mark. This is an identical situation to those occupied by plants growing at Duartbeg in Eddrachillis further up the coast to the north, and on Eilean Mór, just to the south in West Ross.

*Carex arenaria* L.
Seisg Ghainmhich
**Sand Sedge**

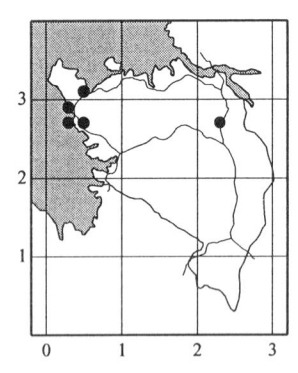

5 tetrads. Frequent at Clachtoll, Clashnessie and Stoer in its traditional habitat of dunes and sandy coastal grassland. Along the road running north from Skiag Bridge there is a colony of this sedge on the verge, extending for 5m. Presumably brought in with sand, it is surviving there alongside a very well-salted road, 6km. from the sea. First recorded in 1886 at Achmelvich Bay by A.Gray.

*Carex disticha* Huds.
Seisg Ruadh
**Brown Sedge**

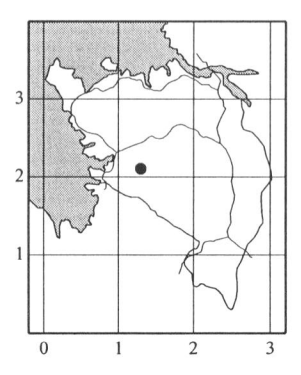

1 tetrad. Very rare, growing in small quantity in a marsh alongside the footpath from Glencanisp Lodge to Suilven. First recorded in 1955 at Achmelvich by J.Anthony; this record is not in his Flora.

*Carex remota* L.
Seisg Sgarta
**Remote Sedge**

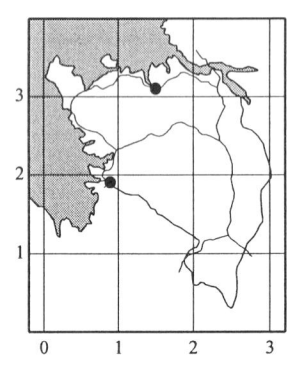

2 tetrads. Growing beneath hazel on the bank of the River Kirkaig, about 1km. from the mouth, and in wet woodland on both sides of the river at Glenleraig. First recorded, in *Anthony*, as 'Lochinver, Stoer, 1944, A.J.W.'.

*Carex ovalis* Gooden.
Seisg Ughach
**Oval Sedge**

56 tetrads. Found very commonly on roadside verges and in other lowland, grassy places. Also, although less often, in marshes. First recorded in 1907 at Inchnadamph by G.C.Druce.

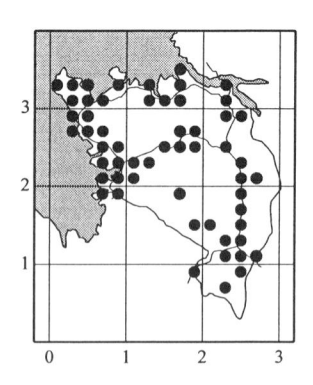

*Carex echinata* Murray
Seisg Reultach
**Star Sedge**

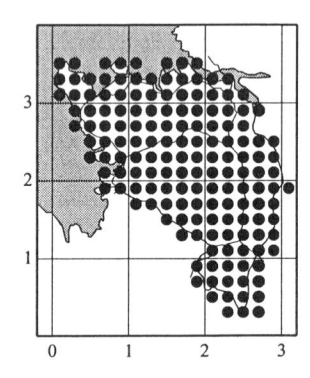

156 tetrads. Widespread and abundant in ground that is permanently wet: mires, bogs and wet heath. First recorded in 1894 at 'Knockan Crags' by G.C.Druce.

*Carex dioica* L.
Seisg Aon-cheannach
**Dioecious Sedge**

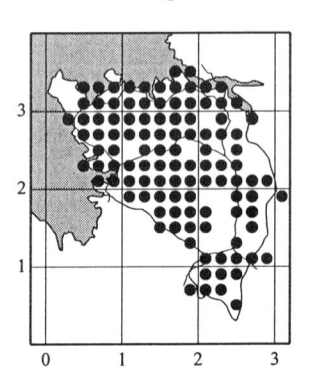

108 tetrads. Widespread and common. In open situations with silty soil, or in stony flushes,

where the conditions are not too acid. First recorded in 1894 at 'Knockan Crags' by G.C.Druce.

*Carex curta* Gooden.
Seisg Bhàn
**White Sedge**

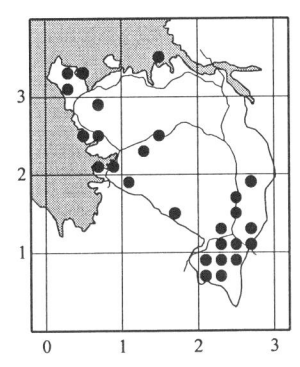

26 tetrads. A puzzling distribution for a reputedly acid-loving species, as the concentration of sites in the south-east would suggest at least some affinity with limestone. Closer examination of these shows that four are on peat over limestone, six on alluvium and others just on 'peat'. Topography may provide a reason, but there seems to be a factor, which we are not appreciating, responsible for its absence from other areas of bog. All our records are lowland. First recorded in 1887 on the 'W. side of Coniveall (108), at over 2500ft.' by E.S.Marshall.

*Carex lasiocarpa* Ehrh.
Seisg Choilleanta
**Slender Sedge**

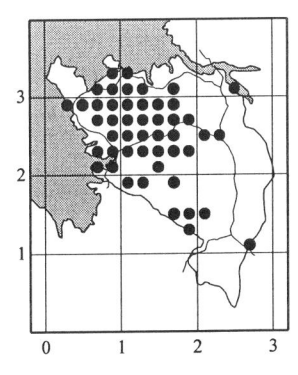

48 tetrads. Frequent as emergent vegetation in lochs and also in marginal reedswamp. This sedge can be the dominant species in extensive mires; under these conditions the plants do not flower as freely as do those in water, but their fine, whippy leaves are unmistakable. First recorded in 1886 in Lochan Feòir by A.Gray.

*Carex rostrata* Stokes
Seisg Shearragach
**Bottle Sedge**

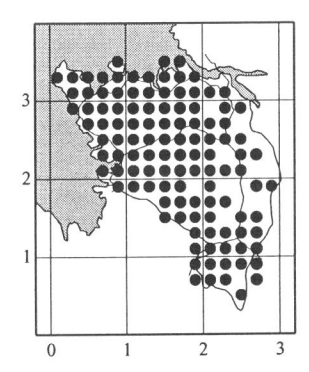

114 tetrads. Widespread and abundant as an emergent in lochs, rivers and bog pools. Grows also in wet heath, bogs, marshes and ditches. Flowers freely in open water but less so in terrestrial situations. First recorded, as '*C.ampullacea*', in 1886 in Lochan Feòir by A.Gray.

*Carex vesicaria* L.
Seisg Bhalganach
**Bladder-sedge**

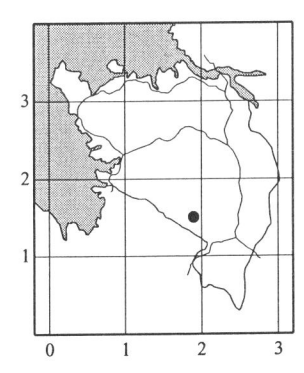

1 tetrad. Found only once during present survey, in a patch of reedswamp at the mouth of the Allt na Braclaich, where it runs into the western end of Cam Loch. First recorded in 1977 at Loch Urigill by E.D.Allen *et al.*

*Carex sylvatica* Huds.
Seisg Choille
**Wood-sedge**

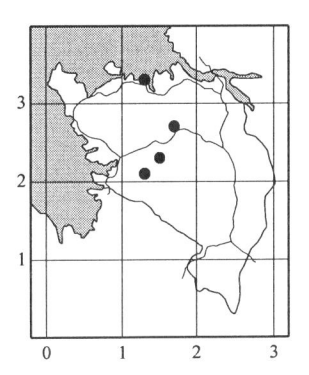

4 tetrads. Occurs in very small quantity in each of its four sites: under hazels at Nedd, beside the Allt an Tiaghaich, on the edge of the burn running out of Loch Bad a'Chigean and in deciduous woodland south-east of Loch Druim Suardalain. First recorded in 1992 at the last-named site.

*Carex capillaris* L.
Seisg Ghrinn
**Hair Sedge**

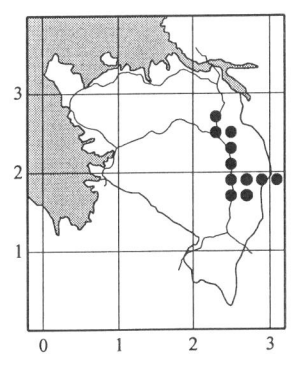

11 tetrads. Confined to limestone areas where it grows in grassland, around outcrops and on ledges. First recorded in 1886 by A.Gray, with the comment 'Loch Assynt (frequent)'. G.C.Druce's 1907 record is more informative, 'Abundant on the limestone, descending to 200ft. near Ardvrick Castle'. Also recorded in 1995 on 'dry slope, Clachtoll' by B.Flannigan, and in 1958 at 'Stoer Bay' by B.Flannigan and R.B.Knox. We have not seen it on the coast, perhaps because of the heavy grazing there.

*Carex flacca* Schreb.
Seisg Liath-ghorm
**Glaucous Sedge**

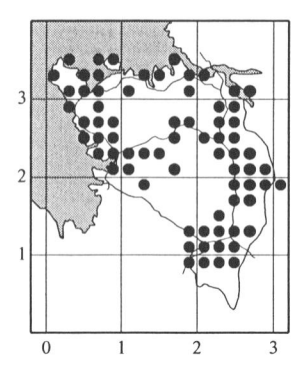

71 tetrads. In limestone and coastal grasslands, but also scattered over the gneiss, in areas of mineral enrichment. First recorded in 1894 at 'Knockan Crags' by G.C.Druce. He later (1907) recorded a 'slender form' on Canisp.

*Carex panicea* L.
Seisg a'Chruithneachd
**Carnation Sedge**

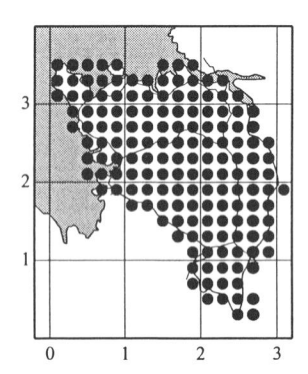

157 tetrads. The most widespread and abundant sedge in Assynt, although *C. echinata* runs it a close second. Grows everywhere: wet and dry heath, grasslands, mires, bogs, stony loch shores and, in a dwarf form, high on the hills. First recorded in 1894 on 'Knockan Crags' by G.C.Druce.

*Carex laevigata* Sm.
Seisg Mhìn
**Smooth-stalked Sedge**

1 tetrad. The first and only record was made in 1992, in the extensively wooded grounds of Glencanisp Lodge, to which it may have been introduced during planting.

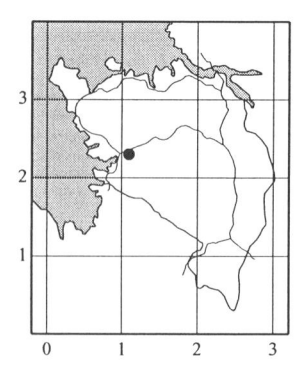

*Carex binervis* Sm.
Seisg Fhèith-ghuirm
**Green-ribbed Sedge**

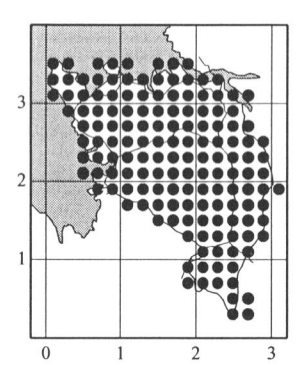

150 tetrads. Widespread and abundant in wet and dry heath, rocky valleys, banks, gorges and cliff ledges. Rarely found in open grassland. First recorded in 1887, 'on the heaths above Loch Assynt', by E.S.Marshall.

*Carex distans* L.
Seisg Fhada-mach
**Distant Sedge**

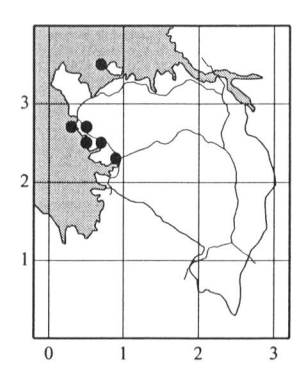

6 tetrads. Found at Port Alltan na Bradhan and on Oldany Island on wet rocks, at Clachtoll in coastal grassland within the spray zone, at Baddidarach and Achmelvich in saltmarsh. First recorded in 1987 at the last-named site.

*Carex extensa* Gooden.
Seisg Anainn
**Long-bracted Sedge**

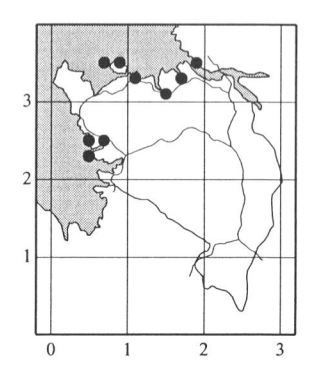

9 tetrads. Occasional in salt marshes and on wet rocky shores. First recorded in 1956 in saltmarsh at the head of Loch Nedd by A.O.Chater.

*Carex hostiana* DC.
Seisg Odhar
**Tawny Sedge**

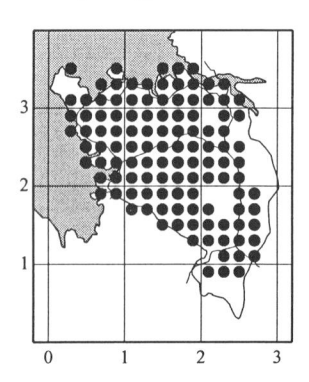

115 tetrads. Widespread but thinly distributed. In wet heath, wet flushes and rough grassland, favouring base-rich habitats. First recorded in 1886, as '*C. fulva*', from 'Tumore – Loch Inver' by A.Gray.

*Carex* x *fulva* Gooden.
*C. hostiana* x *C. viridula*

7 tetrads. Uncommon; in basic flushes, usually near *C. hostiana*.

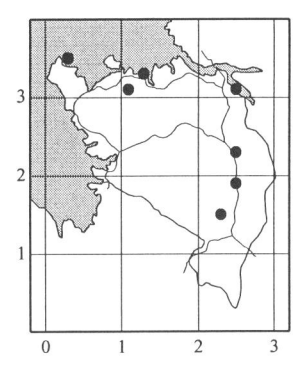

First recorded in 1996 near Calda House, Inchnadamph.

*Carex viridula* Michx.
**Yellow-sedge**

ssp. *brachyrrhyncha* (Celak.) B.Schmid
(*C. lepidocarpa* Tausch)
Seisg Bhuidhe Fhad-chuiseagach
**Long-stalked Yellow-sedge**

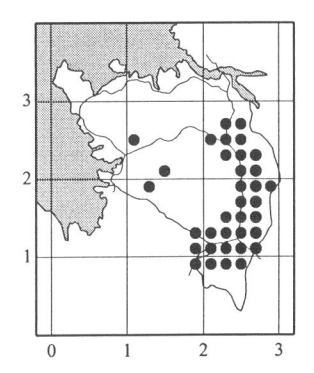

35 tetrads. Common in base-rich habitats: flushes, wet grassland, and on stony edges of lochs and burns. Var. *scotica* (E.W.Davies) B.Schmidt was described new to the British Isles from Inchnadamph in 1952 (as ssp. *scotica*). First recorded in 1943, 'south of Lochinver towards Lady Constance Bay' by A.J.Wilmott and M.S.Campbell; also found at Clashnessie by A.J.Wilmott in 1944.

ssp. *oedocarpa* (Andersson) B.Schmid
(*C. demissa* Hornem)
Seisg Bhuidhe Choitcheann
**Common Yellow-sedge**

153 tetrads. One of the commonest and most widespread sedges in wet and open places: on loch shores, burn edges and flushes, from sea level to high in the hills. First recorded as '*C. chrysites*' in 1890 'by Loch Assynt' by E.S.Marshall and F.J.Hanbury.

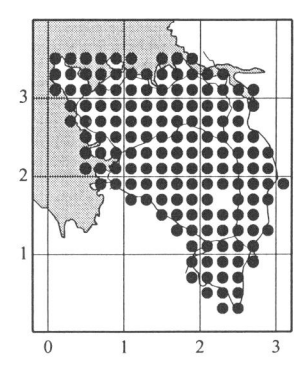

ssp. *viridula*
(*C. serotina* Mérat)
Seisg nam Measan Beaga
**Small-fruited Yellow-sedge**

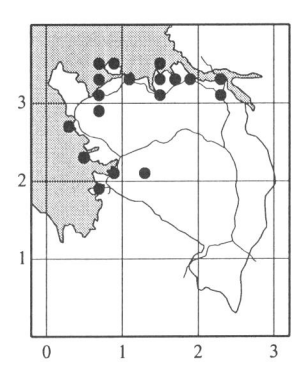

18 tetrads. Almost entirely coastal in distribution, in salt marshes and in wet and stony places at the back of the shore. Two localities are inland, one on the stony margin of Loch Poll Dhaidh and the other on the north shore of the river, just downstream of Loch na h-Airigh Fraoich. First recorded in 1956 in saltmarsh at the head of Loch Nedd by A.O.Chater.

*Carex pallescens* L.
Seisg Gheal
**Pale Sedge**

81 tetrads. It is difficult to say what determines the distribution of this species, which may be common in, or absent from, apparently similar areas. It does not require the damp sheltered woodland conditions associated with it further south in the British Isles, because of the high atmospheric moisture and rainfall of this area. It occurs in dry and wet heath, marshes, grassland, and roadsides. First recorded in 1886 on Loch Assynt by A.Gray.

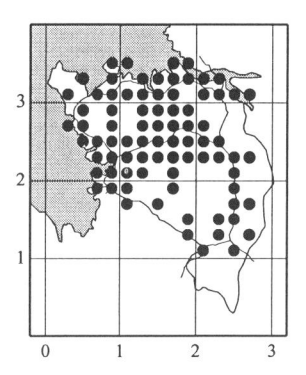

*Carex caryophyllea* Latourr.
Seisg an Earraich
**Spring-sedge**

Not seen during the present survey. First recorded in 1956 from NC21 and 22 by A.C.Crundwell; again from NC22 by M.McC.Webster in 1961.

*Carex pilulifera* L.
Seisg Lùbach
**Pill Sedge**

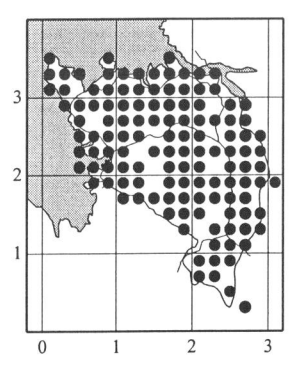

124 tetrads. Widespread in dry places, growing in rock crevices and dry heath. When growing on rock it is easily recognised even in winter, because of its tufts of curved leaves and stems, the latter persisting after the utricles have fallen. First recorded in 1894 on 'Knockan Crags' by G.C.Druce.

161

*Carex limosa* L.
Seisg na Mòna
**Bog-sedge**

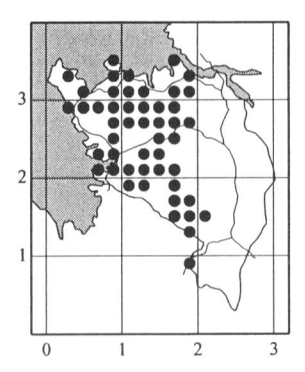

49 tetrads. Occasional in shallow muddy or silty water, in mires, bog pools and amongst reedswamp on the edges of lochs. The plants lie along the surface of the water with the ends of their leaves curving upwards, making them identifiable even in the non-flowering state. Quite often in the same places as *C. lasiocarpa*; the distribution maps are similar. First recorded in 1890 'in two bogs near Lochinver' by E.S.Marshall and F.J.Hanbury.

*Carex aquatilis* Wahlenb.
Seisg Uisge
**Water Sedge**

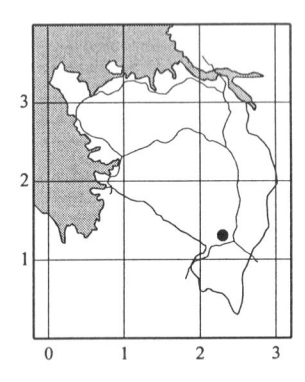

1 tetrad. Found in two places in reedswamp bordering the Ledmore River. First recorded in 1992 near the river mouth, and in 1997 about a kilometre upstream.

*Carex nigra* (L.) Reichard
Gainnisg
**Common Sedge**

153 tetrads. One of the most abundant sedges, growing in a wide range of wet habitats, but particularly common beside water. Also in marshes and flushes and on roadside verges. First recorded in 1887 at Lochan Feòir by E.S. Marshall.

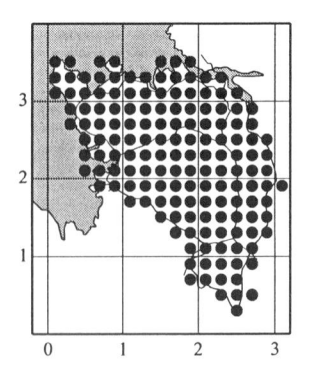

*Carex bigelowii* Torr. ex Schwein.
Dùr-sheisg
**Stiff Sedge**

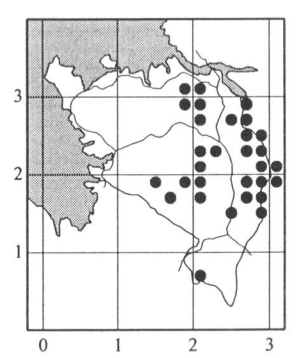

30 tetrads. Frequent on the higher hills, down to just over 400m. In screes, flushes and montane heath. First recorded, as '*C.stricta*', in 1886, on '*Coinne mheal*, Quinag' by A.Gray.

*Carex pauciflora* Lightf.
Seisg nan Lusan Gann
**Few-flowered Sedge**

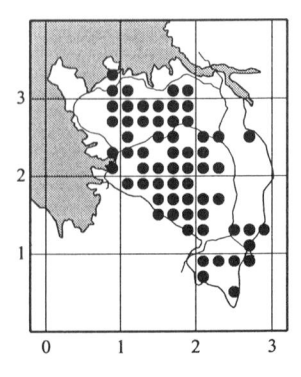

62 tetrads. In bogs, widespread but not common. Found on really wet ground, often in association with *Rhynchospora alba*. Frequently seen growing on *Sphagnum,* on the edge of bog pools and mires. First recorded in 1907 'at base of Quinag' by G.C.Druce.

*Carex rupestris* All.
Seisg na Creige
**Rock Sedge**

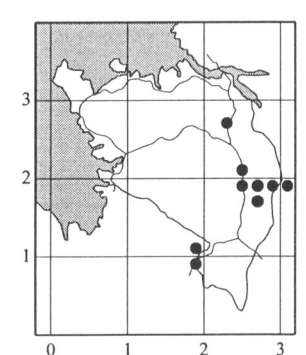

9 tetrads. Confined to limestone cliffs and outcrops, down to 300m. Shy-flowering, but recognisable in the vegetative state by its persistent curled leaves which have earned it the name of 'pig's-tail sedge'. First recorded in 1890 on 'Limestone cliffs, in a valley about three miles on the Altnagealgach side of Inchnadamph, [Allt nan Uamh?], in great abundance, at 600 to 800 feet' by E.S.Marshall and F.J.Hanbury.

*Carex pulicaris* L.
Seisg na Deargainn
**Flea Sedge**

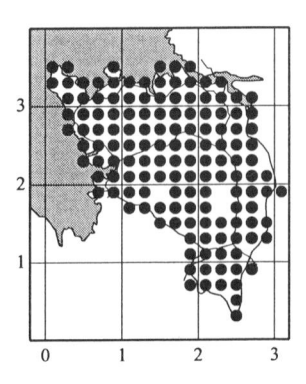

141 tetrads. Widespread and common where there are mesotrophic conditions in wet grassland, wet heath and bog. First

recorded in 1894 on 'Knockan Crags' by G.C.Druce.

POACEAE

*Nardus stricta* L.
Riasg
**Mat-grass**

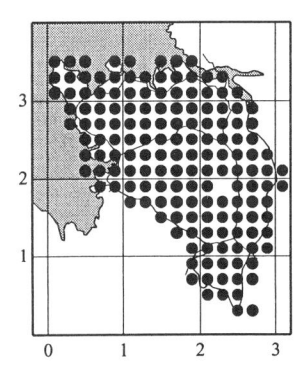

156 tetrads. Abundant and sometimes dominant in dry heath and acid grassland, from sea level to the tops of hills. First recorded in 1767 near Inchnadamph by J.Robertson.

*Festuca arundinacea* Schreb.
Fèisd Ard
**Tall Fescue**

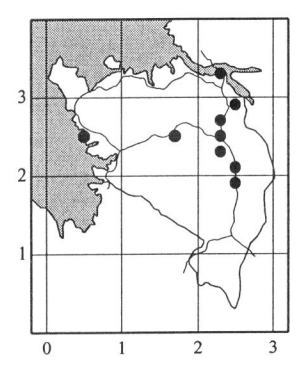

9 tetrads. Found very occasionally along roadside verges, where it has probably been introduced, and once in the grassland by Ardvreck Castle. First recorded in 1988 near Inchnadamph (roadside and 'banks of islets in Traligill Burn') and at Achmelvich by P.J.O.Trist.

*Festuca altissima* All.
Fèisd Choille
**Wood Fescue**

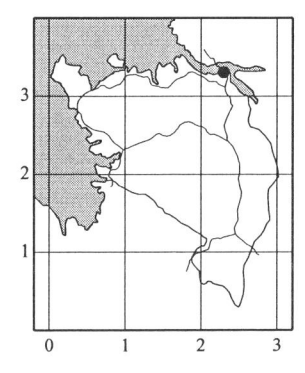

1 tetrad. In the wooded Creag an Spardain ravine, under north-east facing crags, a number of clumps of the fescue grow in a tall herb/tall fern community with much *Luzula sylvatica*. This is its only station in Assynt and the most northerly in Britain. First recorded in the 1970s by R.E.C.Ferreira.

*Festuca rubra* L.
Fèisd Ruadh
**Red Fescue**

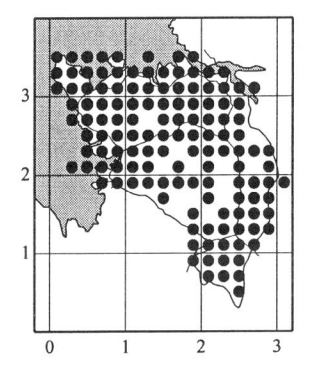

131 tetrads. Widespread and common in all types of grassland, coastal and hill, calcareous and acid. First recorded in 1767 near Inchnadamph by J.Robertson.

There are records for two subspecies:

ssp. *juncea* (Hack.) K.Richt.

4 tetrads. Grassland, including roadside verges, near the sea at Achmelvich, Clashnessie and Unapool. First recorded in 1995 at the last-named site.

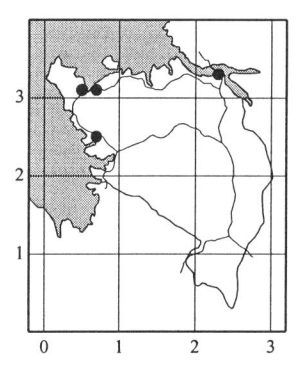

ssp. *arctica* (Hack.) Govor.

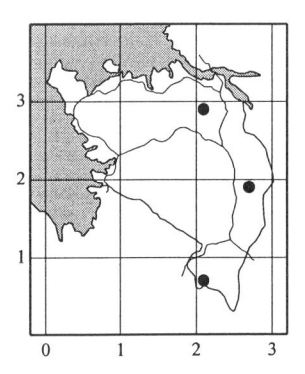

3 tetrads. In the hills: on the stony shore of Loch Mhaolach-coire, in acid grassland on Quinag and on wet scree at Coire an Lochan. First recorded in 1994 at the third site.

*Festuca ovina* L.
Feur Chaorach
**Sheep's-fescue**

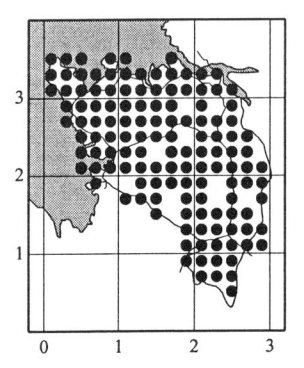

130 tetrads. Widespread and common in grassland and dry heath, but not found on the tops of the hills. First recorded in 1767 near Inchnadamph by J.Robertson.

*Festuca vivipara* (L.) Sm.
Feur Chaorach Bèo-bhreitheach
**Viviparous Sheep's-fescue**

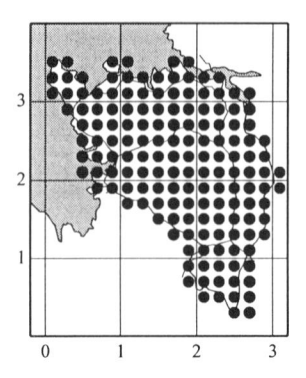

155 tetrads. The most widespread sheep's-fescue, growing in heath and grassland of all kinds, from sea level to high on the hills. First recorded in 1767 near Inchnadamph by J.Robertson.

*Festuca filiformis* Pourr.
(*F. tenuifolia* Sibth.)
Feur Chaorach Mìn
**Fine-leaved Sheep's-fescue**

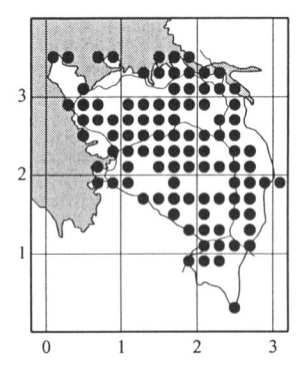

98 tetrads. Widespread and common in grassland and dry, rocky places. Less frequent around the coast than *F. ovina*. First recorded in 1943 near Lochinver by U.K.Duncan.

*Lolium perenne* L.
Breòillean
**Perennial Rye-grass**

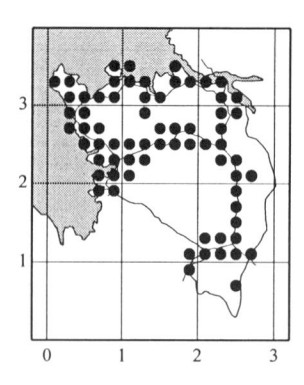

68 tetrads. Not a native grass and its distribution is closely tied in with the influence of man. Occasional on roadside verges, waste places and grassland away from the road which is near old dwellings or fanks. The grass presumably arrived there either as seed on the feet of stock, or as feed at some time in the past. First recorded during the 1950s survey for the first *Atlas*.

*Lolium multiflorum* Lam.
Breòillean Eadailteach
**Italian Rye-grass**

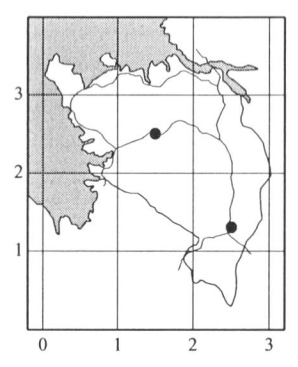

2 tetrads. Found only rarely, as a garden weed at Loch Assynt Lodge and Ledmore. Presumably an accidental introduction. First recorded in 1944 as a 'garden weed' at Lochinver, by A.J.Wilmott and M.S.Campbell.

*Vulpia bromoides* (L.) Gray
Fèisd Aimrid
**Squirreltail Fescue**

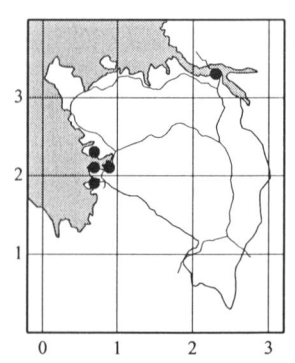

5 tetrads. A rare grass, growing on outcrop rock at Badnaban, Inverkirkaig and Strathan, on a trackside at Ardroe and a roadside at Unapool. First recorded as '*F. sciuroides*' in 1894 'near Elphin and Ledbeg' by G.C.Druce. Recorded for NC12 in the first *Atlas*.

*Cynosurus cristatus* L.
Coin-fheur
**Crested Dog's-tail**

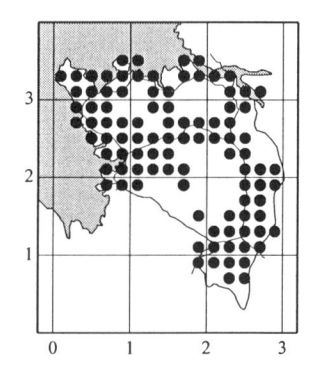

97 tetrads. Widespread and common in the more fertile grasslands, but not in montane habitats. First recorded in 1767 near Inchnadamph by J.Robertson.

*Puccinellia maritima* (Huds.) Parl.
Feur Rèisg Ghoirt
**Common Saltmarsh-grass**

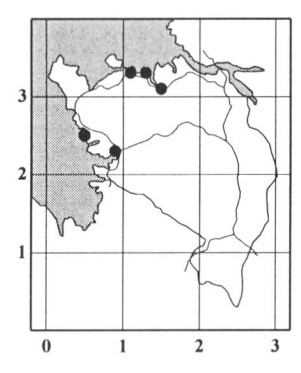

5 tetrads. Occasional in saltmarshes at Achmelvich, Baddidarach, Duart Loch, Loch Nedd and Lochan na Leobaig, but rarely flowers, so may be under-recorded. A *Puccinellia*, presumed to be this species, was recorded in 1956 in 'saltmarsh at the head of Loch Nedd' by A.O.Chater, and in 1972, P.Adam recorded *P. maritima* from 'Loch Nedd saltmarsh'.

*Briza media* L.
Conan Cumanta
**Quaking-grass**

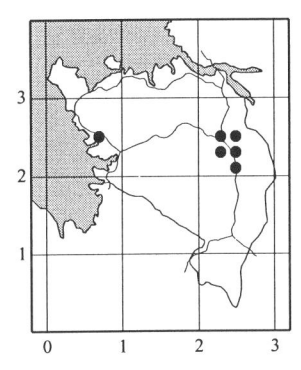

6 tetrads. Apart from a curious outlier on the verge near Cnoc a'Bhainne, all the records come from grassland or roadside verges on limestone: near Inchnadamph, Achmore, Ardvreck Castle and Allt a'Chalda Beag. First recorded in 1949 at Ardvreck Castle by M.McC.Webster. Also mapped for NC21 in the first *Atlas*.

*Poa annua* L.
Tràthach Bliadhnail
**Annual Meadow-grass**

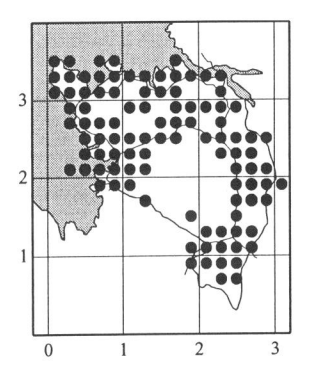

100 tetrads. Widespread and common in grassland of many types, waste and disturbed ground, verges and tracks, where it will tolerate dry conditions. Although not usually on high rocky ground, it does occur in the hills in bryophyte flushes. First recorded in 1767 near Inchnadamph by J.Robertson.

*Poa trivialis* L.
Tràthach Garbh
**Rough Meadow-grass**

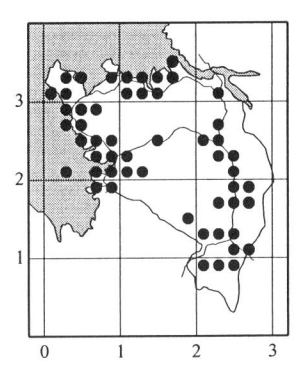

53 tetrads. Occasional in damp places at low altitudes: ditches, marshes, cultivated ground and shaded places where there is long grass, such as in old fanks. First recorded in 1767 near Inchnadamph by J.Robertson.

*Poa humilis* Ehrh. ex Hoffm.
(*P. subcaerulea* Sm.)
Tràthach Sgaoilte
**Spreading Meadow-grass**

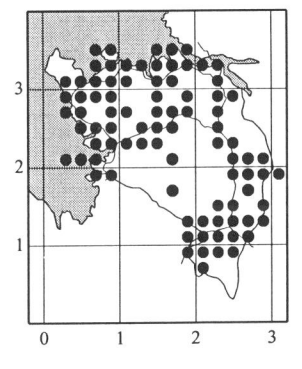

86 tetrads. Frequent in light, disturbed ground on roadsides and particularly noticeable on tops of old walls. Also in hill grasslands to an altitude of c.500m. in the Bealach Traligill. First recorded in 1894 on 'Knockan Crags' by G.C.Druce.

*Poa pratensis* L.
Tràthach Mìn
**Smooth Meadow-grass**

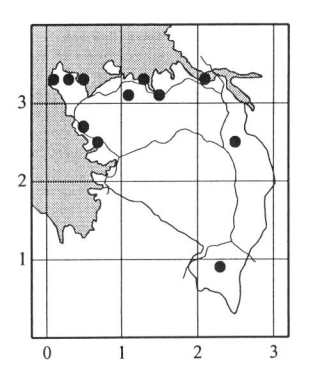

11 tetrads. This species may have been under-recorded in the early stages of the survey, but even so, it is not a common grass, occurring only in coastal and limestone grasslands.

First recorded in 1897 at Inchnadamph by E.S.Marshall.

*Poa angustifolia* L.
Tràthach na Duilleige Caoile
**Narrow-leaved Meadow-grass**

Not seen during the present survey. First recorded in 1890 as '*P. pratensis* var. *angustifolia*...near Inchnadamph on limestone', by E.S.Marshall and F.J.Hanbury.

*Poa glauca* Vahl
Tràthach Liath-ghorm
**Glaucous Meadow-grass**

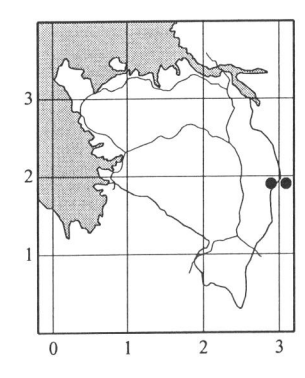

2 tetrads. Very rare. Found by G.P.Rothero on the Fucoid Beds, in two places in Bealach Traligill. First recorded in 1899 on 'Canisp, at 2000 feet' by C.E.Salmon.

*Poa nemoralis* L.
Tràthach Coille
**Wood Meadow-grass**

Not seen during the present survey. First recorded in 1899 on 'Beinn-an-Fhurain' by C.E.Salmon, as 'var. *glaucantha*'; we are not sure to which modern taxon this should be ascribed. *P. nemoralis* was recorded from NC02 during the 1950s survey for the first *Atlas*.

*Dactylis glomerata* L.
Garbh-fheur
**Cock's-foot**

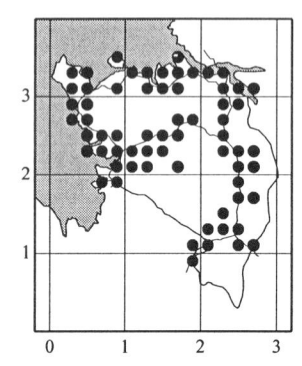

65 tetrads. Frequent in the more fertile grassland where this is affected by man's activities, particularly roadside verges. Plants in the hills away from tracks have probably been brought in as seed by grazing animals. First recorded in 1767 near Inchnadamph by J.Robertson.

*Catabrosa aquatica* (L.) P.Beauv.
Feur-sùghmhòr
**Whorl-grass**

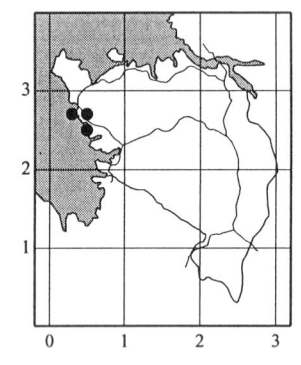

3 tetrads. Rare. Occurs on the damp, sandy margins of small water courses at Achmelvich and in wet hollows or along the high tide mark at Clachtoll. First recorded in 1886 at Clachtoll by A.Gray; also, in 1982, 'in wet area by burn across machair, Stoer', by P.J.O.Trist.

*Catapodium marinum* (L.) C.E.Hubb.
Feur Gainmhich
**Sea Fern-grass**

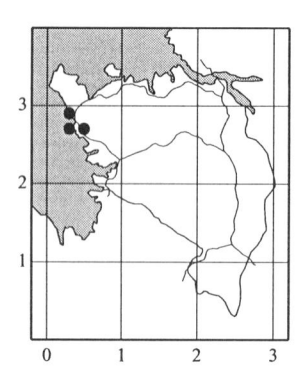

3 tetrads. Rare coastal grass found at Clachtoll and Stoer in sandy grassland close to the sea. First recorded as '*Desmazeria loliacea*', in 1943 on 'coastal rocks, Clachtoll' by A.J.Wilmott and M.S.Campbell.

*Glyceria fluitans* (L.) R.Br.
Mìlsean Uisge
**Floating Sweet-grass**

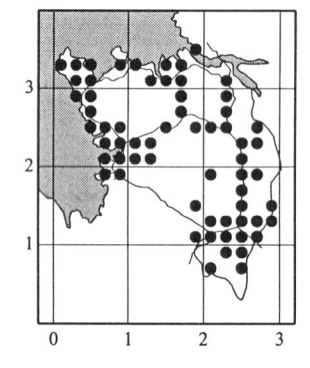

63 tetrads. The only *Glyceria* found in Assynt. Occurs in a range of habitats, both terrestrial and aquatic. Found in ditches and muddy margins of rivers and pools. Also in lochs, pools and slow-moving burns, where the leaves lie along the surface of the water. First recorded in 1890 at '1200 feet in a loch near Inchnadamph' by E.S.Marshall and F.J.Hanbury.

*Melica nutans* L.
Meilig an t-Slèibhe Critheanach
**Mountain Melick**

22 tetrads. An uncommon grass growing mainly on rock faces in gorges and in rocky woodland. Associated on the gneiss with basic or ultra-basic dykes. First recorded in 1897 near Lochinver by E.S.Marshall.

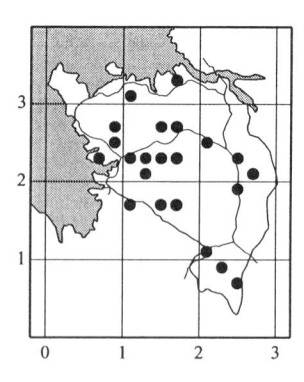

*Helictotrichon pubescens* (Huds.) Pilg.
Feur Coirce Clumhach
**Downy Oat-grass**

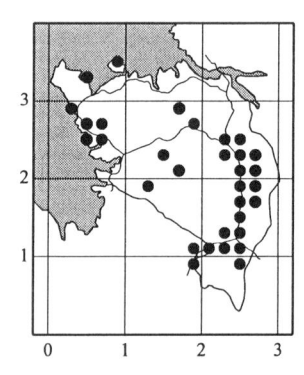

32 tetrads. Occasional, mainly in grasslands near the coast and in limestone areas. First recorded in 1894 at 'Knockan Crags' by G.C.Druce.

*Helictotrichon pratense* (L.) Besser.
Feur Coirce Lòin
**Meadow Oat-grass**

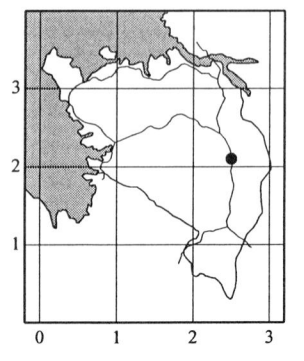

Flowering plants and ferns

1 tetrad. The first and only record was made in 1993, on the gravels of the River Loanan.

*Arrhenatherum elatius* (L.) P.Beauv. ex J. & C.Presl
Feur Coirce Brèige
**False Oat-grass**

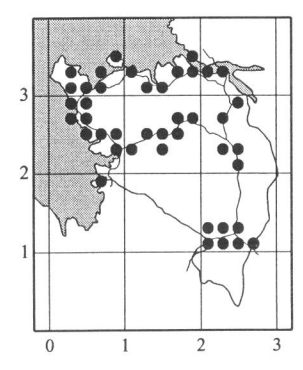

42 tetrads. Almost entirely confined to roadsides, waste and disturbed ground. First recorded in 1894 'near Elphin and Ledbeg' by G.C.Druce.

[*Trisetum flavescens* (L.) P.Beauv.
Feur Coirce Buidhe
**Yellow Oat-grass**

Not seen during the present survey. First reported as '*Avena flavescens*' in 1767 near Inchnadamph by J.Robertson. Anthony has 'Lochinver, 1886, A.G[ray]', but Gray's paper does not localise it to Assynt.]

*Koeleria macrantha* (Ledeb.) Sch. (*K. cristata* auct non (L.)Pers.
Cuiseag Dhosach
**Crested Hair-grass**

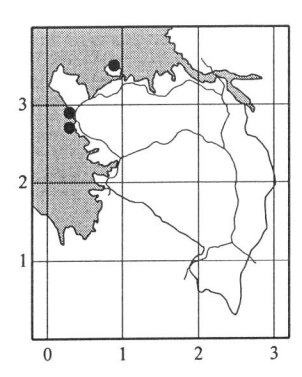

3 tetrads. Rare. We have found it only in coastal grasslands, at Clachtoll, Stoer and Oldany Island. First recorded in 1767, as '*Aira cristata*', near Inchnadamph by J.Robertson.

*Deschampsia cespitosa* (L.) P.Beauv.
**Tufted Hair-grass**

ssp. *cespitosa*
Cuiseag Airgid

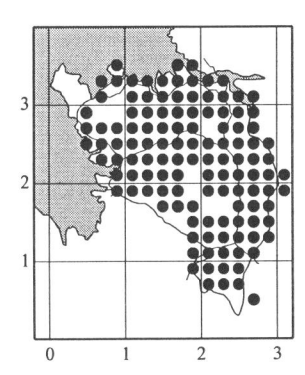

121 tetrads. Widespread and common in marshes, ditches, roadsides and waste places. First recorded during the 1950s survey for the first *Atlas*.

ssp. *alpina* (L.) Hook.
Mòin-fheur Ailpeach

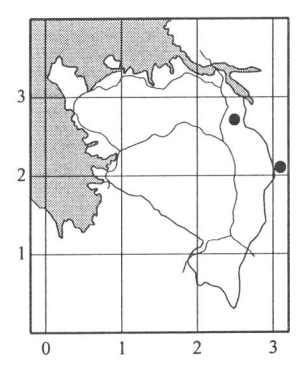

2 tetrads. First recorded in 1995 by G.P.Rothero, in a basic flush at 450m. on Glas Bheinn and in 1998 at 940m. on the north ridge of Conival with *Luzula arcuata*.

*Deschampsia setacea* (Huds.) Hack.
Mòin-fheur Bogaich
**Bog Hair-grass**

12 tetrads. Grows on the very edge of the water, in lochs which have some shallow peaty/muddy margins. This species was probably under-recorded in the first stage of the survey, before we learned to check the lemmas of tufts of fine grass at the water's edge for their characteristically horned and toothed apices.
First recorded in 1993 by D.A. Pearman from Loch an Ordain (Achmelvich).

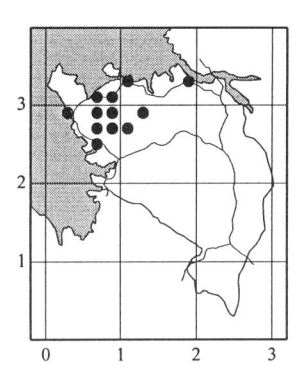

*Deschampsia flexuosa* (L.) Trin.
Mòin-fheur
**Wavy Hair-grass**

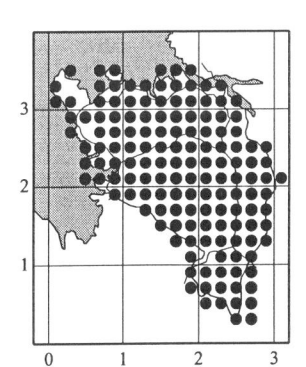

146 tetrads. Widespread and common in dry situations; amongst heather, on banks, in gorges and in open woodland. An attractive grass with its delicate panicles and soft, dense, fine tussocks. First recorded in 1767 near Inchnadamph by J.Robertson.

*Holcus lanatus* L.
Feur a'Chinn Bhàin
**Yorkshire-fog**

130 tetrads. Widespread and abundant in most types of grassland in open situations, although not on the tops of the highest hills.

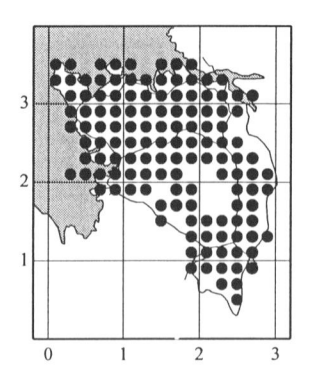

First recorded during the 1950s survey for the first *Atlas*.

*Holcus mollis* L.
Mìn-fheur
**Creeping Soft-grass**

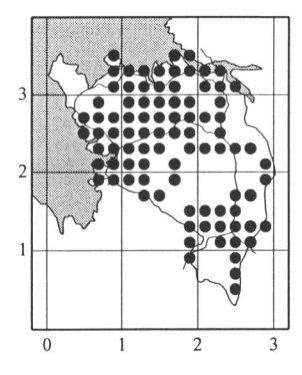

89 tetrads. Common in acid grassland where there is at least a little shade. This may be provided by bracken cover, a small patch of scrub or, of course, woodland. First recorded during the 1950s survey for the first *Atlas*.

*Aira caryophyllea* L.
Sìdh-fheur
**Silver Hair-grass**

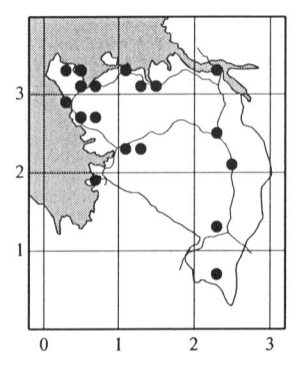

18 tetrads. Considering the wide range of habitats in which it occurs, river gravels, verges, walls, banks and dry heath, this is a surprisingly uncommon grass. First recorded during the 1950s survey for the first *Atlas*.

*Aira praecox* L.
Cuiseag an Earraich
**Early Hair-grass**

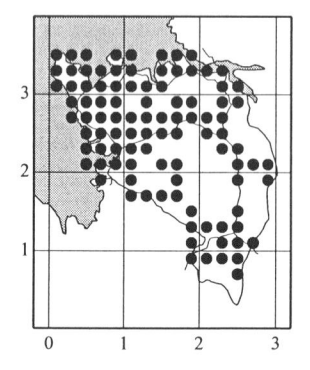

98 tetrads. Common on dry banks, rocks, old walls, tracksides, disturbed ground and in short grassland. First recorded during the 1950s survey for the first *Atlas*.

*Anthoxanthum odoratum* L.
Borrach
**Sweet Vernal-grass**

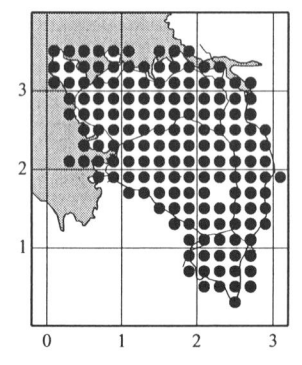

158 tetrads. Widespread and abundant in all types of grassland, lowland and montane, coastal and inland. The 'hole in the map' is the barren tetrad on the quartzite on the south-eastern slope of Canisp, where a determined search might still find it. First recorded in 1767 near Inchnadamph by J.Robertson.

*Phalaris arundinacea* L.
Cuiseagrach
**Reed Canary-grass**

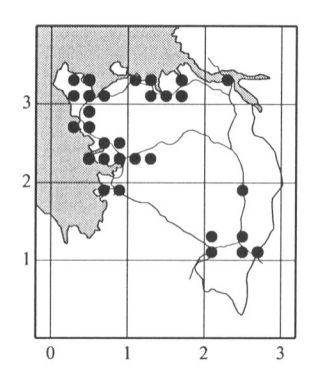

30 tetrads. Occasional in marshy grassland, wet scrub and particularly in ditches. First recorded in 1886 'Elphin – Inchnadamph' by A.Gray.

*Agrostis capillaris* L.
(*A. tenuis* Sibth.)
Freothainn
**Common Bent**

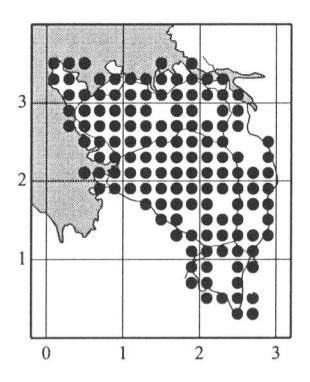

136 tetrads. Widespread and abundant in dry grassland of all kinds. First recorded in 1890 at Kylesku by E.S.Marshall and F.J.Hanbury.

*Agrostis castellana* Boiss. & Reut.
**Highland Bent**

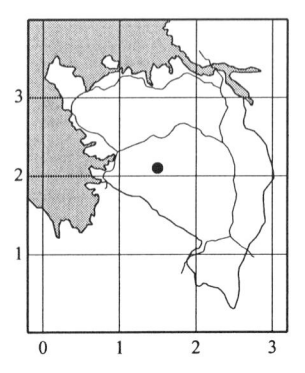

1 tetrad. The first and only record for Assynt was made in 1996, from acid grassland beside the path to Suilven, 0.5 km. west of Suileag. The identification has been confirmed. This is a curious situation for a grass which is said to be introduced for amenity purposes.

*Agrostis stolonifera* L.
Fìoran
**Creeping Bent**

134 tetrads. Widespread and very common in wet habitats, on wet mud at the water's edge, in saltmarshes, marshes and ditches. First recorded in 1767 near Inchnadamph by J.Robertson.

*Agrostis canina* L.
Fioran Mìn
**Velvet Bent**

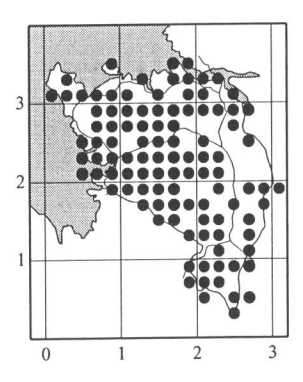

104 tetrads. Widespread and common in wet grassland, wet heath and marshes. *A. canina* s.l. was first recorded in 1767 near Inchnadamph by J.Robertson.

*Agrostis vinealis* Schreb.
Fioran Badanach
**Brown Bent**

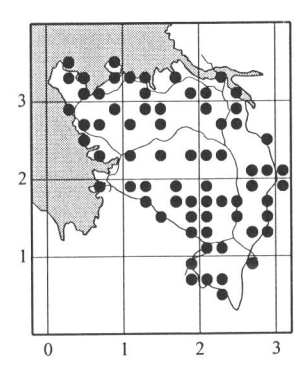

69 tetrads. Frequent, but in much drier situations than the preceding species. Dry heath, acid grassland, rocky banks and sometimes on outcrop rock. First recorded as a species in 1988 during the B.S.B.I. Monitoring Survey.

*Calamagrostis epigejos* (L.) Roth
Cuilc-fheur Coille
**Wood Small-reed**

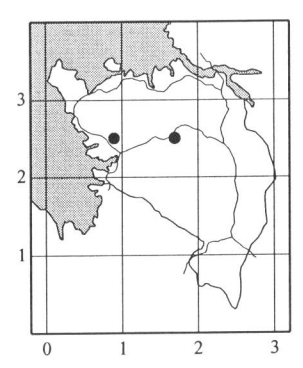

2 tetrads. Very rare. Found by R.E.C.Ferreira in the 1980s in woodland south of Creag Ruigh a'Chàirn, on the north side of the burn. During the present survey it was discovered on the roadside edge of woodland, on the north-western side of the road running along Loch Assynt, at a point about 1.5km. s.w. of Tumore Lodge. This species may occasionally go undetected because, if no flowers are present, it is not easy to distinguish from the ubiquitous *Molinia*. First recorded in 1958 at 'Stoer' by B.Flannigan; also, according to *Anthony*, at Oldany, but there are no further details of this record.

*Ammophila arenaria* (L.) Link
Muran
**Marram**

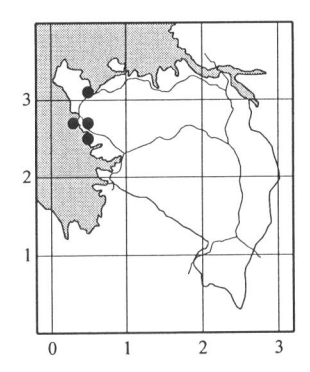

4 tetrads. On sand dunes at Achmelvich, Clachtoll and Clashnessie. First recorded, from NC02, 03 and 13, in the 1950s survey for the first *Atlas*

*Alopecurus pratensis* L.
Fiteag an Lòin
**Meadow Foxtail**

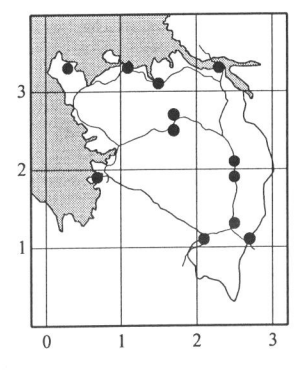

12 tetrads. Occasional in grassland, where its presence is probably due to the influence of man and his stock. Mostly on verges or grassland near fanks. First recorded in 1894 'near Elphin and Ledbeg' by G.C.Druce.

*Alopecurus geniculatus* L.
Fiteag Cham
**Marsh Foxtail**

19 tetrads. Occasional in wet open places; ditches, verges, tracksides and occasionally a mire

169

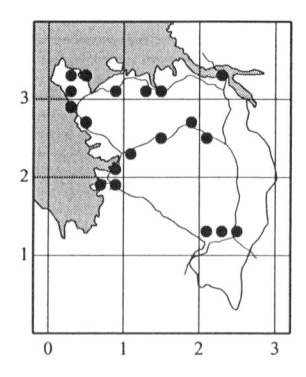

or flush. First recorded in 1894 'near Elphin and Ledbeg' by G.C.Druce.

*Phleum pratense* L.
Feur Cait
**Timothy**

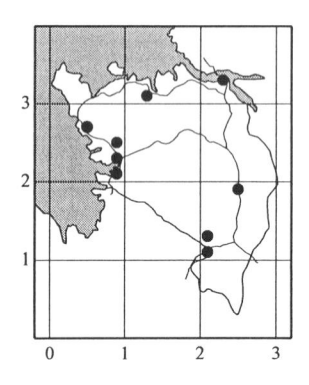

9 tetrads. Occasional on roadsides and in waste places. Probably originally introduced to the area as a fodder grass, or in hay. First recorded in NC21 during the 1950s survey for the first *Atlas*.

*Phleum bertolonii* DC.
Feur Cait Beag
**Smaller Cat's-tail**

Not seen during the present survey. First recorded during the 1950s survey for the first *Atlas*, from NC21.

*Bromus commutatus* Schrad.
Bròmas Lòin
**Meadow Brome**

Not seen during the present survey. First recorded in 1894 in 'cornfields…Elphin and Knockain' by G.C.Druce.

*Bromus hordeaceus* L.
**Soft-brome**

ssp. *hordeaceus*
(B. mollis L.)
Bròmas Bog

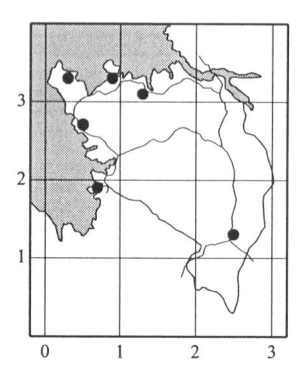

6 tetrads. Most of the records are from man-influenced habitats, road and tracksides or stock-feeding areas. Presumably originally introduced with fodder. First recorded as '*B. mollis*' in 1943 at Lochinver by U.K.Duncan.

*Bromus lepidus* Holmb.
Bròmas Bog Caol
**Slender Soft-brome**

Not seen during the present survey. First recorded in 1959 from NC22.

*Bromopsis ramosa* (Huds.) Holub
(Bromus ramosus Huds.)
Bròmas Giobach
**Hairy Brome**

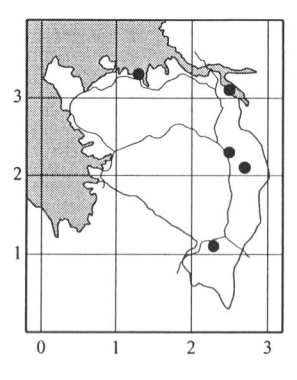

5 tetrads. Rare in rocky situations, wooded, or shaded in some other way: under trees on the banks of the Allt a'Chalda Mòr and the River Traligill, on a wooded cliff at Duart, in rocky woodland at Liath Bhad and on the banks of Na Luirgean. First recorded in 1908 'near Inchnadamph; very rare' by E.S.Marshall and W.A.Shoolbred. Also mapped from NC02 in the first *Atlas*.

*Anisantha sterilis* (L.) Nevski
Bromàs Aimrid
**Barren Brome**

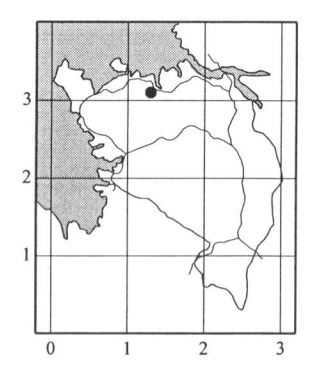

1 tetrad. One record, from Nedd, where it appeared briefly in a garden. First recorded in 1956 at Inchnadamph by G.Halliday.

*Brachypodium sylvaticum* (Huds.) P.Beauv.
Bròmas Brèige
**False Brome**

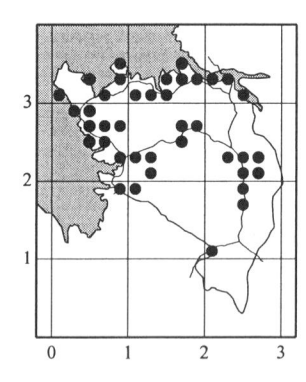

39 tetrads. Frequent in woodland or other shaded rocky places, on burn banks, cliffs and in gorges. First recorded in 1943 on 'rock ledge between Achadhantuir and Feadan; Clashnessie' by A.J.Wilmott and M.S.Campbell.

*Elymus caninus* (L.) L.
(Agropyron caninum (L.) Beauv.)
Glas-fheur Calgach
**Bearded Couch**

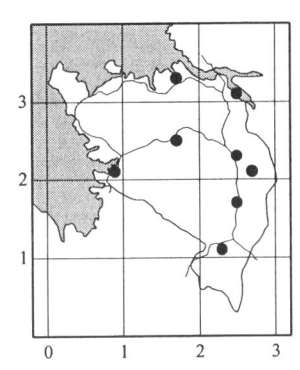

8 tetrads. An uncommon grass of woodland and damp, or even wet, rocky places. One record is from the wall of a sluice at the south-western end of Loch Assynt. First recorded in 1890 by E.S.Marshall and F.J.Hanbury, who said of it 'a state with unusually long awns grows about Lochinver, and on the limestone about Inchnadamph'.

var. *donianus* (Buch.-White) Melderis
**Don's Twitch**

Not identified during the present survey. This short-awned variety was first noted locally in 1887 'near Inchnadamph' by E.S.Marshall. It was re-discovered in 1951 by John Raven after some detective work, and found by him to be frequent all down the lower reaches of the Traligill. It was once regarded as a distinct species, or at least subspecies, but is now known to form 'fertile hybrids with var. *caninus*, that show every degree of intermediacy' (Stace 1997), hence its demotion.

*Elytrigia repens* (L.) Desv. ex Nevski
(Agropyron repens (L.) Beauv).
Feur a'Phuint
**Common Couch**

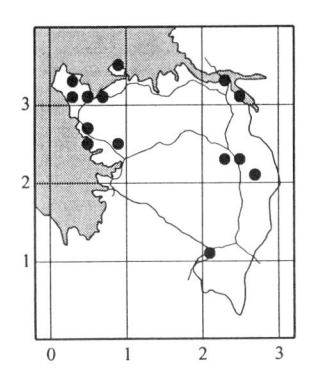

14 tetrads. Fortunately this grass does not live up to its name in Assynt and although it may occur in cultivated ground, it is not the scourge of gardens that it can be further south. First recorded in 1886, as '*Triticum repens*', at Elphin by A.Gray.

*Elytrigia juncea* (L.) Nevski
(Agropyron junceum (L.) P.Beauv.)
Glas-fheur
**Sand Couch**

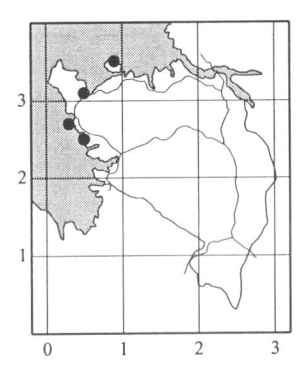

4 tetrads. In blown sand and dunes at Achmelvich, Clachtoll, Clashnessie and Oldany. First recorded at Achmelvich during the 1950s survey for the first *Atlas*.

*Leymus arenarius* (L.) Hochst.
(Elymus arenarius L.)
Taithean
**Lyme-grass**

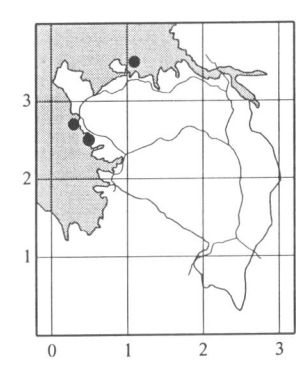

3 tetrads. Rare. Found at Achmelvich in quantity in the fore-dunes, its classic habitat. Also in crevices in an east-facing ridge of rock at Clachtoll and on Meall Dearg at Culkein Drumbeg, where five plants were found in otherwise bare rock. First recorded during the 1950s survey for the first *Atlas*.

*Danthonia decumbens* (L.) DC.
(Sieglingia decumbens (L.) Bernh.)
Feur Monaidh
**Heath-grass**

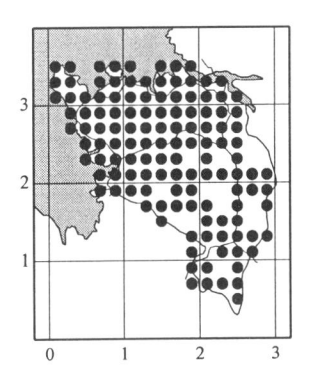

126 tetrads. Widespread and common in damp heathy places, less frequent in limestone areas. First recorded in 1767 near Inchnadamph by J.Robertson.

*Molinia caerulea* (L.) Moench
Fianach
**Purple Moor-grass**

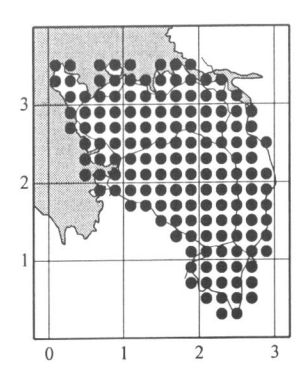

154 tetrads. Widespread and, in suitably flat or gently sloping wet places, dominant. Grows most luxuriantly in this wet climate particularly in flat valleys beside rivers, where it can produce ankle-breaking tussocks over a considerable area. First recorded

in 1767 near Inchnadamph by J.Robertson.

*Phragmites australis* (Cav.) Trin. ex Steud.
(Phragmites communis Trin.)
Cuilc
**Common Reed**

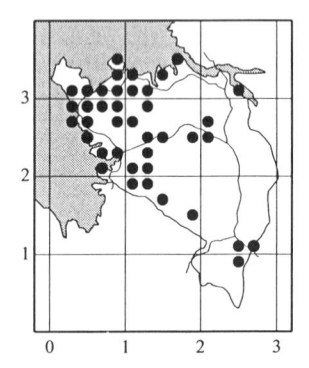

40 tetrads. A striking feature of many lowland lochs, where its stands may be dominant in the shallow water. Also found in mires and drier habitats, such as wet heathland and around small patches of willow scrub, where there is not even seasonal standing water. First recorded in 1767 near Inchnadamph by J.Robertson.

SPARGANIACEAE

*Sparganium erectum* L.
Seisg Rìgh
**Branched Bur-reed**

6 tetrads. Uncommon, in reedswamp alongside rivers. Apparently confined to the extreme southern tip of the parish, where it grows on the Ledmore River in several places, on the Ledbeg River and on the Abhainn a'Chnocain. Also on the Allt nam Meur flowing into Loch Urigill and in a sheltered bay on that loch. Material from that bay was identified as ssp. *neglectum*. First recorded in 1886 at Elphin by A.Gray.

*Sparganium emersum* Rehmann
Seisg Rìgh Madaidh
**Unbranched Bur-reed**

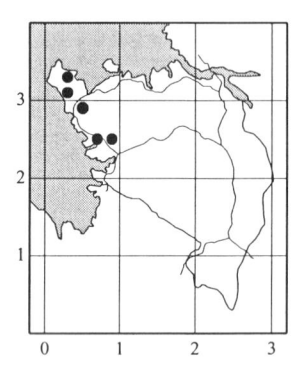

5 tetrads. Uncommon, in the reedswamp on the edge of lochs. Found only in fertile lochs in the north-west, where there is a noticeable richness of vegetation: Loch an Achaidh, Loch an Aigeil, Loch na Claise, Loch Fasg an t-Seana Chlaidh and an un-named loch east of Achmelvich Bay. First recorded in 1955 at Clashmore (perhaps Loch na Claise) by C.D.K.Cook, and from Cam Loch (NC21) during the Scottish Loch Survey.

*Sparganium angustifolium* Michx.
Seisg Rìgh air Bhog
**Floating Bur-reed**

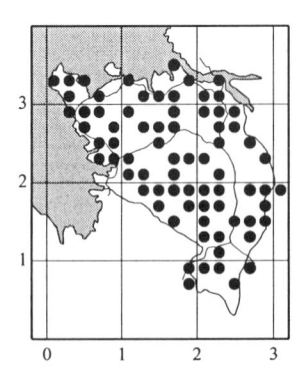

74 tetrads. Widespread and common in rivers, burns and lochs, sometimes high in the hills. The long narrow leaves may extend to a metre or more on the surface of the water. First recorded, as '*S. affine*', in 1886 at Elphin by A.Gray.

*Sparganium* x *diversifolium* Graebn.
*S. angustifolium* x *S. emersum*

Not seen during the present survey. First recorded in 1955 at Clashmore (perhaps Loch na Claise) by C.D.K.Cook.

*Sparganium natans* L.
(S. minimum Wallr.)
Seisg Rìgh Mion
**Least Bur-reed**

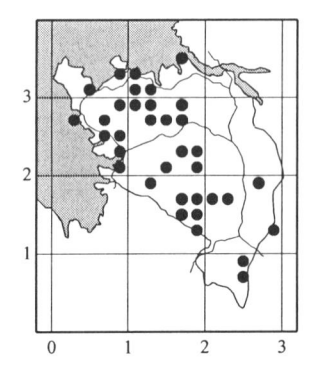

35 tetrads. Occasional in its required habitats of sheltered bays, backwaters and shores, where there is exposed silty mud or very shallow water. The short, bright lettuce-green leaves lying on the mud are easily spotted. First recorded in 1943 at Baddidarach by A.J.Wilmott and M.S.Campbell. Also recorded in 1958 on the River Loanan at Inchnadamph by J.Anthony.

LILIACEAE

*Tofieldia pusilla* (Michx.) Pers.
Bliochan Albannach
**Scottish Asphodel**

3 tetrads. It has been found mainly in the vicinity of the River Traligill. On the Allt na Glaic Móire, which is a southern tributary of the River Traligill, are four sites, three on small islands in the burn and one on rock above the

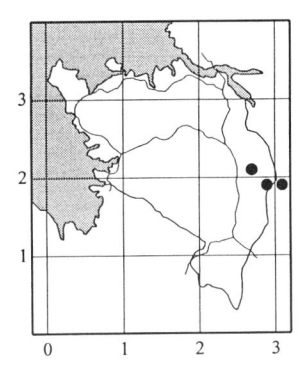

waterfall, all at 100-200m. It grows in stony flushes on the eastern side of Cnoc Eilid Mhathaid, where 500 plants were noted at 350m. Two further sites, in similar habitats, are on Bealach Traligill at over 500m. First recorded in 1890 near Inchnadamph, by E.S.Marshall and F.J.Hanbury.

*Narthecium ossifragum* (L.) Huds.
Bliochan
**Bog Asphodel**

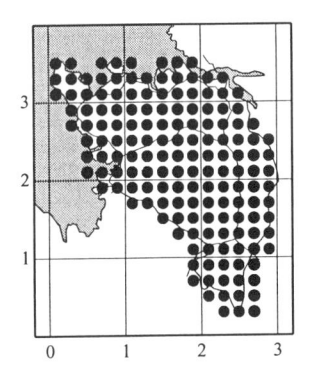

157 tetrads. Widespread and abundant in wet peaty places throughout the parish, making a sheet of colour in bogs and wet heaths in summer. First recorded in 1767, as '*Anthericum ossifragum*' near Inchnadamph by J.Robertson.

*Paris quadrifolia* L.
Aon-dhearc
**Herb-Paris**

Not seen during the present survey. Reported in 1923 'on an islet in Loch Awe, near Inchnadamph' by G.C.Druce, with the comment 'an interesting locality'. *Anthony* has this record as '1895', but it does not appear in Druce's paper of that date.

*Scilla verna* Huds.
Lear-uinnean
**Spring Squill**

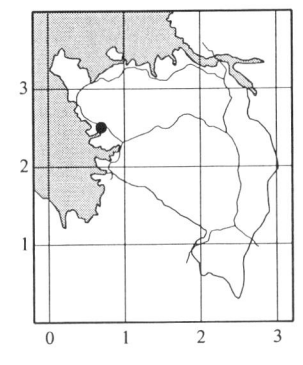

1 tetrad. First recorded in 1996 growing in short, rocky turf, where there is blown sand over gneiss, on the Achmelvich common grazings, just outside the boundary of Mrs I. Ritchie's croft. She has been aware of the population for 18 years; it appears to stay much the same size from year to year, neither declining significantly nor spreading to adjacent patches of apparently similar turf.

*Hyacinthoides non-scripta* (L.) Chouard ex Rothm.
(*Endymion non-scriptus* (L.) Garcke)
Bròg na Cuthaig
**Bluebell**

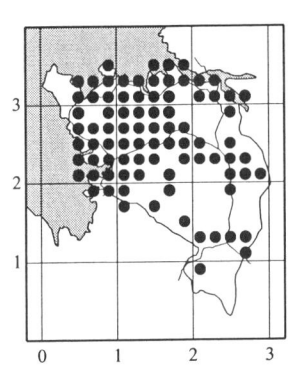

84 tetrads. Frequent except on the higher hills. As well as in the expected habitats of woodland, scrub and shady banks, bluebells thrive beneath bracken and heather; less often in open grassland. First recorded in 1767 near Inchnadamph by J.Robertson.

*Allium ursinum* L.
Creamh
**Ramsons**

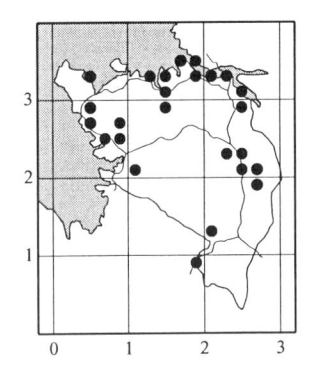

25 tetrads. Frequent in the damp rocky woodlands of the north coast. Can also be found in the shelter provided by a gorge, boulder scree or riverside rocks. First recorded in 1767 near Inchnadamph by J.Robertson.

*Narcissus pseudo-narcissus* L.
Lus a'Chrom-chinn Fiadhaich
**Daffodil**

First recorded in 1886 at Ardvreck Castle by A.Gray. It was probably planted there at least 200 years ago and still persists.

IRIDACEAE

*Iris pseudacorus* L.
Seileasdair
**Yellow Iris**

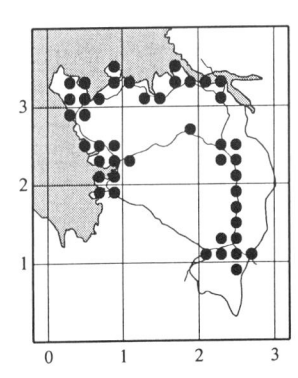

44 tetrads. Common in lowland marshes, riversides, ditches, and wet roadside verges. Salt tolerant and a frequent component of marshes at the back of the shore.

173

First recorded in 1767 near Inchnadamph by J.Robertson.

*Crocosmia* x *crocosmiiflora* (Lemoine) N.E.Br.
Fochann Innseanach
**Montbretia**

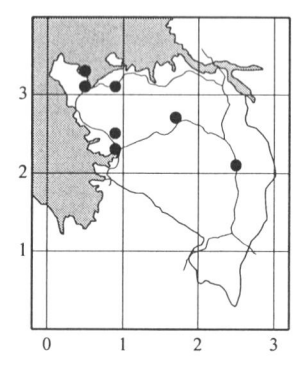

7 tetrads. Naturalised occasionally in marshes, where it may come to dominate quite a large area, at watersides and, in one place, on a roadside among other garden throw-outs. First recorded at Drumbeg during the 1950s survey for the first *Atlas*.

ORCHIDACEAE

*Cephalanthera longifolia* (L.) Fritsch
Eileabor Geal
**Narrow-leaved Helleborine**

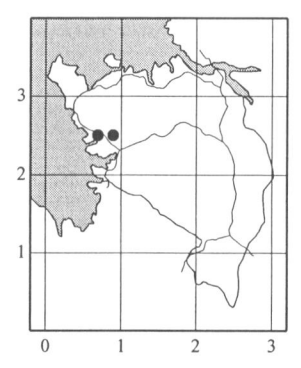

2 tetrads. The most northerly station for this orchid in Britain is the birch/hazel/aspen woodland on the hillside along the Achmelvich road. The helleborines are mostly in that part of the wood near to the road, where the light level is good. The population is thriving there, with over 150 spikes noted by the Inverness Botany Group in 2000. Reported from scrub along Canisp Road, Lochinver in 2000. First recorded in 1943 'between Achadhantuir and Feadan' by A.J.Wilmott and M.S.Campbell. There is also a 1962 record by D.E.Kimmins, from 'roadside near bridge' at the mouth of the River Kirkaig. *Anthony* also has it for Inverkirkaig.

*Epipactis atrorubens* (Hoffm.) Besser
Eilabor Dearg
**Dark-red Helleborine**

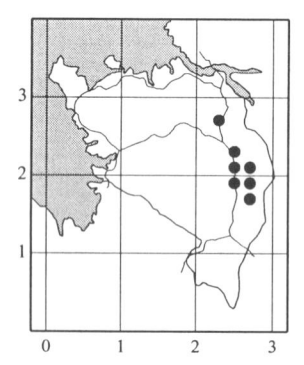

7 tetrads. Rare. Confined to limestone outcrops and cliffs. Found in both the old exclosures on the N.N.R. at Inchnadamph and on the cliffs of Creag Sròn Chrùbaidh. It has been recorded from cliffs near the Inchnadamph Hotel, on ledges above the south-western bank of the River Traligill, upstream of Glenbain, and from Creag nan Uamh. Slightly apart from the main Inchnadamph cluster is the Lairig Unapool site, where the plant grows in crevices and grykes of an isolated limestone outcrop. Depending on the grazing pressure there, the number of spikes can vary greatly; only one was recorded in 1993, but more than 20 in 1998. First recorded in 1827 'on limestone rocks in Assynt' by R.A.Graham. Graham used the name '*Epipactis latifolia*', which is a synonym for *E. helleborine*, but the context indicates that he was referring to *E. atrorubens*. The first reference to Inchnadamph is in 1886 by A.Gray.

*Epipactis helleborine* (L.) Crantz
Eileabor Leathann
**Broad-leaved Helleborine**

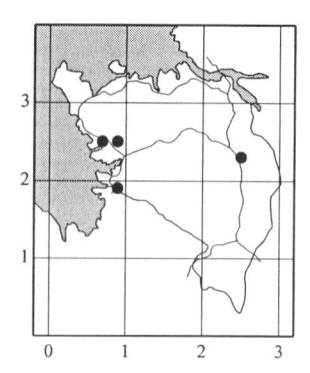

4 tetrads. Very rare. Found in the woodland along the Achmelvich road (see *C. longifolia*), on wooded rocky banks of the Allt a'Chalda Mòr, in a conifer plantation beside the drive up to Achins Bookshop at Inverkirkaig and in deciduous woodland nearby. First recorded in 1943 'between Achadhantuir and Feadan' by A.J.Wilmott and M.S.Campbell.

*Neottia nidus-avis* (L.) Rich.
Mogairlean Nead an Eòin
**Bird's-nest Orchid**

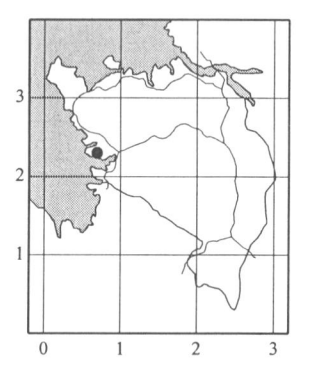

1 tetrad. Found in 1981 by R.E.C.Ferreira, in the woodland on the south side of Loch Dubh. The wood extends up the hillside, but the orchids are beneath hazel on the flat ground at the bottom of the slope. Recorded in 1954 in NC13 by M.E.D.Poore, but not mapped in the first *Atlas*.

*Listera ovata* (L.) R.Br.
Dà-dhuilleach Coitcheann
**Common Twayblade**

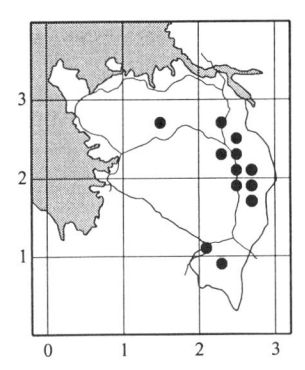

12 tetrads. Occasional in limestone grassland in the Inchnadamph and Elphin areas and also on the Lairig Unapool outcrop. Found, exceptionally, on the gneiss, on wooded cliffs south of Loch Doirean Rairidh. First recorded during the 1950s survey for the first *Atlas* from NC02, as well as 21 and 22. In his card index Anthony lists records from Achmelvich and Stoer.

*Listera cordata* (L.) R.Br.
Dà-dhuilleach Monaidh
**Lesser Twayblade**

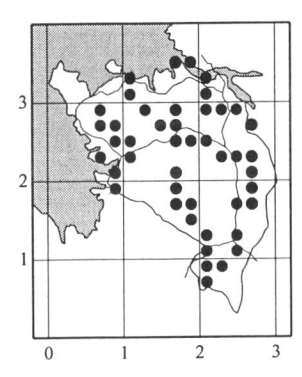

45 tetrads. Widely distributed but not common and, as it is so small and well hidden, probably under-recorded. Grows on moss, particularly *Sphagnum*, under heather and on mossy boulders in woodland. Always in very sheltered situations. First recorded in 1886, just for 'Assynt', by A.Gray.

*Hammarbya paludosa* (L.) Kuntze
Mogairlean Bogaich
**Bog Orchid**

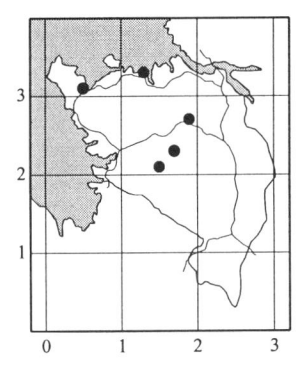

5 tetrads. Apparently rare, but any plant whose flowering spike is green and whose height varies from 2-10cm., is almost certainly under-recorded in a landscape such as this! It grows at Clashnessie, on gravelly sands by a rivulet below the falls, and at Poll Bhuidhe, on bare ground in a *Schoenus* mire. The other three records are from *Sphagnum*: at the edge of runnels at Suileag, beside the Bealach Leireag path and on the stony shore of an un-named loch west of Loch a'Ghlinnein. First recorded in 1890 at Kylesku and near Lochinver by E.S.Marshall and F.J.Hanbury.

*Platanthera chlorantha* (Custer) Rchb.
Mogairlean an Dealain-dè Mòr
**Greater Butterfly-orchid**

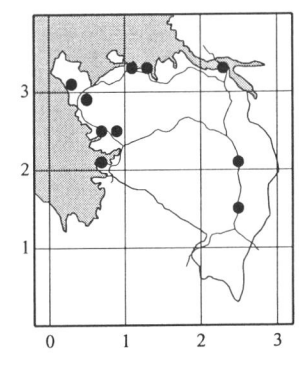

10 tetrads. Found usually in ones and twos, as is the case in a rich pasture at Inchnadamph. At Clashmore, an unpromising-looking field which had had the grazing reduced in 2000, produced that year a goldmine of at least 11 Greater and 30 Lesser Butterfly-orchids. No doubt other places exist where the presence of the orchids will go undetected until a chance alteration in grazing pattern occurs. The sites have all been grassland with the exception of an island in Loch Awe, where the plants were growing under trees. First recorded in 1890, as '*H. chloroleuca*' by E.S.Marshall and F.J.Hanbury, with the remarks 'plentifully at Inchnadamph and Lochinver…an ornament of open grassy meadows'. Not any longer!

*Platanthera bifolia* (L.) Rich.
Mogairlean an Dealain-dè Beag
**Lesser Butterfly-orchid**

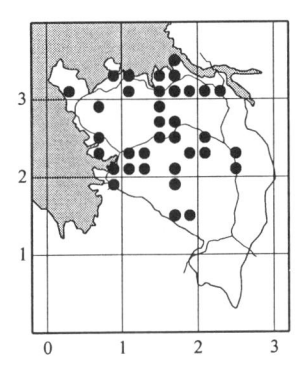

35 tetrads. Occasional in damp places in the hills amongst heather and often near *Schoenus* flushes. Sometimes in grassland (see preceding species). First recorded in 1767, as '*Orchis bifolia*', near Inchnadamph by J.Robertson.

*Pseudorchis albida* (L.)
Á.&D.Löve
(*Leucorchis albida* (L.) E.Mey.)
Mogairlean Bàn Beag
**Small-white Orchid**

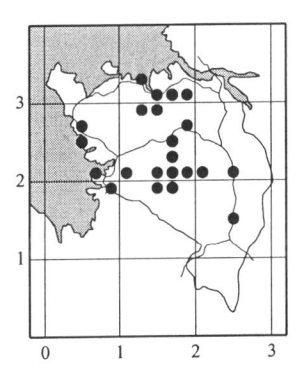

22 tetrads. The scattered sites are mostly in the open amongst short heather, often on the edge of sheep paths, but the plant can also occur on rocky outcrops high in the hills. A large form has been recorded on a wooded island in Loch Awe, presumably in response to the shady conditions there. First recorded in 1767 as '*Orchis albida*' near Inchnadamph by J.Robertson and noted to be 'abundant in Traligill Glen' by G.C.Druce in 1907.

*Gymnadenia conopsea* (L.) R.Br.
Lus Taghte
**Fragrant Orchid**

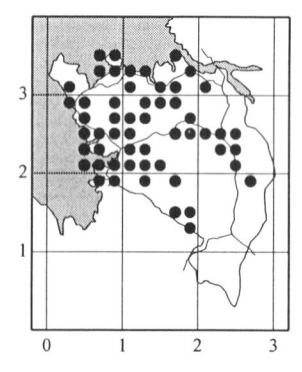

54 tetrads. Frequent on roadsides, dry heath and hill grassland, especially on the gneiss. First recorded during the 1950s survey for the first *Atlas*. Anthony has only ssp. *conopsea*, although current thinking would suggest ssp. *borealis* as the most likely taxon.

*Gymnadenia* x *Platanthera*
*Gymnadenia* x *Dactylorhiza*

These inter-generic hybrids were photographed by L.Tucker alongside the Allt a'Chalda Mór in 1998, but have not been seen since.

*Coeloglossum viride* (L.) Hartm.
Mogairlean Losgainn
**Frog Orchid**

18 tetrads. Very local in grassland, usually on base-rich soils, the greatest concentration of records being in the vicinity of Inchnadamph and in coastal grassland. Also grows in the hills on grassy ledges. First recorded in 1886, just for 'Assynt', by A.Gray. In 1894 G.C.Druce recorded it as 'not infrequent' from Knockan Crags.

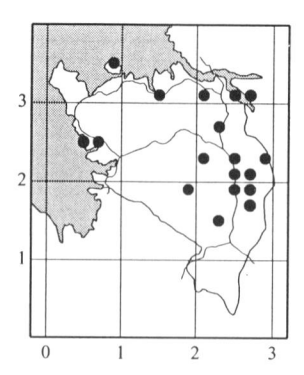

*Dactylorhiza fuchsii* (Druce) Soó
(Dactylorchis fuchsii (Druce) Verm.)
Urach Bhallach
**Common Spotted-orchid**

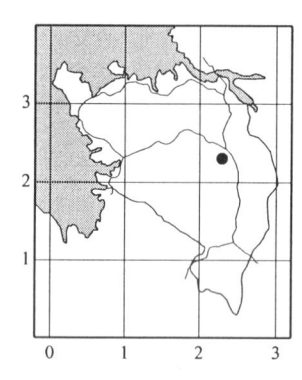

1 tetrad. Very rare. In limestone grassland on the Ardvreck Castle peninsula. First recorded during the 1950s survey for the first *Atlas* for NC03, 21 and 22. Recorded for NC13 in 1981 by P.H.Gamble. 'Ssp. *okellyi*' (now accorded only varietal rank) was recorded, prior to 1970, from NC22 in the *Critical Supplement*.

*Dactylorhiza maculata* (L.) Soó
ssp. *ericetorum* (E.F.Linton)
P.F.Hunt & Summerh.
(Dactylorchis maculata (L.) Verm.)
Mogairlean Mòintich
**Heath Spotted-orchid**

144 tetrads. Widespread and common throughout the parish.

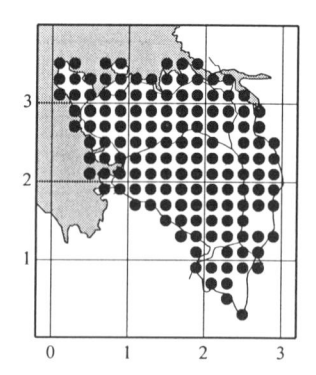

Found in wet and dry heath, open woodland and grassland, from high in the hills down almost to sea level. First recorded in 1767 near Inchnadamph by J.Robertson

*Dactylorhiza incarnata* (L.) Soó
(Dactylorchis incarnata (L.) Verm.)
Mogairlean Lèana
**Early Marsh-orchid**

First recorded, as the species only, in 1890 by E.S.Marshall and F.J.Hanbury, 'about Inchnadamph, but very scarce'. Four subspecies have been recorded during the present survey:

ssp. *incarnata*

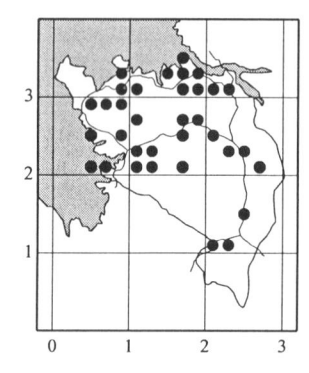

34 tetrads. Scattered in damp grassland, *Schoenus* flushes and wet heath, on neutral or base-rich soils. First recorded in 1990 at Achadh Mór.

ssp. *coccinea* (Pugsley) Soó

2 tetrads. Rare; in flushed grassland below a crag near the sea at Clachtoll and in a wet area in dune grassland at Stoer.

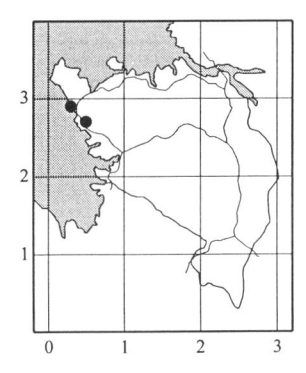

First recorded from the second of these sites in 1997.

ssp. *pulchella* (Druce) Soó

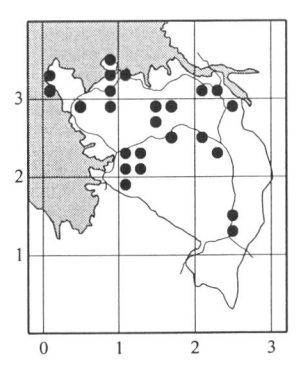

24 tetrads. Occasional in damp peaty places and *Schoenus* mires. First recorded in 1990 near Loch Eileanach.

ssp. *cruenta* (O.F.Müll.) P.D.Sell

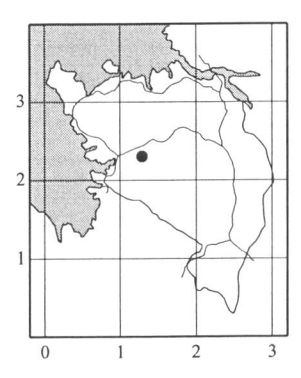

1 tetrad. This Red Data Book plant is a recent addition to the Assynt flora. It was discovered in 1998 by R.E.C.Ferreira and A.Scott. When re-located, a single plant was found beside a small pool, on the western edge of an extensive mire, just south of Loch Poll an Nigheidh. In that area of the mire *Carex lasiocarpa* was the dominant species. See also under *D. lapponica*.

*Dactylorhiza lapponica* (Hartm.) Soó
Mogairlean Lapach
**Lapland Orchid**

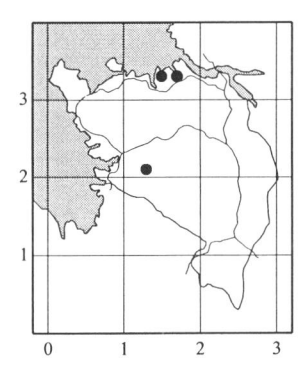

3 tetrads. The southern station for this Red Data Book species is in the same mire as *D. incarnata* ssp. *cruenta*, although in the adjoining tetrad. It grows in the south-eastern corner of the mire, which is dominated at this point by *Schoenus*. Of the two northern tetrads the main populations are in the western one, where it grows in *Schoenus* flushes and where *Schoenus* tufts edge a runnel. Several small stands of *Platanthera bifolia* and *D. incarnata* ssp. *incarnata* are not far away. The eastern site had only one spike, again in a *Schoenus* mire. First recorded in 1986 by R.E.C.Ferreira, from the western of the two northern tetrads.

*Dactylorhiza purpurella* (T. & T.A.Stephenson) Soó
(Dactylorchis purpurella (T. & T.A.Stephenson) Verm.)
Mogairlean Purpaidh
**Northern Marsh-orchid**

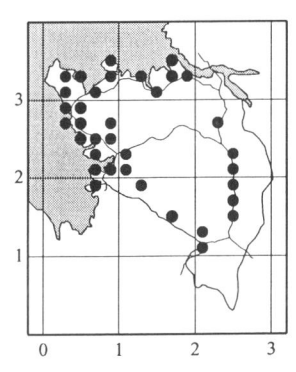

35 tetrads. Frequent in damp grassy places, on roadsides and in marshes. First recorded in 1943 'on machair, Achmelvich Bay' by A.J.Wilmott and M.S.Campbell.

*Dactylorhiza* x *formosa* (T. & T.A.Stephenson) Soó
*D. purpurella* x *D. maculata* ssp. *ericetorum*

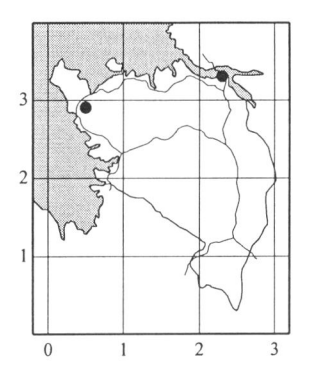

2 tetrads. First recorded in 2001 by L.Tucker from grassland around the building housing the 'Maryck Memories of Childhood' at Unapool. Coincidentally, it was found two days later at Stoer, on the common grazings just outside the grounds of Smithy House.

*Orchis mascula* (L.) L.
Moth-ùrach
**Early-purple Orchid**

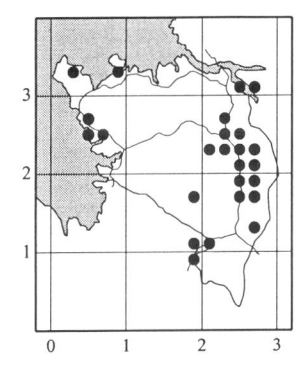

25 tetrads. On limestone or at least base-rich grassland, often amongst rocks. There are a few coastal records, but most are centred on the limestone areas of Inchnadamph and Elphin. An uncommon white-flowered form has been seen on the footpath leading to Port Alltan na Bradhan and at Achmelvich. First recorded in 1890 at Inchnadamph by E.S.Marshall and F.J.Hanbury.

# BRYOPHYTES

# INTRODUCTION

## The physical background

The climate, landform and geology of Assynt have been dealt with in the introduction to the vascular plant flora but a short account of the features that particularly affect bryophytes is necessary. Assynt has a large total annual precipitation but more important for the bryophytes, particularly those described as 'oceanic', is the total number of wet days or conversely, the absence of any prolonged period of drought. Residents in the area will not be surprised that much of Assynt has in excess of 200 wet days per year (Ratcliffe 1968), with a 'wet day' defined as one with at least 1mm. of rain falling. The ameliorating effect of the relatively warm sea means that Assynt as a whole has fewer than 60 frosty days and fewer than 20 days with snow lying (Page 1982), although these sea-level figures need to be adjusted for the higher ground in the east of the parish.

To give an expression to this type of climate, an index of 'oceanicity' (see Averis 1991 and Page 1982), can be derived from the number of wet days per annum divided by the annual temperature range (mean maximum July temperature minus the mean minimum February temperature). Plotting the areas with the highest index gives a narrow band of 'wet mildness' (or 'cool wetness') that extends down the coast of Highland Scotland, including the Hebrides, and the extreme west of Ireland, and which includes all of Assynt. This index has proved a useful indicator of the distribution of oceanic bryophyte species. This very wet climate has also had a marked effect on bryophytes in another way, through the formation of mire areas where mosses, particularly *Sphagnum*, are often dominant.

Bryophytes are mostly small plants and do not compete well with the larger flowering plants and ferns except where conditions are favourable. The landforms of Assynt with a preponderance of rocky slopes, crags, boulders and ravines provide a wonderful assortment of niches where bryophytes can become established and form long-lived communities. The effects of glaciation are everywhere apparent but particularly so on north and east-facing slopes where the resultant crags and screes provide an important habitat for the hepatic heath described below. The run-off from the melting of the glaciers with an increasingly mild and wet climate, over a landscape largely devoid of vegetation and with prodigious quantities of rock debris gave rise to the, sometimes spectacular, ravines which seam the area. These now form very important sites for bryophytes. Another very recent landform with a distinctive bryophyte community is wind-blown shell-sand which has a patchy distribution from Achmelvich round to Oldany.

The effects of the underlying rocks are as apparent for the bryophytes as they are for the flowering plants, though the physiological mechanism through which the rock type operates to determine presence or absence is even less well-understood. The large expanse of limestone and the associated Fucoid Beds and the run-off from the calcareous areas have a nationally important bryophyte flora with a number of species limited to this zone. The sandstone is largely rather acid and often has a limited flora although the gritty texture of the rock is ideal, particularly for the smaller liverworts.

As with the flowering plants, it is the complex mineralogy of the gneiss that poses most of the problems. Some facies of the gneiss are strongly base-rich and have a number of 'calcicole' species while others are absent (though frequent enough on the limestone). *Schoenus nigricans* flushes on the gneiss have a limited number of species, *Scorpidium scorpioides*, *Drepanocladus revolvens*, *Aneura pinguis* and *Blindia acuta*, but the expected *Philonotis fontana*, *Dicranella palustris*, *Jungermannia exertifolia* ssp. *cordifolia* and *Bryum pseudotriquetrum* are rarely seen. Some large gneiss crags are almost completely devoid of bryophytes except at the base and in some seepage lines. The autecological studies necessary to provide an explanation for these distribution patterns have not been carried out. However it would seem likely that it is the balance of a range of minerals including pyroxene, hornblende and plagioclase and their breakdown products that are critical, along with the depletion of several trace elements in the gneiss in general (Johnstone and Mykura 1989).

## Assynt bryophytes in context

In order to put the bryophyte flora of Assynt into context it seems sensible to discuss briefly the importance of the Scottish Highlands as a whole. Britain has an internationally important bryophyte flora; we have approximately 70% of the European bryophytes as compared to only 18% of the European flowering plants. In general, the diversity of bryophytes in Britain increases as you move north and west because of the more diverse geology, more and bigger hills, higher rainfall and less pollution. Within this diverse bryophyte flora, the most important of the bryophyte communities, in global terms, are those which are often described as 'Atlantic' or 'oceanic'.

The climatological and geomorphological features already described for Assynt are common to much of western Scotland. The combination of equable temperatures and consistently high humidity occurs only in a very few areas of the globe, principally on temperate oceanic margins and in high montane zones closer to the equator. In Europe, this zone is limited to the extreme western margins of the continent, where very small areas occur on the coasts of France and Spain and again in south-west Norway. The largest areas are on the western margins of the British Isles, in a band stretching from south-west Ireland to north-west Scotland.

In addition to the climatic factors, buffering from changes in humidity can also be enhanced by a reasonably continuous tree canopy, deeply incised river valleys and by very rocky terrain. The latter two conditions are common in the west of Scotland as a result of glacial and fluvio-glacial processes; the burns tend to be steep and have ravine sections and there are rock-falls and scree slopes, not least those associated with the extensive raised-beaches around the coast. Broadleaf tree cover was probably fairly continuous at one time from Kintyre to Torridon and extensive in favoured spots north of this, probably including much of Assynt.

These various factors have given rise to distinctive and, in global terms, very rare, moss and liverwort communities. Several of the species involved have their only European sites in the British Isles, while many other species that are reasonably common here are extremely rare elsewhere. Perhaps more important is that, though there may be isolated records for most of the species from elsewhere in Europe, it is principally in the British Isles that there are extensive and distinct communities containing these species.

The west coast of Scotland has the largest area of this kind of habitat in Britain and has the largest number of typical oceanic species as well as many of the most important populations. The available habitat in Assynt, as in much of Scotland, has been much reduced in the last few millennia as most of the species, as well as requiring an oceanic climate, require the further buffering from changes in temperature and humidity provided by a broadleaf tree cover. As the area under permanent broadleaf tree cover has decreased, so the populations of the oceanic species have become increasingly fragmented. This means that any semi-natural woodland with populations of oceanic bryophyte species is scientifically important, both nationally and internationally. A much more detailed analysis of the processes giving rise to this rich woodland flora is given by Averis (1991) and Hodgetts (1993) has a good account of the species involved.

These woodland and ravine communities of bryophytes have a better representation further south than Assynt, with woodlands of larger size, a greater diversity of species and more rarities. However, some sites in Assynt are extremely rich in oceanic species and are of international importance. Outside of the woodlands, there is one oceanic community that has some of its best Scottish sites in Assynt and in similar areas in West Ross just to the south. This is the liverwort community under ericaceous shrubs which has been variously described as 'oceanic' heath, 'oceanic-montane' heath, liverwort-rich heath or the 'mixed Northern Atlantic hepatic mat' (Ratcliffe 1968). This community is described more fully below but, essentially, it is a community of rocky slopes with a north or north-easterly aspect, consisting of *Calluna vulgaris* or, at higher altitudes, *Vaccinium myrtillus*, and with a bryophyte layer below which includes a number of large, leafy liverworts with an extremely restricted and disjunct global distribution. The best sites for this community, in the bigger hills in the east of Assynt, are, again, of international importance.

# BRYOPHYTE COMMUNITIES

## Common species

The object of this section is to describe the typical species of a broad range of habitats in Assynt and also to give some account of the rare or unusual plants. Assynt has some very rare species and several interesting bryophyte communities but the nature of the ground and climate means that a large number of plants occur very widely and it seems sensible to deal with these first. Plants like *Hylocomium splendens*, *Pleurozium schreberi*, *Hypnum cupressiforme*, *Racomitrium lanuginosum*, *Dicranum scoparium*, *Diplophyllum albicans* and to a lesser extent *Racomitrium fasciculare*, *Thuidium tamariscinum*, *Rhytidiadelphus squarrosus*, *Rhytidiadelphus loreus*, *Isothecium myosuroides*, *Sphagnum capillifolium*, *Frullania tamarisci* and *Scapania gracilis* can find a niche almost anywhere there is suitable substrate and are usually the most common species in a wide range of habitats. These and some other common species are only mentioned where they form an important, structural part of a community.

## Coast

Assynt has a long coastline, most of which is rocky and

some of which is very exposed. Ubiquitous on all rocky shores within reach of salt spray is *Schistidium maritimum* and the dark-green cushions of this moss occur high up on the sea-blasted Stoer headland. A little higher up the shore, in crevices and on ledges the most frequent species are usually *Trichostomum crispulum* and *Trichostomum brachydontium* and a rarity here is *Tortella flavovirens*, recorded from just two places. Again in the zone above the *Schistidium maritimum*, on the tops of rocks, cushions of *Ulota phyllantha*, with distinctive brown gemmae, are common. Where water seeps down, small patches of *Cratoneuron filicinum* are frequent, and where the rocks are very sheltered, as along the north-facing shore of Eddrachillis Bay, some of the oceanic liverworts can occur, especially *Radula aquilegia*. These sheltered rocky shores often grade into woodland.

Some banks above the sea have good stands of *Weissia controversa* and, rarely, its close relative, *Weissia perssonii* and, on soil in crevices in one rocky slab, there is a good population of *Riccia subbifurca*. *Frullania tamarisci* occurs in abundance both on rocks and in the turf and *Frullania teneriffae* is probably more common on slabby rocks by the coast than elsewhere. In similar sites, and often mixed with *Frullania fragilifolia* and with the same distinctive aroma, there are occasional stands of *Frullania microphylla*. Out on the Stoer peninsula again, in flushed grassland above the cliff edge, are two stands of the 'hyper-oceanic' moss *Myurium hochstetteri*. This plant, one of our most beautiful mosses, is most frequent in the Outer Hebrides and has only a handful of mainland localities; elsewhere it is only known from the Azores, Madeira and the Canary Isles.

The areas of shell-sand form a distinctive coastal community. The mobile sand is often colonised by *Syntrichia ruraliformis* and in amongst the marram *Homalothecium lutescens* is often common. Where the dunes are more stable and in other sandy grassland, *Rhytidiadelphus triquetrus* is locally abundant, often with *Hypnum lacunosum* var. *lacunosum*, *Scapania aspera* and more rarely *Enthodon concinnus* and *Ditrichum gracile*. At Achmelvich there are limited populations of a few bryophytes typical of shell-sand areas in Scotland, including *Distichium inclinatum*, *Encalypta vulgaris* and *Amblyodon dealbatus*.

## Rivers and lochs

The riparian flora of many of the rivers and burns is not particularly diverse but often abundant. On the more acid rocks there is a fairly constant community consisting of *Racomitrium aciculare*, *Scapania undulata*, *Marsupella emarginata* and *Brachythecium plumosum* in varying amounts, often accompanied by *Hyocomium armoricum* on rocks above normal water level. On somewhat more base-rich rocks or run-off, *Racomitrium aciculare* is still common but there may also be stands of *Rhynchostegium riparioides* and in sheltered places, *Thamnobryum alopecurum*. On the sandstone of Quinag, *Rhynchostegium riparioides* is accompanied on two burns by *Rhynchostegium alopecuroides*, a scarce European endemic in its most northerly world locality. *Schistidium rivulare* can be frequent on the upper surfaces of rocks and in one place there is a large population of the nationally scarce *Schistidium agassizii*. With the latter species there are also stands of the Scottish endemic *Bryum dixonii* and this also occurs on the River Inver close to Lochinver. In ravines and where there is some woodland cover the yellow-brown mats of *Hygrohypnum eugyrium* are common. *Fontinalis antipyretica* is relatively infrequent and often restricted to sheltered zones at the margins of rivers or in flushes.

On the faces of boulders in burns and particularly in ravines, but avoiding the limestone, is an oceanic liverwort community of international importance. The micro-habitat is usually the steep faces of rocks in or close by the burn, constantly humid, probably regularly inundated but free from scouring during spate. The most common species here is *Lejeunea patens*, often with *Metzgeria conjugata*, and sometimes with *Lejeunea lamacerina* and *Lejeunea cavifolia*. In the better sites there will also be *Radula aquilegia*, *Drepanolejeunea hamatifolia*, *Harpalejeunea molleri*, *Aphanolejeunea microscopica* and the exquisite *Colura calyptrifolia*. This community is most frequent in wooded ravines but it can persist in more open sites, especially near the coast.

Where the rocks are calcareous, particularly in the Tralligil and the Allt nan Uamh, *Hygrohypnum luridum* is very common, often with *Rhynchostegium riparioides* and more rarely *Amblystegium tenax*. Both rivers also have a good population of the scarce *Hygrohypnum duriusculum*, the rounded leaves of this species rather easily confused with the more abundant *Rhynchostegium riparioides*. Also on the bigger rivers with some base-rich run-off and also on loch margins with inflow from the limestone, there are large, straggling, black patches of *Cinclidotus fontinaloides*. Similarly, *Homalia trichomanoides* has a few sites on lochs and rivers where it will be regularly inundated with base-rich water.

The bryophyte flora of the loch margins also varies markedly with the underlying rocks but consistently provides one of the most interesting habitats in Assynt. On the more acid rocks, the flora is similar to that of the burns, with *Racomitrium aciculare* and *Brachythecium plumosum* prominent often with some *Thamnobryum alopecurum* and *Fontinalis antipyretica*. Other frequent species include *Hygrohypnum eugyrium*, *Bryum capillare*, *Isothecium myosuroides*, *Hypnum cupressiforme*, *Grimmia curvata* and *Racomitrium fasciculare*. The most interesting community usually occurs on basic gneiss boulders and has most of the species above with smaller amounts of *Schistidium strictum*, *Pterogonium gracile*, *Orthotrichum rupestre*, *Antitrichia curtipendula*, *Pterigynandrum filiforme* and a variety of *Grimmia* species of which the most frequent are *Grimmia curvata*, *Grimmia hartmannii*, *Grimmia trichophylla* and *Grimmia funalis*, with rare stands of *Grimmia longirostris*, *Grimmia decipiens* and the Red Data Book species *Grimmia ovalis*.

The softer margins of the lochs can also have an interesting flora, particularly where there is some flushing from above. On more acid substrates common species are *Pellia epiphylla*, *Scapania undulata*, *Marsupella emarginata*, *Scapania irrigua* and *Jungermannia gracillima* often with some *Sphagnum* species. On open gravel, *Pohlia drummondii* can be frequent and in one place there is a small stand of the Scottish endemic, *Pohlia scotica*. Where there is some flushing, the nationally scarce liverwort *Odontoschisma elongatum* can be abundant, often growing in a rather glutinous, algal layer. Another rarity in this habitat is *Haplomitrium hookeri*, usually as scattered upright stems with other bryophytes on open gravel. During one dry spell, *Fossombronia foveolata* was found to be abundant on sandy loch margins in two areas and it is probably more widespread. Loch margins on the limestone have a very rich flushed turf which can include *Plagiomnium elatum*, *Plagiomnium ellipticum* and rarer plants like *Campyliadelphus elodes*, *Cinclidium stygium* and *Pseudobryum cinclidioides*.

There is very little of what can be described as fen on any of the loch margins despite the frequent occurrence of stands of emergent vegetation. Possibly the best area is by Loch na Claise at Balchladich which has populations of *Drepanocladus polygamus*, *Calliergon giganteum*, *Calliergon cordifolium*, *Plagiomnium elatum* and *Rhizomnium pseudopunctatum*. Most lochs on the gneiss have *Sphagnum* associated with the emergent vegetation, usually *Sphagnum denticulatum*, occasionally with *Cladopodiella fluitans* and *Drepanocladus exannulatus*.

## Woodland

An account of the extent and character of woodland in Assynt is given in the introduction to the vascular plant flora. Much of the rather open birch woodland of recent origin is very dull bryologically and all interest centres on those limited areas of older woodland and on woodland over rocky slopes and in ravines. The woodland floor tends to be dominated by common species like *Hylocomium splendens*, *Rhytidiadelphus loreus*, *Dicranum majus* and *Thuidium tamariscinum*, often in considerable abundance, forming hummocks over low rocks and capping the larger boulders. On wetter banks *Sphagnum capillifolium* is often abundant, sometimes with *Sphagnum subnitens*, *Sphagnum girgensohnii* and *Sphagnum quinquefarium*, often with *Plagiothecium undulatum* and, in crevices, *Hookeria lucens*. On the faces of the rocks and crags in some woodlands, a community develops containing *Isothecium myosuroides*, *Scapania gracilis*, *Hymenophyllum wilsonii*, *Plagiochila spinulosa* and less frequently *Plagiochila punctata*. The extent of this community is the best indicator of a relatively undisturbed 'oceanic' woodland that will have other species of interest.

In these better woodlands, large, pure hummocks of the fine fronds of *Hylocomium umbratum* can develop over boulders and tree stumps, swelling cushions of *Mylia taylori* are frequent, often with *Bazzania tricrenata*, and less commonly *Lepidozia cupressina*. On wetter rocks and crags *Saccogyna viticulosa* is usually abundant and on drier faces the dark-green cushions of *Dicranum scottianum* often occur. Of the rarer species in this habitat, *Geocalyx graveolens* has one site on a wooded crag and *Sphenolobopsis pearsonii* has two sites. *Plagiochila killarniensis* favours more open sites, usually on somewhat basic rocks and has a handful of sites, but the tiny *Plagiochila exigua* was only found in the woodland at Creag an Spardain. At the base of some crags and on boulders near burns the *Lejeunea* community, described above in the section on rivers, can occur.

In some of the higher birch woodland, particularly in ravines where there is an under-story with some heather, a few of the 'oceanic-montane' bryophytes start to appear. The most frequent are *Anastrepta orcadensis* and *Herbertus aduncus* ssp. *hutchinsiae*, but there are also records for *Dicranodontium uncinatum*, *Dicranodontium asperulum*, *Paraleptodontium recurvifolium*, *Plagiochila carringtonii* and *Mastigophora woodsii*. In the limited areas of richer rocks as at Duart or Meallard and on the limestone, more base-demanding bryophytes like *Eurhynchium striatum* are abundant, often with *Rhytidiadelphus triquetrus*. Rocks and tree bases can be covered with *Isothecium alopecuroides* and less frequently *Homalothecium sericeum* but the less common oceanic species are usually absent from the rocks; the *Lejeunea* species that occurs here is normally *Lejeunea cavifolia*.

The epiphytic community on the trees is often not very diverse. In the best, old birch woodland many of the larger trees will have cushions of *Scapania gracilis* and *Plagiochila punctata* extending up the trunk, usually with *Dicranum scoparium* and *Hypnum andoi* and less frequently *Dicranum fuscescens*. In a few sites this community contains the scarce, oceanic liverwort *Leptoscyphus cuneifolius*. *Ulota crispa* and *Ulota bruchii* are frequent on the smaller branches of both birch and rowan and the latter often has *Ulota drummondii* and *Ulota phyllantha* as well, usually with abundant *Frullania tamarisci*. This '*Ulota* community' is probably best-developed on hazels and willows in sheltered woodland near the coast where it usually has, in addition to the species already mentioned, *Frullania dilatata*, *Radula complanata* and *Metzgeria furcata* and occasionally *Zygodon conoideus*, *Harpalejeunea molleri* and rarely *Ulota calvescens*.

The epiphytic community on more open, nutrient-rich bark trees like elder, ash and elm is poorly developed in Assynt. Suitable trees are few in number and are often rather exposed, sometimes having little other than abundant *Ulota phyllantha*. A few trees have good stands of *Zygodon viridissimus* var. *viridissimus*, *Zygodon rupestris* is less frequent and there are only single records for species like *Orthotrichum stramineum*, *Orthotrichum affine* and *Orthotrichum pulchellum*.

The bryophyte community on rotting logs is also rather poorly developed in Assynt, possibly because birch, the most frequent tree, breaks down rather rapidly on the ground. *Riccardia palmata* and *Nowellia curvifolia* are both widespread and often abundant, usually with *Cephalozia bicuspidata*, *Cephalozia lunulifolia*, *Scapania umbrosa*, *Lophozia ventricosa* and *Lepidozia reptans*. *Tritomaria exsectiformis*, usually common in this habitat in western woodlands, was only recorded twice. Rotting logs are also the usual habitat for *Dicranodontium denudatum* in Assynt, often with *Campylopus flexuosus*. The tiny, rare liverwort *Calypogeia suecica* has just one site on a log in woodland at Creag an Spardain.

## Heath, mire and flushes

Assynt has vast tracts of heathy ground, with some form of wet heath the most frequent, grading into dry heath on better drained ground with a southerly aspect. The dry heath is usually rather dull bryologically often having little more than an abundance of *Hylocomium splendens*, *Pleurozium schreberi* and *Hypnum jutlandicum*. Where the ground is more rocky there will usually be more *Racomitrium lanuginosum* and occasionally liverworts like *Barbilophozia floerkei* with *Pohlia nutans*, *Campylopus flexuosus* and *Campylopus introflexus* on barer ground.

The wet heath can be equally dour but there is usually a greater variety of species, including *Racomitrium lanuginosum*, *Hylocomium splendens* and *Pleurozium schreberi*. *Sphagnum* species are much less frequent than in the mires but *Sphagnum capillifolium*, *Sphagnum papillosum* and *Sphagnum denticulatum* are usually to be found with smaller amounts of *Sphagnum tenellum* and rarely *Sphagnum molle*. Where there is some flushing, the distinctive 'bottle-brush' stems of *Breutelia chrysocoma* occur, sometimes in huge patches. Where the heath is somewhat degraded or where there is a thin covering over slabby bed-rocks, large cushions of *Campylopus atrovirens* become conspicuous, usually with the equally impressive purple liverwort *Pleurozia purpurea*. The latter species is so common in the western Highlands that it is easy to forget that it is extremely rare in Europe as a whole and has a highly disjunct world distribution. In wet heath, as in other mire areas, deer and sheep dung often has tufts of *Splachnum sphaericum* while the much rarer *Splachnum ampullaceum* seems limited to the larger droppings of the deer.

Areas of deeper peat with mire vegetation are widespread in Assynt but many of these have been exploited for fuel, particularly those close to roads and habitation, past and present. However, there are large populations of *Sphagnum capillifolium*, *Sphagnum papillosum*, *Sphagnum denticulatum*, *Sphagnum fallax* and *Sphagnum cuspidatum* and these seem to regenerate readily after active extraction of peat has ceased. There are smaller populations of *Sphagnum subnitens*, *Sphagnum compactum* and *Sphagnum tenellum*, the latter two often associated with some degradation of the mire, and this is also the case with the few sites for *Sphagnum molle* and *Sphagnum strictum*. An open area of peat in a mire also provided the one recent site for the impressive moss *Campylopus shawii*. This plant is locally abundant on Skye and the Outer Hebrides but very rare on the mainland and is only known elsewhere from the extreme south-west of Ireland, the Azores and some Caribbean islands.

In the least disturbed areas *Sphagnum capillifolium* and *Sphagnum papillosum* form significant hummocks and these better mires may also have lawns of *Sphagnum magellanicum* and a development of a pool system with floating masses of *Sphagnum denticulatum* and *Sphagnum cuspidatum*. Where there are flushed runnels in the mire, often with stands of *Scorpidium scorpioides*, *Sphagnum teres* can form significant stands and in these richer mires there are rare stands of *Sphagnum contortum* and *Sphagnum subsecundum*. Scattered around Assynt, usually some distance from habitation, are a few undisturbed mires, usually with a good cover of *Calluna vulgaris* and *Eriophorum vaginatum*, with large hummocks of *Sphagnum fuscum* and, less frequently, *Sphagnum austinii*. Most of these are on flat ground perched on a watershed, the best being on Mointeach na Totaig north-east of Loch Urigill, but some are valley mires as at the head of a tributary of Abhainn a' Chnocain.

Most hummocks of *Sphagnum capillifolium* and *Sphagnum papillosum* have small liverworts creeping through them. By far the most common of these is *Odontoschisma sphagni* and this may also occur on decaying hummocks and bare peat. *Mylia anomala* is also quite frequent but *Mylia taylori* also occurs in *Sphagnum* hummocks as well. Species of *Kurzia* are frequent but fertile material seems sparse and it was not possible to identify most stands to species and so the maps are rather sparse. *Cephalozia bicuspidata*, *Cephalozia lunulifolia* and *Calypogeia fissa* occur with some regularity and there are scattered stands of *Calypogeia sphagnicola*. There is only one record for

*Cephalozia connivens* and two for *Cephalozia loitlesbergeri* but these species were certainly overlooked elsewhere. The best populations of the more interesting species tend to be associated with mires that have good hummock development, particularly where there are stands of *Sphagnum austinii* and *Sphagnum fuscum*.

Areas of wet ground that are flushed from above are also abundant in Assynt with their flora varying according to the base status of the ground water. Many flushes are overwhelmingly acidic and are dominated by *Sphagnum* species, particularly *Sphagnum denticulatum*, *Sphagnum palustre* and *Sphagnum fallax* often with *Aulacomnium palustre*. On the gneiss there are many flushes that are picked out by stands of *Schoenus nigricans* where *Campylium stellatum* var. *stellatum*, *Scorpidium scorpioides*, *Aneura pinguis*, *Blindia acuta* and less frequently *Drepanocladus revolvens* are abundant along with some *Sphagnum* species. These flushes, with their peculiar chemistry, are seemingly shunned by all other bryophytes.

Other flushes are at least moderately calcareous and have typical species like *Philonotis fontana*, *Dicranella palustris*, *Brachythecium rivulare*, *Bryum pseudotriquetrum*, *Scapania undulata*, *Fissidens adianthoides*, *Fissidens osmundoides* and infrequently *Jungermannia exertifolia* ssp. *cordifolia*. On the limestone *Palustriella commutata* var. *commutata* and *Palustriella commutata* var. *falcata* can be abundant sometimes with *Philonotis calcarea* and in one case *Tritomaria polita*. Both on the limestone and in higher flushes, particularly on Conival, there are stands of *Philonotis seriata* and many montane flushes also have large patches of *Scapania uliginosa*.

Other rarities associated with flushes include *Amblyodon dealbatus* on the limestone and *Calliergon trifarium*, *Pohlia wahlenbergii* var. *glacialis*, *Sphagnum platyphyllum* and *Scapania degenii* on the gneiss in Coire Gorm. Finally, areas of flushed grassland are quite common and *Ctenidium molluscum* is almost ubiquitous here often with *Fissidens adianthoides* and occasionally *Fissidens osmundoides*, *Thuidium delicatulum* and, in coarser vegetation, *Dicranum bonjeanii*.

## Crags, scree and ravines

Much of the account of bryophytes in woodland applies to many crags and ravines on the lower ground with a northerly aspect, as these tend to have at least some woodland cover. Wooded ravines tend to be the richest sites for bryophytes because of the reasonably constant humidity, rock exposure (often with some variety of base-status), and freedom from disturbance. All the woodland species occur here and frequently there is an admixture of plants with a more montane distribution. On crags and in scree much will depend upon aspect, with north-facing slopes generally having a much more diverse flora than those that face the sun.

Constant species on crags include almost all the ubiquitous bryophytes listed above, particularly on ledges, and most crags away from the limestone have plants like *Amphidium mougeottii*, *Anoectangium aestivum*, *Racomitrium heterostichum*, *Racomitrium aquaticum*, *Bartramia pomiformis*, *Bartramia ithyphylla*, *Fissidens* species, *Isothecium myosuroides* var. *brachythecioides*, *Andreaea rupestris*, *Polytrichum piliferum* and *Campylopus atrovirens*. The flora will depend very much on base-status with *Homalothecium sericeum*, *Isothecium alopecuroides*, *Tortella tortuosa*, *Ctenidium molluscum*, *Preissia quadrata*, *Plagiochila porelloides*, *Gymnostomum aeruginosum* and *Trichostomum brachydontium* all more or less common on the more base-rich rocks. At the other end of the scale, some crags on the gneiss and the quartzite are almost devoid of any bryophytes except on the larger ledges, and the same is true to a lesser extent, of the more exposed sandstone.

The gneiss, as ever, is the least predictable of the rocks because of the small-scale variation in base-status and the numerous igneous intrusions. The 'better' gneiss is often signalled by stands of *Grimmia funalis* and *Grimmia trichophylla* in open sites, *Neckera crispa* where more sheltered, or *Racomitrium ellipticum* where water trickles down. A large number of such gneiss crags proved to have populations of *Glyphomitrium daviesii*, an interesting find as this is a plant that was thought to be virtually confined to the Tertiary volcanic rocks in Britain (Birks in Hill et al. 1992). It is an extremely oceanic species only known outside the British Isles from Iceland, the Faeroes, Norway, Madeira and the Azores.

The bryophytes on scree slopes are also very dependant on aspect and geology. South-facing block-scree is usually rather dull and may have little other than *Racomitrium lanuginosum*, *Racomitrium heterostichum*, *Racomitrium fasciculare* and *Diplophyllum albicans*, sometimes with *Hedwigia stellata*, *Ulota hutchinsii* and, on the sheltered side of larger rocks, *Scapania gracilis*. North-facing scree, particularly where the blocks are large, often has a flora like that of the rocky woodland with an abundance of the larger carpet-forming mosses including occasionally *Hypnum callichroum*, much *Scapania gracilis* and *Isothecium myosuroides*, and quite frequently *Hymenophyllum wilsonii* and *Plagiochila spinulosa*. *Mylia taylori* and *Bazzania tricrenata* are usually common, often with *Sphagnum capillifolium*, *Anastrepta orcadensis* and less frequently *Herbertus aduncus ssp hutchinsiae*, providing a clear link with the more montane hepatic community described below. On the rocks, *Dicranum fuscescens* can be frequent often with *Lophozia ventricosa* and rarely *Lophozia sudetica* but *Barbilophozia floerkei* is surprisingly infrequent.

A number of other interesting species have their only sites in these habitats at relatively low level. The predominantly eastern moss *Cyodontium jenneri* has two sites on dry, acid crags on the gneiss, the newly described *Leiocolea fitzgeraldii*, endemic to the British Isles, has two sites on wet gneiss and *Leiocolea heterocolpos* was found once on a ledge of a gneiss crag. *Trichostomum hibernicum* and *Molendoa warburgii*, both apparently endemic to the British Isles, have sites in the more base rich ravines while *Aulacomnium androgynum*, so common elsewhere in Britain, has just two sites on sheltered but extremely acid rocks.

## The limestone

The limestone is extremely rich and has an abundance of the more common calcicole bryophytes so I will concern myself mostly with the species of interest. *Ctenidium molluscum* is abundant in much of the grassland with *Enthodon concinnus* in drier places and usually with *Ditrichum gracile*. Plants collected suggest that *Thuidium recognitum* is more frequent in the grassland than *Thuidium philibertii* but neither are at all common. *Neckera crispa* can occur in the 'Dryas heath', sometimes with *Scapania aspera*, but the latter is much more frequent on boulders and crags. On the north side of Cnoc Eilid Mhathain there is an odd community which has large patches of *Herbertus stramineus* in the heathy grassland over the limestone. Where the grassland is a little damper there are a few sites for *Brachythecium glareosum*. Flushed turf by a burn near Knockan also provides the habitat for *Moerckia hibernica* and the Red Data Book liverwort *Leiocolea gillmanii* in their one Assynt site.

The tops of low rocks in the grassland is the preferred habitat for *Ditrichum flexicaule*, here looking very distinct from *Ditrichum gracile*, being straight-leaved, dark green in colour and, at least in the winter months, with fragile apical shoots with reduced leaves. Also on these lower, easy-angled rocks are cushions of *Tortella densa*, so easily overlooked as *Tortella tortuosa* which is abundant, often with *Didymodon ferrugineus*. Species of *Schistidium* are common on all the boulders; by far the most common seems to be *Schistidium apocarpum* sensu stricto, but *Schistidium robustum* also occurs, as does the nationally scarce species *Schistidium trichodon* in the scree near Creag nan Uamh. Also in this same community on the broken outcrops above Lairig Unapool is a population of the nationally rare *Didymodon icmadophila*.

On the damper limestone the frequent patches of *Orthothecium rufescens* can provide a magnificent display, set off by the grey rocks. On damp ledges in two places there are stands of *Conardia compacta*, the Cambrian limestone providing the only sites north of the Great Glen for this rare plant. On dripping rock faces in caves and sheltered spots by rivers and often covered by an unpleasant algal slime, there are several populations of *Seligeria trifaria* in by far the most northerly of the two Scottish localities. In the Traligill valley, similar wet places can have very large cushions of *Hymenostylium insigne*, a Red Data Book species with a disjunct world distribution. One of the most interesting finds of the current survey was a small patch of *Hypnum bambergeri,* on a wet ledge on a crag at the very moderate altitude of 180m. This is a rare moss, previously thought to be limited to the Central Highlands where it usually occurs high up on strongly calcareous mica-schist, although there are sites on limestone in the Ben Alder range and Glencoe.

The more montane limestone and the larger crags also have a number of species of interest. Creag Sròn Chrùbaidh is often rather dry but has *Syntrichia princeps* and *Grimmia orbicularis* in one area and scattered stands of *Bryum elegans* and a good range of other calcicoles. The smaller crags near Stronchrubie have some *Leucodon sciuroides* and there is much more on the stone walls in the fields below. Creag nan Uamh and its associated scree are rather better and there are good populations of *Pseudoleskeella catenulata* and *Pseudoleskeella rupestris* on the crag near the cave and on the big boulders below there is a small population of *Encalypta alpina*. The latter also occurs sparingly on the highest limestone in the Bealach Traligill with *Encalypta ciliata* in its only Assynt site. *Encalypta* species (other than *Encalypta streptocarpa*) and a few other limestone bryophytes, like *Anomodon viticulosus*, are surprisingly rare in Assynt given the limestone exposure. The 'high limestone' of Cnoc Eilid Mhathain also has some *Pseudoleskeella catenulata* and good stands of *Mnium thomsonii* and *Scapania aequiloba*.

## Upland

Much of the species that would normally be considered 'upland' have found a place under one or other of the habitats already covered. The occurrence of 'alpine' species at low altitude in the north west of Scotland is a well-known phenomena and is as true for bryophytes as for flowering plants. As explained below, the higher ground in Assynt did not receive the attention it deserves during this survey so this account is rather sketchy. The highest ground, that on the western flanks of Conival, was not covered at all, but older records of mosses like *Aulacomnium turgidum* and *Kiaeria falcata* suggest that more survey work here may prove productive. Unfortunately the excellent late snow-lie vegetation on the north side of Conival is not in Assynt. The sandstone hills are spectacular but provide little area at high altitude and the high ridges of Breabag and Beinn Uidhe are composed mainly of a rather uncompromising quartzite.

The best development of the species-rich ledges that are a feature of the better hills further south, occur where the gneiss reaches its highest altitude on Sàil Gorm on

Quinag, on the north side of Glas Bheinn and on one facies of the sandstone on the north side of Canisp. Here there are ledges with *Distichium capillaceum*, *Ditrichum gracile*, *Leiocolea bantriensis*, *Molendoa warburgii*, *Plagiobryum zieri*, *Orthothecium intricatum* and *Hypnum hamulosum* and rocks with *Grimmia torquata*, *Grimmia curvata* and *Cololejeunea calcarea*. Less common species include *Amphidium lapponicum*, *Isopterygiopsis muelleriana*, *Tortula subulata* var. *graeffii* and *Orthothecium rufescens*.

Most other interesting montane crag and fell field species have only isolated occurrences. *Ditrichum zonatum* var. *zonatum* has a few scattered sites on open soil in exposed sites, usually in a mat with more common species like *Oligotrichum hercynicum*, *Diplophyllum albicans* and *Nardia scalaris*. *Arctoa fulvella* was only seen on a couple of occasions, both sites being on the exposed sandstone rocks on the ridges of Quinag. *Racomitrium macounii* subsp. *alpinum* was seen once in the Bealach Traligill but this is an easy species to overlook. *Plagiothecium platyphyllum* occurs in a few acid flushes high on Quinag and Glas Bheinn and *Kiaeria blyttii* is quite frequent in the larger areas of block scree.

The most important upland community is the heath that develops at the base of crags, on steep slopes and over scree on north or north-east facing slopes. This heath, already described briefly above, usually has an open canopy of *Calluna vulgaris* or at higher altitudes, *Vaccinium myrtillus*, often with an abundance of *Sphagnum capillifolium* or *Racomitrium lanuginosum* in the ground layer. Mixed in with this are reasonably widespread plants like *Scapania gracilis*, *Anastrepta orcadensis*, *Pleurozia purpurea*, *Mylia taylori*, *Herbertus aduncus* ssp. *hutchinsiae* and *Bazzania tricrenata* and varying amounts of the rare liverworts *Scapania ornithopodioides*, *Scapania nimbosa*, *Mastigophora woodsii*, *Plagiochila carringtonii*, *Bazzania pearsonii*, *Anastrophyllum donianum* and *Anastrophyllum joergensenii*. These liverworts have an extraordinary world distribution (see below)

That such an assemblage of disjunct species is present in north-west Britain is remarkable and there are excellent examples in Assynt, particularly in the hills in the east and on Quinag. All these species would appear to require cool, montane, high rainfall sites and it is worth noting that in their Himalaya sites they tend to occur in moist juniper-rhododendron scrub above the natural tree line and it is possible that they are derived from similar dwarf shrub vegetation at or near the tree line in Scotland that has now been largely lost (Long, pers. comm.).

*Anastrophyllum donianum* and *Anastrophyllum joergensenii* tend to be limited to the higher ground, often where there may be some development of snow patches as in Coire Gorm. *Scapania ornithopodioides* is less demanding and is quite widespread in suitable habitat but *Scapania nimbosa* is rather patchy and often absent from seemingly suitable sites. *Bazzania pearsonii* is probably locally frequent but can be rather hard to distinguish from the abundant *Bazzania tricrenata*. Usually most abundant of all are *Plagiochila carringtonii* and *Mastigophora woodsii*, frequently forming large cushions and sometimes extending down into the higher woodland. This type of oceanic-montane heath is covered by the National Vegetation Classification H21b (*Mastigophora woodsii* – *Herbertus aduncus* ssp. *hutchinsiae* sub-community of the *Calluna vulgaris* – *Vaccinium myrtillus* – *Sphagnum capillifolium* heath) or, higher up, H20c (the *Bazzania tricrenata* – *Mylia taylori* sub-community of the *Vaccinium myrtillus* – *Racomitrium lanuginosum* heath) and Rodwell (1991), gives a much fuller treatment.

There are other interesting species associated with this kind of habitat. The predominantly woodland species *Plagiochila spinulosa* and *Dicranodontium denudatum* occasionally occur here but much more frequent, particularly on Quinag, is *Dicranodontium uncinatum* which can form fine carpets in the turf below the crags. Also on Quinag, with *Mastigophora woodsii* and *Plagiochila carringtonii*, there are two populations of *Campylopus setifolius*, an uncommon oceanic species in Britain and elsewhere only known from northern Spain.

| Species | ex-UK distribution |
| --- | --- |
| *Scapania ornithopodioides* | W. Ireland, Norway, Faeroes, Himalaya, W. China, Japan, Taiwan, Philippines, Hawaii |
| *Scapania nimbosa* | W. Ireland, Nepal, Sikkim, W. China (Yunnan) |
| *Mastigophora woodsii* | W. Ireland, Faeroes, N.W. America, E. Himalaya |
| *Plagiochila carringtonii* | W. Ireland, Faeroes, Nepal |
| *Bazzania pearsonii* | W. Ireland, E. & S.E. Asia, N.W. America |
| *Anastrophyllum donianum* | Faeroes, S.W. Norway, Tatra, W. Tibet, Sikkim, Nepal, Bhutan, Yunnan, Alaska, W. Canada |
| *Anastrophyllum joergensenii* | S.W. Norway, Sikkim, Nepal, Bhutan, Yunnan |

Another species with a disjunct distribution which often occurs close to these plants is the moss *Paraleptodontium recurvifolium*, outside of the British Isles only known from British Columbia and Alaska.

## HISTORY OF BRYOPHYTE RECORDING

The history of recording in Assynt reflects the magnetic attraction of the limestone, with only a relatively small number of records from sites away from the calcareous band and the mountains close to it. Assynt has never had a resident bryologist and visitors have always had to contend with its relative isolation which, even in these days of good roads, tends to deter all but the few. This means that a small number of intensive visits have produced the vast majority of the records prior to this survey. For example, of the 2733 records in the Biological Records Centre (B.R.C) database that can be localised to Assynt, over 1000 originate from three visits by John Birks and others in 1966, 1967 and 1972.

Written accounts of visits to Assynt with bryological intent are very sparse and a full history of recording would require much time trawling through the herbaria of the early collectors. Given the time-scale of the project and the, probably, limited interest of the information gleaned, this approach was not adopted. Almost all of this information has come from the B.R.C database and from a few papers. It is remarkable that in MacVicar's paper *The Distribution of Hepaticae in Scotland* (1910) there are only two records that he assigns to sites in Assynt, though there are a number of records from Sutherland.

The earliest records from Assynt are those collected by James Robertson in the vicinity of Inchnadamph in 1767 (Henderson and Dickson 1994), background to which is given in the companion section for the vascular plants. These include a number of common species like *Sphagnum palustre*, *Polytrichum commune*, *Hylocomium splendens* (as *Hypnum proliferum*), *Racomitrium lanuginosum* (as *Bryum hypnoides*), *Homalothecium sericeum* (as *Hypnum sericeum*), *Lophocolea bidentata* (as *Jungermannia bidentata*) and *Aulacomnium palustre* (as *Mnium palustre*). There are plants of more interest including *Fissidens bryoides* (as *Hypnum bryoides*) and *Marchantia polymorpha* sensu lato, that are uncommon in Assynt, and three species not now in the Assynt list, *Leptobryum pyriforme* (as *Bryum pyriforme*), *Brachythecium velutinum* (as *Hypnum velutinum*) and *Drepanocladus aduncus* (as *Hypnum aduncum*). We do not know who named these plants and there are no voucher specimens and given the taxonomic changes it seems sensible to merely mention them here.

The earliest record in the B.R.C. is that of *Pleurozia purpurea* from Canisp in 1833, credited to W.H. Campbell, with the specimen in the herbarium in the Royal Botanic Gardens (R.B.G.) in Edinburgh. This was probably collected during the expedition to Sutherland by Robert Graham in 1833, when a Mr Campbell was certainly a member of the party. However there is no direct mention of a visit to Canisp in the report (Graham 1833) and so some doubt must remain about the source. There will almost certainly be other records from around this date in the Greville herbarium in the R.B.G. and probably elsewhere.

The richness of the bryophyte flora was first revealed during a visit to the area by the noted English bryologists H.N.Dixon, W.E. Nicholson and E.S. Salmon from the 17th to the 24th of July in 1899 (Nicholson 1900). By then, Dixon had already 'traversed the country from Lairg to Altnaharra, whence Ben Clibreck and Ben Hope were ascended, then went on by Tongue and Erriboll to Durness, exploring from this centre Smoo Cave, Cape Wrath and the Far-out Head (sic)'. Based at Inchnadamph, they went over Conival and Ben More Assynt, Canisp, Quinag, Glas Bheinn, Beinn an Fhurain and covered the limestone at Inchnadamph and the Allt nan Uamh, an impressive itinerary for a weeks' botanising. Nicholson's paper in the Journal of Botany is, in effect, a list of the first records for Assynt of a considerable number of moss species. A number of their montane records are given as '108... Ben More Assynt' and it is difficult to know whether they were aware that the eponymous mountain is neither in Assynt nor v.c.108, a problem of interpretation which has persisted down the years.

They found many of the interesting species on the limestone and were enthusiastic about its 'southern flora', finding that the hill ground 'is very poor in northern or arctic species as compared with...the Grampians'. South of Inchnadamph they found *Syntrichia princeps* (as *Tortula princeps*) in 'immense, rounded cushions', presumably on Creag Sròn Chrùbaidh where it still occurs. In Glen Dubh on rocks in the Traligill they found *Gymnostomum calcareum* (as *Weisia calcarea*) and also described a robust form of *Hymenostylium recurvirostrum* (as *Weisia curvirostris*), now elevated to species status as *Hymenostylium insigne*. In the Allt nan Uamh they recorded *Pseudoleskeella catenulata*, noting the form that is now recognised as a species, *Pseudoleskeella rupestris*. Other notable finds on the limestone include *Mnium thomsonii*, *Cinclidium stygium*, *Thuidium recognitum*, *Conardia compacta* (as *Amblystegium compactum*) and *Leucodon sciuroides*.

Away from the limestone, Nicholson remarks on a form of *Rhynchostegium riparioides* (as *Eurhynchium rusciforme*) with 'slender, julaceous, brown branches' in a waterfall on Quinag, where it still occurs today and is now considered a species, the nationally scarce *Rhynchostegium alopecuroides*. Also on Quinag they found *Arctoa fulvella*, again the only site found for this plant during the present survey. Again on Quinag and also on Conival they recorded *Bryum mildeanum* but these collections have been re-determined as *Bryum riparium*, the only Assynt records for this species. Other interesting montane species they recorded include *Isopterygiopsis muelleriana* on Beinn Uidhe, *Herzogiella striatella* on Conival and *Bryum muehlenbeckii*, *Calliergon trifarium* and *Tetraplodon angustatus* on Glas Bheinn.

During the 1899 visit, as Nicholson remarks later, 'the Hepatics were left severely alone' (Nicholson 1923) and it was not until 1921 than he returned to remedy this omission, this time in the company of another celebrated bryologist, H.H. Knight. In his account, Nicholson remarks on the relative poverty of the limestone in liverworts compared with the mosses but is enthusiastic about the slopes of Beinn an Fhurain and the 'damp wood by the southern shore of Loch Assynt'. It seem reasonable to assume that this is the wood at An Coimhleum. The most important single discovery of this trip was undoubtedly that of *Anastrophyllum joergensenii*, new to the British Isles, on Beinn an Fhurain. They also made the first West Sutherland records of other members of the of oceanic-montane heath community including *Herbertus aduncus* ssp. *hutchinsiae*, *Mastigophora woodsii*, *Anastrophyllum donianum*, *Bazzania tricrenata*, *Bazzania pearsonii*, *Plagiochila carringtonii* (as *Jamesoniella Carringtonii*), *Scapania ornithopodiodes* and *Scapania nimbosa*.

In 'the wood by Loch Assynt' they found an excellent selection of the liverwort species of oceanic woodland most of which were re-found during the current survey. They provide the first West Sutherland records for *Plagiochila spinulosa*, *Plagiochila punctata* and the much less common *Plagiochila exigua* (as *Plagiochila tridenticulata*). They also found a good population of *Leptoscyphus cuneifolius* 'mostly on *Frullania* growing on rock' but during the current survey it was found only on trees here. On 'moist, shaded rocks' they found *Colura calyptrifolia*, *Drepanolejeunea hamatifolia*, *Aphanolejeunea microscopica*, *Lejeunea patens* and *Radula aquilegia*, happily all still frequent in the woodland.

They did find some hepatic interest on the limestone, including *Scapania cuspiduligera* (as *Scapania Bartlingii*) with *Scapania aequiloba* by the Traligill, *Porella platyphylla* and *Radula lindenbergiana* on Creag na Uamh and *Harpanthus flotovianus* in the upper part of Gleann Dubh. Nicholson also managed to find several montane species which have eluded this survey in *Marsupella sphacelata*, *Marsupella alpina* (as *Gymnomitrion alpinum*) and *Marsupella adusta*.

Apart from these two very productive visits described by Nicholson, records in B.R.C. suggest few other visits before World War II. J.B. Duncan visited several times in his long bryological career but added little of interest in Assynt except *Grimmia orbicularis*, though further examination of his herbarium may reveal more useful records. E.C. Wallace visited in 1939 and several times later, recording *Hymenostylium insigne*, *Leucodon sciuroides*, *Antitrichia curtipendula* and new records for *Eurhynchium pumilum* by Inchnadamph and, in 1957, *Taxiphyllum wissgrillii* at Knockan, by far its most northerly site in Britain. In 1946, C.W. Muirhead, more noted as a vascular plant botanist, found *Campylopus shawii* by the Allt a' Bhathaich on Quinag and C.D. Pigott found *Rhytidium rugosum* on cliffs at Inchnadamph in 1950, remarkably the only record.

In 1951, Alan Crundwell made the first of several visits spanning over 40 years and added *Bazzania trilobata*, *Entosthodon attenuatus*, probably from An Coimhleum, *Metzgeria leptoneura* from Quinag and *Weisia controversa* from the limestone. In 1956, he also added *Kiaeria falcata* from Conival and, in 1959, *Trichocolea tomentella* from Elphin. E.F. Warburg visited Inchnadamph in 1952 recording *Gyroweisia tenuis* and *Thuidium philibertii* from the limestone. During the 1950s and 60s, D.A. Ratcliffe made his inevitable, interesting contributions to the records of the area with *Harpanthus flotovianus* from Beinn an Fhurain, *Hygrohypnum duriusculum* from the Allt nan Uamh in 1959, *Myurium hochstetteri* on Stoer in 1961 and *Campylopus setifolius* from Creag na h-Iolaire (Bealach Leireag) in 1966.

In 1960 the British Bryological Society (B.B.S.) summer field meeting was based in Ullapool and made three excursions to Assynt, all to the limestone (Crundwell 1961). At Knockan interesting species included *Reboulia hemisphaerica*, *Pseudoleskeella rupestris* (as *Pseudoleskeella catenulata* var. *acuminata*) and *Hypnum hamulosum*. Despite the attention it had already received, the Inchnadamph area provided a number of interesting new finds including *Weissia rutilans*, *Thuidium recognitum*, *Ulota calvescens*, *Rhynchostegiella teneriffae* (as *Rhynchostegiella teesdalei*) and *Porella cordeana*. Finally, the trip to the Allt nan Uamh and Creag nan Uamh yielded *Seligeria trifaria* (as *Seligeria tristicha*), *Encalypta alpina*, *Rhodobryum roseum*, *Schistidium trichodon* and *Anomobryum julaceum* var. *concinnatum*.

In 1966 H.J.B. Birks made the first of his three visits (1966, 1967 and 1972) with H.H. Birks and J. Dransfield which, while contributing only a few new records, confirmed a number of the older rarities and visited areas away from the limestone, like Stoer. The new records made on these trips are *Meesia uliginosa* from the Allt nan Uamh in 1966, *Aulacomnium turgidum* from Conival and *Sphenolobopsis pearsonii*, probably from An Coimhleum, in 1967 and *Harpanthus scutatus* by Loch Nedd in 1972. Also in 1966, S. Ward made the remarkable find of *Anastrophyllum saxicola* on Meallan Liath Mor and Paddy Coker found *Oncophorus virens* by the Traligill; neither species has been re-found.

Jean Paton made several new records during a visit in 1969 and provided a large number of records for more common species away from the limestone. Interesting liverworts recorded by her include new records for *Lophozia opacifolia* and *Haplomitrium hookeri* from Loch nan Cuaran and the only Assynt record for *Barbilophozia lycopodioides* from Inchnadamph. David Long has made several visits, first in 1973 with David Chamberlain and again in 1981 and 1982. His were the first records of the interesting flora at Achmelvich including *Amblyodon dealbatus*, *Riccardia incurvata*, *Scapania cuspiduligera* and *Distichium inclinatum*. From the Inchnadamph area came records for the Red Data Book species *Dicranella grevilleana* and also *Hypnum lindbergii*, *Amphidium lapponicum* and *Mnium thomsonii*. In 1978 the two distinguished American bryologists W.B. Schofield and R.M. Schuster found *Barbilophozia kunzeana* in a mire above Creag Sròn Chrùbaidh and *Andreaea megistospora* on rocks near Unapool, the only West Sutherland records for these species. Finally, Ray Woods found *Weissia perssonii* out on the Stoer peninsula in 1985.

# PRESENT SURVEY

The survey work on the flora of the flowering plants and ferns was already well underway before there was any notion of a bryophyte add-on. The idea of attempting to include the bryophytes of Assynt in the planned flora grew out of the records made by the British Bryological Society field meeting in Lochinver in 1992 (Rothero 1993) and the preparations for that meeting. Over the next few years I made regular visits, usually during the winter months, based in Nedd but exploring much of the parish. Driven by the indefatigable enthusiasm of I.M.E., my natural indolence in recording was given a severe jolt and it became apparent that a useful body of records was being built up. Once the decision was made to include bryophytes in the flora, never a conscious one on my part (!), the frequency and duration of my visits to Assynt increased and a deliberate attempt was made to cover as much of the parish as was possible during the time-scale of the vascular flora project.

With such a large number of tetrads (164) and recording limited to, at best, two weeks per year over the last nine years, it is inevitable that most tetrads (99) were only visited once, and a number (48) were not visited at all. Of the 99 tetrads, some were rapidly surveyed en route to somewhere else and some have the merest handful of records. This must be borne in mind when reading the species accounts as well as when using the distribution maps. They are in no way comparable to those of the vascular plant flora as will be apparent if the description of that survey is examined. Even so, this body of records is unique in the Highlands at the moment, and, though more survey time and time to research various herbaria would have produced a more complete work, the opportunity to publish it, warts and all, along with the vascular plant flora, could not be missed.

There is a bias in the recording. The well-known bryophyte flora of the limestone areas did not receive the proportionate coverage it deserves; it has been reasonably well-covered by others. Most survey work took place outside of the summer months, so, despite my personal interest in the montane flora, there is scope for much more recording and interesting finds in the higher hills. With so much wild country to cover, very little attention was paid to the more 'anthropogenic' habitats provided by the settlements, the roadsides and what little cultivation there is. This is reflected in the paucity of records for a good number of ruderal species. I humbly admit to not being an aficionado of the genus *Sphagnum* and this is glaringly obvious in the species accounts.

Apart from the initial work by the B.B.S. in 1992, virtually all the records that are mapped are mine, the only exception being a number of interesting records from David Long, the earliest dating from 1973. Earlier records of species that were either not re-found during this survey or that are rare in Assynt are mentioned in the text but are not mapped. The Assynt bryophyte database used to produce the maps now holds some 13,600 records of 156 liverwort and 345 moss taxa. In addition to this there are some 14 liverworts and 20 mosses reliably recorded for Assynt which were not re-found on this survey. For the bryophytes there is no equivalent to *Anthony* so it is very difficult to be certain how many new species the survey added to the Assynt list but some 34 species recorded during the survey were new for West Sutherland as a whole.

## The future

As I hope I have made clear from the preamble, there is

considerable scope for more bryophyte recording in Assynt. There are large gaps in the maps that it would be satisfying to fill. The limestone certainly has more interesting species to be found, with many of the smaller crags and burns having had few if any visits. More attention to the mire areas by someone with more expertise in *Sphagnum* and an enthusiasm for *Cephalozia* species would be worthwhile, and there must be more than one site for *Campylopus shawii*. The higher hills have had only cursory attention and offer the possibility of good plants in spectacular scenery. And then there is the rest of Sutherland.

## PLAN OF SPECIES ACCOUNT

The bryophyte accounts are in systematic order beginning with the liverworts. This order and the taxonomy follows the most recent bryophyte *Census Catalogue* published by the British Bryological Society (Blockeel and Long 1998). There have been numerous changes in moss nomenclature since the publication of the standard British moss flora (Smith 1978) and, where appropriate, I have included a synonym. We are fortunate in having an excellent new British liverwort flora (Paton 1999) which should be consulted for the liverwort taxonomy and for much fuller accounts of the species. Those tetrads which were not visited are marked 'X'; in one or two cases there may be an odd record from one of these tetrads so there may be a dot superimposed on the 'X'.

*The following species was inadvertently omitted from p.236 of the Species Account:

*Bryum riparium* I.Hagen
Recorded, as *B. mildeanum,* from both Quinag and Conival by Dixon and Nicholson in 1899 (Whitehouse 1963).

# SPECIES ACCOUNT

## LIVERWORTS

*Haplomitrium hookeri* (Sm.) Nees

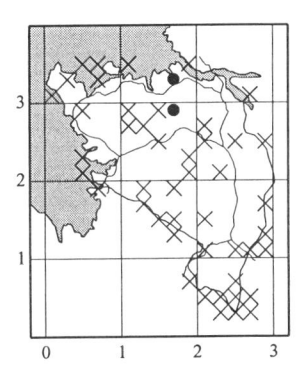

2 tetrads. Rare but possibly overlooked on stony or gravelly loch margins; Loch an Leothaid and Loch Ardbhair. There is a further record from Loch nan Cuaran, Beinn Uidhe by Paton in 1969.

*Mastigophora woodsii* (Hook.) Nees

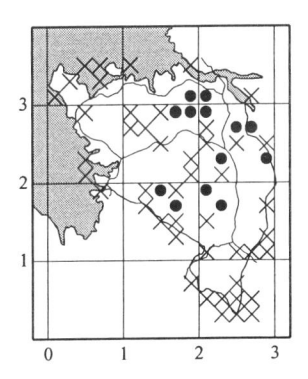

13 tetrads. This oceanic-montane liverwort is very local but can be abundant in suitable habitat on the higher ground, under humid, dwarf-shrub heath and heathy scree on rocky, north or north-east -facing slopes and also occurs rarely in wooded ravines.

*Herbertus stramineus* (Dumort) Trevis

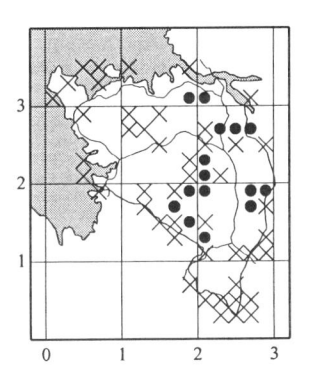

15 tetrads. This oceanic-montane liverwort is unusual in this group in requiring rather base-rich conditions. It occurs on montane ledges on the gneiss, in grassy turf on the limestone and is locally abundant on the north side of Cnoc Eilid Mhathain.

*Herbertus aduncus* (Dicks.) Gray ssp. *hutchinsiae* (Gottsche) R.M.Schust.

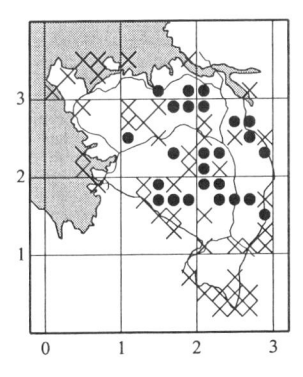

25 tetrads. One of the most constant liverworts in the hepatic community under oceanic-montane heath on rocky north and north-east facing slopes; it can be locally abundant in striking, orange-brown mats.

*Blepharostoma trichophyllum* (L.) Dumort.

16 tetrads. This inconspicuous plant grows through other bryophytes on damp ledges on crags and in ravines. It is most frequent on the more base-rich rocks but can occasionally occur on acid, peaty soil.

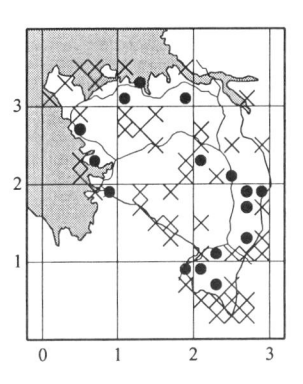

*Trichocolea tomentella* (Ehrh.) Dumort.

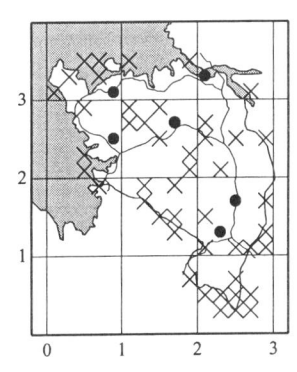

6 tetrads. This beautiful species is rather sparse in Assynt given its relative frequency in the South-West Highlands. It occurs on wet, rocky banks in ravines where there has been a build-up of silt in a drainage line.

*Kurzia pauciflora* (Dicks.) Grolle

There are unlocalised records from the Inchnadamph area by Birks et al. in 1966 and 1967.

*Kurzia sylvatica* (A.Evans) Grolle

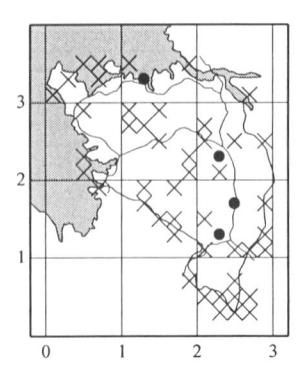

4 tetrads. Rare; occurs in *Sphagnum* hummocks and as thin mats on peaty banks; Allt Mhic Mhurchaidh Gheir, An Coimhleum, Duart and the east end of Cam Loch.

*Kurzia trichoclados* (Müll.Frib.) Grolle

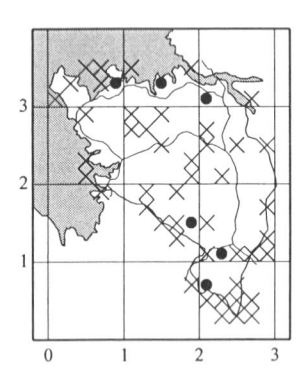

6 tetrads. The most frequent of the three *Kurzia* species, usually occurring as deep, aromatic cushions under heather on steep peat, capping vertical rocks.

*Lepidozia reptans* (L.) Dumort.

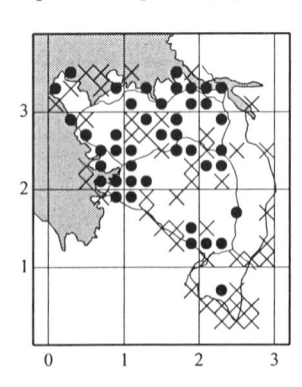

43 tetrads. Common and locally abundant in humid sites on peaty banks, on rocks in woodland, sheltered crags and as an epiphyte on tree boles, particularly birch and alder.

*Lepidozia pearsonii* Spruce

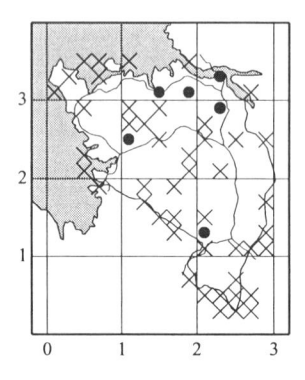

6 tetrads. Seemingly rare in Assynt but possibly overlooked. It usually grows through mats of large bryophytes like *Sphagnum capillifolium*, *Dicranum majus* and *Plagiothecium undulatum* on the steep sides of ravines. It is difficult to distinguish from attenuated *Lepidozia reptans* in the field.

*Lepidozia cupressina* (Sw.) Lindenb.

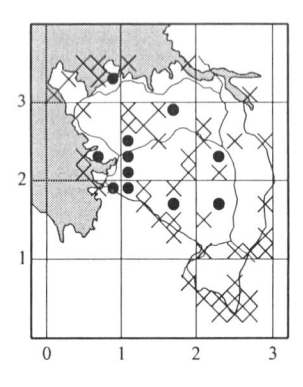

11 tetrads. This handsome plant forms swelling cushions on rocks and tree bases in sheltered, rocky woodland and on north-facing crags, particularly those close to the sea.

*Bazzania trilobata* (L.) Gray

5 tetrads. Surprisingly rare in Assynt and rarely in great quantity, this large liverwort occurs in humid rocky woodland, on heathy banks, particularly near the sea and rarely in block scree.

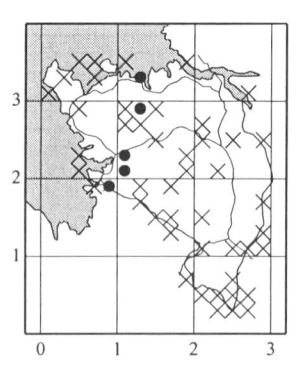

*Bazzania tricrenata* (Wahlenb.) Lindb.

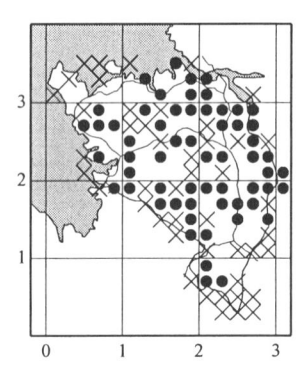

59 tetrads. This variable plant is a common constituent of the hepatic flora underneath oceanic heath with *Mylia taylori*, *Herbertus aduncus* ssp. *hutchinsiae* and species of *Scapania*, but also occurs widely in ravines, rocky woodland, scree and on crags.

*Bazzania pearsonii* Steph.

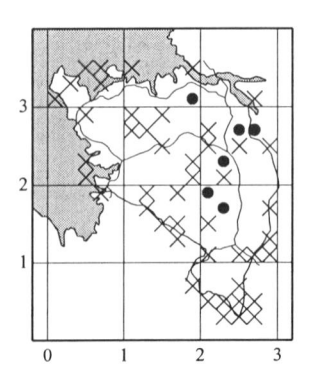

6 tetrads. This oceanic-montane liverwort is rare but can be locally plentiful, usually in dwarf shrub heath on rocky north and north-

east facing slopes, as in Coire Gorm on Glas Bheinn and Sàil Gorm on Quinag, typically with *Bazzania tricrenata*, *Plagiochila carringtonii* and *Mastigophora woodsii*. It is not easy to distinguish from the common *Bazzania tricrenata* in the field.

*Calypogeia fissa* (L.) Raddi

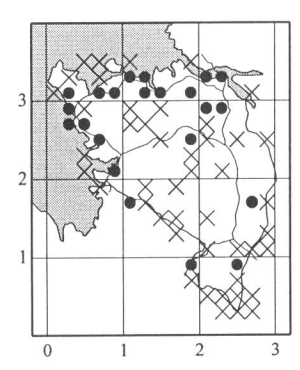

22 tetrads. Widespread but usually in small quantity, on wet peaty banks and growing through *Sphagnum* in mires.

*Calypogeia muelleriana* (Schiffn.) Müll.Frib.

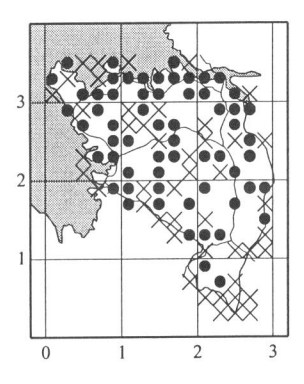

61 tetrads. Very common on shaded, peaty banks and on acid rocks and less tolerant of constant wetness than *Calypogeia fissa*.

*Calypogeia azurea* Stotler & Crotz

1 tetrad. Only recorded on this survey from a very wet, vertical quartzite crag in the Bealach Traligill. There is an older record from Inchnadamph by Paton in 1960.

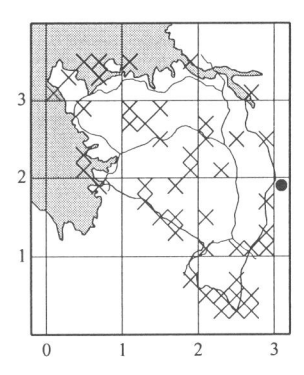

*Calypogeia sphagnicola* (Arnell & J.Perss.) Warnst. & Loeske

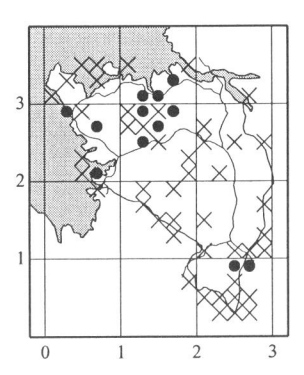

12 tetrads. Uncommon but possibly overlooked as it invariably grows through *Sphagnum* hummocks with other small liverworts in the less disturbed mire areas.

*Calypogeia suecica* (Arnell & J.Perss.) Müll.Frib.

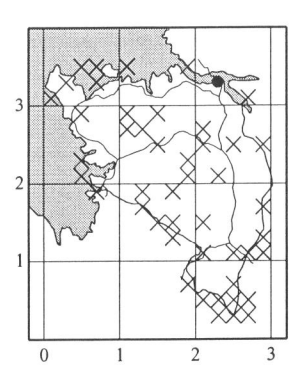

1 tetrad. Very rare with only one site on a rotting log in woodland at Creag an Spardain, found by Blockeel in 1992.

*Calypogeia arguta* Mont. & Nees

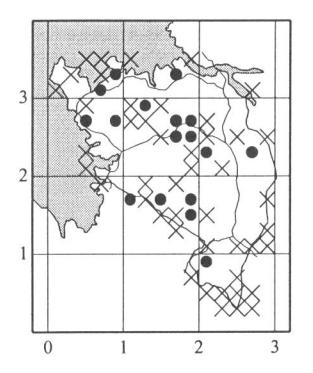

17 tetrads. A widespread and locally frequent species of steep, shaded soil banks, often in woodland and more tolerant of basic conditions than other species in the genus.

*Cephalozia bicuspidata* (L.) Dumort.

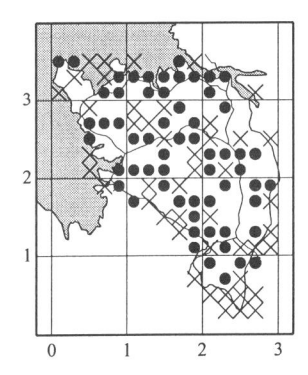

60 tetrads. A common species of damp disturbed soil in a variety of habitats and extending to the tops of the hills; particularly frequent on stony paths but also occurring on peat and rotting logs.

*Cephalozia leucantha* Spruce

There are three localised records for this species in Assynt, from Inchnadamph (W.E.Nicholson in 1921) and Allt nan Uamh (B.B.S. in 1960) and, more recently, from Little Assynt by Paton in 1969.

*Cephalozia lunulifolia* (Dumort.) Dumort.

25 tetrads. Frequent and probably under-recorded; most common on

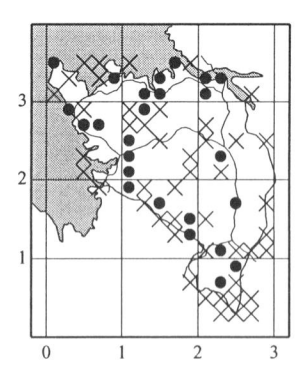

peaty banks, often with *Kurzia trichoclados* but also occurs on rotting logs and on *Sphagnum* in mires.

*Cephalozia loitlesbergeri* Schiffn.

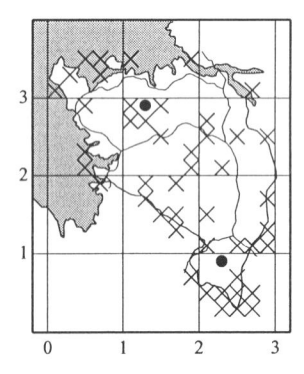

2 tetrads. Recorded only twice, from hummocks of *Sphagnum austinii* in a good mire at the head of a tributary of Abhainn a'Chnocain near Elphin and from a mire south of Loch na Loinne, but very easy to overlook. There is an old record from Inchnadamph by Nicholson in 1923.

*Cephalozia connivens* (Dicks.) Lindb.

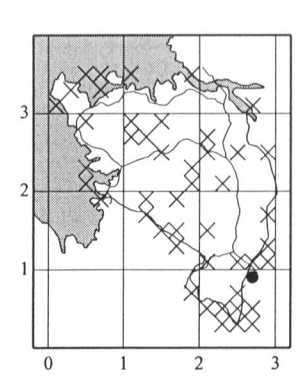

1 tetrad. Surprisingly rare but almost certainly overlooked; the only record is from *Sphagnum* hummocks in an excellent mire, Mointeach na Totaig, surrounded by forestry above Loch Urigill; there is an unlocalised 1966 record from Inchnadamph.

*Nowellia curvifolia* (Dicks.) Mitt.

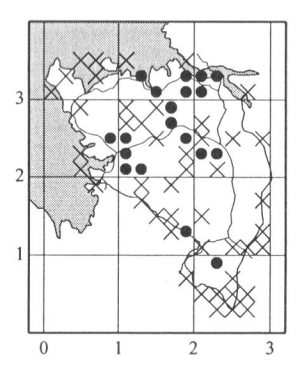

19 tetrads. Not infrequent on logs in woodland, usually with *Metzgeria palmata* and *Scapania umbrosa*, but limited by the paucity of suitable habitat as birch logs often fall to pieces before colonisation can occur.

*Cladopodiella fluitans* (Nees) H.Buch

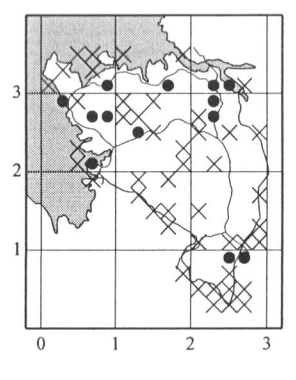

13 tetrads. An uncommon but very locally abundant species, usually growing with *Sphagnum* species at the margin of pools, often where there is some moderate base-enrichment.

*Hygrobiella laxifolia* (Hook.) Spruce

4 tetrads. Surprisingly rare in Assynt given the extent of suitable habitat on wet rocks

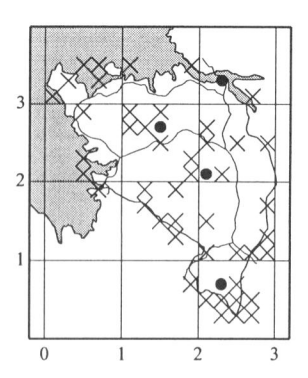

in ravines and crags.

*Odontoschisma sphagni* (Dicks.) Dumort.

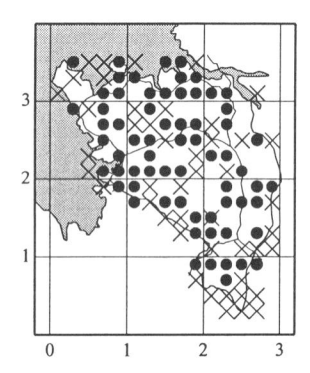

66 tetrads. Common and often abundant on *Sphagnum* hummocks and occasionally on bare peat.

*Odontoschisma denudatum* (Mart.) Dumort.

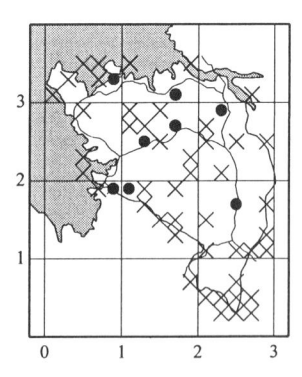

8 tetrads. Scattered records from peat banks and cuttings but probably under-recorded.

*Odontoschisma elongatum* (Lindb.) A.Evans

19 tetrads. One of the interesting

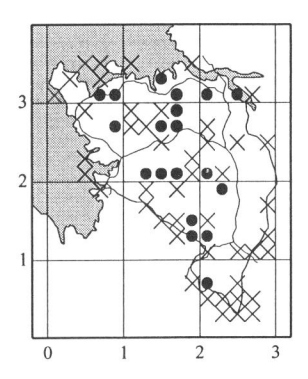

finds of this survey was that this nationally scarce species is a frequent constituent of the community of loch margins, growing over gneiss gravel where there has been some silt deposition and where there is some flushing.

*Cephaloziella hampeana* (Nees) Schiffn.

Recorded from the Little Assynt area by Paton in 1969.

*Cephaloziella divaricata* (Sm.) Schiffn.

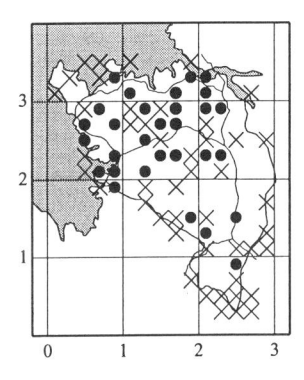

30 tetrads. Frequent and sometimes locally abundant, growing through other bryophyte cushions on rocks and trees, particularly in open, but humid, sites.

*Anthelia julacea* (L.) Dumort.

25 tetrads. Frequent in the more montane areas, growing in wet, acid, stony flushes and sometimes in extensive stands; it also occurs lower down on wet stony paths.

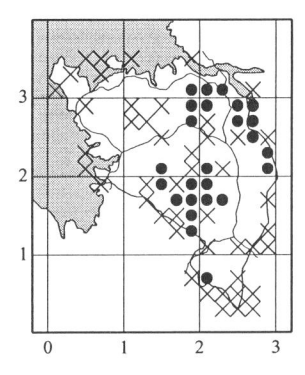

*Anthelia juratzkana* (Limpr.) Trevis

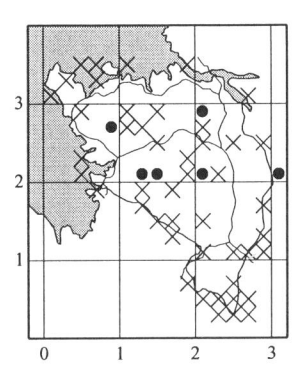

6 tetrads. Only recorded on a few occasions in limited quantity in dry exposed liverwort or lichen crusts; there will be further stands on the higher hills where it has been recorded from the Conival ridge by Long in 1982.

*Barbilophozia kunzeana* (Huebener) Müll.Frib.

Recorded from a mire above the Stronechrubie cliffs by the distinguished American bryologists W.B.Schofield and R.M.Schuster in 1978.

*Barbilophozia floerkei* (F.Weber & D.Mohr) Loeske

31 tetrads. Widespread on exposed rocks and scree but nowhere very common despite the abundance of seemingly suitable habitat.

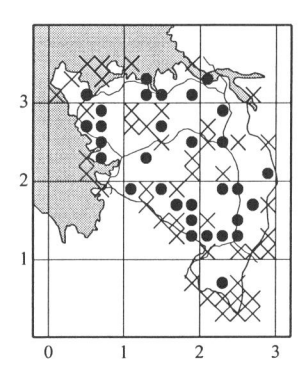

*Barbilophozia atlantica* (Kaal.) Müll.Frib.

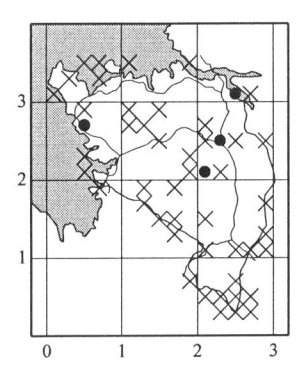

4 tetrads. Rare and in small quantity, with other bryophytes on dry, exposed, acidic and usually south-facing rocks; Con a'Chreag by Unapool, Creag Liath by Clachtoll, by a tributary of the Allt Sgiathaig and near Meallan Liath Beag.

*Barbilophozia attenuata* (Mart.) Loeske

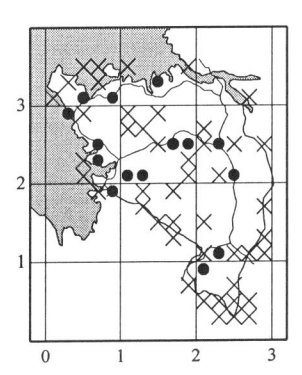

15 tetrads. Surprisingly sparse in Assynt given its frequency on the west coast further south; usually mixed with other bryophytes on

rocks and trees in humid woodland.

*Barbilophozia lycopodioides* (Wallr.) Loeske

There is a record from the Inchnadamph area by Paton in 1969.

*Barbilophozia barbata* (Schmidel ex Schreb.) Loeske

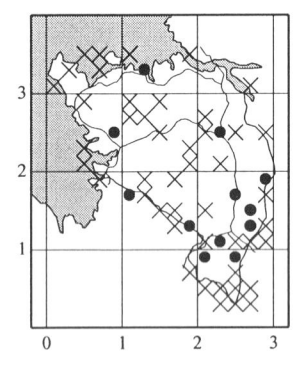

12 tetrads. Widespread but uncommon, usually occurring in a bryophyte mat at the base of boulders or, more rarely, on their upper surface; there is a cluster of records from the limestone but it is not a noted calcicole.

*Anastrepta orcadensis* (Hook.) Schiffn.

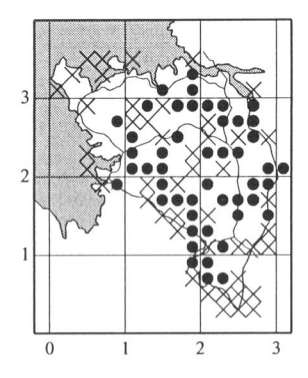

48 tetrads. One of the most frequent of the oceanic-montane hepatics and common on heathy banks, often with *Sphagnum capillifolium*, and in rocky woodland.

*Lophozia ventricosa* (Dicks.) Dumort.

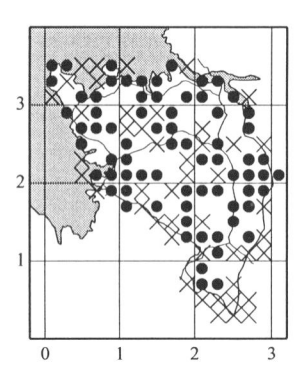

72 tetrads. One of the most common liverworts, occurring in a wide variety of habitats but particularly frequent on peat over rocky outcrops, acid crags in woodland, rotting logs and tree bases.

*Lophozia sudetica* (Nees ex Huebener) Grolle

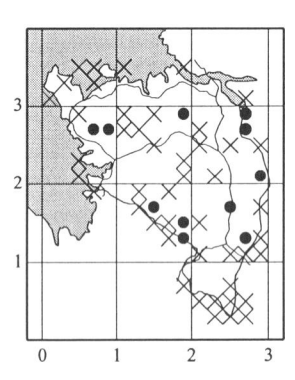

11 tetrads. A scattering of records but usually in small quantity on the tops of rocks in scree and on crags; more work on the higher ground would produce more records.

*Lophozia excisa* (Dicks.) Dumort.

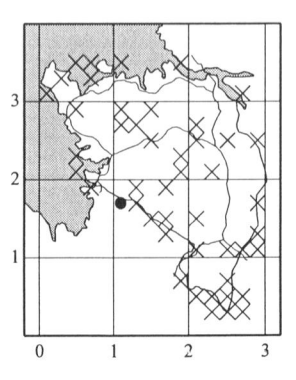

1 tetrad. Very rare, with just one record from a south-facing rock outcrop in the Kirkaig ravine.

*Lophozia obtusa* (Lindb.) A.Evans

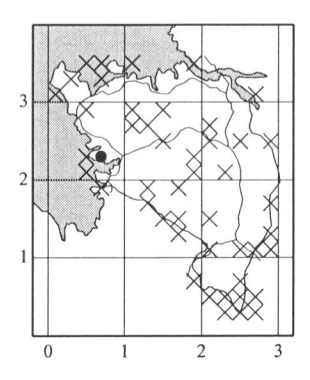

1 tetrad. This rare and distinctive liverwort is recorded from just one site at an unusually low altitude, on a boulder in wooded, block scree above Loch Dubh at Ardroe.

*Lophozia incisa* (Schrad.) Dumort.

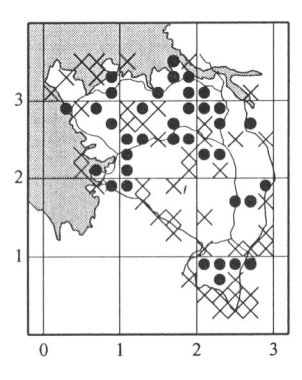

37 tetrads. Widespread and frequent in a variety of damp, acid habitats including peaty banks, rotting logs and open, drier areas in mires; usually in small quantity.

*Lophozia opacifolia* (Culm.) Meyl.

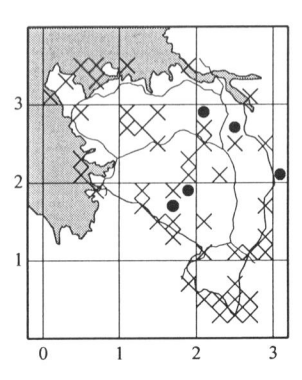

5 tetrads. Rare on damp, organic soil in rocky crevices on top of the higher hills; only recorded from Conival, Glas Bheinn, Quinag, Canisp and Suilven, with an older record from Beinn an Fhurain by Paton in 1960.

*Lophozia bicrenata* (Schmidel ex Hoffm.) Dumort.

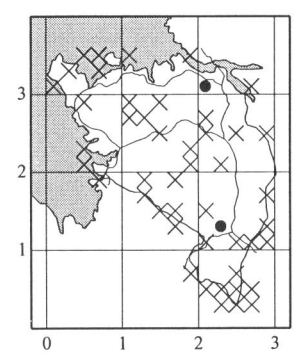

2 tetrads. Rare but possibly overlooked; both sites are on shallow, damp organic soil over rocks at low levels: Cam Loch at south-east end and Airigh na Beinne. Also recorded by Paton at Little Assynt in 1960.

*Leiocolea gillmanii* (Austin) A.Evans

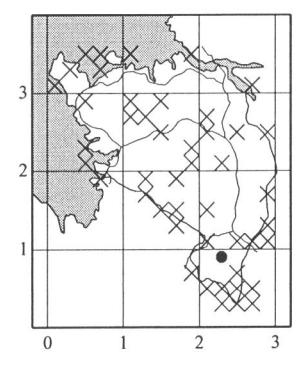

1 tetrad. This nationally rare species has one site in Assynt, in flushed calcareous grassland by a tributary of the Abhainn a'Chnocain above Knockan where it occurs with *Moerckia hibernica*.

*Leiocolea bantriensis* (Hook.) Jörg.

20 tetrads. Most frequent and often abundant on damp, sheltered limestone ledges or in calcareous flushes; this plant also has scattered sites in ravines or on montane crags on the more base-rich gneiss.

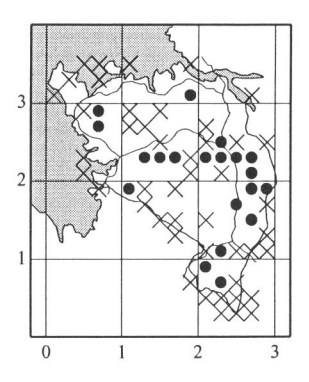

*Leiocolea fitzgeraldiae* Paton & A.R.Perry

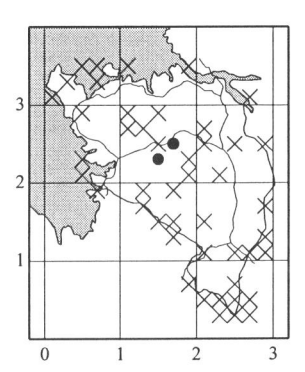

2 tetrads. This distinctive but only recently described species is known from two sites on irrigated basic gneiss crags on the south side of Loch Assynt: Allt an Tiaghaich and Cnoc Aird na Seilge.

*Leiocolea alpestris* (Schleich. ex F.Weber) Isov.

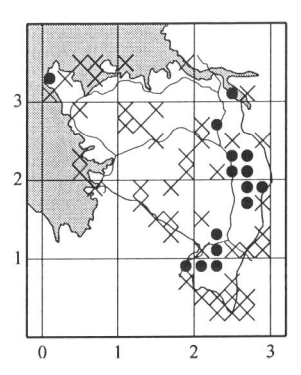

15 tetrads. Fairly frequent on the limestone and the Fucoid Beds where it occurs on thin, damp soil on ledges of crags; there is an isolated locality on the Torridonian at Stoer.

*Leiocolea heterocolpos* (Thed. ex Hartm.) H.Buch

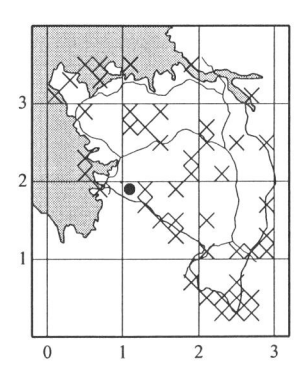

1 tetrad. The single locality, the only West Sutherland site, is on a ledge of a low, basic gneiss crag north of Loch Gleannan na Gaoithe, where it occurs with *Riccardia chamedryfolia*, *Selaginella selaginoides* and *Conocephalum conicum*.

*Leiocolea badensis* (Gottsche) Jörg.

Recorded from the limestone at Inchnadamph by Paton in 1969.

*Leiocolea turbinata* (Raddi) H.Buch

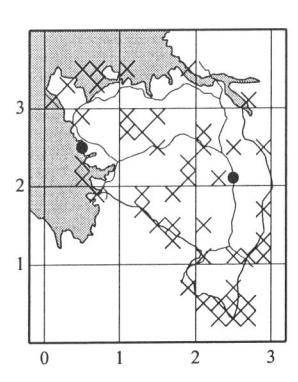

2 tetrads. Two sites, one on the dunes at Achmelvich and the other on the limestone at Inchnadamph. The rarity of this species, and of *Leiocolea badensis*, is hard to explain given the expanse of limestone in the east of the parish.

197

*Gymnocolea inflata* (Huds.) Dumort.

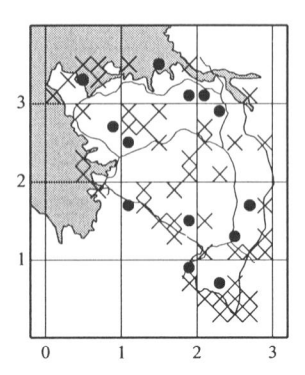

13 tetrads. Widespread but uncommon; usually occurring either as cushions on the edge of peaty banks or as mats on wet peat over stony ground.

*Sphenolobopsis pearsonii* (Spruce) R.M.Schust

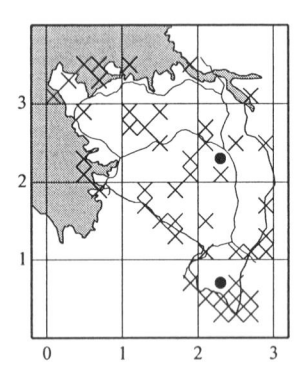

2 tetrads. A rare oceanic species forming thin mats on sheltered, vertical faces on quartzite crags in two sites, An Ciomhleum, on the south side of Loch Assynt and Allt an Achaidh.

*Anastrophyllum minutum* (Schreb.) R.M.Schust.

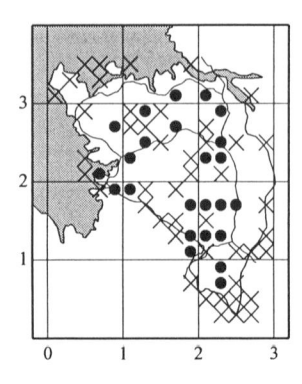

24 tetrads. Widespread and sometimes quite frequent, particularly on the vertical, peat capping on north-facing rocks overhung by 'leggy' heather.

*Anastrophyllum hellerianum* (Nees ex Lindenb.) R.M.Schust.

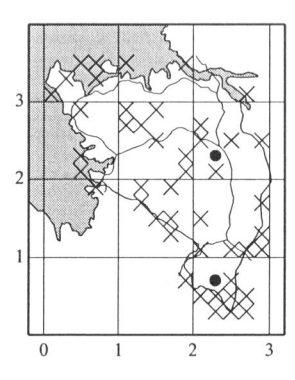

2 tetrads. This tiny liverwort usually grows amongst algae on the vertical bark of trees but also occurs in an algal slime on steep sheltered rocks: An Coimhleum on the south side of Loch Assynt and Allt a'Phollain.

*Anastrophyllum saxicola* (Schrad.) R.M.Schuster

There is a record (and voucher specimen in **BBSUK**) for this nationally rare species from Meallan Liath Mor, collected by Ward in 1966. This is an extraordinary record as all other British sites for this 'continental' liverwort are in a small area of the eastern Cairngorms. There is no doubt that, apart from its oceanic climate, the quartzite scree in the Canisp area offers much, seemingly suitable, habitat but an extensive search in the area indicated by the map reference was unsuccessful.

*Anastrophyllum donnianum* (Hook.) Steph.

10 tetrads. One of the most distinctive of our large, oceanic-montane hepatics; it is locally frequent in heathy ground and in block scree on the higher hills, often where snow accumulates, but descending quite a long way down on the north side of Quinag and north of Glas Bheinn.

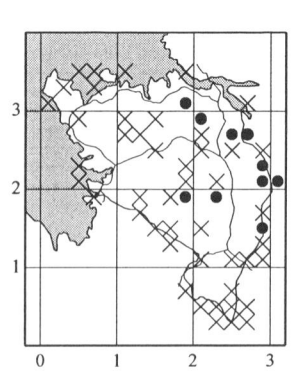

*Anastrophyllum joergensenii* Schiffn.

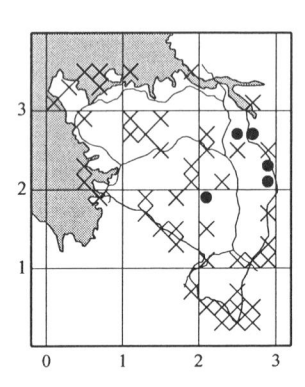

5 tetrads. In similar habitat to *Anastrophyllum donianum* and almost always growing with it making identication somewhat problematic, particularly as it is always much less frequent; there are sites on Canisp, Beinn an Fhurain, near Loch nam Caorach, Coire Gorm and elsewhere on Glas Bheinn, with good populations on the last two hills.

*Tritomaria exsectiformis* (Breidl.) Loeske

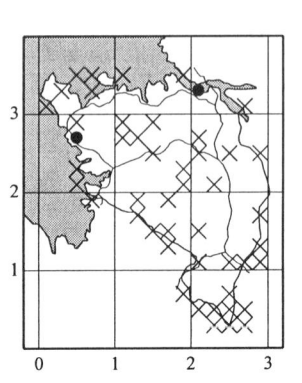

2 tetrads. Inexplicably rare in Assynt given its frequency further south, but this may be due to the scarcity of its preferred habitat of rotting logs: Alltana'bradhan and Tòrr a'Ghamhna.

*Tritomaria quinquedentata* (Huds.) H.Buch

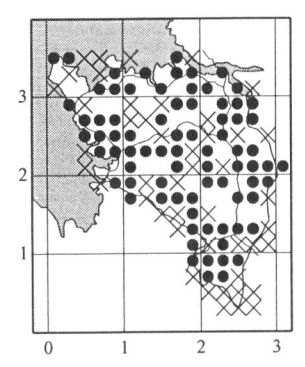

78 tetrads. At home on both the gneiss and the limestone, this species is widespread and locally frequent, often forming patches, particularly on loch-side boulders and sheltered basic crags.

*Tritomaria polita* (Nees) Jörg.

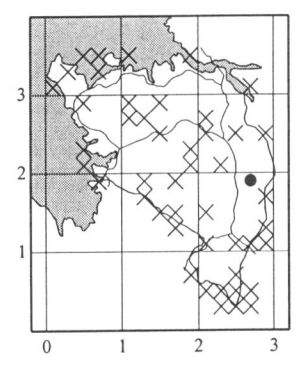

1 tetrad. Very rare; this interesting calcicole is only recorded from the flushes south of Loch Mhaolach-choire on the limestone, one of only two sites in West Sutherland.

*Mylia taylori* (Hook.) Gray

74 tetrads. Common and often abundant in swelling, purple-red to yellow-green cushions in rocky woodland, moorland and block scree on north-facing or sheltered slopes and also in mires.

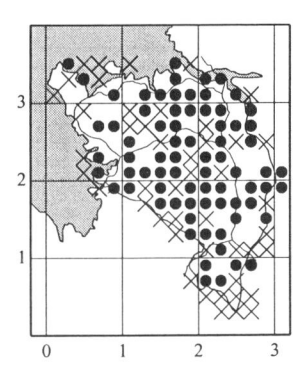

*Mylia anomala* (Hook.) Gray

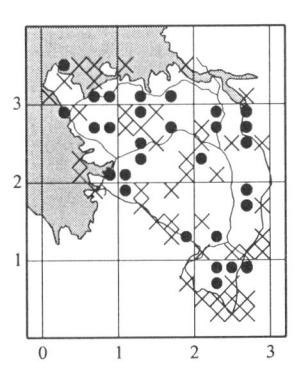

29 tetrads. Widespread but usually in small isolated stands in *Sphagnum* hummocks in the least disturbed mire areas, where it may occur with *Mylia taylori*.

*Jungermannia atrovirens* Dumort.

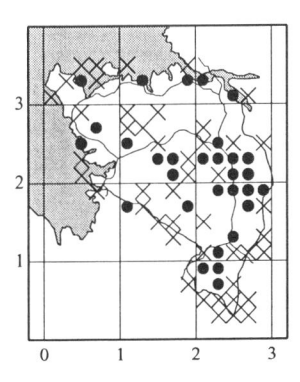

30 tetrads. Widespread and locally frequent in damp places on the limestone but also fairly common on wet, moderately basic rocks elsewhere.

*Jungermannia pumila* With.

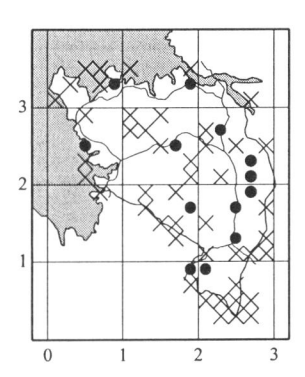

13 tetrads. Much less common than *Jungermannia atrovirens* and more base-demanding and so most stands are associated with wet limestone rocks, usually in burns or ravines.

*Jungermannia exsertifolia* Steph. ssp. *cordifolia* (Dumort.) Váňa

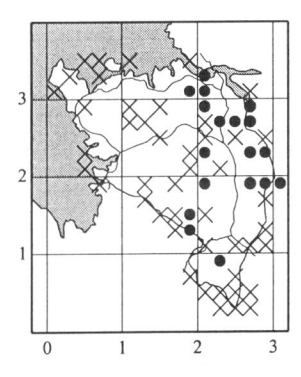

18 tetrads. A number of scattered localities, mostly in flushes on the limestone or on the Torridonian and notably absent from the *Schoenus* flushes so typical of the gneiss.

*Jungermannia sphaerocarpa* Hook.

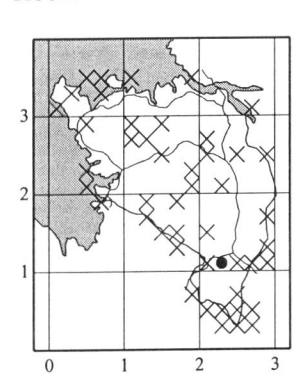

1 tetrad. This inconspicuous liverwort has one record, from the ravine of the river below Loch Urigill.

*Jungermannia gracillima* Sm.

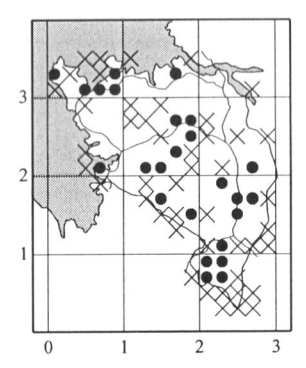

25 tetrads. A frequent coloniser of bare ground and so most often recorded on gravelly loch margins and stony paths; almost certainly under-recorded.

*Jungermannia hyalina* Lyell

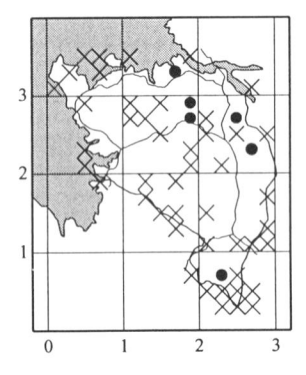

6 tetrads. Surprisingly infrequent given the expanse of suitable habitat in the form of wet rocks and gravel by burns and in ravines; it may not like the chemical composition of the gneiss.

*Jungermannia paroica* (Schiffn.) Grolle

4 tetrads. A similar story to *Jungermannia hyalina*, which may strengthen the supposition about the chemistry of the gneiss. Found in Gleann Dorcha, the ravine at north end of Loch Leitir Easaidh, Meallan Liath Beag and Uamh an Tartair.

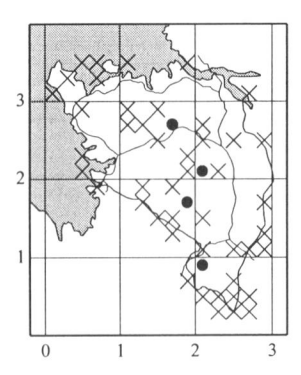

*Jungermannia obovata* Nees

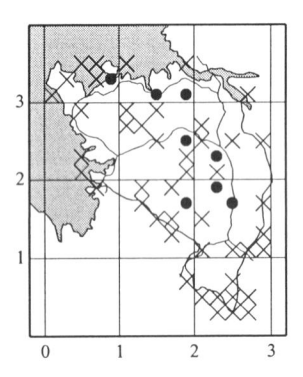

8 tetrads. Uncommon in widely scattered sites, all on wet acid rocks in burns or ravines.

*Jungermannia subelliptica* (Lindb. ex Kaal.) Levier

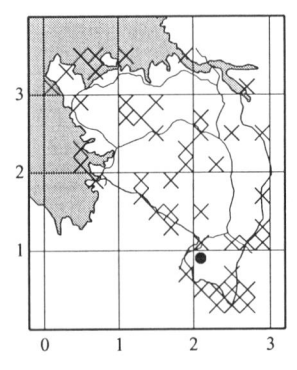

1 tetrad. In similar places to *Jungermannia obovata* but more restricted to base-rich habitats; the one site is from wet soil over the limestone behind Knockan.

*Nardia compressa* (Hook.) Gray

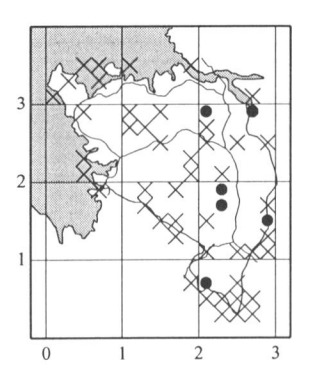

6 tetrads. An uncommon plant in the far north of Scotland and in Assynt limited to a few sites on acid rocks in burns where it can form large stands.

*Nardia scalaris* Gray

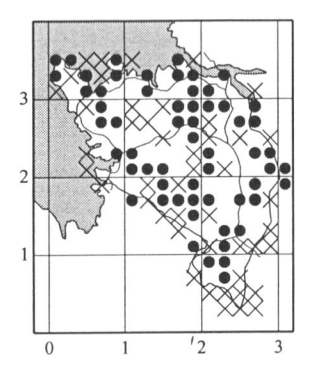

59 tetrads. A common, almost ubiquitous, species on disturbed mineral soil, often forming extensive mats at the sides of tracks, in ravines, scree slopes and loch margins.

*Nardia geoscyphus* (De Not.) Lindb.

There is an old record for this species from 'near Inverkirkaig' by Poore in 1955 and B.R.C. has assigned this to v.c.108, but the site could equally well be referred to v.c.105.

*Marsupella emarginata* (Ehrh.) Dumort.

79 tetrads. A very common species on wet or periodically

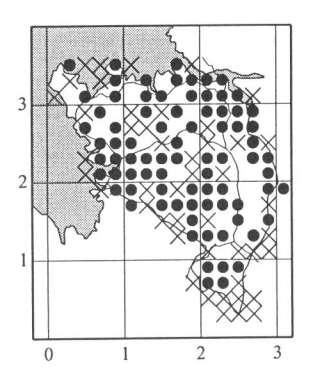

irrigated rocks in a wide variety of acid habitats. No effort was made to record the varieties, but var. *emarginata* is by far the most frequent with var. *aquatica* probably quite common and var. *pearsonii* much less so.

*Marsupella sphacelata* (Gieseke ex Lindenb.) Dumort.

Not seen during this survey but recorded from Allt Poll an Droighinn by Nicholson in 1923 and from Sàil Gharbh by Schuster in 1978.

*Marsupella funckii* (F.Weber & D.Mohr) Dumort.

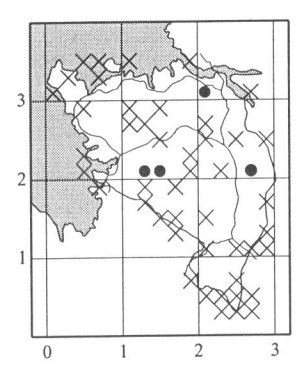

4 tetrads. Rare but possibly overlooked; all the sites are on the firm organic soil developed at the sides of old paths and stands on the Suilven path are quite frequent.

*Marsupella sprucei* (Limpr.) Bernet

2 tetrads. Apparently rare, but more time spent surveying in the higher hills would have produced more sites; both sites are on the tops of rocks in scree in a mat with *Andreaea* species: Fuaran nan Each and Bealach Traligill.

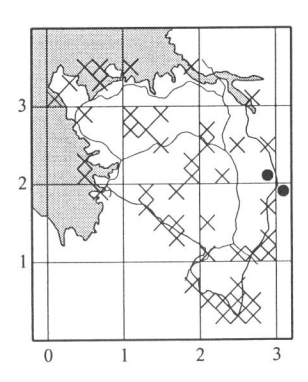

*Marsupella adusta* (Nees) Spruce

Recorded from the Inchnadamph area by Paton in 1969.

*Marsupella alpina* (Gottsche ex Limpr.) Bernet

Recorded from 'Ben Fhurain' by Nicholson in 1921 and he made an earlier (1899) record from Inchnadamph area.

*Gymnomitrion concinnatum* (Lightf.) Corda

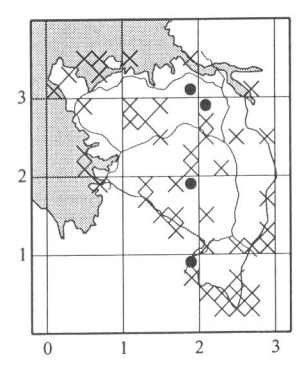

4 tetrads. The rarest of the three *Gymnomitrion* species occurring in Assynt, usually growing as cushions on exposed rocks on high ground, as on Quinag and Canisp, with a lower site at Knockan.

*Gymnomitrion obtusum* Lindb.

15 tetrads. Widely scattered in the more montane areas where its grey-green cushions can be locally abundant on the steep faces of sheltered crags and large rocks in scree.

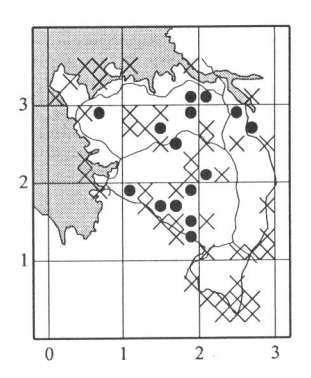

*Gymnomitrion crenulatum* Gottsche ex Carrington

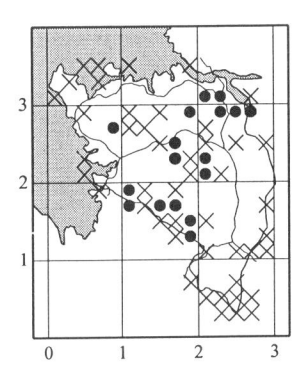

17 tetrads. As widespread as *Gymnomitrion obtusum* but always occurring as much smaller stands, forming small, tight, red cushions on more exposed rocks.

*Douinia ovata* (Dicks.) H.Buch

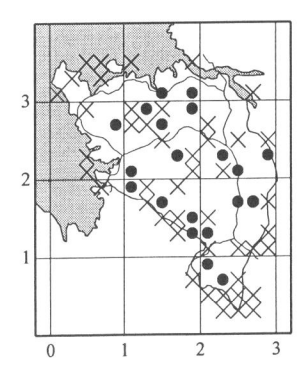

20 tetrads. An attractive little liverwort which forms thin, grey-green patches on dry, acid rocks in sheltered places, usually in rocky woodland or on large rocks in scree.

*Diplophyllum albicans* (L.) Dumort

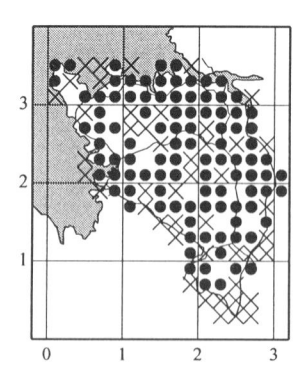

101 tetrads. Unsurprisingly, one of the most common and abundant liverworts in Assynt, growing on most substrates apart from bare limestone and extending to the tops of the hills.

*Diplophyllum obtusifolium* (Hook.) Dumort.

There is an old record in B.R.C. for the N.N.R. at Inchnadamph with no other information; this should probably be disregarded.

*Scapania compacta* (Roth) Dumort

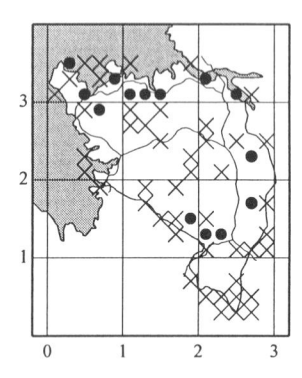

14 tetrads. Scattered round Assynt usually at low levels, with most sites being associated with sunny coastal rocks or boulders on loch margins.

*Scapania cuspiduligera* (Nees) Müll. Frib.
1 tetrad. Only recorded recently by Long in 1982 from the calcareous dunes at Achmelvich but there is an earlier record (1921) from the Inchnadamph limestone by Nicholson.

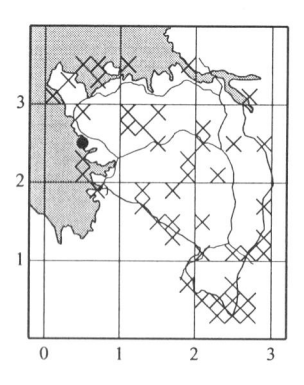

*Scapania scandica* (Arnell & H.Buch) Macvicar

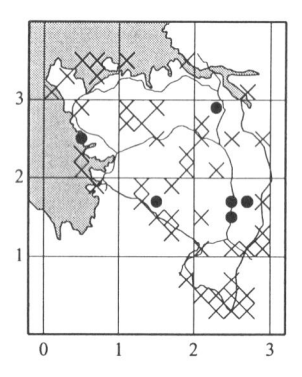

6 tetrads. In similar 'pioneer' habitats to *Diplophyllum albicans* and *Nardia scalaris* but much less common and easily overlooked.

*Scapania umbrosa* (Schrad.) Dumort

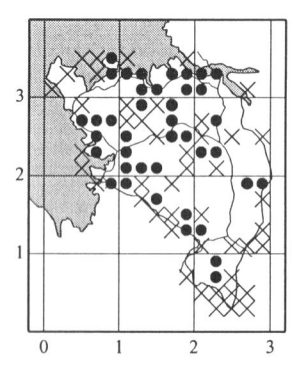

40 tetrads. Most common on low, sloping rocks and rotting logs in woodland but also occurring on bare peat under long heather.

*Scapania nemorea* (L.) Grolle

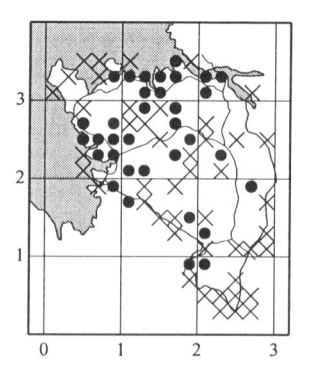

34 tetrads. The 'greasy', green cushions, usually with red gemmae, are widespread but local on rocks and logs in humid sites in rocky woodland and ravines.

*Scapania irrigua* (Nees) Nees

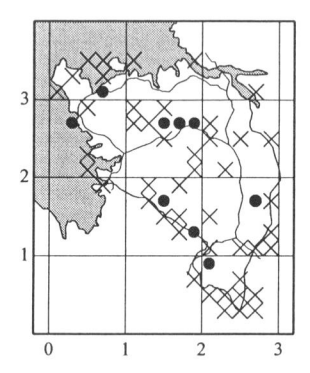

9 tetrads. Most frequent as part of the community on the gravel shores of the larger lochs and probably under-recorded.

*Scapania degenii* Schiffn. ex Müll. Frib.

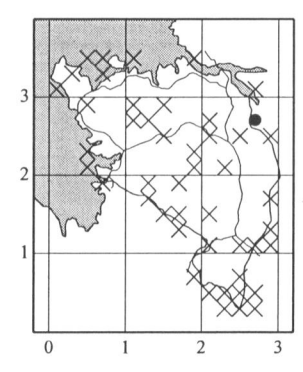

1 tetrad. This fine species has just one isolated site in Assynt, the only one in West Sutherland; it occurs as a healthy population in

one flush below basic gneiss crags in Coire Gorm on the north side of Glas Bheinn.

*Scapania undulata* (L.) Dumort.

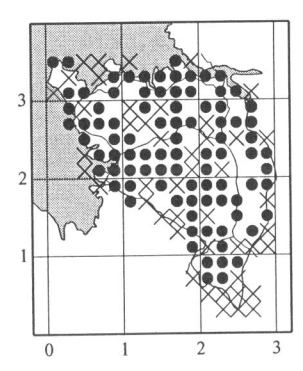

89 tetrads. Common and usually abundant on wet rocks in burns, on loch margins and other irrigated rocks, usually with *Marsupella emarginata* and *Racomitrium aciculare*; also in upland flushes, sometimes with *Scapania uliginosa*.

*Scapania subalpina* (Nees ex Lindenb.) Dumort.

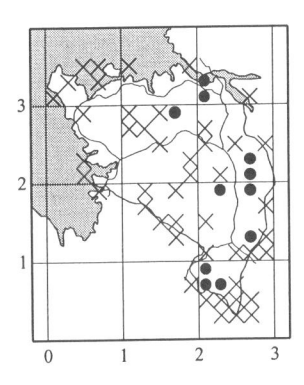

11 tetrads. An uncommon plant of bigger burns, where it usually occurs as silt-encrusted cushions on rocks or gravel, above normal water-levels, but where it is regularly inundated.

*Scapania uliginosa* (Sw. ex Lindenb.) Dumort.

10 tetrads. In similar habitats to *Scapania undulata* but limited to montane sites where it is most frequent in acid flushes, sometimes forming extensive patches.

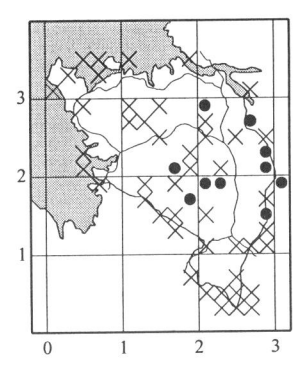

*Scapania aequiloba* (Schwägr.) Dumort.

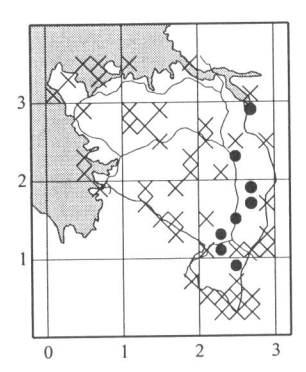

8 tetrads. Apparently strictly limited to the limestone or the Fucoid Beds where it occurs in sheltered crevices; difficult to distinguish from the more common *Scapania aspera* in the field and possibly under-recorded.

*Scapania aspera* M. & H.Bernet

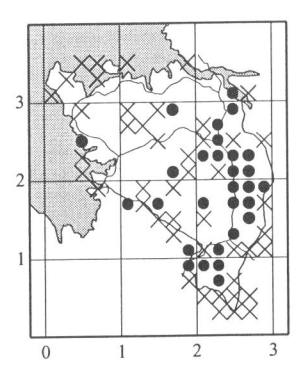

29 tetrads. Common and often abundant on the limestone and occurring more sparingly on base-rich gneiss.

*Scapania gracilis* Lindb.

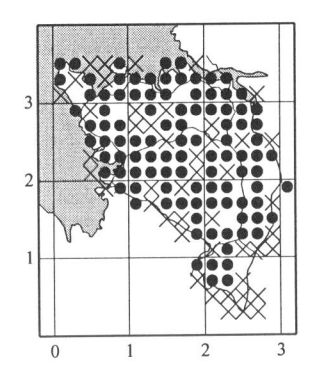

102 tetrads. The most frequent of the species of *Scapania* occurring, often in great quantity, on sheltered rocks and trees and also in block scree and the bryophyte-rich community under tall heather on humid slopes.

*Scapania ornithopodioides* (With.) Waddell

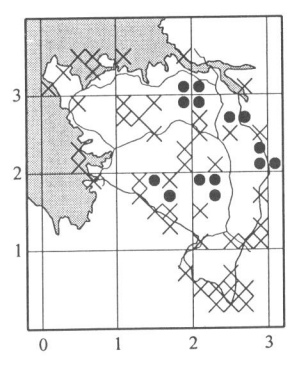

14 tetrads. This large and beautiful, dull-red liverwort is a frequent constituent of the oceanic-montane hepatic mat that occurs under dwarf-shrub heath, in scree and at the base of crags on north or north-east facing slopes on the bigger hills.

*Scapania nimbosa* Taylor ex Lehm.

7 tetrads. Another fine and easily recognised species and part of the same community as *Scapania ornithopodioides* but much less common and apparently more strictly limited to montane sites:

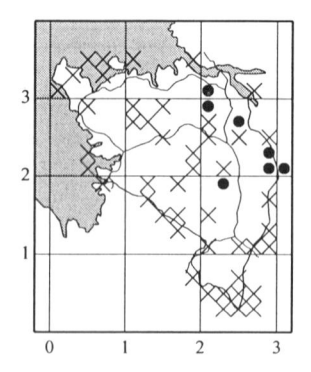

Sàil Gorm and Bàthaich Cuinneige on Quinag, Coire Dearg on Glas Bheinn, Fuaran nan Each, Na Tuadhan, above Lochan na Caorach and on Meallan Liath Mor.

*Leptoscyphus cuneifolius* (Hook.) Mitt.

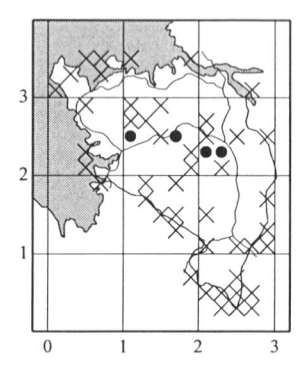

4 tetrads. This tiny liverwort is a rare epiphyte forming small intricate patches on the trunks of older birches in rocky woodland on shaded slopes; never in great quantity in Assynt and almost invariably occurring with *Plagiochila punctata*: An Coimhleum, woodland east of Manse Loch, Aird na Seilge and Garbh Dhoire.

*Lophocolea bidentata* (L.) Dumort.

44 tetrads. A common species of grassy places and occasionally occurring on logs and rocks in woodland.

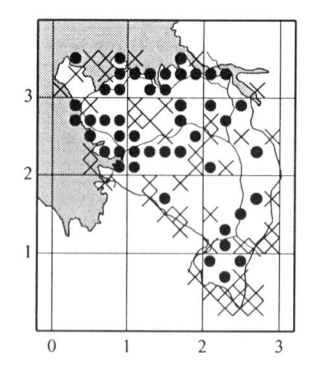

*Chiloscyphus polyanthos* (L.) Corda

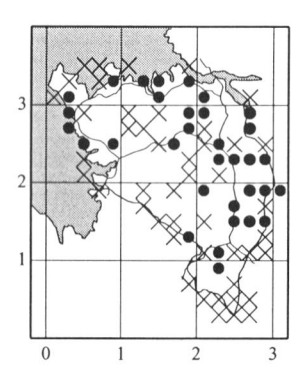

33 tetrads. Widespread but usually in small quantity and most frequent on the limestone, occurring on stones in burns and flushes. Regrettably, no attempt was made to distinguish *Chiloscyphus pallescens* but it is likely that most plants are *Chiloscyphus polyanthos*.

*Geocalyx graveolens* (Schrad.) Nees

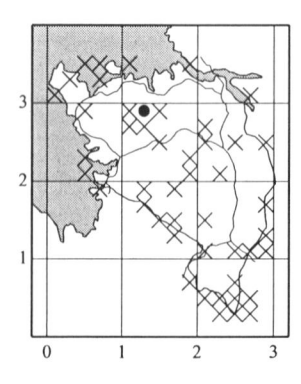

1 tetrad. This nationally rare liverwort has one site in Assynt, on a low cliff on the south side of Loch na Loinne, an unusual, inland site for what is almost exclusively a coastal plant in Britain. *Geocalyx graveolens* is included on Schedule 8 of the Wildlife and Countryside Act.

*Harpanthus flotovianus* (Nees) Nees

This species of montane flushes was recorded from Beinn an Fhurain by Ratcliffe in 1952.

*Harpanthus scutatus* (F.Weber & D.Mohr) Spruce

This species was recorded from a peaty bank by Loch Nedd by Birks in 1972, the only West Sutherland locality, though it occurs on peat and rocks on the West Ross side of the River Kirkaig.

*Saccogyna viticulosa* (L.) Dumort.

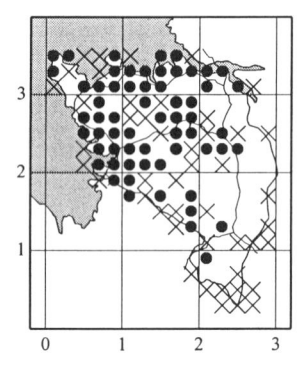

61 tetrads. A widespread and locally abundant oceanic liverwort of steep, wet rocks and soil in woodland and ravines and occasionally loch margins.

*Plagiochila carringtonii* (Balf.) Grolle

21 tetrads. Another large and distinctive member of the oceanic-montane hepatic community, forming yellow-green cushions under heather, on crag ledges and in scree on rocky north and east facing slopes; it extends down to

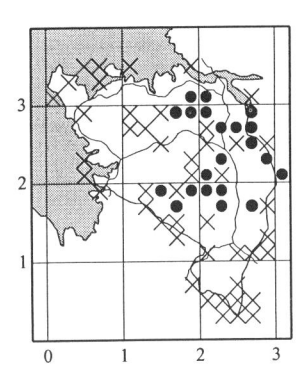

quite low altitudes and occurs rarely in rocky woodland.

*Plagiochila porelloides* (Torr. ex Nees) Lindenb.

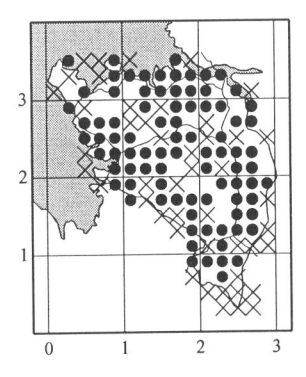

86 tetrads. A common liverwort of periodically irrigated rocks in burns and sides of ravines; also occurring on drier basic rocks and can be abundant on the limestone.

*Plagiochila asplenioides* (L.) Dumort.

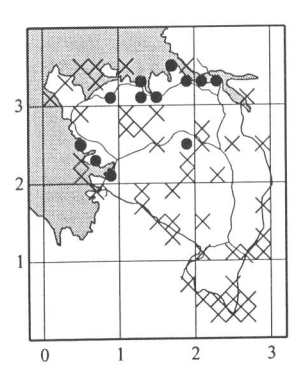

12 tetrads. This large and handsome plant is almost restricted to the richer soils on damp banks in coastal woodland where it can be quite frequent.

*Plagiochila spinulosa* (Dicks.) Dumort.

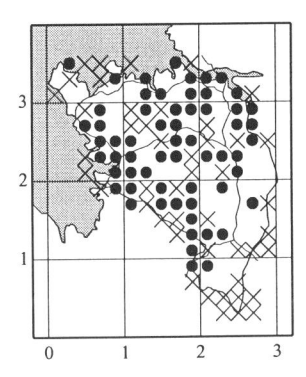

60 tetrads. One of the most frequent constituents, with *Scapania gracilis* and *Hymenophyllum wilsonii*, of the community on rocks in oceanic woodland. It also occurs on rocks in scree and under heather on north- or north-east facing slopes in the hills.

*Plagiochila killarniensis* Pearson

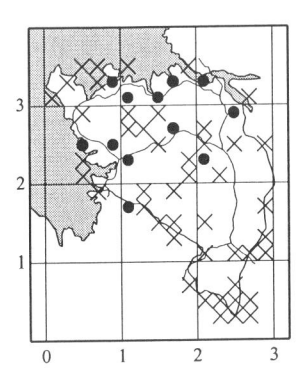

12 tetrads. By far the least common of the larger oceanic *Plagiochila* species and always in small stands; it usually occurs on rather open but humid rocks which are somewhat base-rich and get some sunshine.

*Plagiochila punctata* Taylor

34 tetrads. Occurs in similar habitats to *Plagiochila spinulosa* but is most frequent as cushions on older birch trees in humid woodland with *Scapania gracilis* and, rarely, *Leptoscyphus cuneifolius*.

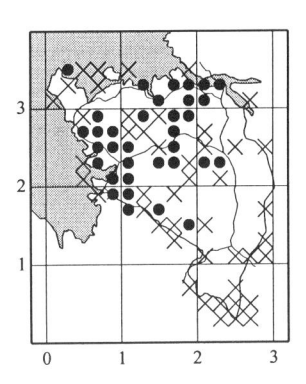

*Plagiochila exigua* (Taylor) Taylor

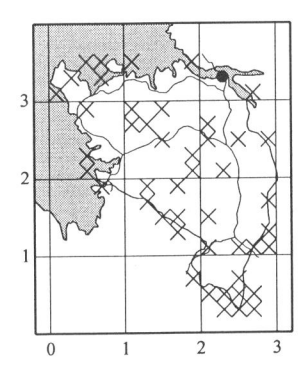

1 tetrad. This tiny liverwort is very rare this far north in Scotland and this survey produced only one record, from Creag an Spardain. There is an older record from woods on the south side of Loch Assynt, probably An Coimhleum, by Birks in 1966.

*Pleurozia purpurea* Lindb.

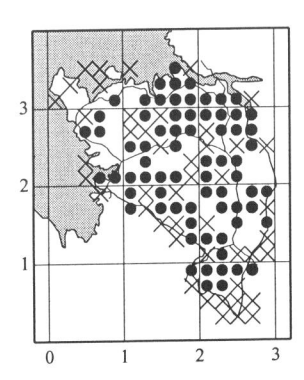

71 tetrads. This common and unmistakable species is frequent on peat in all kinds of wet heath and mires and in scree on north and north-east facing slopes.

*Radula complanata* (L.) Dumort.

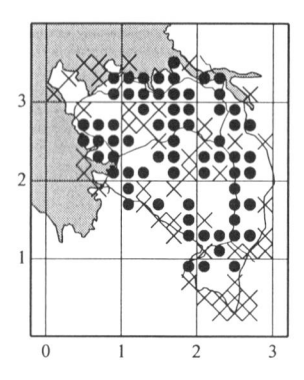

67 tetrads. Common as an epiphyte on hazel and willows in woodland on better soils but also frequent on damp rocks in woodland, ravines and crags, particularly where the rock is somewhat base-rich.

*Radula lindenbergiana* Gottsche ex C.Hartm.

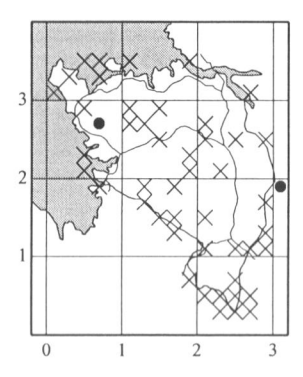

2 tetrads. Rare and only recorded from moderately basic rocks at two sites, Airigh Bheag at Clachtoll and Bealach Traligill. There is an older record from the Allt nan Uamh by Birks et al. in 1966.

*Radula aquilegia* (Hook.f. & Taylor) Gottsche, Lindenb. & Nees

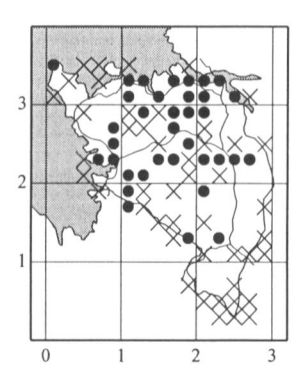

35 tetrads. The attractive green-brown mats of this oceanic species are a frequent feature of damp, sloping rocks in the better ravines and rocky woodland and it also occurs on periodically irrigated coastal rocks and montane crags up to 500m. on Canisp.

*Ptilidium ciliare* (L.) Hampe

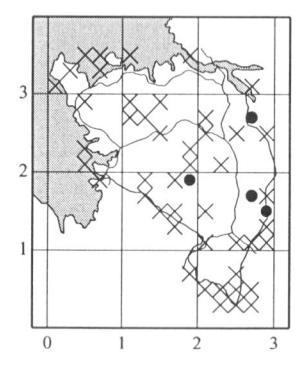

4 tetrads. Apparently rare in Assynt; the wet heaths and block scree are possibly too wet, but there may well be more sites in drier heath on higher ground.

*Ptilidium pulcherrimum* (Weber) Vainio

Recorded from Lochinver by Nicholson in 1921.

*Porella platyphylla* (L.) Pfeiff.

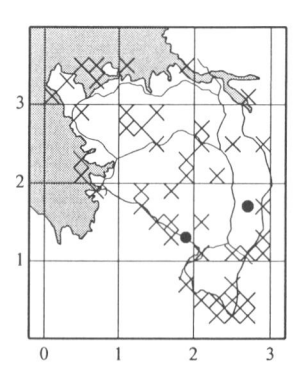

2 tetrads. Rare with just two sites, on the limestone by the Allt na Uamh and on Torridonian on the north shore of Loch Veyatie.

*Porella cordeana* (Huebener) Moore

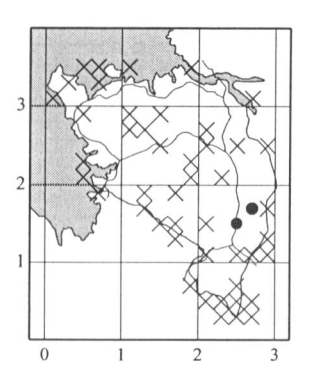

2 tetrads. Rare, occurring in just two places, on damp limestone in the Allt nan Uamh and on a limestone wall at Lyne; there is an older record from trees by the Traligill by Warburg and Crundwell in 1960.

*Porella arboris-vitae* (With.) Grolle

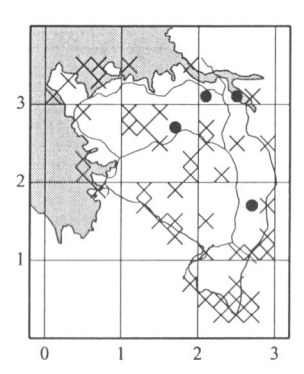

4 tetrads. Inexplicably rare given the extent of base-rich rock in the area: Sàil Gorm on Quinag, Allt nan Uamh, Liath Bhad and a ravine at the north end of Loch Leitir Easaidh.

*Porella obtusata* (Taylor) Trevis

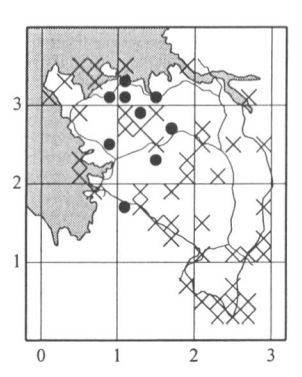

9 tetrads. Surprisingly, the most common of the *Porella* species here, occurring most frequently on rocks just above normal water level in the larger ravines. Some stands, particularly on the Kirkaig River, in Gleann Leireag and on the Oldany River, are extensive.

*Frullania tamarisci* (L.) Dumort.

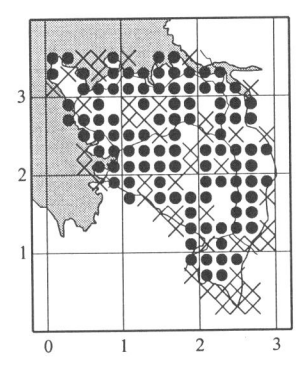

102 tetrads. One of the most common liverworts, occurring as an epiphyte on most trees, on dry rocks of all types and in limestone and coastal turf.

*Frullania teneriffae* (F.Weber) Nees

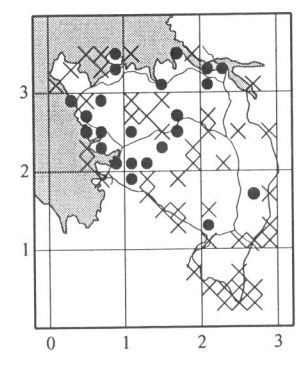

23 tetrads. Widespread, but much less frequent than *Frullania tamarisci*, favouring more open and rocky sites and rare as an epiphyte; all the best sites are on coastal rocks.

*Frullania microphylla* (Gottsche) Pearson

6 tetrads. Not known in West Sutherland before this survey and found in six sites, mostly on the coast, on rocks that are open but not too exposed.

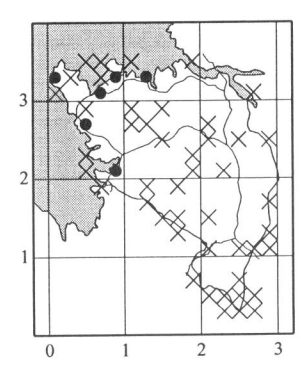

*Frullania fragilifolia* (Taylor) Gottsche, Lindenb. & Nees

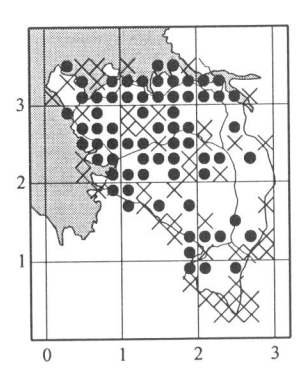

68 tetrads. Widespread but often quite local, although some stands are large; it particularly favours the base-rich facies of the gneiss, especially on south-facing rocks and is rather rare as an epiphyte.

*Frullania dilatata* (L.) Dumort.

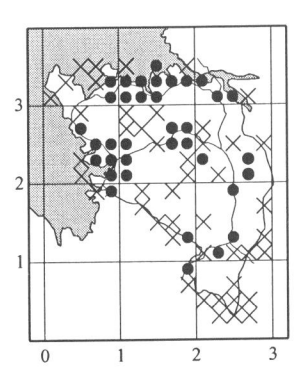

36 tetrads. Common and locally abundant on hazel in some of the better woodlands particularly near the coast, but rare on rocks and absent from much of the area.

*Drepanolejeunea hamatifolia* (Hook.) Schiffn.

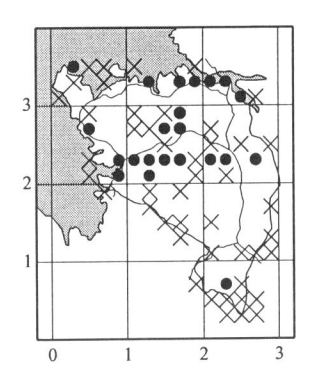

22 tetrads. This tiny oceanic liverwort usually occurs as thin patches on the vertical faces of rocks in burns, where it is humid but where scouring is absent; it can also occur on sheltered humid faces at the base of montane crags and on boulders by lochs where it may also occur as an epiphyte on leggy heather.

*Harpalejeunea molleri* (Steph.) Grolle
(*H. ovata* (Hook.) Schiffn.)

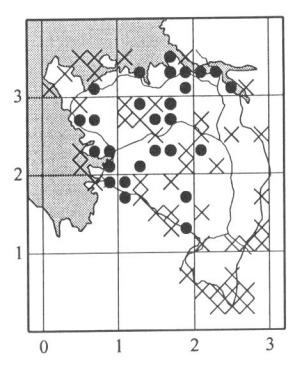

27 tetrads. Similar in its requirements to *Drepanolejeunea hamatifolia* and often growing with it, but more tolerant of basic conditions. It also occurs as an epiphyte on hazel in some areas of coastal woodland and in ravines.

*Microlejeunea ulicina* (Tayl.) A.Evans
(*Lejeunea ulicina* (Taylor) Gottsche, Lindenb. & Nees)

5 tetrads. Remarkably rare in Assynt given its frequency only a little further south; all sites are on

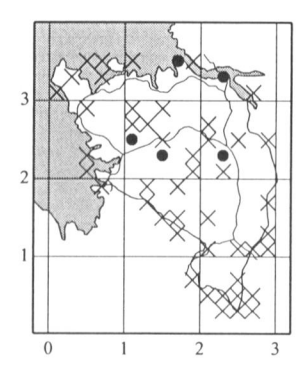

old birches in woodland on north facing slopes: Ardvar, Creag an Spardain, Allt an Tiaghaich, An Coimhleum and Bad a'Bhainne. woodland.

*Lejeunea cavifolia* (Ehrh.) Lindb.

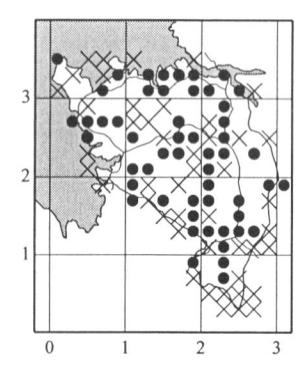

54 tetrads. Widespread and locally frequent on damp rocks particularly where they are base-rich and seems to be the only *Lejeunea* species on the limestone; very occasionally it occurs as an epiphyte on hazel.

*Lejeunea lamacerina* (Steph.) Schiffn.

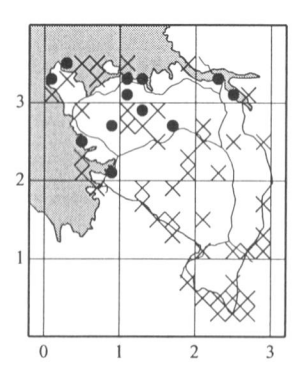

12 tetrads. Scattered sites in the north-west of Assynt on sloping, damp rocks in ravines and on periodically irrigated coastal rocks.

*Lejeunea patens* Lindb.

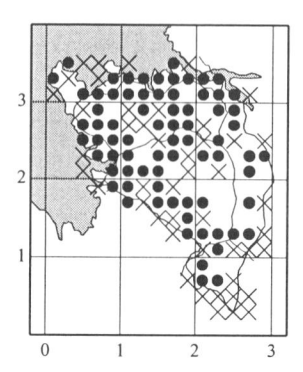

74 tetrads. This is the common *Lejeunea* species in most of the burns and ravines in Assynt, forming distinctive, whitish-green patches on sheltered faces of rocks; it also occurs on rocks on loch margins and on crags but is largely absent from the limestone areas.

*Colura calyptrifolia* (Hook.) Dumort.

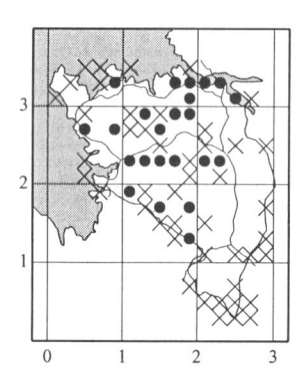

23 tetrads. This delightful little oceanic liverwort is often an associate of *Drepanolejeunea hamatifolia* on the vertical faces of humid rocks in ravines and on sheltered crags.

*Cololejeunea calcarea* (Lib.) Schiffn.

18 tetrads. Another tiny liverwort, somewhat resembling *Drepanolejeunea hamatifolia* but restricted to basic rocks where it forms small dense, green patches; only at all frequent on damp limestone.

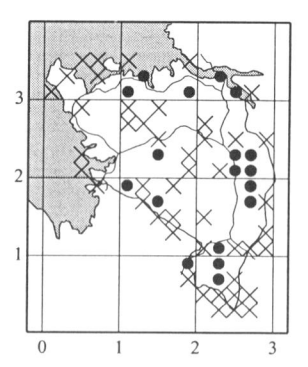

*Aphanolejeunea microscopica* (Taylor) A.Evans

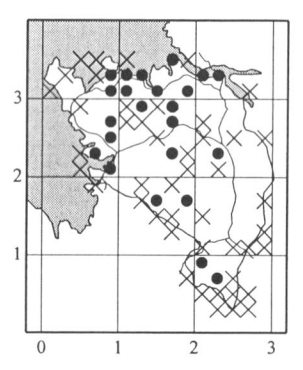

23 tetrads. The last and least of the tiny Lejeunaceae, forming tiny, bright-green patches on damp rocks in ravines, usually where they are wooded.

*Fossombronia foveolata* Lindb.

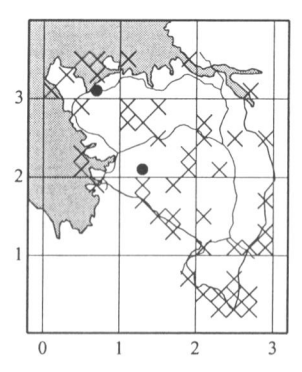

2 tetrads. Rare but probably overlooked, on fine bare silt and sand on loch margins and in flushes; it is frequent on the shore of Loch na h-Airigh Fraoich and at the west end of Loch na Bruthaich.

## Bryophytes

*Pellia epiphylla* (L.) Corda

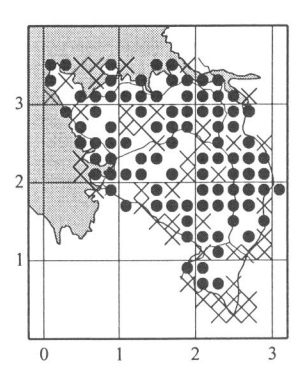

91 tetrads. Very common on acid soil on damp banks, in woodland, at the base of crags, ditches and on marshy ground.

*Pellia neesiana* (Gottsche) Limpr.

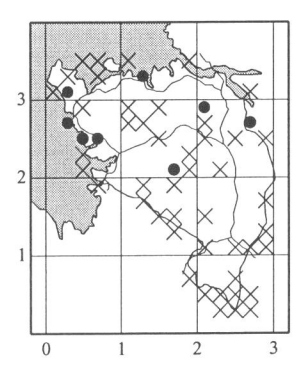

8 tetrads. Apparently rare, but the fertile material necessary to distinguish it from *Pellia epiphylla* is often absent; usually in wetter ground and sometimes forming turfs of upright thalli, often with *Juncus* species on marshy ground.

*Pellia endiviifolia* (Dicks.) Dumort.

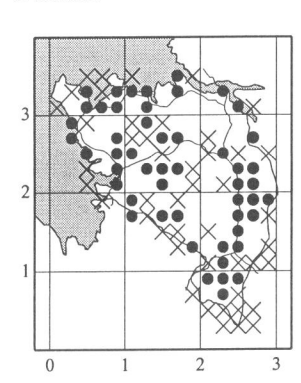

51 tetrads. This common liverwort can be seen as replacing *Pellia epiphylla* in more base-rich habitats where it can form extensive mats.

*Moerckia hibernica* (Hook.) Gottsche

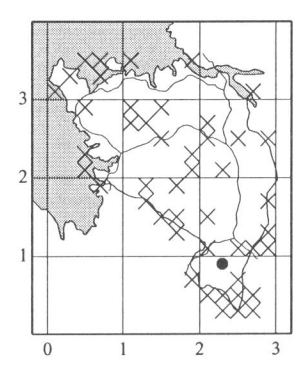

1 tetrad. Very rare and only known from flushed calcareous grassland by a tributary of the Abhainn a'Chnocain, above Knockan, where it occurs with *Leiocolea gilmanii*.

*Blasia pusilla* L.

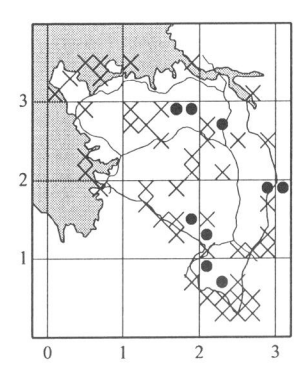

9 tetrads. Uncommon in Assynt, occurring on the gravels of loch margins and burns where patches can be quite large.

*Aneura pinguis* (L.) Dumort.

93 tetrads. A very common species of moderate to strongly base-rich flushes and wet rocks and one of the few constant bryophytes in *Schoenus* flushes on the gneiss.

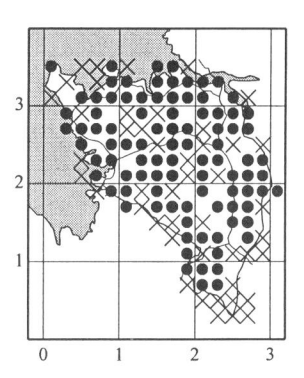

*Riccardia multifida* (L.) Gray

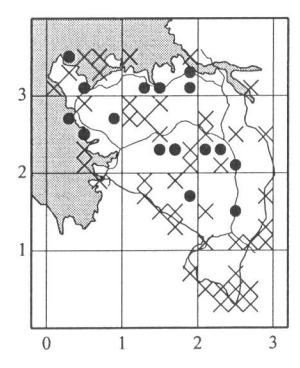

16 tetrads. Scattered sites, usually on wet rocks in ravines or on crags and less commonly in flushes; it does not seem to like the gneiss.

*Riccardia chamedryfolia* (With.) Grolle

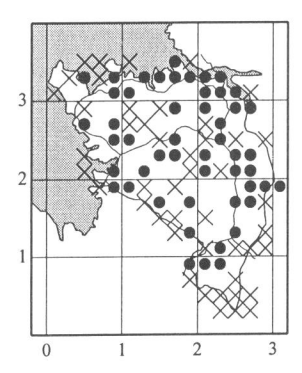

50 tetrads. Widespread in similar habitats to *Riccardia multifida* but much more frequent and preferring at least moderately base-rich substrates.

*Riccardia incurvata* Lindb.

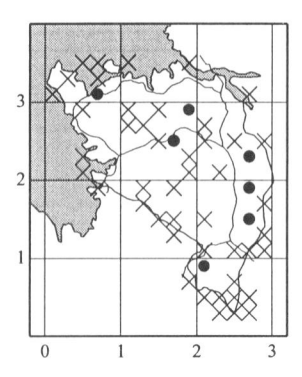

7 tetrads. Widespread but rare and always in small quantity; occurs as patches of thalli in tight moss cushions, often of *Gymnostomum aeruginosum*, in flushes or on wet ledges of base-rich crags.

*Riccardia palmata* (Hedw.) Carruth.

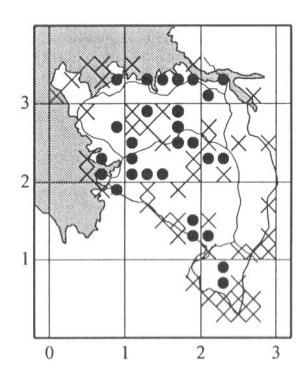

28 tetrads. A common species on rotting logs, often with *Nowellia curvifolia* and *Scapania umbrosa*, and more rarely on damp peat.

*Riccardia latifrons* (Lindb.) Lindb.

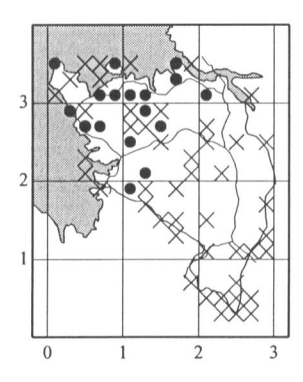

17 tetrads. An uncommon species most frequently found growing through *Sphagnum* hummocks, but also occurring in other bryophyte cushions on wet ledges of base-rich crags and in flushes.

*Metzgeria fruticulosa* (Dicks.) A.Evans

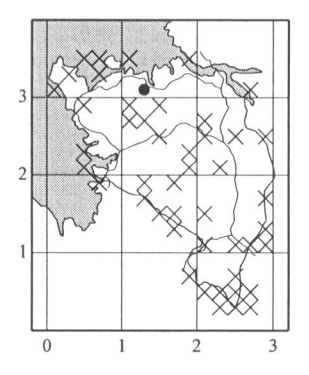

1 tetrad. Very rare, with just one site on an old elder by the roadside in Nedd.

*Metzgeria furcata* (L.) Dumort.

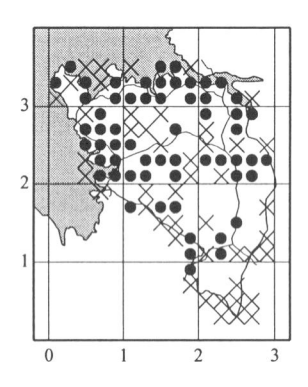

59 tetrads. The flat, yellow-green patches of this liverwort are a common feature as an epiphyte, particularly on hazel, and also on steep, dry faces of moderately basic rocks.

*Metzgeria conjugata* Lindb.

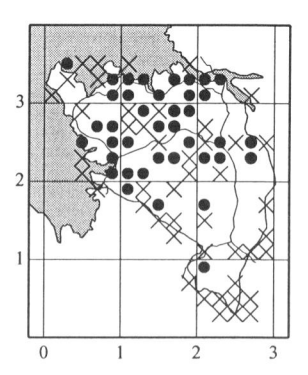

39 tetrads. Superficially similar to *Metzgeria furcata*, this is usually a larger plant, forming more loosely attached patches on rocks in humid sites, particularly on the faces of boulders in ravines.

*Metzgeria leptoneura* Spruce

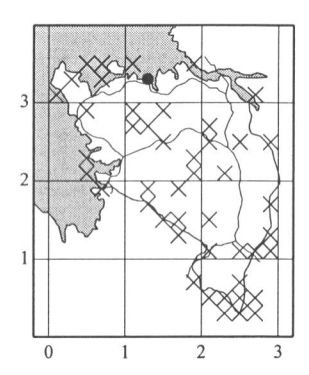

1 tetrad. The only recent record is from the base-rich woodland at Duart by Nedd, but there is an older record from wet rocks, above Lochan Bealaich Cornaidh, Quinag by Crundwell in 1951.

*Apometzgeria pubescens* Kuwah.

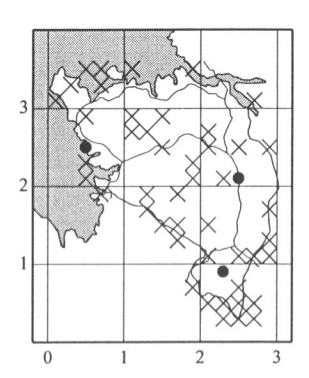

3 tetrads. Surprisingly rare considering the extent of the limestone; recorded from Achmelvich and from the limestone at Stronechrubie and above Knockan. There are several older records from Inchnadamph and the Allt nan Uamh.

*Lunularia cruciata* Adans.

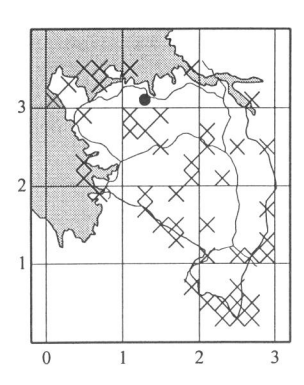

1 tetrad. Occurs as a garden weed at Nedd and was recorded from a garden path, Stoer House, Stoer by Reid in 1960; presumably it occurs in similar sites elsewhere.

*Conocephalum conicum* (L.) Domort.

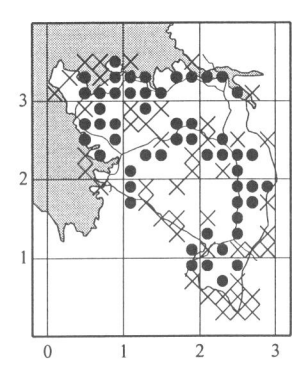

53 tetrads. An unmistakeable thalloid liverwort with a strong smell, often forming large mats on damp, humus-rich soil on banks, at the base of crags and on thin soil over rocks in ravines, almost always where deeply shaded.

*Reboulia hemisphaerica* (L.) Raddi

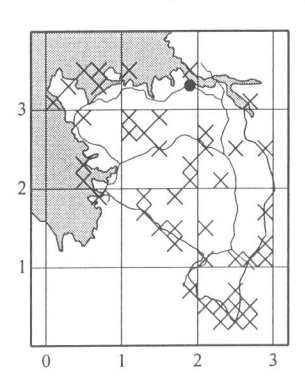

1 tetrad. Very rare but possibly overlooked; occurs on south facing basic gneiss in Gleann Ardbhair. First recorded from the limestone at Elphin by Perry in 1960 and there are two other records from Knockan and the Allt nan Uamh.

*Preissia quadrata* (Scop.) Nees

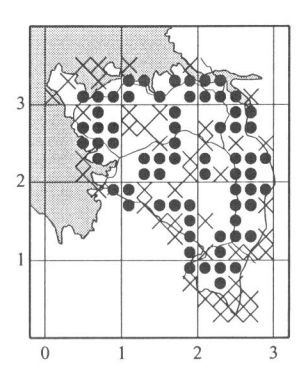

67 tetrads. A common species on the soil on ledges of base-rich crags and occasionally in stony flushes.

*Marchantia polymorpha* L. sensu lato

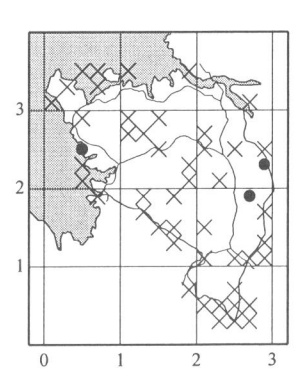

3 tetrads. A rare plant in Assynt occurring in flushed ground at sea level at Achmelvich, up in the hills at Imir Fada and on the limestone at Loch Mhaolach-coire. There is also a record from boulders in the Allt nan Uamh by Perry in 1960.

*Riccia sorocarpa* Bisch.

2 tetrads. Recorded from soil over rocks by the Baddidarach road in Lochinver by Blockeel in 1992, and with the *R. subbifurca* at Clachtoll in 1998 (see below).

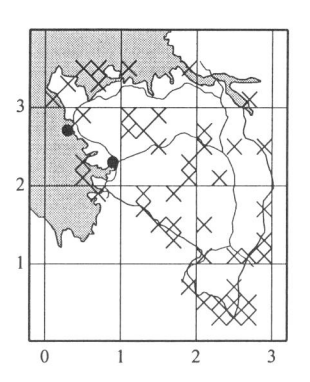

*Riccia subbifurca* Croz.

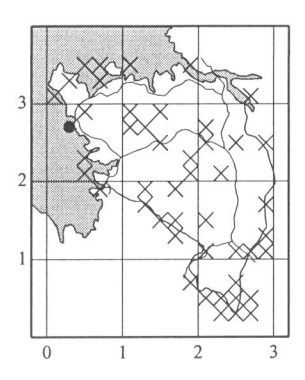

1 tetrad. The only site is on soil in the crevices of an exposed slab of Torridonian sandstone on the sea cliffs at Clachtoll, where it was found in 1998, the first locality for West Sutherland.

211

# MOSSES

*Sphagnum austinii* Sull. ex Aust. (S. imbricatum Hornsch. ex Russow subsp. austinii (Sull. ex Aust.) Flatberg)

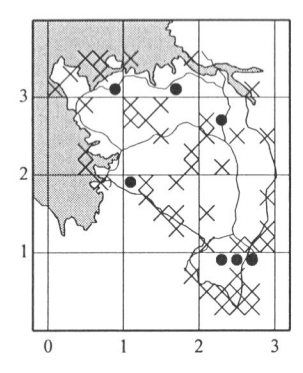

7 tetrads. Scattered localities but generally rare and restricted to the least disturbed mires, usually valley or saddle mires, and often as isolated hummocks; often associated with *Sphagnum fuscum*, and absent from areas of peat cutting or muir-burn.

*Sphagnum papillosum* Lindb.

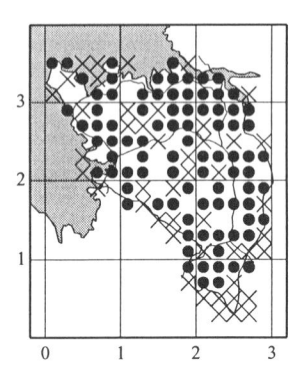

88 tetrads. Common and often abundant in most mires and sometimes forming large hummocks, particularly where the peat is deep.

*Sphagnum palustre* L. var. *palustre*

64 tetrads. Common in wet areas, usually where there is some water movement, often with *Juncus effusus*, *Molinia caerulea* and *Sphagnum fallax*.

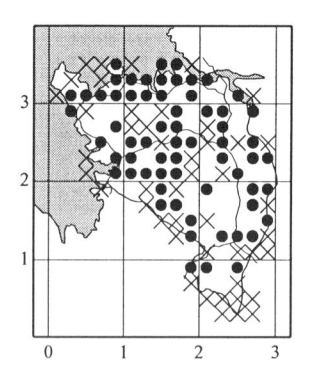

*Sphagnum magellanicum* Brid.

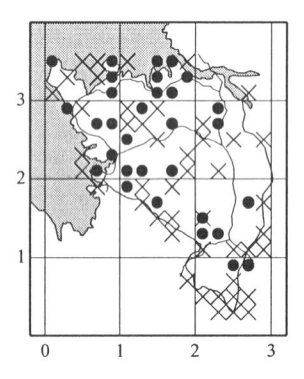

31 tetrads. Widespread in the wetter, least disturbed mires, forming lawns and hummocks, but also in old peat cuttings in some places.

*Sphagnum squarrosum* Crome

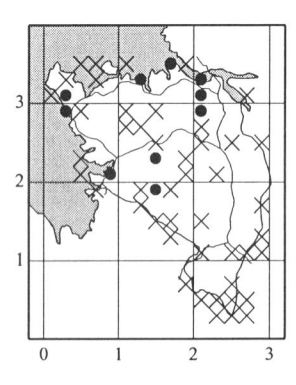

10 tetrads. An uncommon species in Assynt with sites in wet woodland, often under sallows, but also in flushed ground in the hills.

*Sphagnum teres* (Schimp.) Angstr.

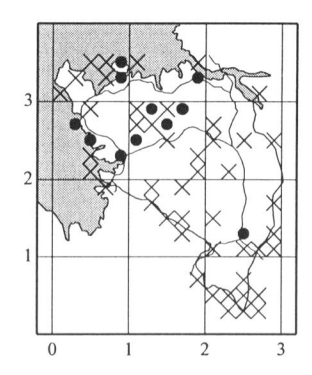

11 tetrads. Widespread but sparse, though occasionally frequent in some flushed mire areas on the gneiss; apparently largely absent from the Torridonian and the limestone.

*Sphagnum fimbriatum* Wilson

Recorded from the Allt nan Uamh by Birks et al. in 1966.

*Sphagnum girgensohnii* Russow

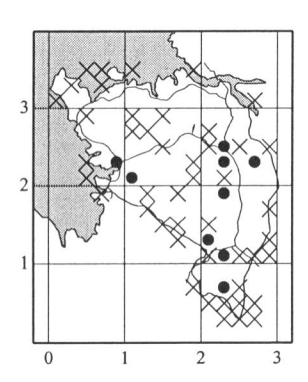

9 tetrads. A few scattered localities on sheltered, flushed banks, often in woodland; possibly under-recorded.

*Sphagnum russowii* Warnst.

There is an unlocalised record from NC22 by Birks et al. in 1967.

*Sphagnum quinquefarium* (Lind. Ex Braithw.) Warnst.

22 tetrads. Widespread but nowhere frequent, occurring on

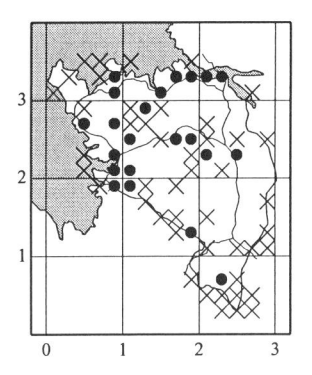

steep banks in rocky woodland, often with a northerly aspect.

*Sphagnum warnstorfii* Russow

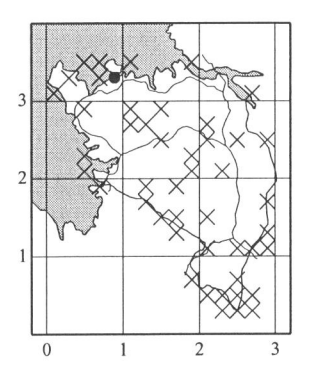

1 tetrad. Only recorded from the flushed margin of a mire at Oldany but possibly overlooked elsewhere.

*Sphagnum capillifolium* (Ehrh.) Hedw.

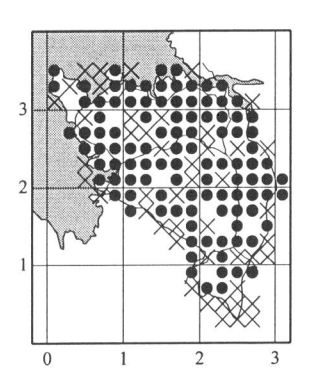

105 tetrads. Very common and usually abundant in both mire and heathland and, with the large oceanic-montane hepatics, forming an important community on steep rocky slopes in the hills. No attempt was made to distinguish between subsp. *capillifolium* and subsp. *rubellum*.

*Sphagnum fuscum* (Schimp.) H.Klinggr.

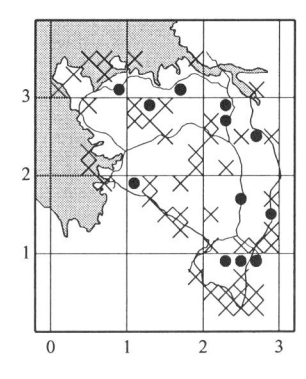

12 tetrads. Scattered sites in the better mire areas, often with *Sphagnum austinii* and sometimes forming substantial hummocks.

*Sphagnum subnitens* Russ. & Warnst. var. *subnitens*

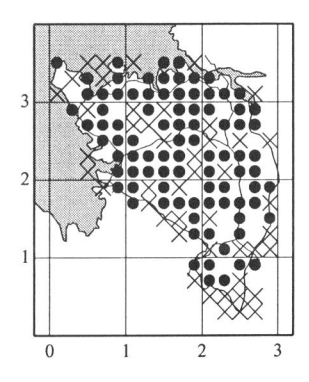

89 tetrads. A common species, occurring in a wide variety of wet habitats including rocky banks, edges of flushes, wet woodland and flushed slopes under heather. All Assynt plants are presumably var. *subnitens*.

*Sphagnum molle* Sull.

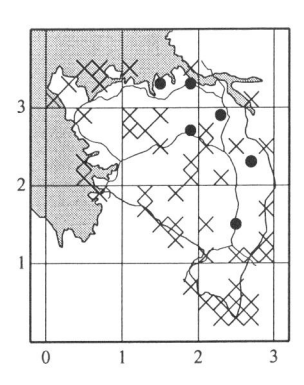

6 tetrads. Only recorded from sites on rather open shallow peat in wet heath or degraded mire; probably over-looked elsewhere. There are older records from Little Assynt and the Point of Stoer.

*Sphagnum strictum* Sull.

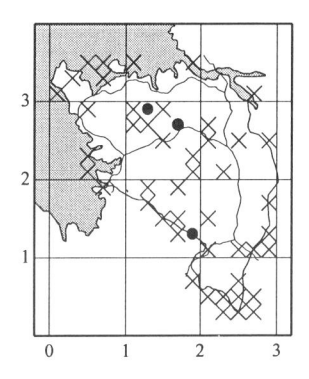

3 tetrads. Confirmed from just three sites on open shallow peat in wet heath but possibly overlooked.

*Sphagnum compactum* Lam. & DC.

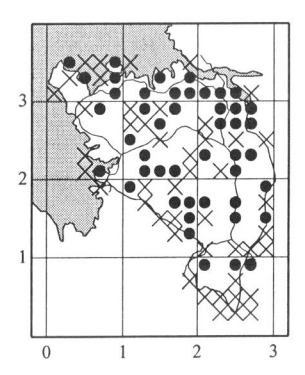

46 tetrads. Widespread, but not usually in any great abundance, on open areas in wet heaths and degraded, peaty ground.

*Sphagnum subsecundum* Nees

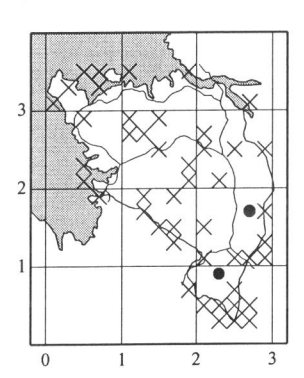

2 tetrads. Very rare in two small mire areas with run-off from base-rich rocks on the upper part of the Allt nan Uamh and on the Cam Alltan at Knockan.

*Sphagnum inundatum* Russow
(S. auriculatum Schimp. var. inundatum (Russow) M.O.Hill)

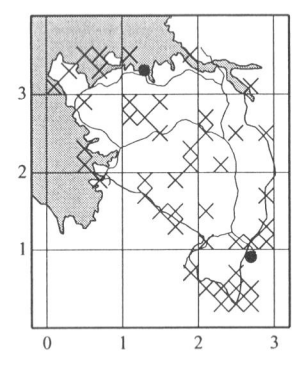

2 tetrads. Only recorded from two localities but almost certainly overlooked elsewhere.

*Sphagnum denticulatum* Brid.
(S. auriculatum Schimp.)

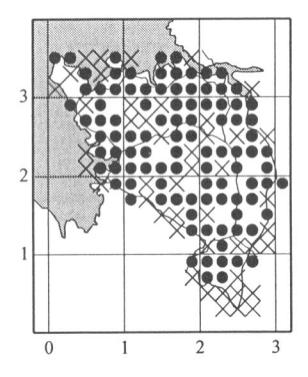

100 tetrads. Common as extensive lawns at the margin of bog pools and in acid flushes, and as swelling mats on flushed rocks in woodland and high into the hills.

*Sphagnum contortum* Schultz

2 tetrads. Rare in base-rich mires, usually with *Campylium stellatum* var. *stellatum* and *Drepanocladus revolvens*: Balchladich and in a bog just south of the River Inver near Lochinver. There is an older record from Knockan by Birks et al. in 1967.

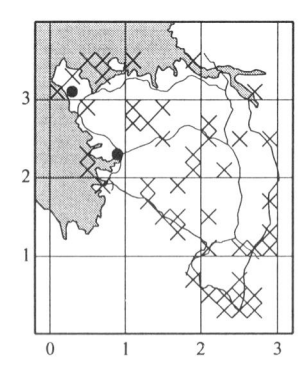

*Sphagnum platyphyllum* (Lindb. ex Braithw.) Warnst.

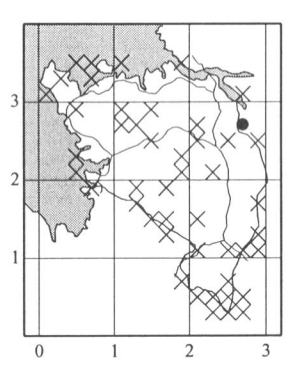

1 tetrad. Only known from one base-rich flush, below gneiss crags in Coire Gorm on the north side of Glas Bheinn.

*Sphagnum tenellum* (Brid.) Bory

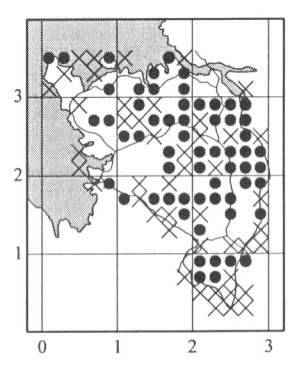

60 tetrads. Occurs in a variety of acid peaty habitats but is most frequent on *Sphagnum*-rich areas of wet heath and, with *Sphagnum capillifolium*, returns quite quickly after burning.

*Sphagnum cuspidatum* Ehrh. ex Hoffm.

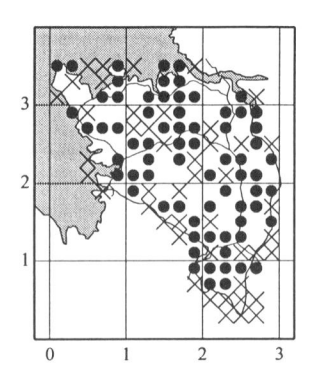

71 tetrads. Very common in bog pools and in wet areas of mires and can be frequent in old peat cuttings.

*Sphagnum fallax* (H.Klinggr.) H.Klinggr. subsp. *fallax*
(S. recurvum P.Beauv. var. mucronatum (Russow) Warnst.)

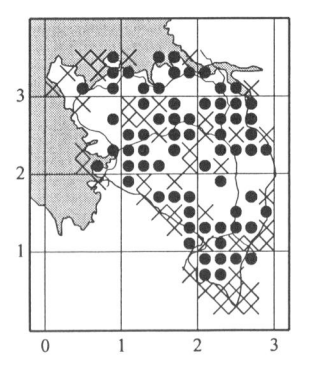

68 tetrads. Common in a variety of wet acid habitats but probably most frequent as carpets with *Juncus effusus* or *Sphagnum palustre* in wet valley bottoms.

*Sphagnum angustifolium*
(C.E.O.Jensen ex Russow) C.E.O.Jensen
(S. recurvum P.Beauv. var. tenue H.Klinggr.)

1 tetrad. Only recorded from the one site in the Cromalt Hills, where it occurs in some abundance

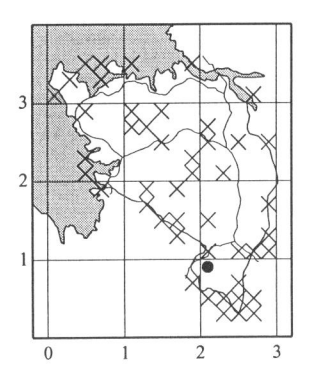

in tall heather on wet heath, but probably overlooked elsewhere.

*Andreaea alpina* Hedw.

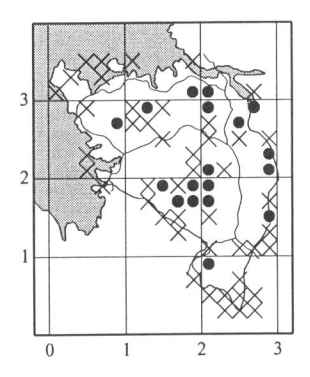

18 tetrads. Frequent on the higher hills but very local at lower altitudes, on wet rocks and shallow mineral soil on ledges of crags, often where the substrate is somewhat base-rich.

*Andreaea rupestris* Hedw.

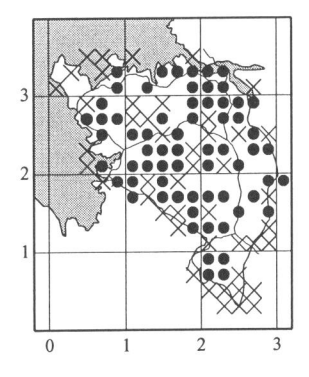

65 tetrads. Common and often abundant both on exposed dry rocks and where water tracks down rock surfaces, but absent from the limestone. No attempt was made to distinguish between var. *rupestris* and var. *papillosa*.

*Andreaea rothii* F.Weber & D.Mohr

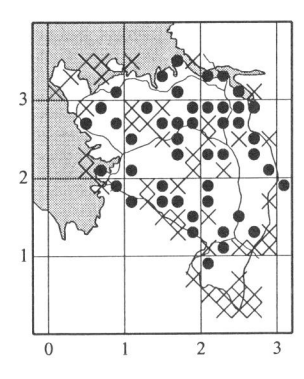

46 tetrads. Generally much less common than *Andreaea rupestris* and limited to acid rocks which are regularly irrigated, where it can form extensive patches. No attempt was made to distinguish between subsp. *rothii* and subsp. *falcata*.

*Andreaea megistospora* B.M.Murray

Collected 'south of Unapool' by the American bryologists Schofield and Schuster in 1978.

*Pogonatum aloides* (Hedw.) P.Beauv.

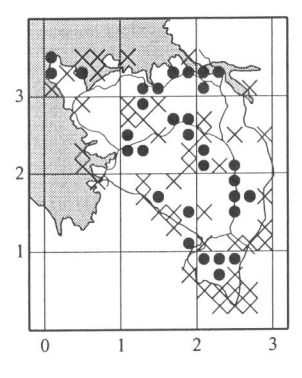

31 tetrads. Widely scattered sites where it is usually an early colonist of mineral soil, on steep eroding banks, root-plates of fallen trees and new tracks.

*Pogonatum urnigerum* (Hedw.) P.Beauv.

26 tetrads. A similar ecology to *Pogonatum aloides*, but possibly persisting longer in closed communities and frequent on disturbed soil in the higher hills.

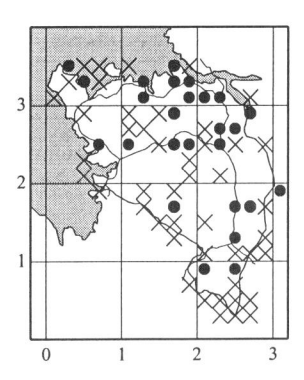

*Polytrichum alpinum* Hedw.

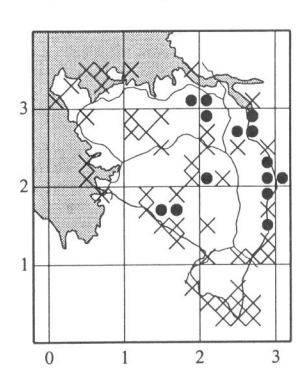

14 tetrads. Sometimes common in drier heathy ground on the higher hills and possibly overlooked elsewhere.

*Polytrichum formosum* Hedw.

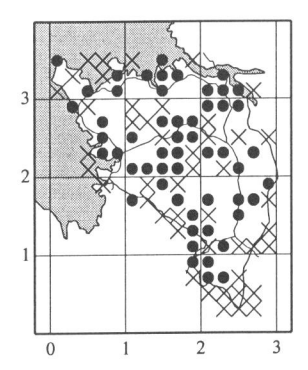

54 tetrads. Common and locally frequent in the better areas of woodland and a characteristic species of the deep bryophyte cap on the tops of boulders in rocky woodland.

*Polytrichum commune* Hedw. var. *commune*

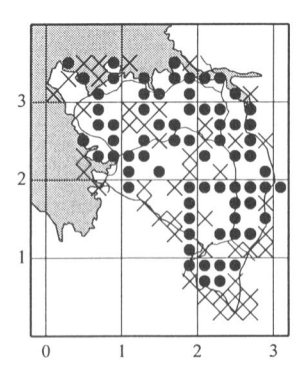

70 tetrads. Common and often abundant in a variety of wet habitats including boggy woodland, wet grassland with *Juncus effusus* and in mires.

*Polytrichum piliferum* Hedw.

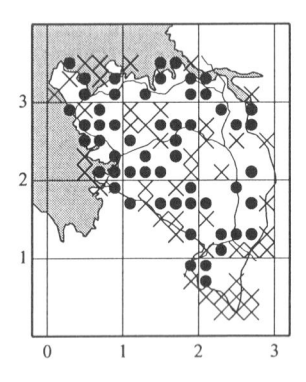

55 tetrads. Common in a variety of well drained sites including the tops of boulders and walls, gravelly soils under heather and gravel by burns and lochs.

*Polytrichum juniperinum* Hedw.

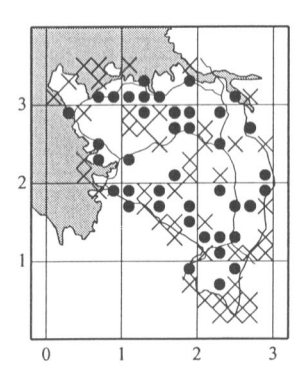

40 tetrads. Frequent in a variety of dry sites on open, gravelly or sandy soils, tops of boulders and walls, and extending to the tops of the hills.

*Polytrichum strictum* Brid. (P. alpestre Hoppe)

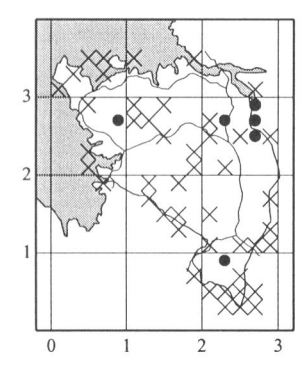

6 tetrads. Surprisingly uncommon despite the extent of seemingly suitable habitat; all sites are in hummocks with *Sphagnum* species in undisturbed mires.

*Oligotrichum hercynicum* (Hedw.) Lam. & DC.

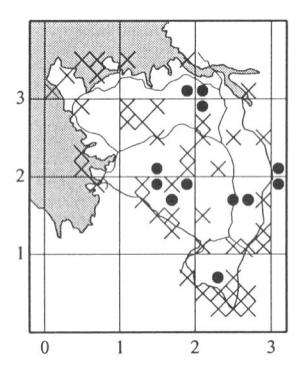

12 tetrads. A coloniser of disturbed, often eroding, gravelly soils in the hills; most frequent on the highest ground but with occasional sites lower down.

*Atrichum undulatum* (Hedw.) P.Beauv.

38 tetrads. A common species of damp sheltered sites on good but not necessarily calcareous soils, often where there has been some disturbance;

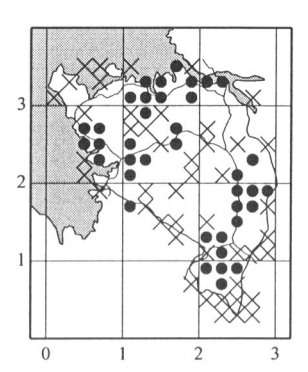

most frequent in woodland but also in ravines and at the base of crags.

*Tetraphis pellucida* Hedw.

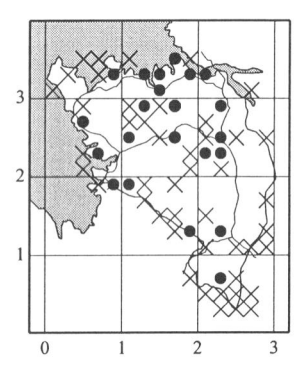

22 tetrads. Widespread but always local, occurring most frequently on peaty soils in shaded, humid sites, especially on the peat capping on crags or boulders in woodland, under 'leggy' heather and very occasionally on rotting logs.

*Diphyscium foliosum* (Hedw.) Mohr

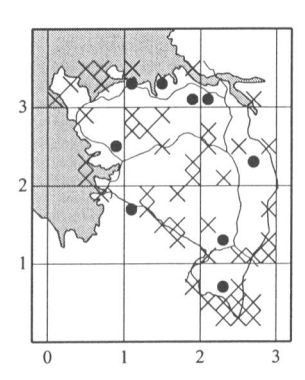

9 tetrads. Uncommon but possibly overlooked when not fruiting; occurs in the bryophyte mat at the top of acidic, steep, often overhanging, earthy banks on

crags, the sides of burns and occasionally the tops of old walls.

*Archidium alternifolium* (Hedw.) Schimp.

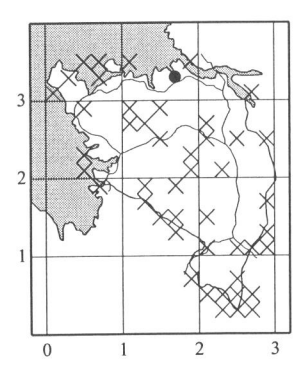

1 tetrad. Very rare but possibly overlooked; only recorded from near the coast at Loch Ardbhair, by Crundwell on the B.B.S meeting in 1992.

*Ditrichum heteromallum* (Hedw.) Britt.

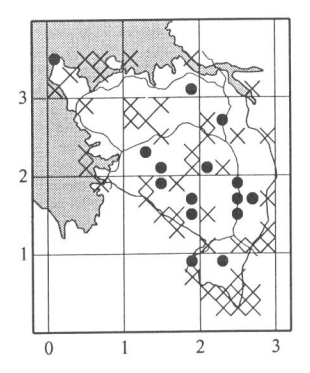

15 tetrads. A colonist of disturbed stony soils where there has been erosion in ravines, newly excavated tracks or on the margin of fine scree in the hills.

*Ditrichum zonatum* (Brid.) Braithw. var. *zonatum*

2 tetrads. Only recorded from the shallow lithosols on the stony terraces west of the summit of Suilven and from similar ground at the head of Coire Gorm on Glas Bheinn;

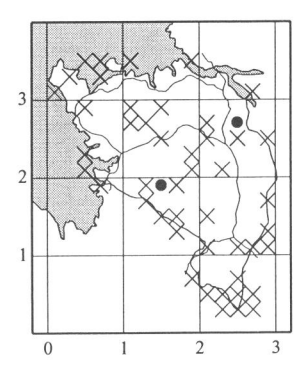

more work on the higher ground will produce further records.

*Ditrichum flexicaule* (Schwägr.) Hampe

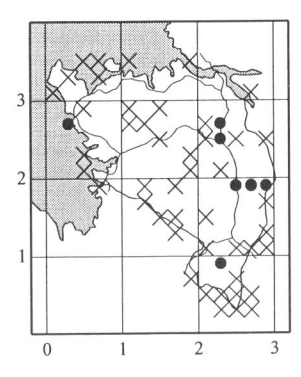

7 tetrads. Uncommon and almost restricted to the limestone where it usually forms tight, dark-green cushions on the tops of low outcrops and boulders.

*Ditrichum gracile* (Mitt.) Kuntze (D. *crispatissimum* (Müll.Hal.) Paris)

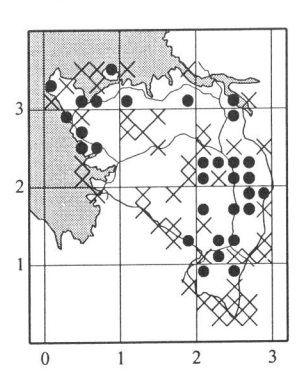

30 tetrads. Frequent and sometimes abundant in flushed grassland on the limestone and on stable sand dunes, with scattered sites on the more base-rich gneiss.

*Distichium capillaceum* (Hedw.) Bruch, Schimp. & W.Gümbel

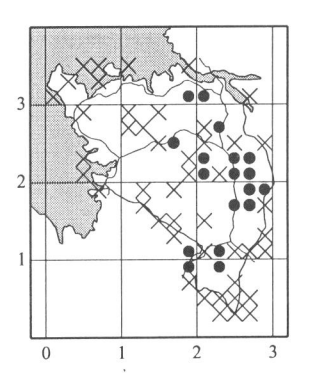

19 tetrads. Common in sheltered crevices on the limestone but rather rare elsewhere on basic ledges on crags.

*Distichium inclinatum* (Hedw.) Bruch, Schimp. & W.Gümbel

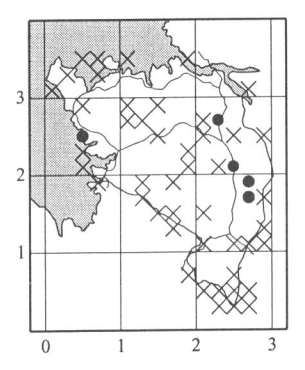

5 tetrads. Rare and limited to damp, shaded ledges on limestone crags, with one site on damp, calcareous sand at Achmelvich.

*Ceratodon purpureus* (Hedw.) Brid.

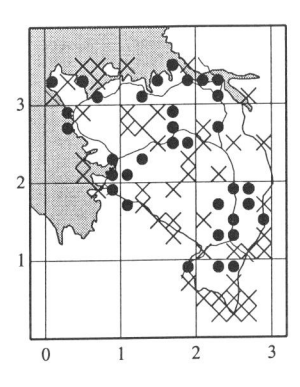

34 tetrads. A widespread species of dry, acid sites where there is little competition and often where

there has been some nutrient enrichment; in Assynt it is particularly frequent on 'bird-perch' sites on rocks and posts.

*Rhabdoweisia fugax* (Hedw.) Bruch, Schimp. & W.Gümbel

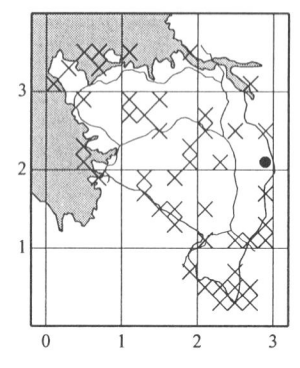

1 tetrad. Very rare and only recorded from sheltered acid crags at Fuaran nan Each. Unfortunately a voucher was not retained so this record, though new for the vice-county, cannot be entered into the v.c.108 list.

*Rhabdoweisia crispata* (Dicks.) Lindb.

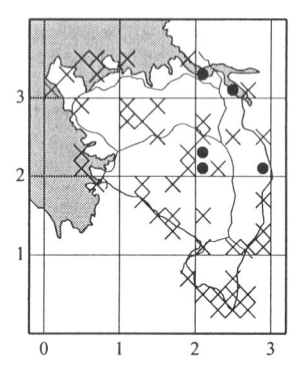

5 tetrads. Rare in sheltered crevices of dry acid crags; recorded from Con a'Chreag, Fuaran nan Each, Garbh Dhoire, Ruigh Chnoc and an older record from the Allt na Ghamhna by Long in 1973.

*Cynodontium jenneri* (Schimp.) Stirt.

2 tetrads. Very rare in crevices of dry, acid crags; recorded from Loch Leathad an Lochain (Inverkirkaig) and Con a'Chreag.

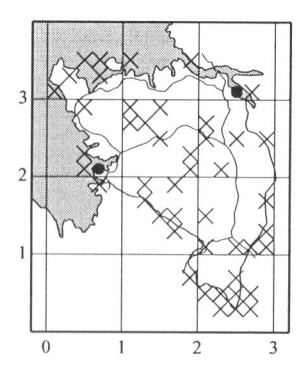

*Oncophorus virens* (Hedw.) Brid.

There is one record for this interesting species from limestone on the bed of an intermittent stream at Inchnadamph by Coker in 1966.

*Dichodontium pellucidum* (Hedw.) Schimp.

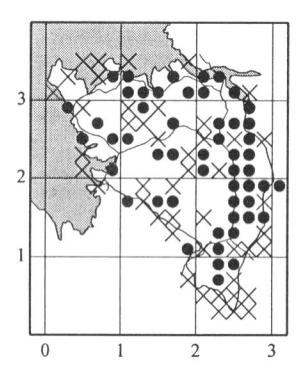

52 tetrads. Common on wet rocks and gravel on the margins of burns and on wet crags in ravines; it is particularly frequent on the limestone. No sporophytes (essential for distinguishing *D. flavescens*) were seen so all records are mapped as *D. pellucidum* sensu lato.

*Dicranella palustris* (Dicks.) Crundw. ex E.F.Warb

38 tetrads. Widespread but usually very local in flushes and on flushed rocks, but noticeably avoiding flushes with *Schoenus nigricans*.

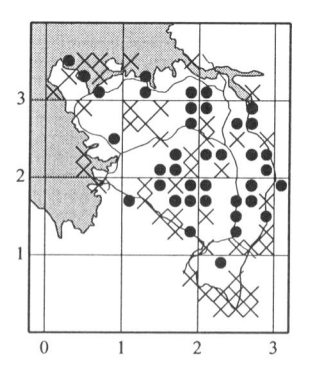

*Dicranella schreberiana* (Hedw.) Dixon

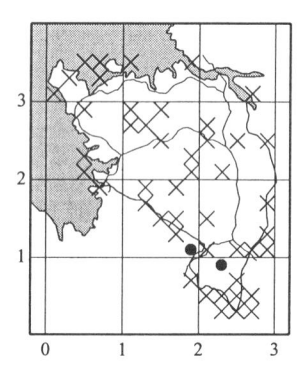

2 tetrads. On disturbed soil which is at least somewhat calcareous; rare but probably overlooked. There is an older record from a ditch by the road at Elphin (Smith, 1960).

*Dicranella grevilleana* (Brid.) Schimp.

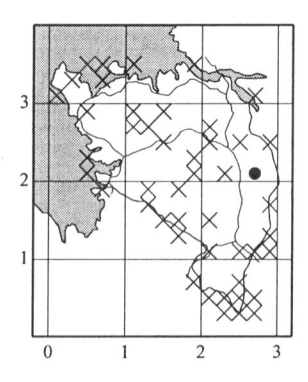

1 tetrad. Very rare and only seen once, on a damp, turfy ledge on limestone in Gleann Dubh by Long in 1981.

*Dicranella subulata* (Hedw.) Schimp.

There is an unlocalised record from NC22 by Dixon and Nicholson which is dated 1900, but was presumably collected on their visit in 1899.

*Dicranella varia* (Hedw.) Schimp.

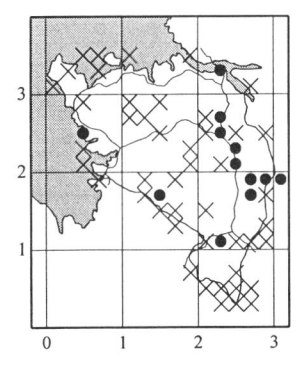

12 tetrads. Widespread but very sparse in the more calcareous areas, occurring on soil on banks in limestone grassland, at the base of damp crags and in calcareous sand dunes.

*Dicranella heteromalla* (Hedw.) Schimp.

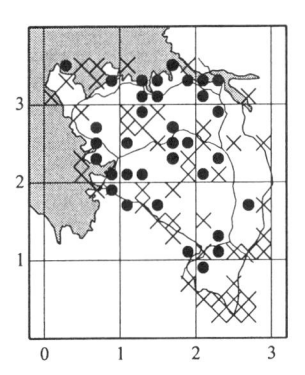

34 tetrads. Fairly common but perhaps not quite as frequent as might have been expected; in a variety of sites on organic soil in woodland and, in particular, on the root-plates of fallen trees.

*Dicranoweisia crispula* (Hedw.) Milde

2 tetrads. Very rare with both sites at comparatively low altitude for this montane species; both sites are on open crags, at Cnoc Aird na Seilge, and on an ultra-basic dyke above the Allt an Tiaghaich (with *Asplenium septentrionale*), both on the south side of Loch Assynt.

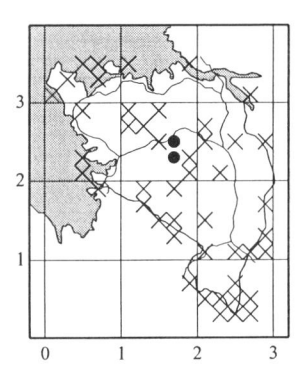

*Arctoa fulvella* (Dicks.) Bruch, Schimp. & W.Gümbel

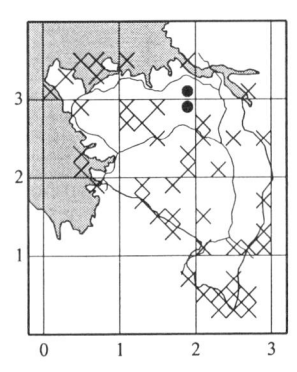

2 tetrads. The sites are on Sàil Gorm and the ridge to the south on Quinag where it occurs on low, exposed Torridonian sandstone. It was also recorded from Quinag in 1899 by Dixon and Nicholson and they also record it from Glas Bheinn.

*Kiaeria falcata* (Hedw.) I.Hagen

The only relatively modern record is from the north-west side of Conival at c. 760m. by Crundwell in 1956; Dixon and Nicholson recorded it from 'Ben Mor Assynt' in 1899 but assigned the record to v.c.108.

*Kiaeria blyttii* (Bruch, Schimp. & W.Gümbel) Broth.

9 tetrads. Occasional as dark green cushions, often with whitish capsules, on large boulders in scree or isolated rocks below crags, usually, but not exclusively, in the higher hills.

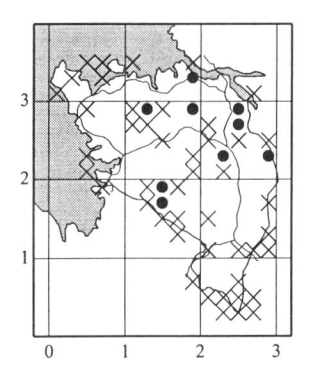

*Kiaeria starkei* (F.Weber & D.Mohr) I.Hagen

The only record is from Ben Mor Assynt by Dixon and Nicholson in 1899; they assign the record to v.c.108, but the more likely habitat on the hill, and where the moss certainly occurs, is in v.c.107.

*Dicranum bonjeanii* De Not.

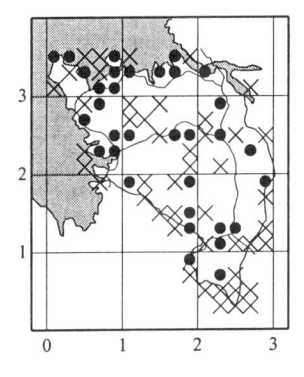

33 tetrads. A characteristic species of flushed grassland, often on the margins of open flushes and on loch shores, but also in rough coastal turf.

*Dicranum scoparium* Hedw.

102 tetrads. Very common and often abundant on firm substrates in a wide variety of habitats, particularly on open rocks and on trees, but also occurring under

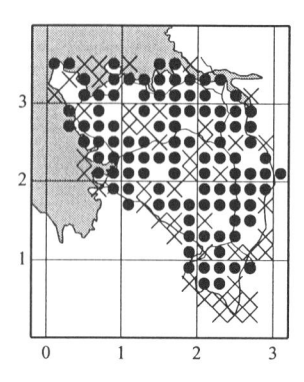

heather in dry and wet heaths and with *Sphagnum* in mires.

*Dicranum majus* Sm.

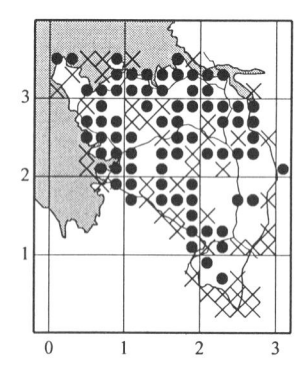

75 tetrads. This handsome species is common and locally abundant in shaded rocky woodland and on ledges of ravines and montane crags, particularly those with a northerly aspect. It can also occur with *Sphagnum capillifolium* and liverworts in oceanic-montane heath.

*Dicranum fuscescens* Sm.

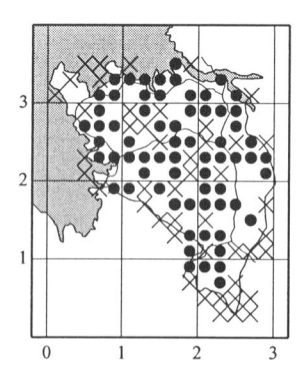

67 tetrads. Common, but much less so than *Dicranum scoparium*; most frequent on open boulders or dry crags and on larger birches; it can be abundant in some areas of block-scree.

*Dicranum scottianum* Turner

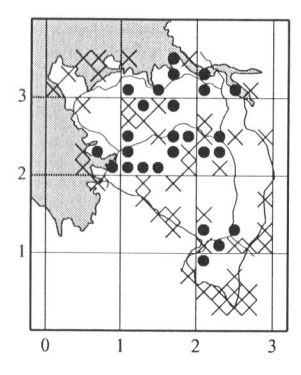

26 tetrads. Widespread but local, with the characteristic sites being sheltered, vertical acid crags where its dark-green cushions can be frequent with *Campylopus flexuosus*, with which it is easily confused. Very rarely it occurs on trees with *Dicranum fuscescens*.

*Dicranodontium uncinatum* (Harv.) A.Jaeger

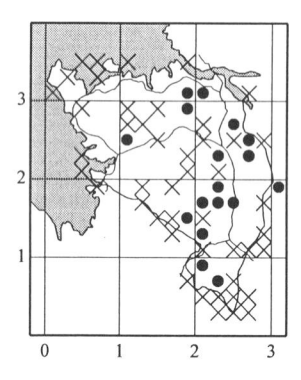

17 tetrads. Occasional in very acid, damp ground at the base of crags in ravines and with heather in montane heath; it is locally abundant on the craggy slopes on the north side of Quinag.

*Dicranodontium asperulum* (Mitt.) Broth.

1 tetrad. Only recorded recently on quartzite in the ravine below Eas Crom in the Cromalt Hills, where it occurs in some quantity, but recorded from Gleann Dubh and woods on the south side of Loch Assynt by Dixon and Nicholson in 1899.

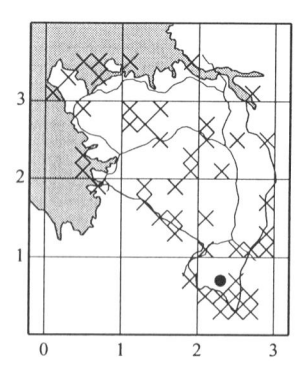

*Dicranodontium denudatum* (Brid.) E.Britton

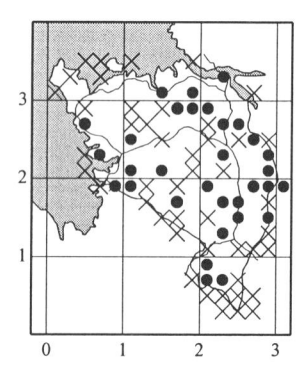

33 tetrads. Widespread but never in great quantity; the most common sites are rocks, or more frequently rotting logs, in woodland and on peat under heather on steep, humid, rocky slopes in the bigger hills.

*Campylopus schimperi* Milde

Recorded by Dixon and Nicholson from Gleann Dubh in 1899.

*Campylopus gracilis* (Mitt.) A.Jaeger
(*C. schwarzii* Schimp.)

4 tetrads. A rare, but possibly overlooked plant, occurring on gravel on the margin of Loch Veyatie, on wet rocks on Meall Diamhain (Breabag), on the north

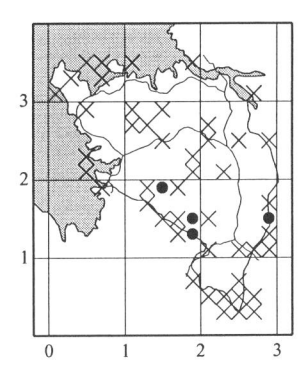

side of Suilven on gneiss and on the south shore of Cam Loch.

*Campylopus fragilis* (Brid.) Bruch, Schimp. & W.Gümbel

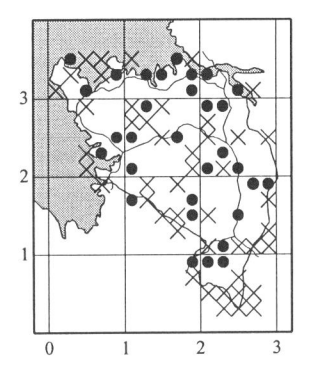

31 tetrads. Widespread but very local and more tolerant of basic conditions than other species of *Campylopus*; most frequent on flushed, species-rich banks in craggy ground at low levels.

*Campylopus pyriformis* (Schultz) Brid.

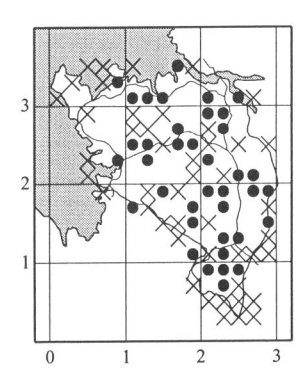

38 tetrads. Widespread and locally abundant on drier peaty ground, often under heather and particularly noticeable on the dry edges of cuttings and peat hags.

*Campylopus flexuosus* (Hedw.) Brid.
(*C. paradoxus* Wilson)

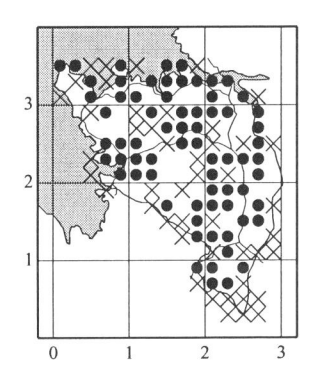

69 tetrads. A very common calcifuge species of bare peat, stumps, rotting logs, boulders and sheltered acid crags.

*Campylopus setifolius* Wilson

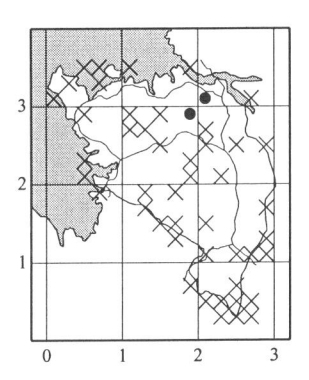

2 tetrads. Very rare and known only from two sites, Creag na h-Iolaire, where it was first found by Ratcliffe in 1966, and from the base of Sàil Garbh on Quinag. In both sites it occurs in rocky montane heath with species like *Mastigophora woodsii* and *Dicranodontium uncinatum*.

*Campylopus shawii* Wilson ex Hunt

1 tetrad. This beautiful, hyper-oceanic species has one of its few mainland localities in Assynt, on open peat in a flushed mire area on the bealach south of Creag na h-Iolaire. There is a further record from the Sanctuary area on Quinag

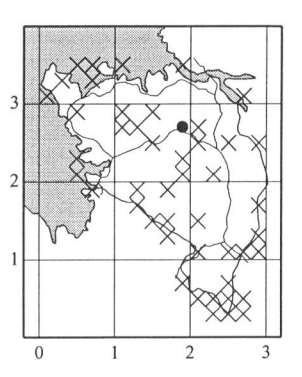

by Muirhead in 1946, but recent searching has failed to re-find this site.

*Campylopus atrovirens* De Not. var. *atrovirens*

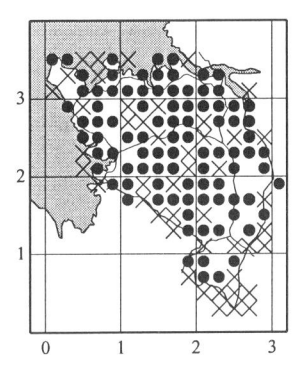

92 tetrads. Common, widespread and often abundant on periodically irrigated rocks and open peat in wet heath and degraded mires, in the latter habitat often with *Pleurozia purpurea*.

*Campylopus atrovirens* De Not. var. *falcatus* Braithw.

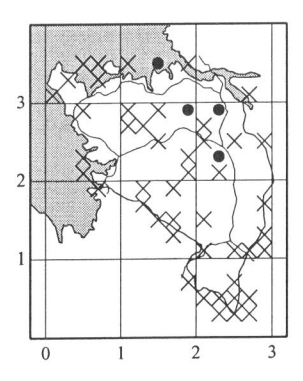

4 tetrads. Rare and always in small quantity on peaty ledges on sheltered crags; recorded from Creagan a'Chait, south of Loch

221

nan Eun, Bealach Leireag and An Sgonnan (Ardvar).

*Campylopus introflexus* (Hedw.) Brid.

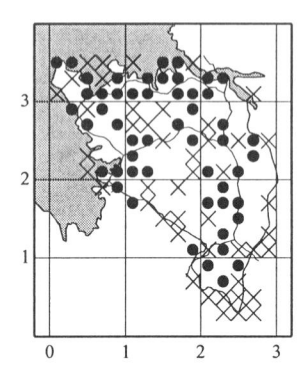

52 tetrads. This non-native species is spreading inexorably over bare peat, particularly where disturbed, and also occurs on logs and at the edge of paths.

*Campylopus brevipilus* Bruch, Schimp. & W.Gümbel

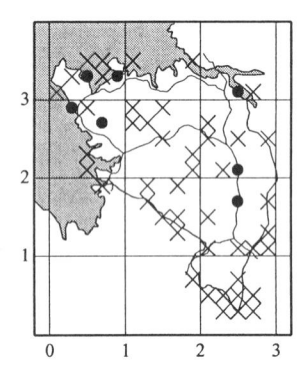

7 tetrads. An uncommon but distinctive species forming yellow-brown cushions in undisturbed areas of mire, usually where there is good hummock development.

*Leucobryum glaucum* (Hedw.) Ångstr.

32 tetrads. Widespread but local and usually in small quantity; most frequent as isolated hummocks in wet heath or mire and surprisingly rare in woodland.

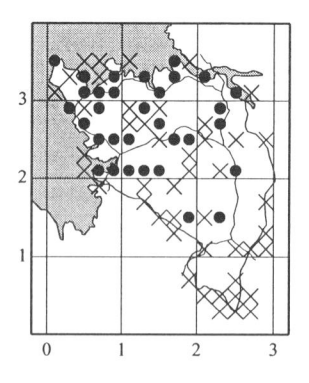

*Fissidens pusillus* (Wils.) Milde

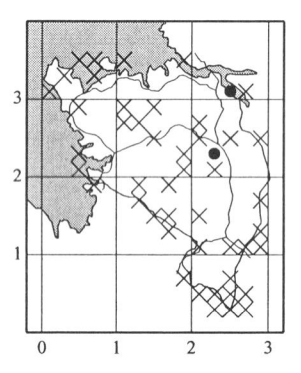

2 tetrads. Rare, but this diminutive species is easily overlooked; the two sites are both on sheltered, damp calcareous rocks, at Ardvreck and Liath Bhad.

*Fissidens bryoides* Hedw.

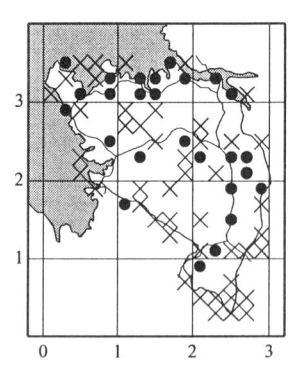

26 tetrads. An infrequent species of well-drained, sheltered sites on open steep soil in woodland and ravines and more open sites on the limestone.

*Fissidens osmundoides* Hedw.

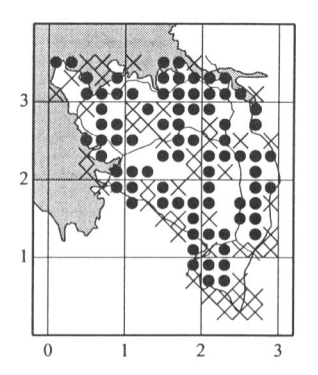

80 tetrads. A common species in Assynt; its distinctive, dense tufts are a feature of wet ledges on crags and in flushes where the run-off is somewhat base-rich.

*Fissidens taxifolius* Hedw.

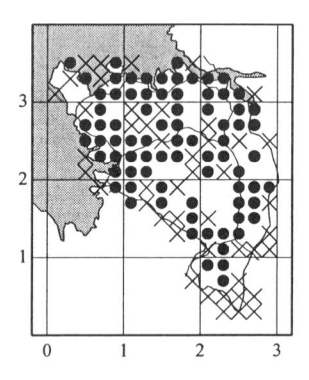

79 tetrads. A common plant of wet soil and rock faces, tolerant of both moderately acid and moderately calcareous substrates. The var. *pallidicaulis*, an unsatisfactory taxon, has been recorded once, from the Falls of Kirkaig.

*Fissidens dubius* P.Beauv. (F. cristatus Wilson ex Mitt.)

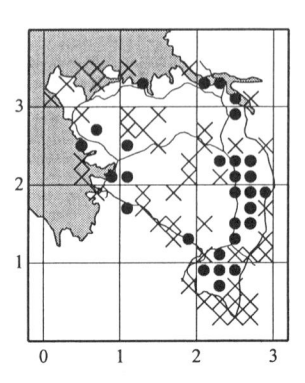

29 tetrads. Common on the limestone and possibly overlooked elsewhere; found on both damp rocks and on soil in limestone grassland, but also occurs on the base-rich gneiss.

*Fissidens adianthoides* Hedw.

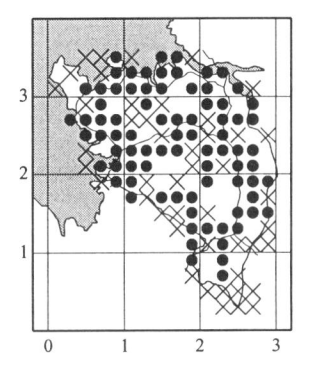

81 tetrads. A common plant of wet, moderately base-rich sites and sometimes forming large, lax tufts; it is probably most frequent on wet rocks and ledges in ravines, but also occurs in flushes and flushed grassland.

*Encalypta streptocarpa* Hedw.

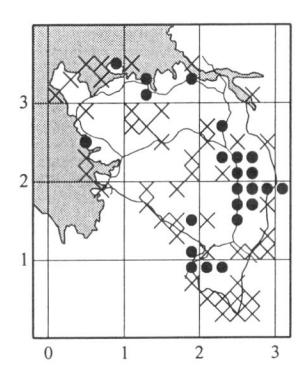

23 tetrads. Common on the limestone and occurs on the sand dunes at Achmelvich, otherwise only on mortared walls.

*Encalypta alpina* Sm.

2 tetrads. This rare calcicole occurs in two places, on the limestone boulders below Creag nan Uamh and on a limestone outcrop in the Bealach Traligill.

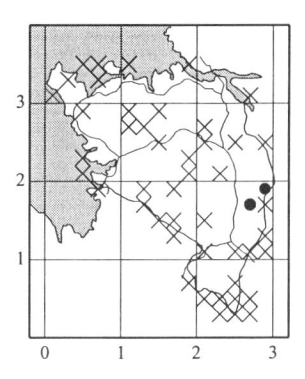

*Encalypta rhaptocarpa* Schwägr.

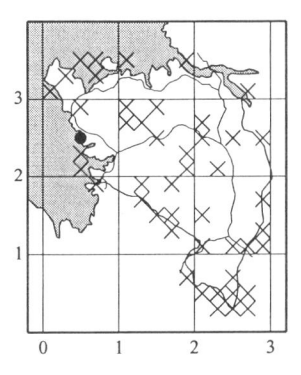

1 tetrad. Only recorded from the calcareous sand dunes at Achmelvich and seemingly absent from the limestone.

*Encalypta ciliata* Hedw.

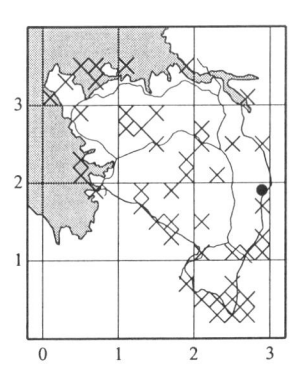

1 tetrad. Very rare; only found on the isolated limestone outcrop in the Bealach Traligill.

*Eucladium verticillatum* (Brid.) Bruch, Schimp. & W Gumbel

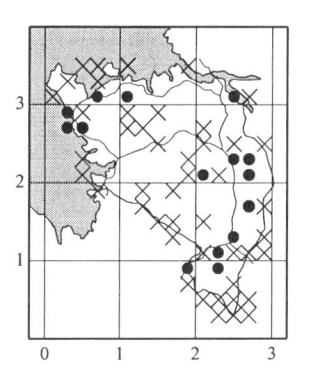

15 tetrads. Locally frequent on steep, wet limestone and occasional on wet basic rocks elsewhere, usually encrusted with lime.

*Weissia controversa* Hedw. var. *controversa*

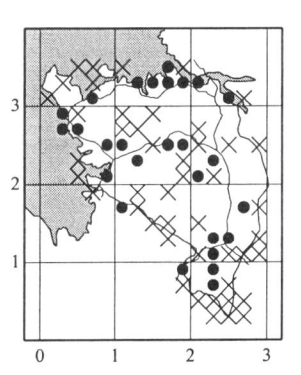

27 tetrads. A wide spread but uncommon plant of moderately basic soil on open, stable banks and most frequent near the sea.

*Weissia perssonii* Kindb.

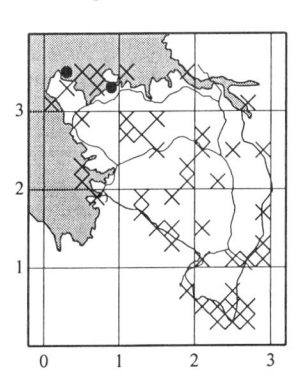

2 tetrads. In similar situations to *Weissia controversa* but much rarer and always coastal; west of

Glac Cùilce (Oldany) and in Geodh' Dearg on Stoer.

*Weissia rutilans* (Hedw.) Lindb.

Recorded from the limestone at Inchnadamph by Appleyard in 1960 and the Allt na Uamh by Birks et al. in 1969.

*Weissia brachycarpa* (Nees & Hornsch.) Jur.
(W. microstoma (Hedw.) Müll.Hal.)

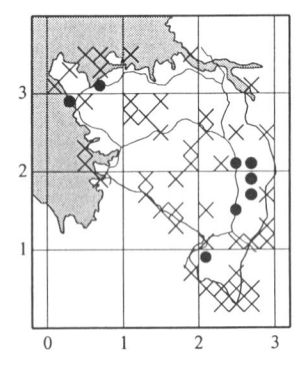

8 tetrads. Occasional and confined to the limestone and calcareous shell-sand where it occurs on open, dry soil on banks; all plants in Assynt are var. *obliqua*.

*Tortella tortuosa* (Hedw.) Limpr.

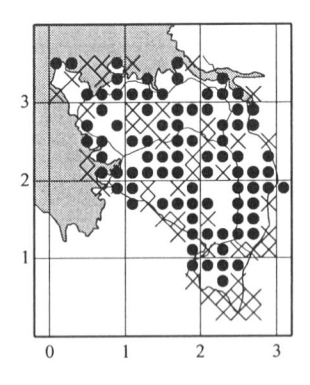

85 tetrads. Common and locally abundant on base-rich gneiss, limestone and occasionally on Torridonian sandstone.

*Tortella densa* (Lors. & Molendo) Crundw. & Nyholm

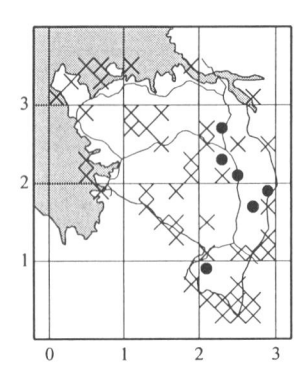

6 tetrads. Rather rare and only on the limestone where it occurs with, and is easily overlooked as, the abundant *Tortella tortuosa*.

*Tortella flavovirens* (Bruch) Broth. var. *flavovirens*

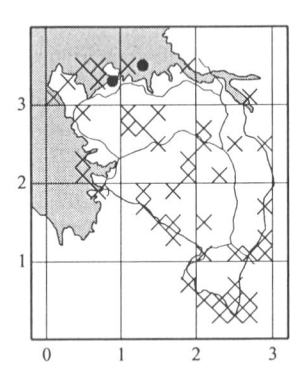

2 tetrads. This coastal species is apparently rare in the north-west of Scotland and is only recorded from two localities in Assynt, Sgeir Liath off Drumbeg and Oldany, in both places on coastal rocks.

*Trichostomum brachydontium* Bruch

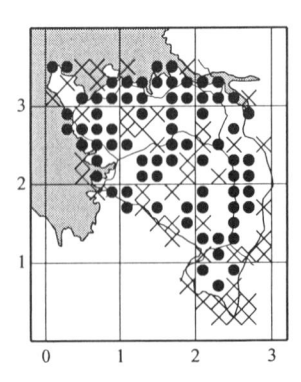

73 tetrads. A common plant of soil on ledges and in crevices of at least moderately base-rich rocks. Particularly frequent on some facies of the gneiss near the coast.

*Trichostomum crispulum* Bruch

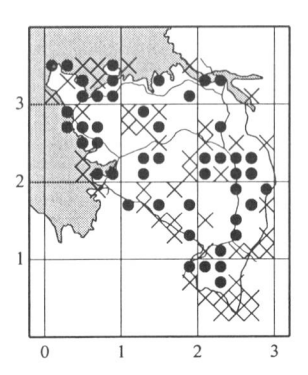

47 tetrads. Widespread in similar habitats to *Trichostomum brachydontium* but generally less frequent, except on the limestone and on rocky places near the coast.

*Trichostomum tenuirostre* (Hook. & Tayl.) Lindb. var. *tenuirostre* (Oxystegus tenuirostris (Hook. & Taylor) A.J.E.Smith)

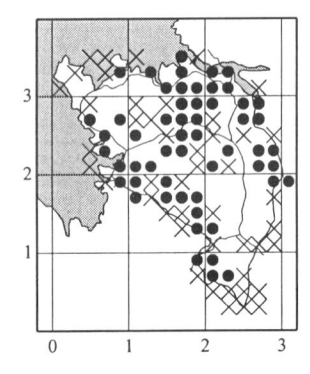

55 tetrads. A common species of wet acid rocks by burns and on sheltered crags, usually in small quantity.

*Trichostomum hibernicum* (Mitt.) Dixon
(Oxystegus hibernicus (Mitt.) Hilpert)

2 tetrads. A very rare moss of steep, flushed banks over base-rich rocks; in two places, Gleann

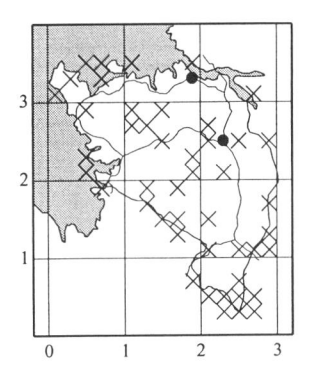

Ardbhair and an incised burn north of Achmore Farm.

*Paraleptodontium recurvifolium* D.G.Long
(*Leptodontium recurvifolium* (Taylor) Lindb.)

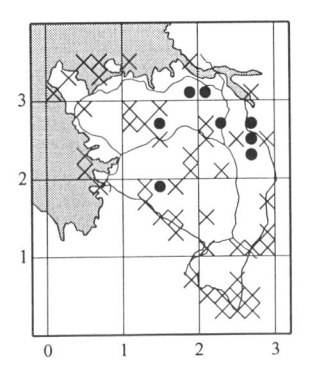

8 tetrads. This interesting oceanic-montane species is very occasional in damp turf at the base of crags in the higher hills but also occurs sparingly lower down, on wet ledges in wooded ravines.

*Pseudocrossidium revolutum* (Brid.) R.H.Zander
(*Barbula revoluta* Brid.)

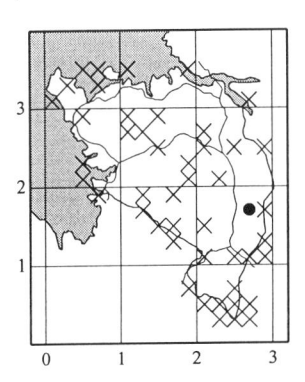

1 tetrad. Recorded once from the limestone in the Allt na Uamh but possibly overlooked on mortared walls elsewhere.

*Bryoerythrophyllum recurvirostrum* (Hedw.) P.C.Chen
(*Barbula recurvirostra* (Hedw.) Dixon)

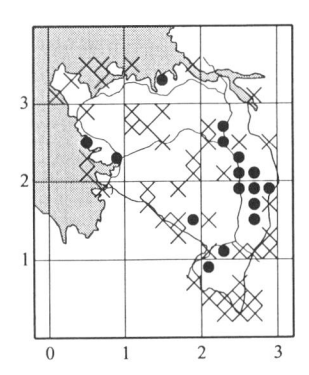

16 tetrads. Occasional on the tops of boulders and on open crags on the limestone but very rare elsewhere.

*Bryoerythrophyllum ferruginascens* (Stirt.) Giacom.
(*Barbula ferruginascens* Stirt.)

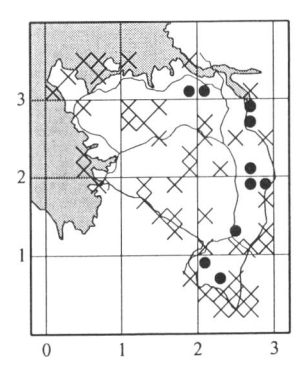

10 tetrads. An uncommon species of damp soil on ledges on base-rich montane crags, with rare occurrences lower down on the limestone.

*Leptodontium flexifolium* (Dicks.) Hampe

1 tetrad. Only recorded once, from the north side of Sàil Gorm on Quinag by Blockeel in 1992.

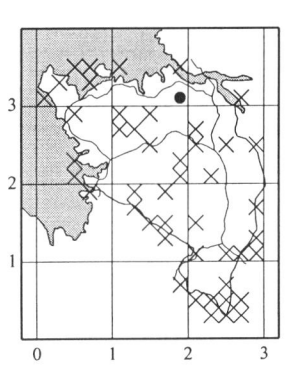

*Hymenostylium recurvirostrum* (Hedw.) Dixon
(*Gymnostomum recurvirostrum* Hedw.)

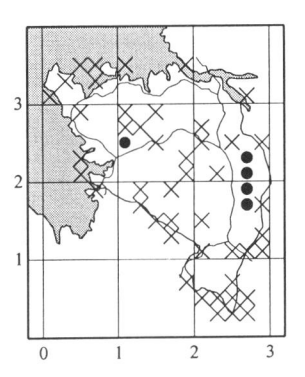

5 tetrads. Scattered sites on wet ledges on sheltered limestone crags, sometimes with *Hymenostylium insigne*, and very rare elsewhere.

*Hymenostylium insigne* (Dixon) Podp.
(*Gymnostomum insigne* (Dixon) A.J.E.Smith)

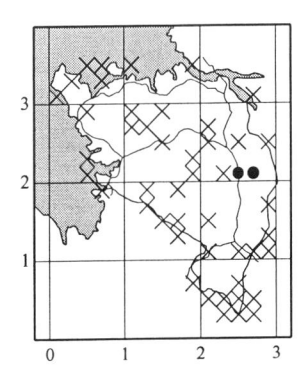

2 tetrads. Rare but very locally abundant on wet limestone crags and rocks by the river in the

Traligill valley; seemingly absent from the limestone elsewhere.

*Anoectangium aestivum* (Hedw.) Mitt.

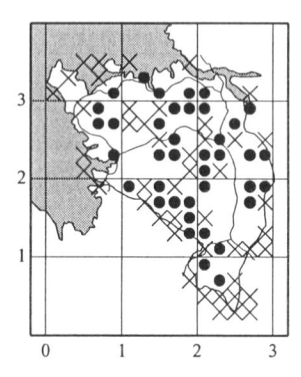

39 tetrads. Widespread and sometimes forming large swelling cushions on damp ledges and in crevices of at least moderately base-rich crags, often with *Preissia quadrata*.

*Gyroweisia tenuis* (Hedw.) Schimp.

Recorded once, from an overhanging bank by the River Traligill by Warburg in 1952, but possibly overlooked elsewhere on the limestone and mortared walls.

*Gymnostomum calcareum* Nees & Hornsch.

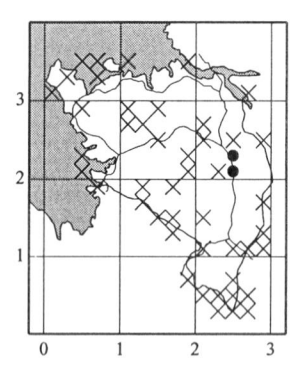

2 tetrads. Occurs sparingly in crevices of the limestone on the crags at Stronechrubie and in Gleann Dubh, where it was first recorded by Dixon in 1899.

*Gymnostomum aeruginosum* Sm.

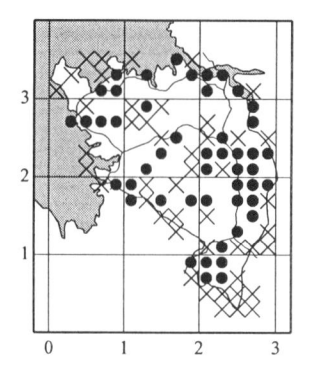

48 tetrads. A common moss of wet calcareous places, often forming large, dark-green cushions in wet places on steep or over-hanging crags.

*Molendoa warburgii* (Crundw. & M.O.Hill) R.H.Zander (Anoectangium warburgii Crundw. & M.O.Hill)

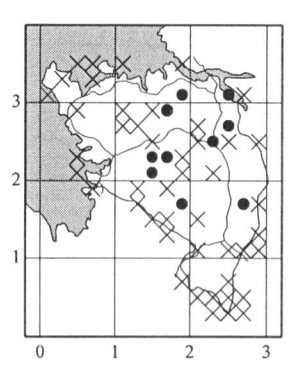

10 tetrads. An uncommon plant, usually forming loose mats of tiny stems in wet open sites on base-rich crags and in ravines, usually associated with the higher hills.

*Barbula convoluta* Hedw.

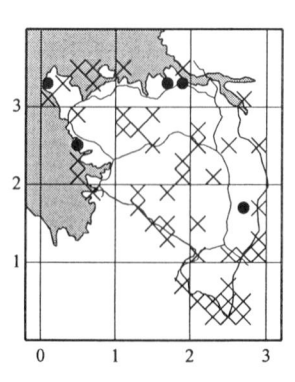

5 tetrads. Apparently uncommon but almost certainly overlooked, on disturbed ground on roadsides and in gardens and the like.

*Barbula unguiculata* Hedw.

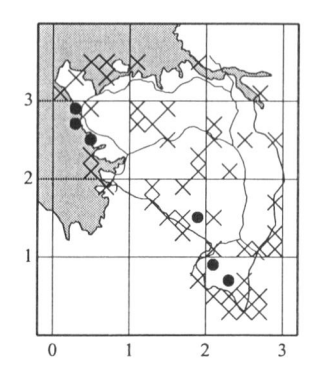

6 tetrads. As with *Barbula convoluta*, apparently uncommon, but more attention to ruderal sites in the parish would probably produce many more records.

*Didymodon icmadophilus* (Schimp. ex Müll.Hall.) K.Saito (Barbula icmadophila Schimp. ex Müll.Hall.)

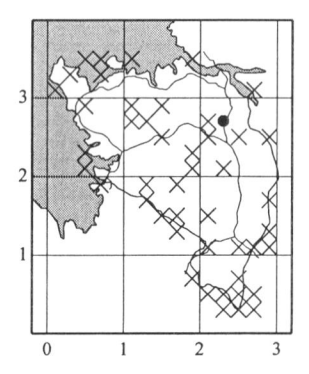

1 tetrad. This nationally rare species occurs on top of the low limestone outcrops at Lairig Unapool with *Ditrichum flexicaule*, the only known locality for West Sutherland.

*Didymodon rigidulus* Hedw. (Barbula rigidula (Hedw.) Mitt.)

10 tetrads. Occasional on the limestone where it usually occurs on the tops of boulders or open

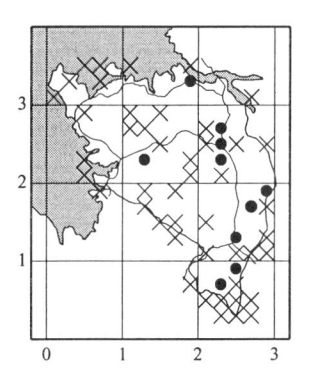

low outcrops, often with *Schistidium* species and *Ditrichum flexicaule*.

*Didymodon vinealis* (Brid.) R.H.Zander

Dubiously recorded during this survey at Achmelvich (B.B.S.) and recorded in the past from Clachtoll and Stoer by Wallace in 1939.

*Didymodon insulanus* (De Not.) M.O.Hill
(Barbula cylindrica (Taylor) Schimp.)

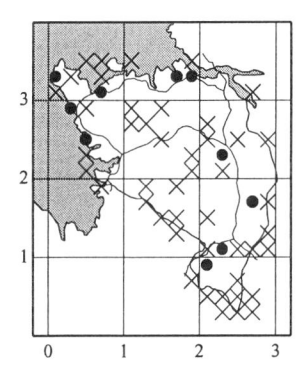

10 tetrads. Uncommon but widespread on base-rich soil and on roadsides; more work in ruderal habitats would almost certainly produce more records.

*Didymodon tophaceus* (Brid.) Lisa
(Barbula tophacea (Brid.) Mitt.)

5 tetrads. Rare on soil on base-rich banks close to the sea with one site on the limestone in the Allt nan Uamh.

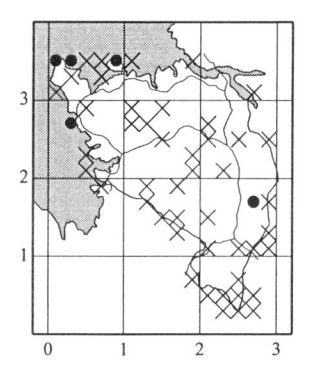

*Didymodon spadiceus* (Mitt.) Limpr.
(Barbula spadicea (Mitt.) Braithw.)

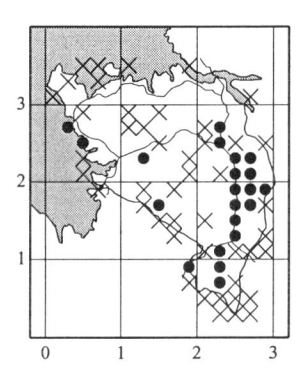

21 tetrads. Quite frequent on wet, gravelly soil by burns or flushes on the limestone, where it can form large cushions, but rare elsewhere.

*Didymodon fallax* (Hedw.) R.H.Zander
(Barbula fallax Hedw.)

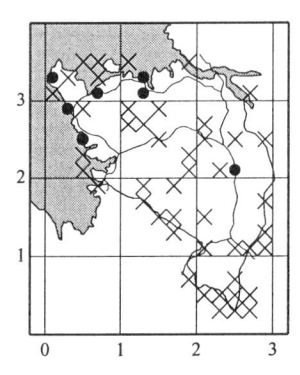

7 tetrads. Uncommon but widespread on sand-dunes, base-rich soil and on roadsides; more work in ruderal habitats would almost certainly produce more records.

*Didymodon ferrugineus* (Schimp. ex Besch.) M.O.Hill
(Barbula reflexa (Brid.) Brid.)

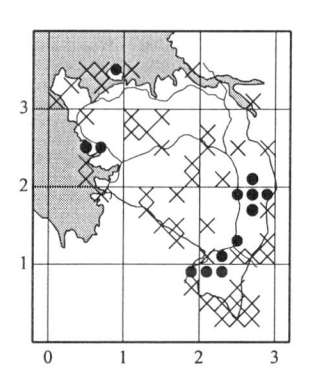

13 tetrads. This attractive little moss with its regular, reflexed leaves is locally common on the limestone, on boulders and open, low crags, and also occurs on shell-sand at Achmelvich and on Oldany Island.

*Tortula subulata* Hedw.

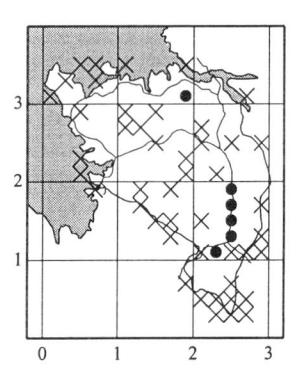

6 tetrads. An uncommon moss of soil on dry, open, flat ledges, confined to the limestone, apart from one site on basic gneiss on Sàil Gorm where it occurs as the var. *graeffii*.

*Tortula muralis* Hedw.var. *muralis*

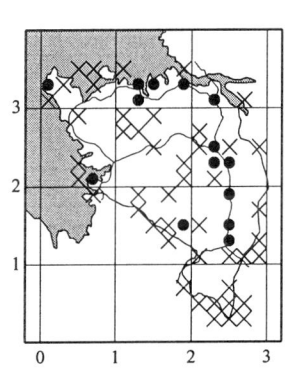

227

14 tetrads. Quite frequent on the limestone along dry cracks and in crevices, and from mortared walls elsewhere; under-recorded in man-made habitats.

*Syntrichia ruralis* (Hedw.) F.Weber & D.Mohr.
(*Tortula ruralis* (Hedw.) P.Gaertn., B.Mey. & Scherb.)

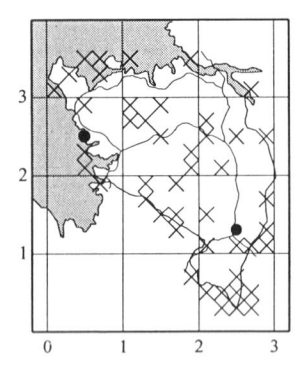

2 tetrads. Apparently very rare on open rocks, with sites on basic gneiss at Achmelvich and on the limestone at Ledbeg, and an older record from Inchnadamph by Wallace in 1939.

*Syntrichia ruraliformis* (Besch.) Cardot
(*Tortula ruralis* ssp. *ruraliformis* (Besch.) Dixon)

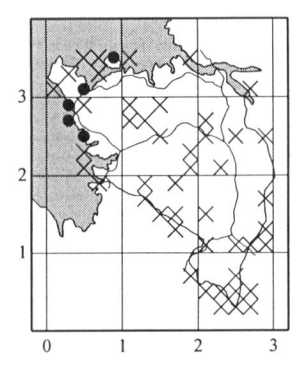

5 tetrads. Limited to the areas of sand-dunes where it is an effective coloniser of open sand; locally abundant at all large shell-sand sites in the area.

*Syntrichia princeps* (De Not.) Mitt.
*Tortula princeps* De Not.)

1 tetrad. Known from only one site on the limestone crags above the road north of Stronechrubie, where it was first recorded by Dixon and Nicholson in 1899; seemingly limited to one overhung bay in the crags where it forms huge cushions.

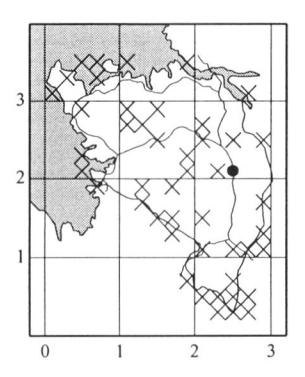

*Cinclidotus fontinaloides* (Hedw.) P.Beauv.

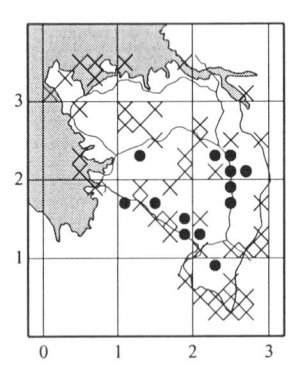

13 tetrads. Frequent on rocks in burns on the limestone and on rocks on the margins of lochs with high base-status, with this effect extending down the River Kirkaig.

*Schistidium maritimum* (Turner) Bruch & Schimp.

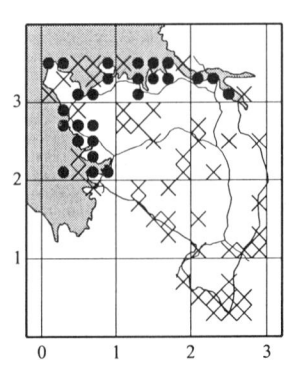

26 tetrads. Common and locally abundant on all the rocky coast and extending up onto the higher ground on Point of Stoer, providing some indication of the drenching with salt-spray that this exposed headland receives.

*Schistidium rivulare* (Brid.) Podp. (*Schistidium alpicola* (Hedw.) Limpr. var. *rivulare* (Brid.) Limpr.)

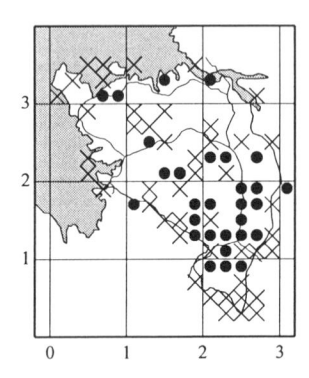

22 tetrads. Occasional but locally frequent on rocks in the larger rivers and burns and also on loch margins.

*Schistidium agassizii* Sull. & Lesq.

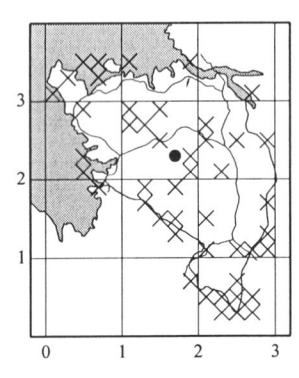

1 tetrad. This rare species has one site in Assynt, on wet, base-rich gneiss rocks below a large waterfall on the Allt an Tiaghaich, where it is abundant in spreading black cushions.

*Schistidium apocarpum* (Hedw.) Br.E sens. lat.

45 tetrads. Common on the limestone and Fucoid Beds and also on base-rich gneiss on loch margins and coastal rocks. In the latter part of the survey some attempt was made to apply Blom's

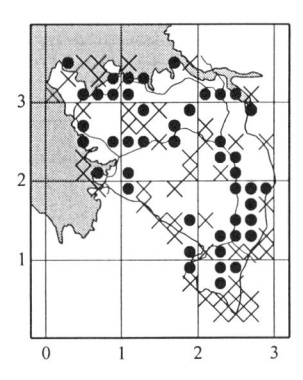

revision of the *Schistidium apocarpum* complex (Blom 1996) to Assynt material other than *Schistidium strictum* and *Schistidium trichodon* which were recorded from the outset. It would appear that *Schistidium apocarpum* s.s. is by far the most common taxon with some 17 tetrads. There is a scattering of records for *Schistidium crassipilum* and *Schistidium robustum* (mostly from the limestone) and single records for *Schistidium frigidum* var. *frigidum* (Ledbeg River at Lyne) and var. *havaasii* (Cnoc Aird na Seilge). The majority of these records have been verified by A.J.E. Smith.

*Schistidium trichodon* (Brid.) Poelt

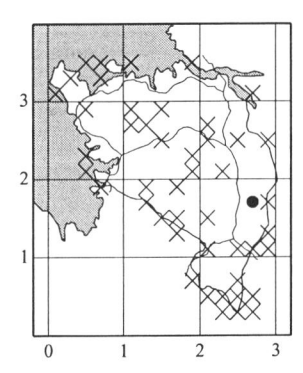

1 tetrad. This rare species occurs in flat, black patches on the upper surfaces of limestone boulders in the scree to the west of Creag nan Uamh, its only locality north of the Great Glen.

*Schistidium strictum* (Turner) Loeske ex Martensson

41 tetrads. Frequent but never in great quantity on inclined surfaces of base-rich rocks, often on loch margins or by burns on the lower ground. All material collected is *Schistidium strictum* but it is possible that plants of *Schistidium papillosum* were overlooked in the field.

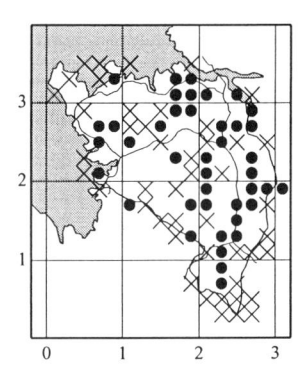

*Grimmia longirostris* Hook. (G. affinis Hornsch.)

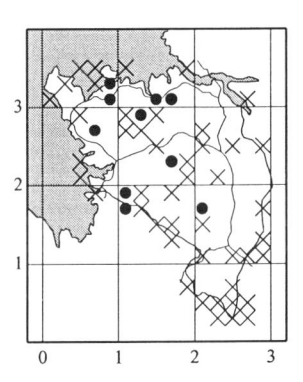

10 tetrads. This scarce species in Britain has a surprising number of records in Assynt on basic gneiss boulders on the margins of lochs; unless fruiting, it is easily overlooked as the much more common *Grimmia funalis*.

*Grimmia ovalis* (Hedw.) Lindb.

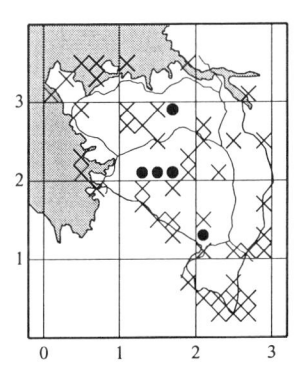

5 tetrads. This rare moss occurs in scattered sites as robust, untidy cushions on base-rich gneiss rocks on loch margins and by burns; Abhainn Bad na h-Achlaise, Abhainn na Clach Airigh, Lochan Buidhe, Loch an Leothaid and Cam Loch. It has apparently declined markedly in Britain over the last 50 years and is rated as Vulnerable in the bryophyte Red Data Book (Church et al. 2001) and these small stands in Assynt would seem to form a significant part of the British population.

*Grimmia pulvinata* (Hedw.) Sm. var. *pulvinata*

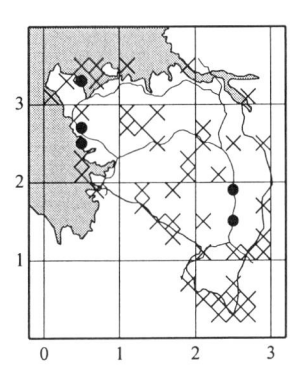

5 tetrads. Surprisingly rare in Assynt with sites on coastal rocks and on the limestone; closer attention to more man-made habitats would probably produce many more records.

*Grimmia orbicularis* Bruch ex Wilson

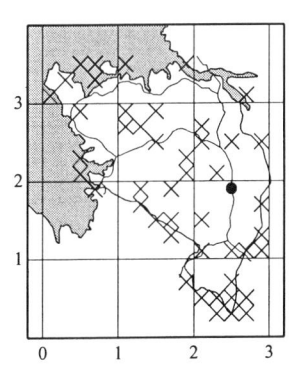

1 tetrad. One site on a dry, exposed ledge on the limestone crags at Stronechrubie but may well occur elsewhere on the limestone.

*Grimmia torquata* Hornsch. ex Grev.

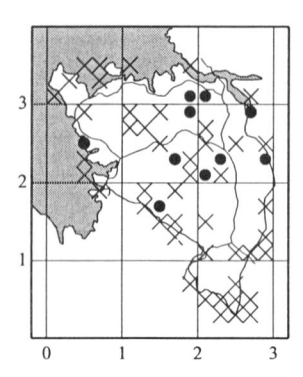

10 tetrads. An uncommon species forming neat, small cushions on sheltered, basic crags but avoiding limestone; it is usually montane (it is frequent on the gneiss on Sàil Gorm on Quinag) but does also occur on the coast at Achmelvich.

*Grimmia funalis* (Schwägr.) Bruch & Schimp.

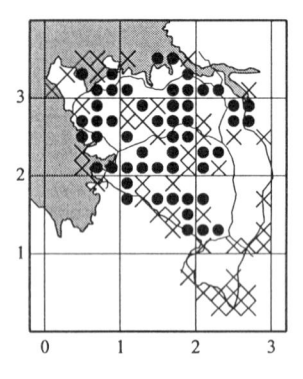

52 tetrads. The neat, hoary cushions of this moss are a consistent feature of exposures of base-rich gneiss on crags, coastal rocks and particularly on boulders on loch margins; interestingly, as with *Grimmia torquata*, it is absent from the limestone.

*Grimmia trichophylla* Grev.

47 tetrads. A common species of dry, exposed rocks and a frequent constituent of the community on loch-side boulders; it is absent from both the most acid gneiss and from the limestone. Several records were made of *G. britannica* but this is now regarded as a form of *G. trichophylla*.

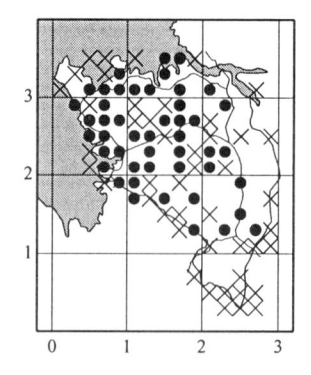

*Grimmia hartmanii* Schimp.

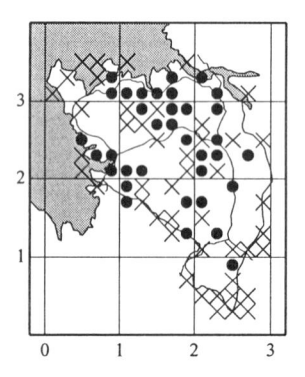

35 tetrads. A widespread but local species, most frequent on boulders on the margin of lochs but also on rocks in ravines and, rarely, in woodland.

*Grimmia decipiens* (Schultz) Lindb.

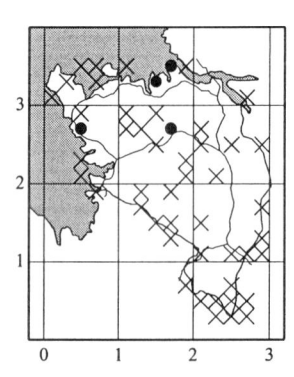

4 tetrads. A rare species in Assynt, forming relatively robust, untidy cushions on base-rich gneiss rocks on loch margins with sites on Loch na Creige Léithe south-east of Clachtoll, Loch Assynt south of Tumore, and on Loch a'Meallard, Ardvar and at Kerrachar.

*Grimmia curvata* (Brid.) De Sloover
(*Dryptodon patens* (Hedw.) Brid.)

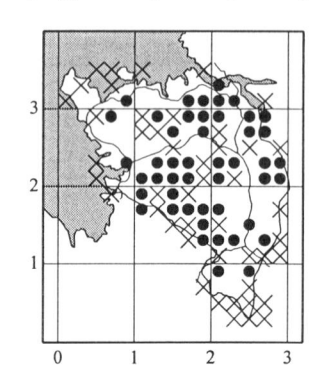

47 tetrads. A frequent and often abundant species, forming large, spreading cushions on moderately base-rich rocks on loch margins and on large boulders in rocky burns.

*Racomitrium ellipticum* (Turner) Bruch & Schimp.

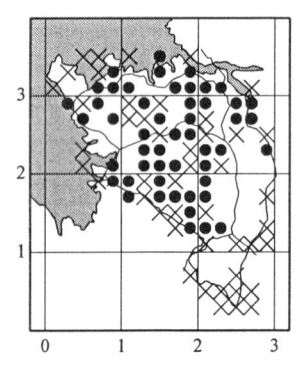

50 tetrads. A common and characteristic species of periodically irrigated gneiss, on crags, on rocks on loch margins and in ravines, where its dark green cushions with short brown capsules can be abundant.

*Racomitrium aciculare* (Hedw.) Brid.

102 tetrads. A ubiquitous species of wet rocks in burns, loch margins and crags, occurring on the most

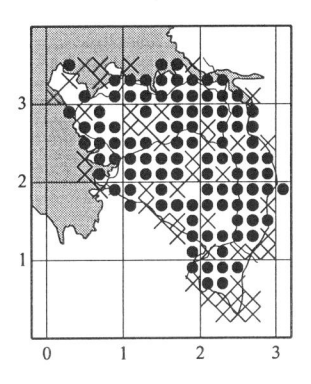

acid quartzite and on limestone.

*Racomitrium aquaticum* (Schrad.) Brid.

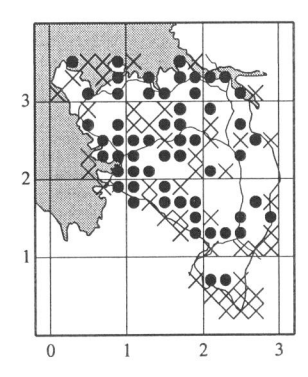

56 tetrads. A common species of acid rocks on crags where water regularly tracks down from above.

*Racomitrium fasciculare* (Hedw.) Brid.

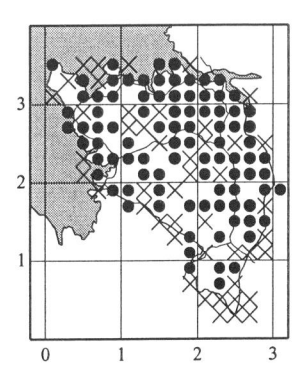

91 tetrads. The spreading, yellow-green patches of this moss are very common on acid rocks but also occur on limestone, usually in relatively sheltered sites and often on low rocks that are periodically irrigated.

*Racomitrium macounii* Kindb. subsp. *alpinum* (E.Lawton) Frisvoll

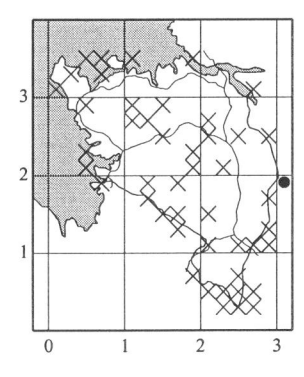

1 tetrad. Only recorded from one site from rocks by a small burn on the slopes of Conival in the Bealach Traligill, the only known locality in West Sutherland.

*Racomitrium sudeticum* (Funck) Bruch & Schimp.

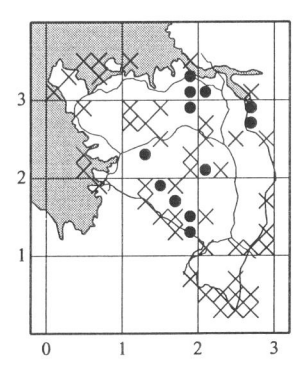

12 tetrads. Uncommon but almost certainly under-recorded; occurs as dense, dark-green cushions on exposed acid rocks and is usually associated with the higher hills.

*Racomitrium heterostichum* (Hedw.) Brid.

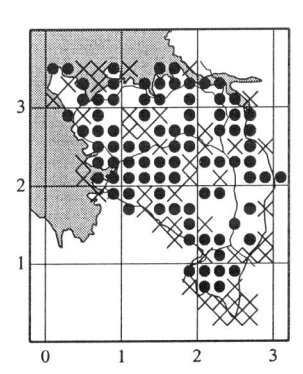

86 tetrads. A very common species of exposed acid rocks from sea level to the tops of the hills, but most frequent on large rocks in montane scree.

*Racomitrium lanuginosum* (Hedw.) Brid.

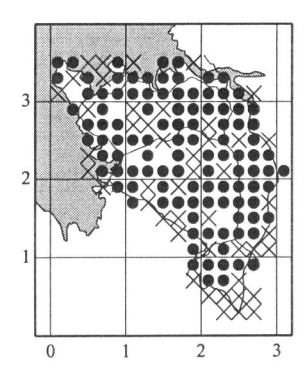

107 tetrads. A ubiquitous and often abundant species of exposed rocks, scree, old walls, on hummocks in bogs and stony places in the high hills where it can form extensive stands.

*Racomitrium ericoides* (Brid.) Brid.

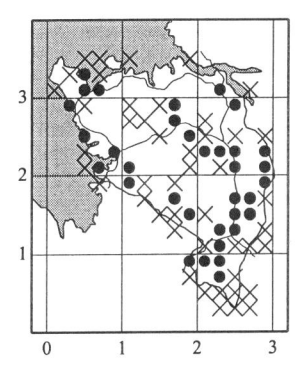

33 tetrads. A widespread species of damp gravelly ground by burns or on loch margins and occasionally on thin soil over limestone.

*Racomitrium elongatum* Frisvoll

5 tetrads. Apparently much less frequent than *Racomitrium ericoides* and only recorded from the limestone where it occurs in

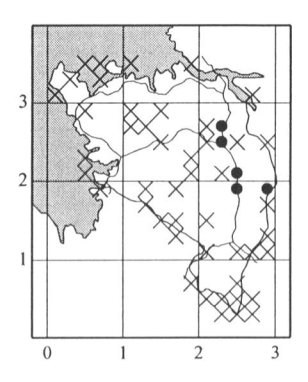

stony grassland; possibly under-recorded.

*Ptychomitrium polyphyllum* (Sw.) Bruch & Schimp.

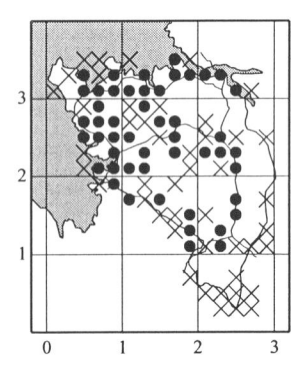

49 tetrads. A common species on open, acid or mildly base-rich rocks on crags, loch margins and in ravines; it is often an early coloniser of newly exposed rock surfaces.

*Glyphomitrium daviesii* (Dicks.) Brid.

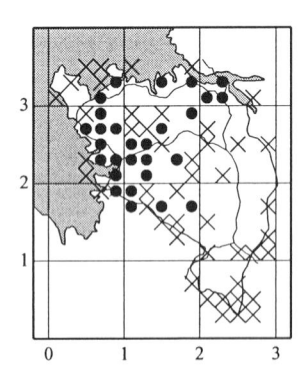

28 tetrads. Widespread but usually in small quantity on sheltered faces of moderately base-rich gneiss crags and boulders. Primarily thought of as a moss of Tertiary basalts in Scotland, the frequency in Assynt of the attractive, dark-green cushions with urn-shaped capsules was one of the surprises of the survey.

*Blindia acuta* (Hedw.) Bruch, Schimp. & W.Gümbel

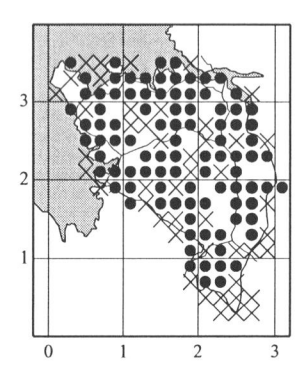

95 tetrads. A very common species, sometimes forming extensive open carpets, on at least moderately base-rich rocks which are regularly irrigated and on gravel in flushes.

*Seligeria pusilla* (Hedw.) Bruch, Schimp. & W.Gümbel

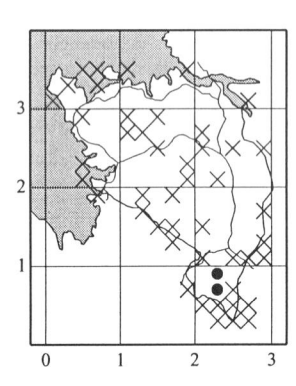

2 tetrads. Recorded from two sites on vertical, damp limestone in the Allt a'Phollain and the Abhainn a'Chnocain above Knockan; hard to see and easily confused with *Seligeria donniana* and possibly under-recorded.

*Seligeria recurvata* (Hedw.) Bruch, Schimp. & W.Gümbel

7 tetrads. Rather rare and seemingly limited to the Fucoid Beds and avoiding the harder limestones; it usually occurs as tiny, open turfs on sheltered, damp rock faces and capsules are abundant.

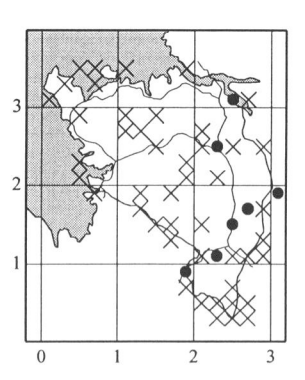

*Seligeria trifaria* (Brid.) Lindb.

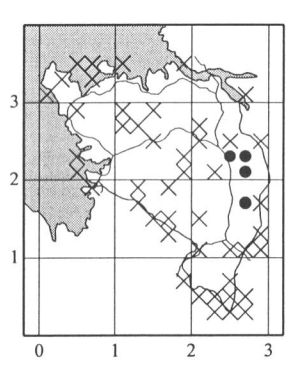

4 tetrads. Rare but can be abundant in limited areas of suitable habitat; it forms open mats of tiny stems in the algal 'scum' that forms on heavily-shaded, wet limestone under overhangs or in caves: River Traligill, Allt Poll an Droighinn and the Allt nan Uamh. The plants in Assynt are *Seligeria trifaria* s.s. rather than *Seligeria patula* and are the most northerly in Britain for this scarce species.

*Seligeria donniana* (Sm.) Müll.Hal

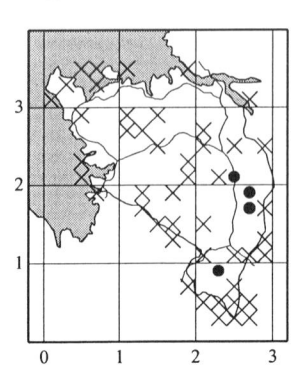

4 tetrads. Rare on damp, vertical

faces of limestone rocks in ravines or at the base of crags; this tiny plant is difficult to spot and may be under-recorded: Inchnadamph, Cam Alltan, Cnoc Eilid Mhathain, Creag nan Uamh.

*Funaria hygrometrica* Hedw.

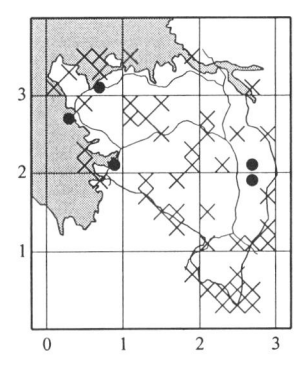

5 tetrads. Only a handful of records for this normally common species probably reflects the neglect of man-made and disturbed habitats during this survey.

*Entosthodon attenuatus* (Dicks.) Bryhn
(*Funaria attenuata* (Dicks.) Lindb.)

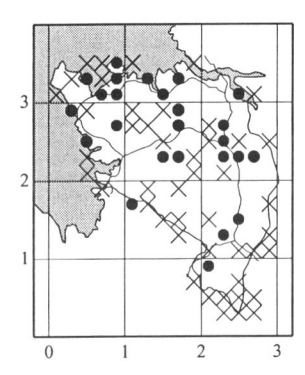

25 tetrads. A widespread but sparse moss of flushed soil on steep banks, often with *Fissidens* species, easily overlooked if not fertile.

*Entosthodon obtusus* (Hedw.) Lindb.
(*Funaria obtusa* (Hedw.) Lindb.)

36 tetrads. In similar sites to *Entosthodon attenuatus* and usually more frequent and occasionally forming extensive turfs with other bryophytes on suitable flushed banks.

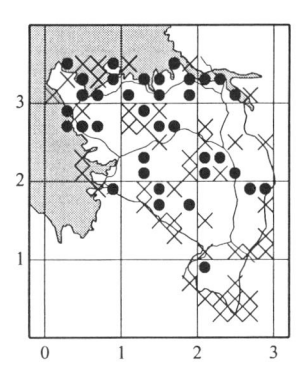

*Ephemerum serratum* (Hedw.) Hampe var. *minutissimum* (Lindb.) Grout

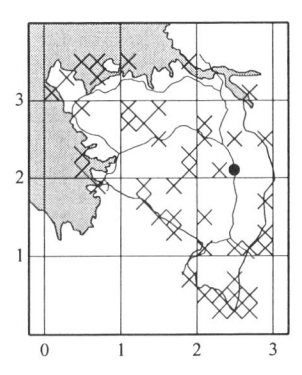

1 tetrad. Recorded from soil on a bank on the limestone by the Allt Poll an Droighinn, new to West Sutherland, by Blockeel on the B.B.S meeting in 1992.

*Tetraplodon angustatus* (Hedw.) Bruch & Schimp.

'A single tuft with male flowers only' of this scarce species was recorded from Glas Bheinn in 1899 by Dixon & Nicholson and it has not been seen in West Sutherland since.

*Tetraplodon mnioides* (Hedw.) Bruch & Schimp.

18 tetrads. Uncommon but widely scattered, occurring as dense cushions, often with abundant reddish sporophytes, on carnivore scats and not seen on herbivore dung.

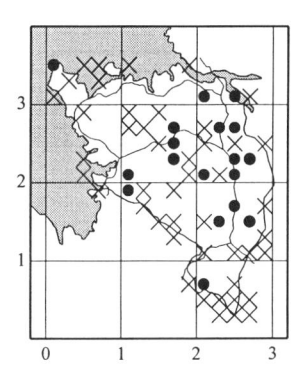

*Splachnum sphaericum* Hedw.

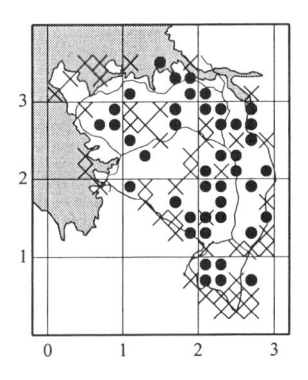

43 tetrads. Widely scattered and locally frequent in areas where deer or sheep gather; only seen on the dung of sheep, deer and cattle in boggy places.

*Splachnum ampullaceum* Hedw.

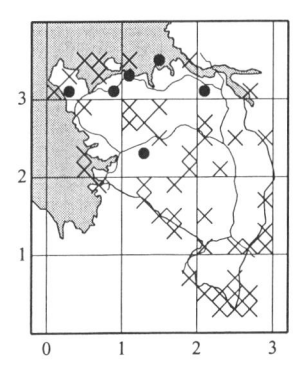

6 tetrads. Rare but possibly overlooked as *Splachnum sphaericum* when without sporophytes; only seen on deer or cattle dung in boggy places: Airigh na Beinne, Balchladich, Oldany River, Loch Poll, An Sgonnan (Ardvar), and Mòinteach na Dubha Chlaise.

233

*Pohlia elongata* Hedw.

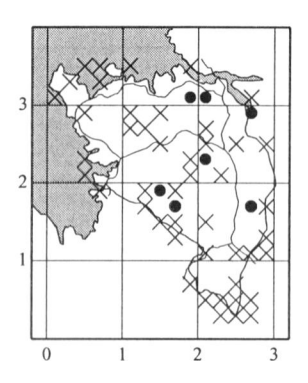

7 tetrads. Rather rare and usually in small quantity on thin organic soil in crevices and ledges of acidic crags; all plants collected were ssp. *elongata* var. *elongata*.

*Pohlia cruda* (Hedw.) Lindb.

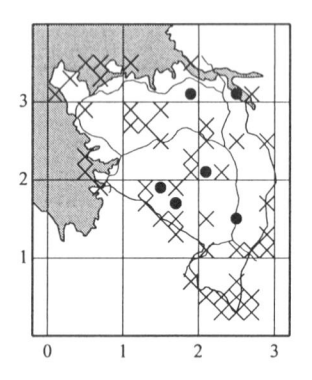

6 tetrads. A rare moss in Assynt growing in shaded crevices on crags but absent from many seemingly suitable sites.

*Pohlia nutans* (Hedw.) Lindb.

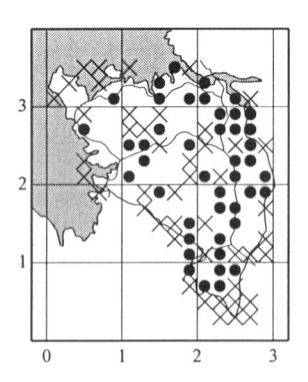

44 tetrads. A common and often plentiful species of organic substrates, especially bare peat, in a variety of acidic habitats but particularly in the drier heathland.

*Pohlia drummondii* (Müll.Hal.) A.L.Andrews

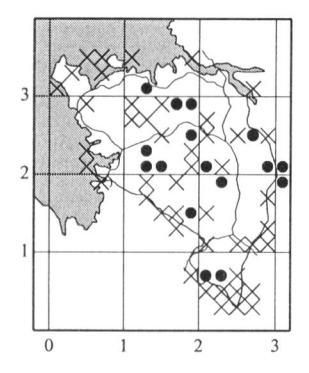

16 tetrads. Widely scattered but nowhere at all frequent, on open, gravelly soil on the margins of burns, lochs, tracks and eroding banks.

*Pohlia scotica* Crundw.

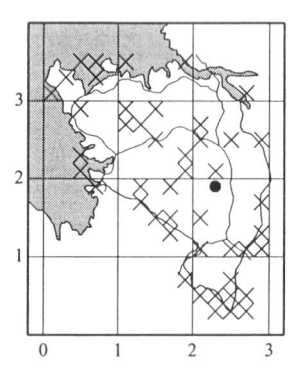

1 tetrad. This nationally rare moss, apparently endemic to Scotland, was found once during the survey on wet gravel on the edge of a small burn at the inflow to Loch nam Meallan Liatha, north of Canisp.

*Pohlia filum* (Schimp.) Martensson

Recorded from a lay-by on the road between Elphin and Ledmore by Paton in 1969.

*Pohlia bulbifera* (Warnst.) Warnst.

2 tetrads. In similar habitats to *Pohlia drummondii* but much less frequent; only recorded from Loch an Leothaid and the Allt an Achaidh (Cromalt), with an older record from a lay-by on the road between Elphin and Ledmore by Paton in 1969.

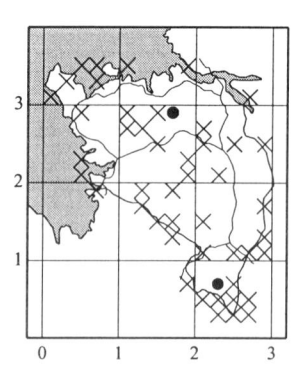

*Pohlia annotina* (Hedw.) Lindb.

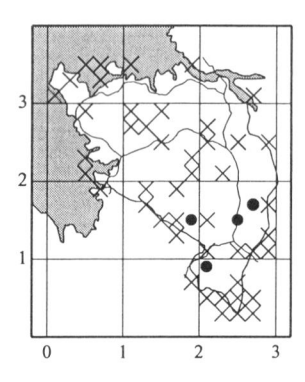

5 tetrads. An uncommon pioneer species, like *Pohlia drummondii*, on open gravelly soils in a variety of damp or periodically irrigated sites.

*Pohlia camptotrachela* (Renault & Cardot) Broth.

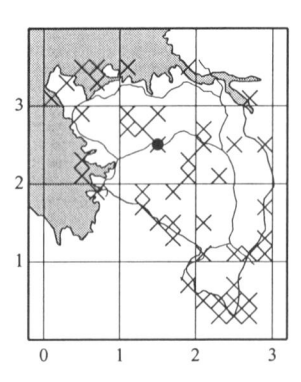

1 tetrad. There is a single record from open ground near the outflow of Loch Assynt made by Pool during the 1992 B.B.S meeting.

*Pohlia flexuosa* Harv.
(Pohlia muyldermansii R.Wilczek & Demaret)

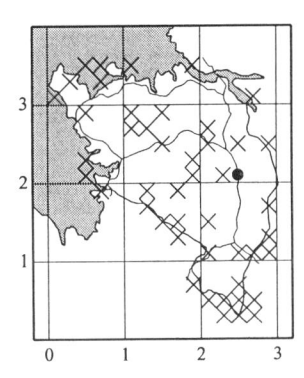

1 tetrad. Only recorded once on this survey, from Inchnadamph by the B.B.S in 1992.

*Pohlia ludwigii* Spreng. ex Schwägr.) Broth.

The only unequivocal Assynt record is from Canisp by Dixon and Nicholson in 1899.

*Pohlia melanodon* (Brid.) A.J.Shaw
(P. carnea (Schimp.) Lindb.)

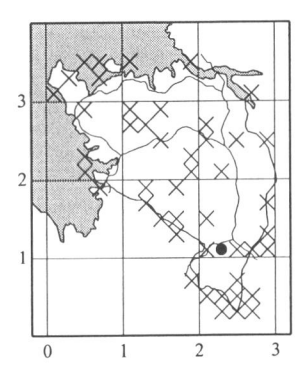

1 tetrad. An uncommon species in the north of Scotland and only recorded once on this survey, from wet basic soil in the ravine at Na Luirgean.

*Pohlia wahlenbergii* (F.Weber and Mohr) A.L.Andrews var. *wahlenbergii*

25 tetrads. Widespread but only at all frequent on the limestone, where it occurs on wet soil by burns and tracks, in ditches and flushes.

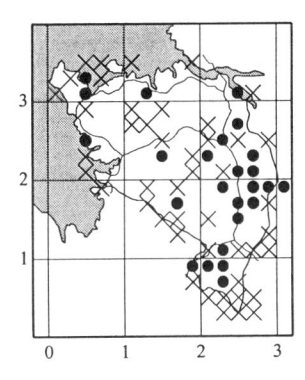

*Pohlia wahlenbergii* var. *glacialis* (Schleich. ex Brid.) E.F.Warb.

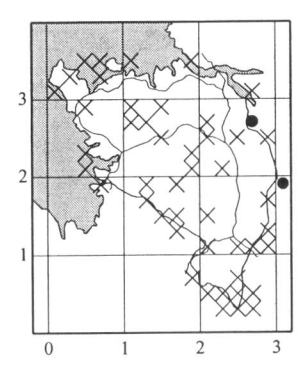

2 tetrads. This striking moss grows in montane flushes and is limited to two sites on the higher hills, in the Bealach Traligill and on Glas Bheinn.

*Plagiobryum zieri* (Hedw.) Lindb.

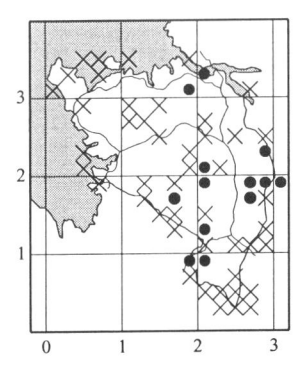

13 tetrads. An uncommon calcicole moss with its best populations in damp crevices of crags on the limestone, but also more sparingly on basic rocks elsewhere.

*Anomobryum julaceum* (P.Gaertn., B.Mey. & Scherb.) Schimp.
(Anomobryum filiforme (Dicks.) Husn.)

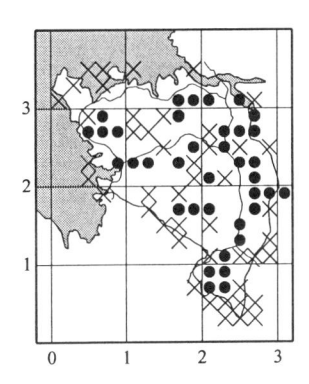

37 tetrads. Widespread and often frequent in small scattered stands, usually in sand or gravel by burns or lochs, in flushes or in crevices of damp basic rocks. All records on this survey were of var. *julaceum* but there is an old record for var. *concinnatum* from the Allt nan Uamh by Warburg and Crundwell in 1960.

*Bryum pallens* Sw.

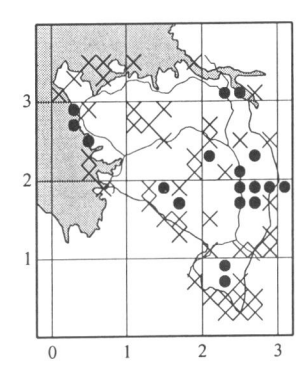

18 tetrads. A widespread but uncommon species of damp soil and gravel, probably most frequent on the limestone.

*Bryum weigelii* Spreng.

1 tetrad. A rare plant in Assynt and only recorded from a flush

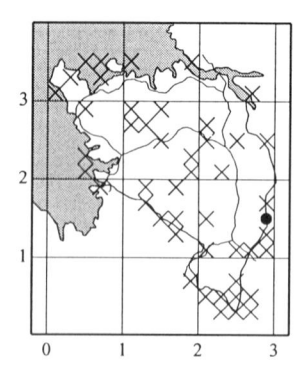

high on Meall Diamhain (Breabag).

*Bryum capillare* Hedw.

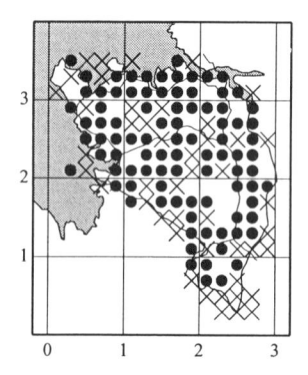

91 tetrads. A very common moss on mildly basic rocks and trees with nutrient-rich bark; it is most frequent on loch-side boulders and exposed coastal rocks. All plants seen on this survey were var. *capillare*.

*Bryum elegans* Nees ex Brid.

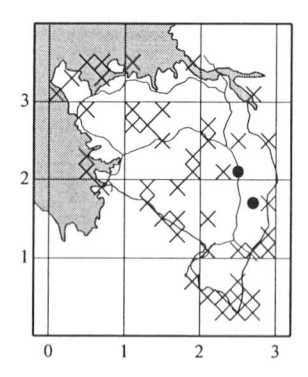

2 tetrads. A rare species forming neat, dense cushions on dry limestone rocks on the cliffs at Stronechrubie and Creag nan Uamh.

*Bryum pseudotriquetrum* (Hedw.) P.Gaertn., B.Mey. & Scherb.

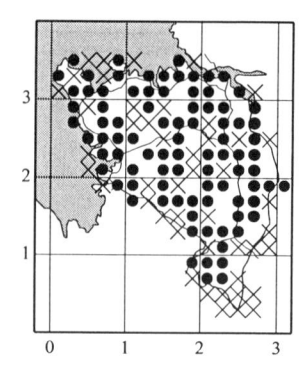

88 tetrads. A very common moss of wet rocks, flushes and wet soil where irrigation is at least mildly basic; the robust, often pinkish cushions can be a conspicuous feature of stony flushes. No attempt was made to distinguish var. *bimum*.

*Bryum argenteum* Hedw.

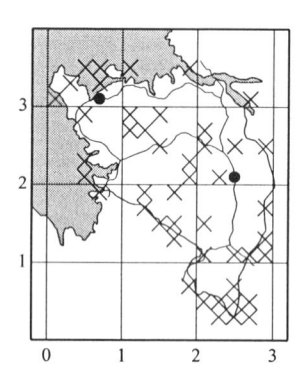

2 tetrads. A ruderal species which is not common but is certainly under-recorded; more work in man-made habitats will produce many more records.

*Bryum dixonii* Cardot ex W.E.Nicholson

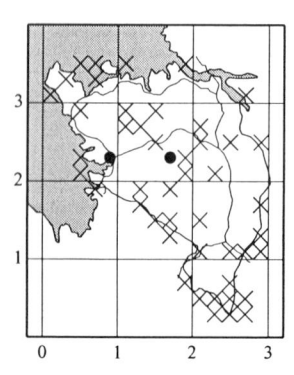

2 tetrads. This, apparently endemic, Scottish moss has two sites in Assynt, on basic gneiss rocks in the River Inver just above Lochinver and on similar rocks just below the big fall on the upper Allt an Tiaghaich where it occurs with *Schistidium agassizii*.

*Bryum bicolor* Dicks.

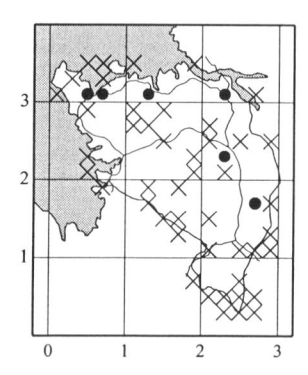

6 tetrads. Uncommon and usually occurring on the soil beside paths or roads; certainly under-recorded from man-made habitats.

*Bryum radiculosum* Brid.

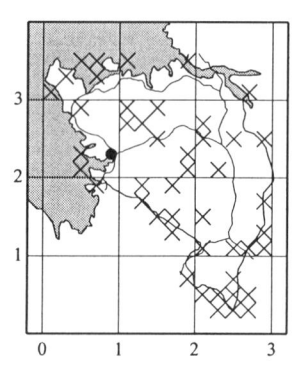

1 tetrad. Only recorded once in Assynt on this survey, from a mortared wall in Lochinver by Blockeel during the B.B.S meeting in 1992; there is an earlier record from Glenbain, Inchnadamph by Yeo in 1987.

*Bryum riparium* I.Hagen: see p.190

*Bryum mildeanum* Jur.

2 tetrads. A rare moss of limestone boulders by the Traligill

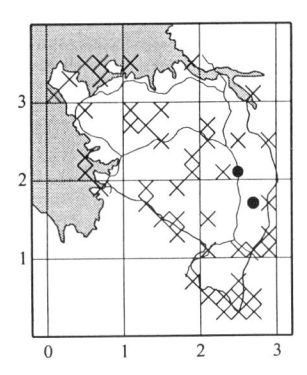

and the Allt nan Uamh; in limited quantity at both sites.

*Bryum muehlenbeckii* Bruch, Schimp. & W.Gümbel

There are old records for this distinctive moss from 'protruding rocks in streams' on Glas Bheinn and Conival by Dixon and Nicholson in 1899 (Nicholson, 1900), and Nicholson saw it again in 1921 in Gleann Dubh.

*Bryum alpinum* With.

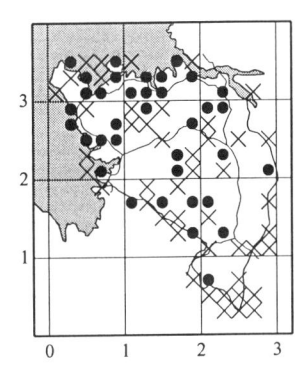

36 tetrads. The beautiful red cushions of this species are a frequent feature of seepage lines on slabby crags, particularly on the gneiss and usually avoiding the limestone.

*Rhodobryum roseum* (Hedw.) Limpr.

2 tetrads. This fine and distinctive moss is rare in Assynt with sites on a bank over calcareous rocks in the ravine of the Na Luirgean river below Loch Urigill and on a bank below a gneiss crag in Gleann Ardbhair. There is an older record from the Allt nan Uamh by Paton in 1960.

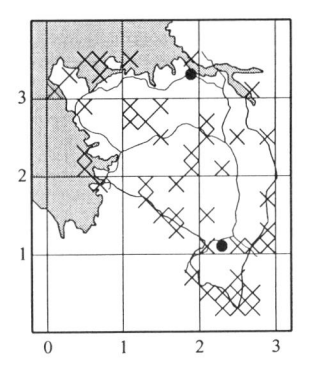

*Mnium hornum* Hedw.

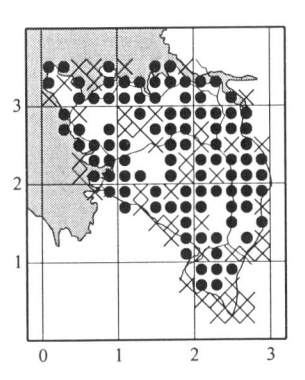

97 tetrads. A very common moss of soil and tree bases in woodland and ravines, and in crevices in scree and at the base of crags.

*Mnium thomsonii* Schimp.

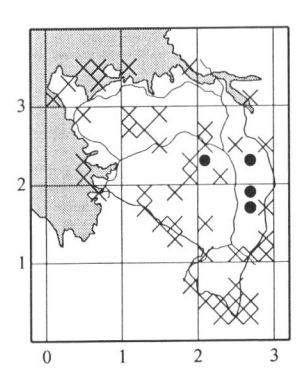

4 tetrads. Similar in appearance to the common *Mnium hornum* but a calcicole, occurring sparingly on dry ledges of limestone crags at Creag nan Uamh, Cnoc Eilid Mhathain, Allt Poll an Droighinn and with one site on the Torridonian on Beinn Garbh.

*Mnium stellare* Hedw.

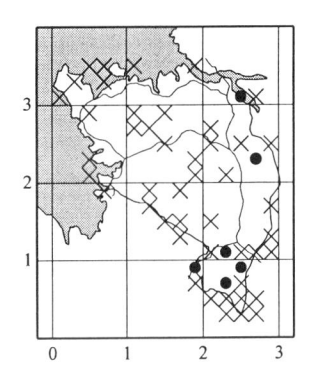

6 tetrads. An uncommon moss with a scattering of records on the limestone or Fucoid Beds where it occurs in dry crevices; there is an earlier record from the shell-sand at Achmelvich.

*Cinclidium stygium* Sw.

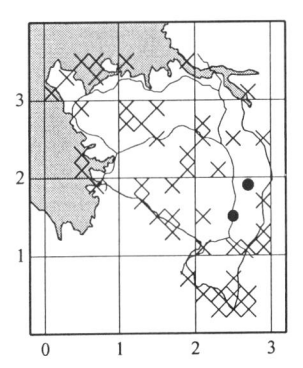

2 tetrads. This fine moss of wet calcareous ground is rare in the north of Scotland; the two Assynt sites are both in flushed turf on loch margins, by Loch Awe and by Loch Mhaolach-coire, where it was first seen by Dixon and Nicholson in 1899.

*Rhizomnium punctatum* (Hedw.) T.J.Kop.

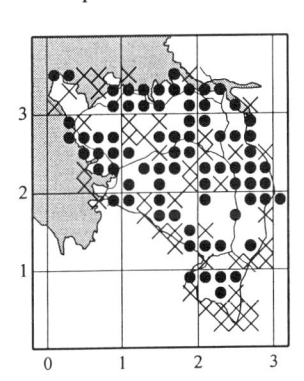

75 tetrads. A common and distinctive plant of various wet habitats, occurring on rock or soil, but perhaps most frequent on gravel by burns.

*Rhizomnium pseudopunctatum* (Bruch & Schimp.) T.J.Kop.

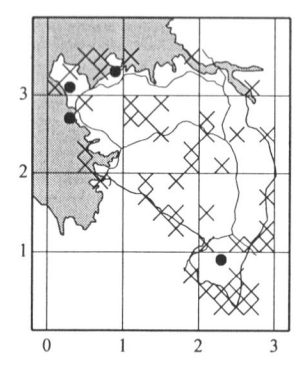

4 tetrads. A rare species in Assynt, occurring in flushed turf or marshy ground where the water is at least moderately base-rich: Balchladich, Cam Alltan, west of Clachtoll and Lon Ruadh (Oldany).

*Plagiomnium cuspidatum* (Hedw.) T.J.Kop.

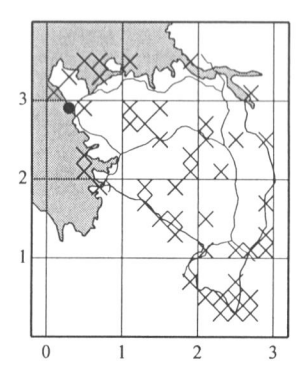

1 tetrad. Only recorded once on this survey, from the coast at Stoer. There is an old record from the Allt nan Uamh by Birks et al. in 1966.

*Plagiomnium elatum* (Bruch & Schimp.) T.J.Kop.

14 tetrads. Widespread in Assynt in but largely absent from the gneiss; it grows in marshy places, sometimes amongst reeds, or in flushed turf and is most frequent on the limestone.

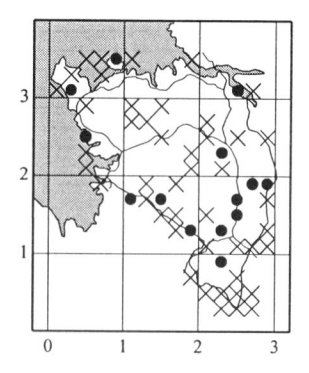

*Plagiomnium ellipticum* (Brid.) T.J.Kop.

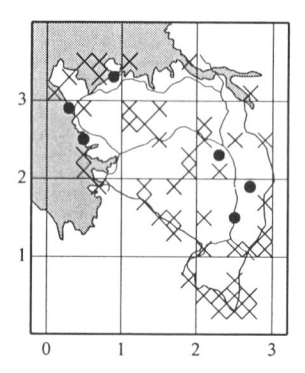

6 tetrads. In similar habitats to *Plagiomnium elatum* but generally much less common.

*Plagiomnium undulatum* (Hedw.) T.J.Kop.

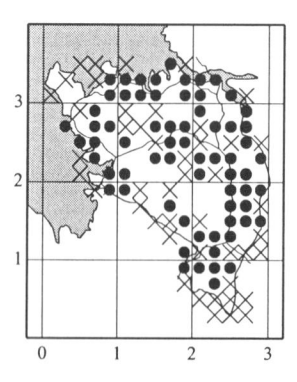

67 tetrads. A very common species of damp and moderately basic soils in grassland, crag bases and ledges, flushed banks and woodland.

*Plagiomnium rostratum* (Schrad.) T.J.Kop.

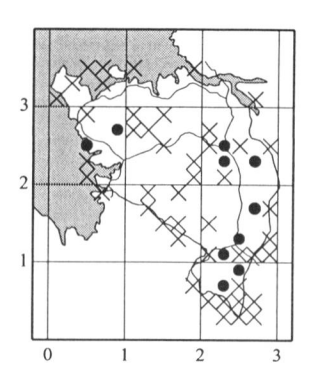

10 tetrads. A moss of good soils, usually in shaded, grassy sites and only at all frequent on the limestone.

*Pseudobryum cinclidioides* (Kindb.) T.J.Kop.

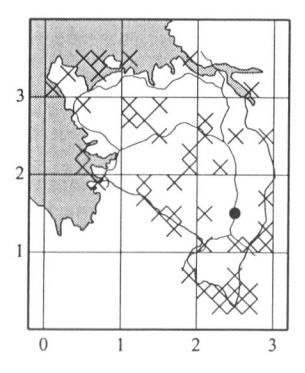

1 tetrad. Apparently very rare in Assynt with just one record from a flush on the shore of Loch Awe where it occurs with *Cinclidium stygium*.

*Aulacomnium palustre* (Hedw.) Schwägr.

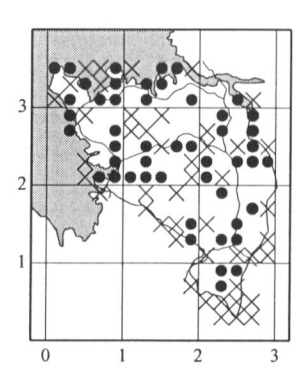

49 tetrads. A common and often abundant plant of a variety of wet,

peaty habitats such as rushy grassland, boggy woodland and with *Sphagnum* in mires.

*Aulacomnium turgidum* (Wahlenb.) Schwägr.

There is a record from the west side of Conival by Birks in 1967; this is a rare but characteristic species of exposed, species-rich turf on mountain tops in the north-west of Scotland.

*Aulacomnium androgynum* (Hedw.) Schwägr.

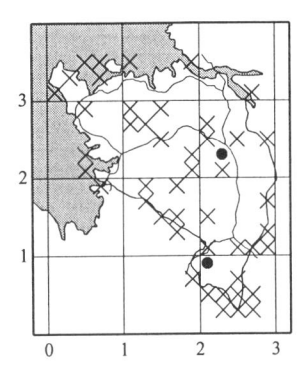

2 tetrads. A very rare moss in the north-west of Scotland, occurring in Assynt in sheltered crevices of quartzite crags at An Coimhleum and below Uamh an Tartair near Knockan.

*Meesia uliginosa* Hedw.

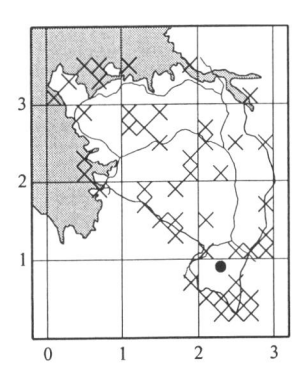

1 tetrad. Given the extent of flushed calcareous ground on the limestone, this is a curiously rare plant in Assynt. Only seen on this survey in flushed turf on the southern shore of Loch Urigill; there is an older record from the Allt nan Uamh by Birks et al. in 1966.

*Amblyodon dealbatus* Bruch & Schimp.

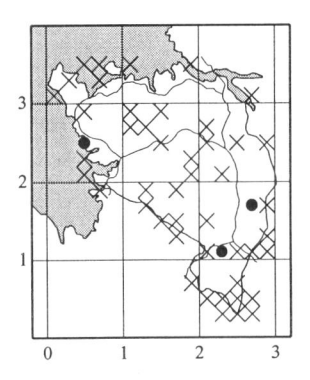

3 tetrads. A rare plant of flushed calcareous ground with sites by the resurgence in the lower Allt nan Uamh, by a waterfall on the Na Luirgean river below Loch Urigill and on damp calcareous dunes at Achmelvich.

*Plagiopus oederianus* (Sw.) H.A.Crum & L.E.Anderson (*P. oederi* (Brid.) Limpr.)

A record in the B.R.C list from the Allt na Uamh by Birks et al. in 1966 requires confirmation.

*Bartramia pomiformis* Hedw.

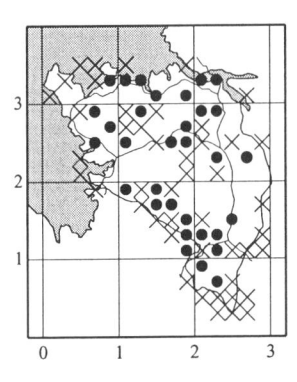

32 tetrads. Widespread and fairly common in crevices of dry acid crags and largely absent from the limestone.

*Bartramia ithyphylla* Brid.

27 tetrads. Very similar in appearance and abundance to *Bartramia pomiformis* but preferring slightly more basic rocks.

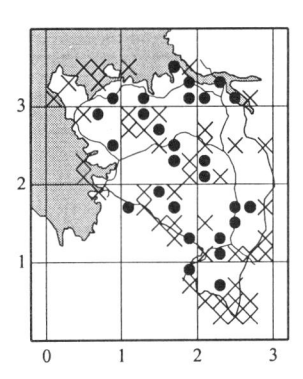

*Philonotis fontana* (Hedw.) Brid.

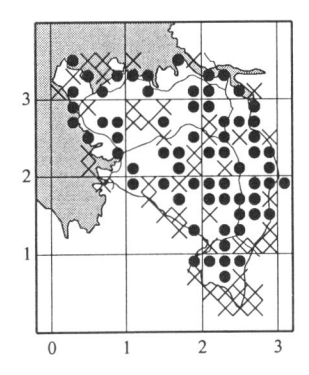

66 tetrads. Common and widespread in a variety of wet and flushed habitats, particularly in the higher hills; it is never abundant on the gneiss and is largely absent from the stony flushes with *Schoenus nigricans*.

*Philonotis seriata* Mitt.

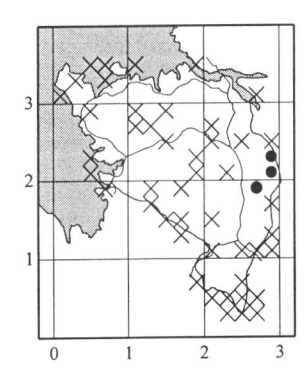

3 tetrads. A rare plant of montane flushes with sites on the western flanks of Conival, Beinn an Fhurain and above Loch Mhaolach-coire.

*Philonotis calcarea* (Bruch & Schimp.) Schimp.

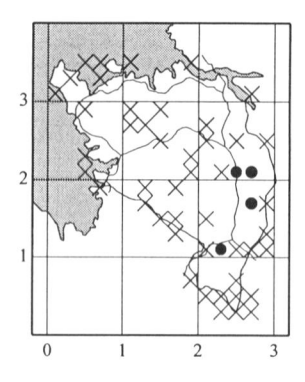

4 tetrads. An uncommon moss of flushes and flushed grassland on the limestone where some stands are extensive: Inchnadamph, Allt nan Uamh, Na Luirgean and by the Traligill.

*Breutelia chrysocoma* (Hedw.) Lindb.

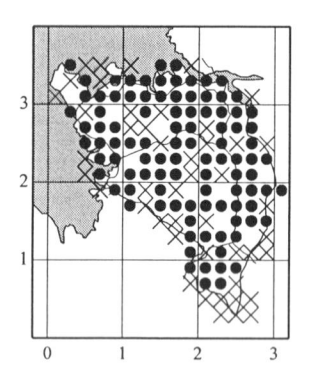

99 tetrads. This striking moss is common and often very abundant in wet grassland, wet heath, on ledges and in ravines, particularly where there is some slight flushing.

*Amphidium lapponicum* (Hedw.) Schimp.

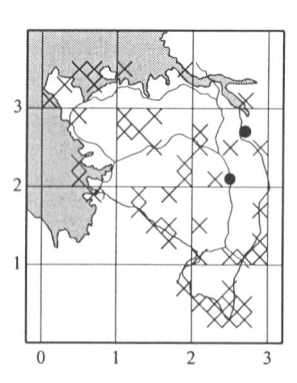

2 tetrads  Seen only once during this survey on base-rich gneiss crags above the outflow burn from Lochan a'Choire Ghuirm on Glas Bheinn; also recorded by Long from the crags above the Inchnadamph Hotel in 1973.

*Amphidium mougeotii* ((Bruch & Schimp.) Schimp.

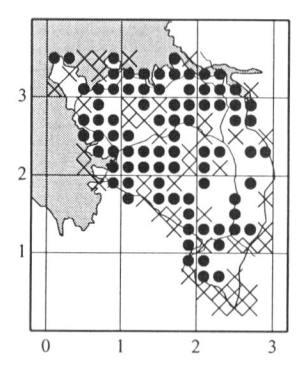

82 tetrads. A common moss of periodically irrigated cracks and crevices in acid or mildly basic crags, where the swelling, mid-green cushions can be a conspicuous feature.

*Zygodon viridissimus* (Dicks.) Brid. var. *viridissimus*

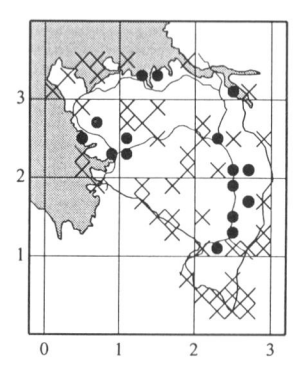

16 tetrads. An uncommon moss occurring as both an epiphyte on the trunks of trees with nutrient-rich bark and on shaded rocks.

*Zygodon viridissimus* var. *stirtonii* (Schimp. ex Stirt.) I.Hagen

1 tetrad. Only recorded from the limestone at Stronechrubie.

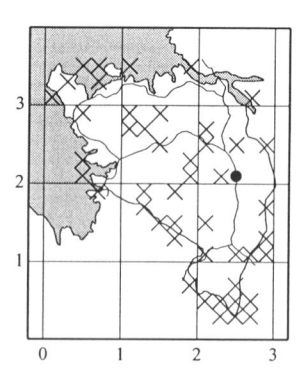

*Zygodon rupestris* Schimp. ex Lor. (Zygodon baumgartneri Malta)

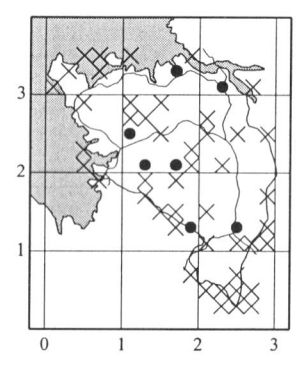

7 tetrads. A very sparse epiphyte on the trunks of trees with nutrient-rich bark, with one odd record from the open rocks of an ultra-basic dyke at Clach Airigh.

*Zygodon conoideus* (Dicks.) Hook. & Tayl.

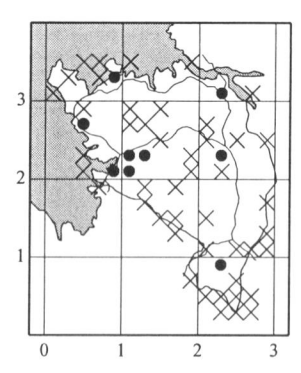

9 tetrads. Another uncommon epiphyte on the trunks of trees with nutrient-rich bark, its sparsity probably reflecting the shortage of such trees in Assynt.

*Orthotrichum affine* Brid.

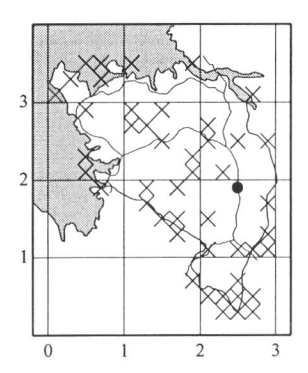

1 tetrad. Only recorded from one site, a large goat willow by the burn above Stronechrubie.

*Orthotrichum rupestre* Schleich. ex Schwägr.

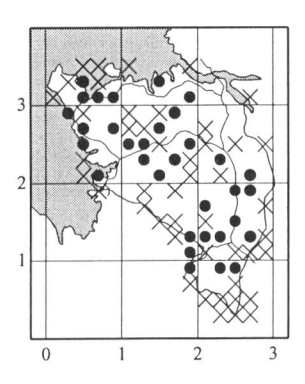

33 tetrads. Widespread but usually in small quantity, typically as part of the community on exposed, moderately basic rocks on loch margins but also on other open crags.

*Orthotrichum anomalum* Hedw.

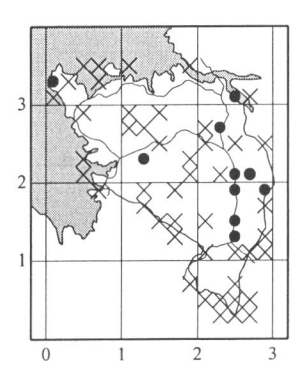

10 tetrads. Quite frequent on limestone boulders and low outcrops but rare elsewhere and seemingly avoiding the gneiss.

*Orthotrichum cupulatum* Brid.

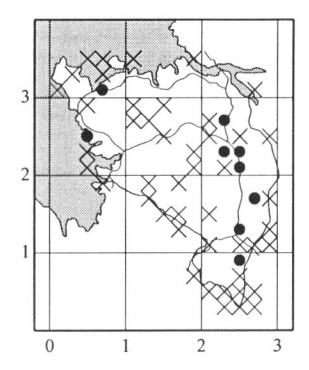

9 tetrads. Uncommon on low outcrops and slabby rocks on the limestone and very rare elsewhere; no attempt was made to distinguish the var. *riparium*.

*Orthotrichum stramineum* Hornsch. ex Brid.

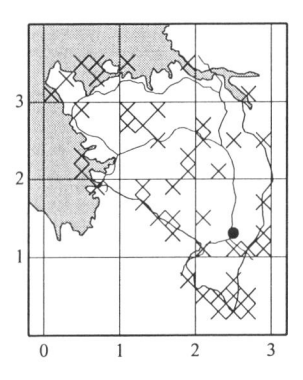

1 tetrad. Very rare; only recorded from a large wych elm by a waterfall on the Ledbeg River at Ledbeg, the only known West Sutherland site.

*Orthotrichum diaphanum* Brid.

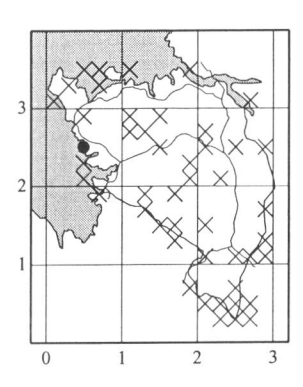

1 tetrad. Recorded by the B.B.S at Achmelvich in 1992 and in the past from elders at Inchnadamph by Nicholson in 1899 and the Allt nan Uamh by the B.B.S in 1960.

*Orthotrichum pulchellum* Brunt.

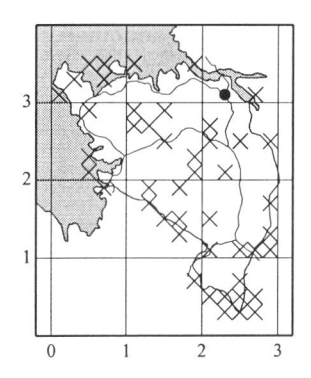

1 tetrad. As with most other epiphytic Orthotricaceae, this is a rare species with just one site on an old elder at Newton; in the past recorded from Inchnadamph by Nicholson in 1899 and by Paton in 1969.

*Ulota drummondii* (Hook. & Grev.) Brid

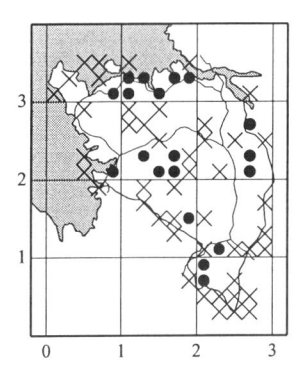

19 tetrads. Widespread but only very locally plentiful, occurring as spreading patches, particularly on rowan and sallow; possibly overlooked as it is only easily identified in the field when it is dry.

*Ulota crispa* (Hedw.) Brid.

62 tetrads. A common epiphyte on all trees, including tiny bushes of *Salix aurita* and occasionally on

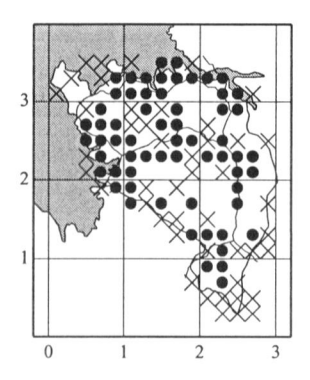

heather, but most frequent on hazel in richer woodland or ravines.

*Ulota bruchii* Hornsch. ex Brid.

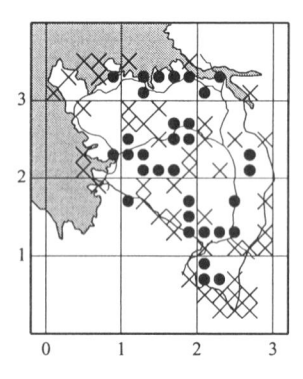

32 tetrads. A common epiphyte in similar situations to *Ulota crispa*, and often an early coloniser; only distinguished from *Ulota crispa* later in the survey and so is much under-recorded.

*Ulota calvescens* Wilson

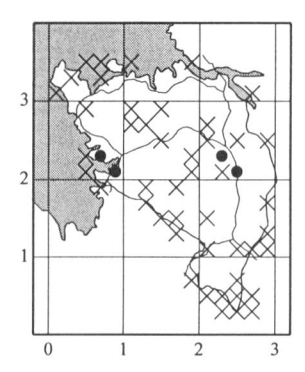

4 tetrads. A rare epiphyte in Assynt with records from hazel, ash and birch; not easily distinguished from *Ulota crispa* in the field, so possibly under-recorded: Inchnadamph (B.B.S.), Ardvreck, Culag Wood and Loch Dubh (Ardroe).

*Ulota hutchinsiae* (Sm.) Hammar

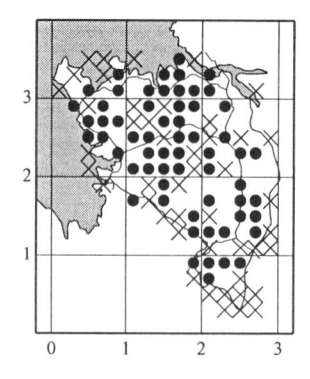

61 tetrads. A common moss on exposed rocks on crags and in scree but particularly frequent and occasionally abundant on rocks on loch margins; absent from the limestone.

*Ulota phyllantha* Brid.

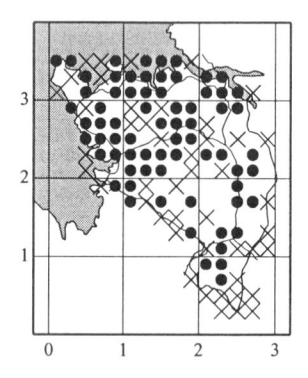

70 tetrads. A very common epiphyte particularly near the coast, where its swelling cushions with conspicuous brown gemmae can be a feature of the most exposed trees; it also frequently occurs on coastal rocks.

*Hedwigia ciliata* (Hedw.) P.Beauv. var. *ciliata*

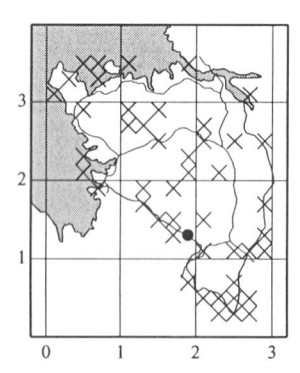

1 tetrad. Only recorded once in Assynt, on a Torridonian sandstone boulder on the margin of Loch Veyatie, where it had sporophytes and occurred in some abundance.

*Hedwigia stellata* Hedenäs

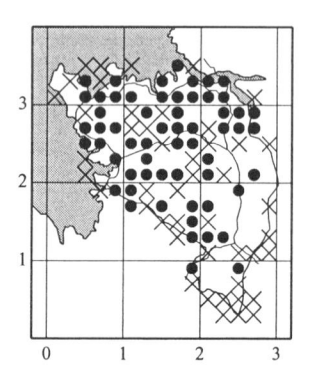

59 tetrads. A common and sometimes abundant species of acid or mildly basic rocks in exposed sites.

*Fontinalis antipyretica* Hedw.

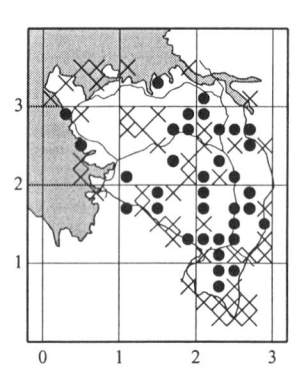

35 tetrads. A common species, growing on rocks in or just above the water on loch margins and the quieter parts of large burns, and more rarely on flushed rocks elsewhere. No attempt was made to distinguish any of the varieties.

*Fontinalis squamosa* Hedw.

9 tetrads. A rare moss in Assynt, usually growing on acid or mildly basic rocks in fast-flowing burns

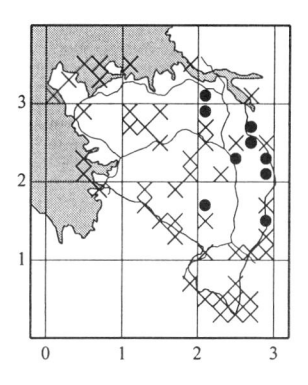

and largely confined to the higher hills in the east of the parish.

*Climacium dendroides* (Hedw.) F.Weber & D.Mohr

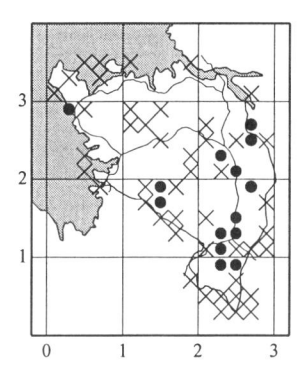

14 tetrads. Occasional in flushed turf on the margins of lochs and also in other damp turf on the limestone.

*Leucodon sciuroides* (Hedw.) Schwägr. var. *sciuroides*

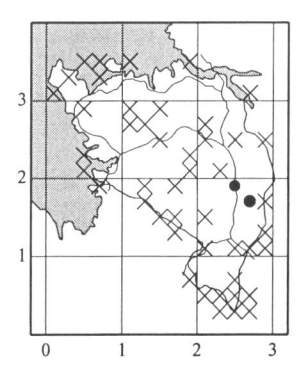

2 tetrads. A rare moss growing in large patches on limestone boulders by the Allt nan Uamh, and on old limestone walls near Stronechrubie where it is quite frequent.

*Antitrichia curtipendula* (Hedw.) Brid.

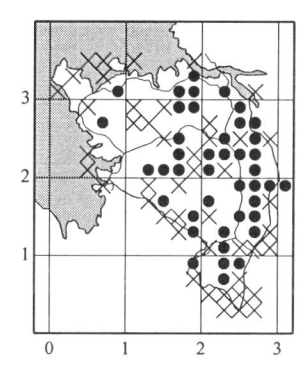

42 tetrads. This fine moss is relatively common in Assynt but never abundant, though some patches can be large; it is most frequent on rocks on the margin of lochs and only rarely occurs on trees in ravines where it occasionally produces sporophytes.

*Pterogonium gracile* (Hedw.) Sm.

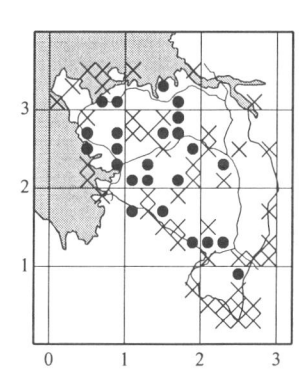

24 tetrads. Widespread in the parish; usually occurring in small quantity and almost exclusively on mildly basic rocks on the margins of lochs.

*Myurium hochstetteri* (Schimp.) Kindb.

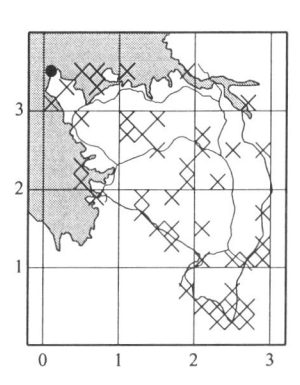

1 tetrad. This beautiful hyper-oceanic moss has one of its few mainland sites on the Point of Stoer; here it occurs in two places in exposed, flushed grassland above the sea-cliffs.

*Neckera crispa* Hedw.

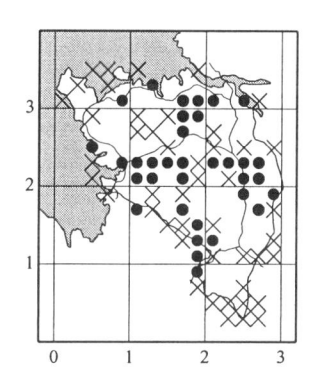

34 tetrads. The pendant mats of this robust moss are a frequent feature of sheltered basic crags, often in ravines, and it also occurs in *Dryas* heath on the limestone.

*Neckera complanata* (Hedw.) Huebener

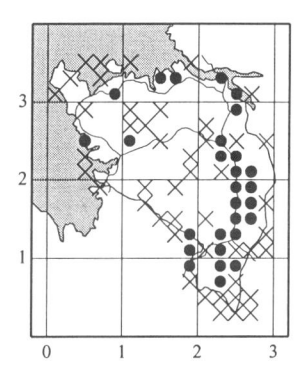

28 tetrads. In similar sheltered, craggy habitats to *Neckera crispa* but seemingly much less frequent away from the limestone; in Assynt it is very rare as an epiphyte.

*Homalia trichomanoides* (Hedw.) Bruch, Schimp. & W.Gümbel

5 tetrads. Rare and invariably in small quantity; it occurs on soil, tree roots or bases and on faces of sheltered rocks that are periodically inundated by burns or

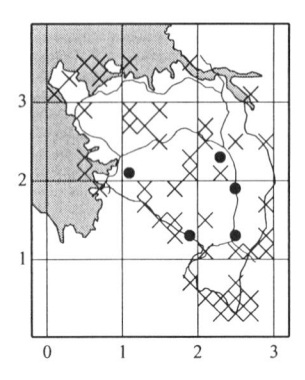

lochs: Loch Druim Suardalain, Allt Mór (Stronechrubie), falls on Ledbeg River, Ardvreck and the north shore of Loch Veyatie.

*Thamnobryum alopecurum* (Hedw.) Gangulee

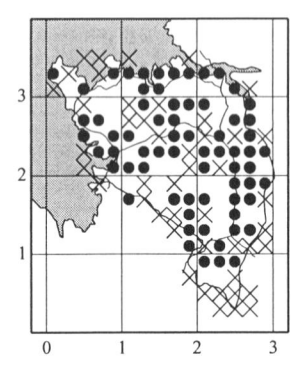

67 tetrads. This robust moss is common in two habitats, on rocks in and by falls in burns and rivers and on sheltered, vertical, basic rock faces and in both it can form extensive patches.

*Pterigynandrum filiforme* Hedw.

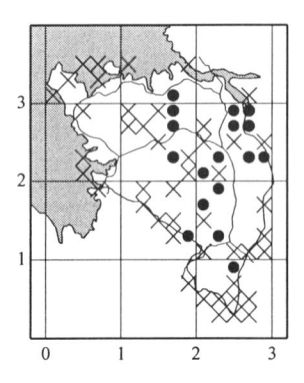

17 tetrads. The neat, brownish-green patches of this inconspicuous moss are an occasional constituent of the community on loch-side boulders, particularly where the rock is somewhat base-rich. No attempt was made to distinguish the var. *majus*, which is probably only a habitat modification.

*Hookeria lucens* (Hedw.) Sm.

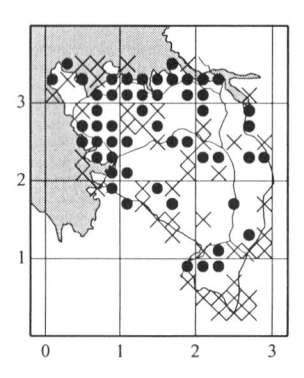

52 tetrads. A common species of damp, acid soils in sheltered sites, particularly in woodland but also in ravines and rarely in crevices at the base of montane crags.

*Pseudoleskeella catenulata* (Schrad.) Kindb.

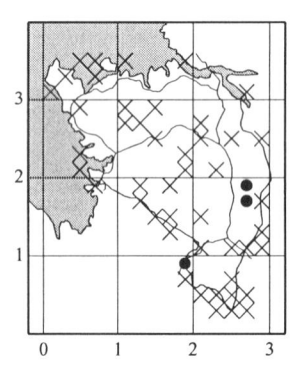

3 tetrads. A rare plant forming dark, olive-green patches on open limestone rocks; it occurs on Cnoc Eilidh Mhathain, on Creag nan Uamh, where it is very locally frequent, and on the crags near the road at Knockan.

*Pseudoleskeella rupestris* (Berggr.) Hedenäs & Söderström

2 tetrads. This nationally rare moss has two sites in Assynt, with *Pseudoleskeella catenulata* on Creag nan Uamh, and on the low limestone crags at Lairig Unapool; it is similar to *Pseudoleskeella catenulata* but prefers rather more sheltered sites. There is also a 1960 record from Knockan Crag.

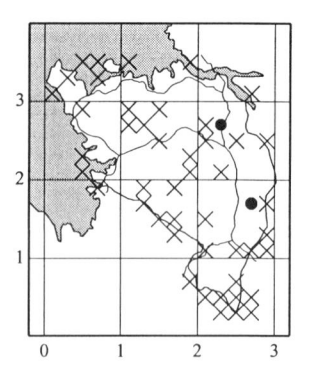

*Anomodon viticulosus* (Hedw.) Hook. & Taylor

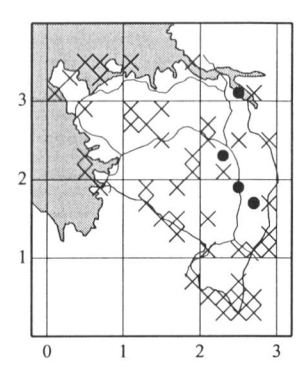

4 tetrads. This is a rare plant in Assynt, which is surprising given the extent of seemingly suitable habitat on the limestone; it occurs in sheltered niches at the base of calcareous crags or large boulders in the Allt nan Uamh, Ardvreck, Liath Bhad and Stronechrubie.

*Heterocladium heteropterum* var. *heteropterum* Bruch, Schimp. & W.Gümbel

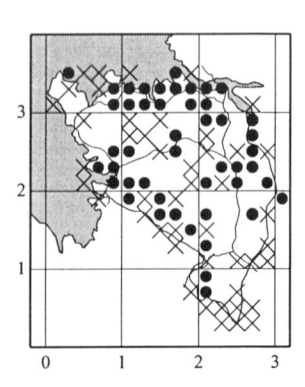

47 tetrads. A common moss

forming dark wefts in sheltered sites on rocks in burns and damp crags.

*Heterocladium heteropterum* var. *flaccidum* Bruch, Schimp. & W.Gümbel

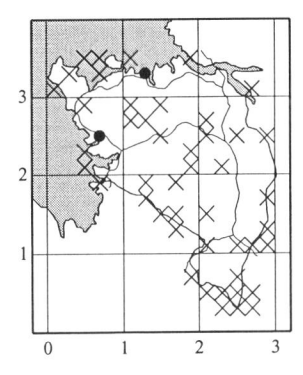

2 tetrads. In similar sites to var. *heteropterum* but much rarer and usually on somewhat basic rock; Duart (Nedd) and Loch Dubh (Ardroe).

*Thuidium tamariscinum* (Hedw.) Bruch, Schimp. & W.Gümbel

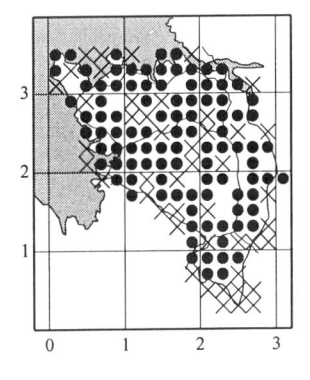

100 tetrads. A common and often abundant species in drier places in rocky woodland, on sheltered banks, montane grassland and it can be dominant in the ground layer under bracken.

*Thuidium delicatulum* (Hedw.) Mitt.

33 tetrads. A frequent moss in flushed stony grassland and on low rocks where there is some seepage; it may well have been overlooked as *Thuidium tamariscinum* elsewhere.

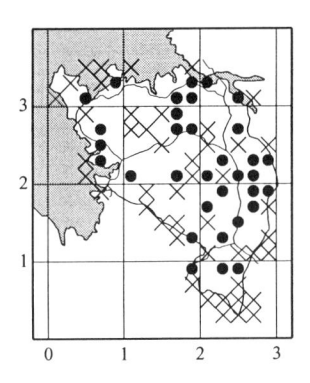

*Thuidium philibertii* Limpr.

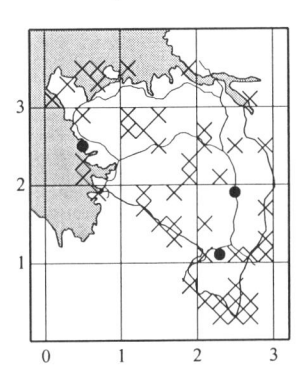

3 tetrads. Seemingly a rare moss in Assynt, occurring in dry, calcareous grassland on the limestone at Knockan and Stronechrubie and on shell-sand at Achmelvich, with an old record from limestone near Cùil Dhubh by Warburg in 1952.

*Thuidium recognitum* (Hedw.) Lindb.

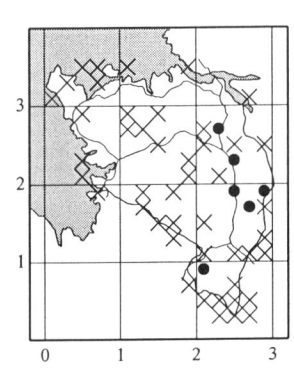

6 tetrads. An uncommon species growing in dry grassland on shallow soils over the limestone; apparently more frequent here than the very similar *Thuidium philibertii*.

*Palustriella commutata* var. *commutata* (Hedw.) Ochyra (Cratoneuron commutatum (Hedw.) Roth.)

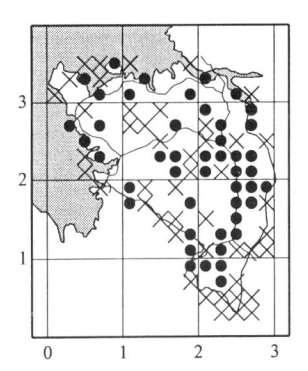

46 tetrads. A characteristic moss of seepage lines and springs on basic rocks, often abundant on the limestone where it forms tufa, but also frequent on the base-rich gneiss.

*Palustriella commutata* var. *falcata* (Brid.) Ochyra

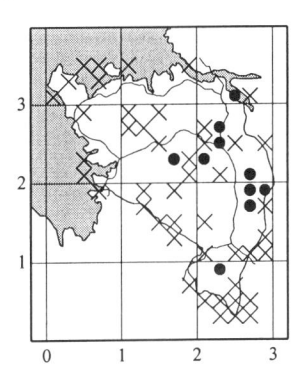

10 tetrads. Locally common in flushes on the limestone, but rare elsewhere and much less common than var. *commutata*.

*Palustriella commutata* var. *virescens* (Schimp.) Ochyra

1 tetrad. A poorly marked taxon recorded from seepage lines below

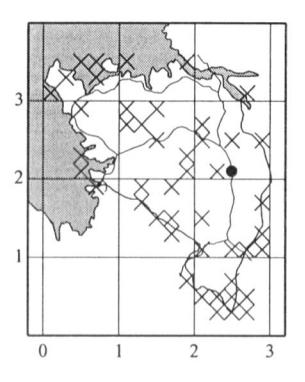

the main Stronechrubie crags.

*Cratoneuron filicinum* (Hedw.) Spruce

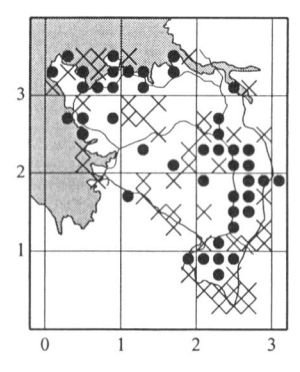

44 tetrads. A frequent moss in a variety of damp basic habitats; perhaps most common in seepage lines on the limestone, but also plentiful on damp coastal rocks.

*Campylium stellatum* var. *stellatum* (Hedw.) J.Lange & C.Jens.

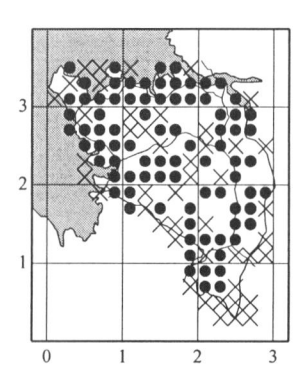

84 tetrads. A common species of wet places that are at least moderately base-rich; often abundant in marshy grassland, at the base of wet crags and is one of the few bryophytes that flourishes in *Schoenus* mires.

*Campylium stellatum* var. *protensum* (Brid.) Bryhn.

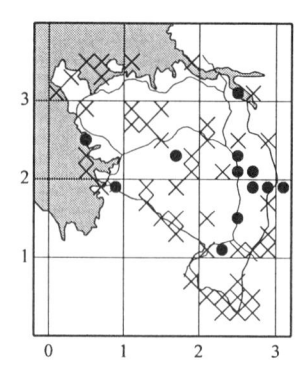

12 tetrads. Usually in drier places than var. *stellatum* and only at all frequent on damp ledges on limestone crags.

*Campyliadelphus chrysophyllus* (Brid.) Kanda
(*Campylium chrysophyllum* (Brid.) J.Lange)

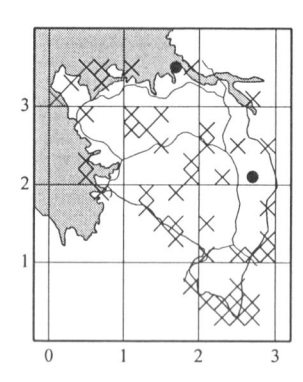

2 tetrads. Rare on ledges on base-rich crags; Ardvar and the Traligill glen.

*Campyliadelphus elodes* (Lindb.) Kanda
(*Campylium elodes* (Lindb.) Kindb.)

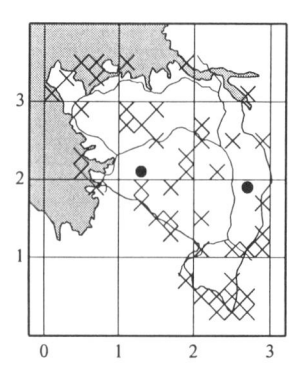

2 tetrads. A rare moss of wet calcareous soil occurring in flushed turf by Loch Mhaolach-coire and on ledges on basic gneiss in a ravine on the Abhainn Bad na h-Achlaise.

*Amblystegium serpens* (Hedw.) Bruch, Schimp. & W.Gümbel

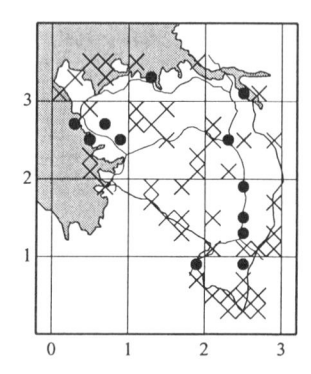

12 tetrads. Widespread but distinctly uncommon, growing as flat patches on damp, basic rocks and also in damp coastal turf.

*Amblystegium tenax* (Hedw.) C.E.O.Jensen

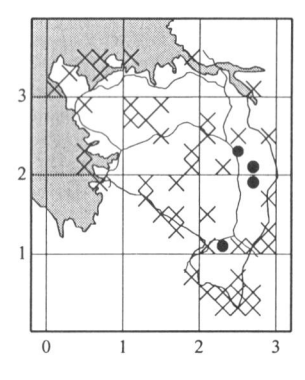

4 tetrads. A rare moss of boulders in burns on the limestone where it occurs as untidy, partially denuded patches; recorded from the Allt a' Chalda Mór, Na Luirgean, Cnoc Eilid Mhathain and the Traligill.

*Conardia compacta* (Müll.Hal.) H.Rob.
(*Amblystegium compactum* (Müll.Hal.) Austin)

2 tetrads. This rare moss has two sites in Assynt, both on damp soil on heavily shaded ledges on the

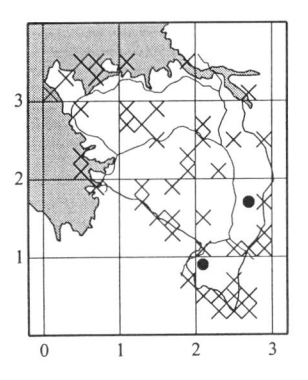

limestone, one in caves on Creag na Uamh and the other near Uamh an Tartair at Knockan.

*Warnstorfia fluitans* (Hedw.) Loeske
(Drepanocladus fluitans (Hedw.) Warnst. Hedw.)

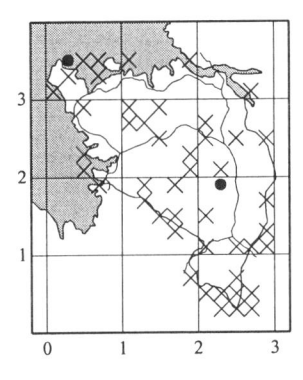

2 tetrads. Recorded from peaty pools in a mire area on the Point of Stoer and the edge of Loch nam Meallan Liatha.

*Warnstorfia exannulata* (Bruch, Schimp. & W.Gümbel) Loeske
(Drepanocladus exannulatus (Bruch, Schimp. & W.Gümbel) Warnst.)

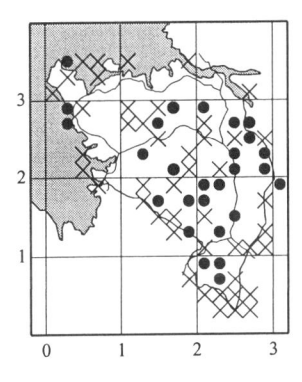

27 tetrads. Widespread over the parish in flushes and more base-rich mire pools, but not at all common, with the best populations restricted to the higher ground in the east.

*Drepanocladus polygamus* (Bruch, Schimp. & W.Gümbel) Hedenäs
(Campylium polygamum (Bruch, Schimp. & W.Gümbel) J.Lange & Jensen)

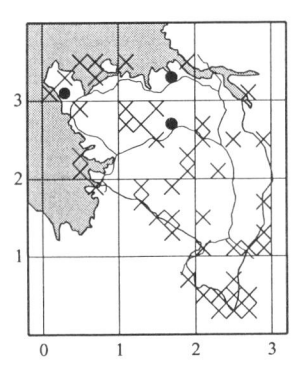

3 tetrads. Rare but possibly overlooked in wet, coastal grassland and in poor fen at the margin of lochs: Loch Ardbhair, Balchladich and Loch Bad a'Chigean.

[*Drepanocladus aduncus* (Hedw.) Warnst.

Despite records in B.R.C. and thus in the *Bryophyte Atlas* there is no verified record of this taxon in Assynt or elsewhere in v.c.108.]

*Drepanocladus revolvens* (Sw.) Warnst.

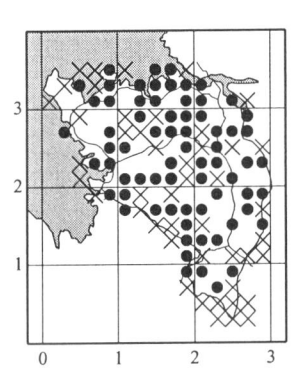

69 tetrads. Common in stony flushes and in *Schoenus* mires, often with *Scorpidium scorpioides*, and more occasional on the margin of flushed peaty pools.

*Drepanocladus cossonii* (Schimp.) Loeske

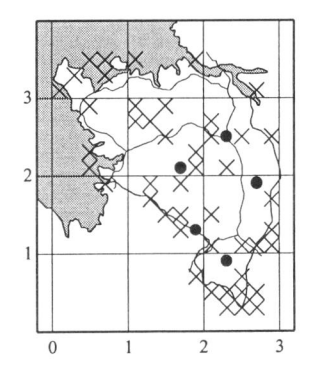

5 tetrads. Apparently rare in Assynt, in similar, but more obviously base-rich, sites as *D. revolvens*; *D. cossonii* was distinguished from *D.revolvens* only during the latter part of the survey and is recorded from a tributary of the Allt Sgiathaig, in a flush at Clach Airigh, at Loch Mhaolachcoire, on the shore of Loch Veyatie and at three sites on the Cam Allt.

*Sanionia uncinata* (Hedw.) Loeske
(Drepanocladus uncinatus (Hedw.) Warnst.)

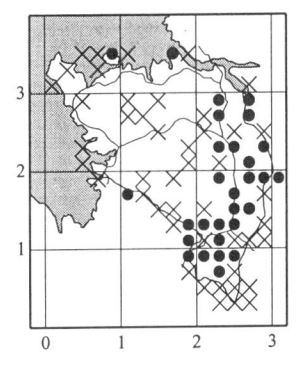

29 tetrads. Locally frequent on the calcareous Cambrian rocks, and occasionally on willows in the east of the parish but very rare on the gneiss.

*Hygrohypnum ochraceum* (Turner ex Wilson) Loeske

13 tetrads. An uncommon riparian

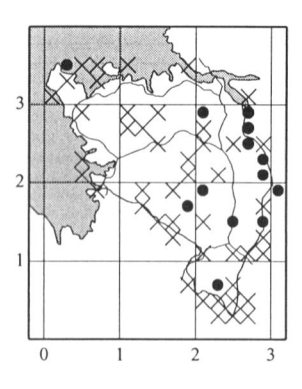

moss, growing on rocks in burns or on the margin of lochs; most records are on the higher ground in the east of the county and usually on acid rocks.

*Hygrohypnum luridum* (Hedw.) Jenn. var. *luridum*

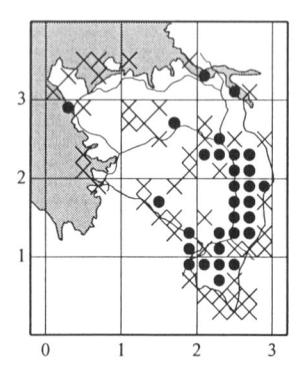

30 tetrads. Common and often abundant on wet rocks on the limestone and Fucoid Beds with rare sites on basic rocks elsewhere. There is a curious small form on damp rocks on the limestone with non-falcate leaves and julaceous stems rather reminiscent of *Pterigynandrum filiforme*.

*Hygrohypnum eugyrium* (Schimp.) Broth.

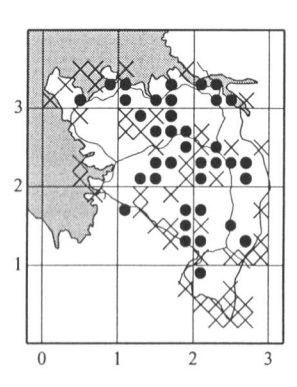

38 tetrads. The yellow-green mats of this moss, often with a brownish tinge, are a frequent feature of flat rocks in burns, particularly in ravines, and on loch margins.

*Hygrohypnum duriusculum* (De Not.) Jamieson
(H. dilatatum (Wils. ex Schimp.) Loeske)

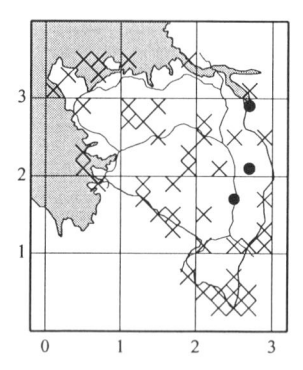

3 tetrads. This nationally scarce species has at least three sites in Assynt but is easily overlooked as *Rhynchostegium riparioides*; it is locally frequent on rocks in the lower part of the Allt nan Uamh, occurs in similar places on the Traligill and there is a small population on rocks by Loch Bealach a'Bhuirich.

*Scorpidium scorpioides* (Hedw.) Limpr.

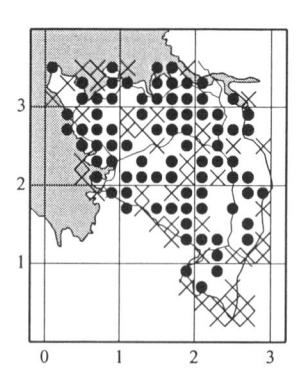

81 tetrads. Common and locally abundant in stony flushes, on loch margins and in wet ground by more basic mire pools; one of the few bryophytes to flourish in *Schoenus nigricans* mires.

*Calliergon stramineum* (Brid.) Kindb.

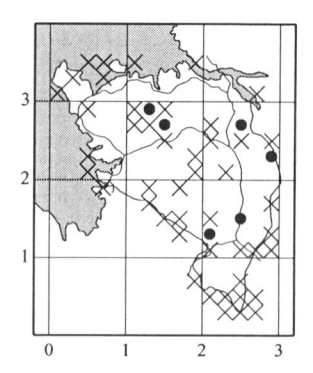

6 tetrads. Surprisingly rare with a few, well-scattered sites in wet, basic grassland and mires.

*Calliergon trifarium* (Web. & Mohr) Kindb.

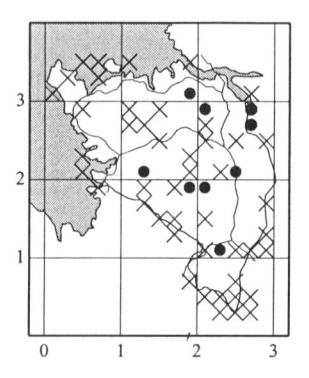

9 tetrads. A distinctive and uncommon arctic-alpine moss of base-rich stony flushes with most sites on the higher ground and which is, perhaps surprisingly, scarce on the limestone.

*Calliergon cordifolium* (Hedw.) Kindb.

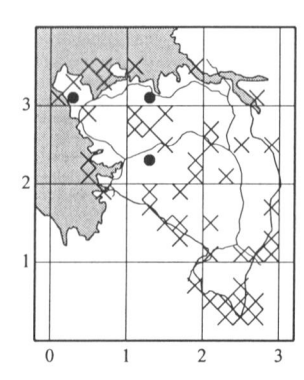

3 tetrads. Seemingly a rare plant in Assynt with scattered sites at

Balchladich, Nedd and by the Allt nan h-Airbhe.

*Calliergon giganteum* (Schimp.) Kindb.

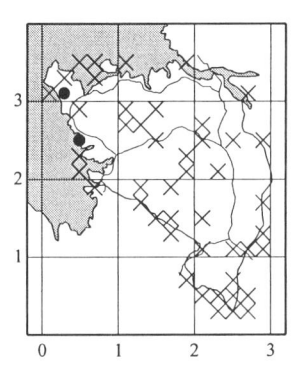

2 tetrads. Very rare and limited to the margins of Loch na Claise at Balchladich, recorded by Stern and a lochan at Achmelvich, both on the B.B.S meeting in 1992.

*Calliergon sarmentosum* (Wahlenb.) Kindb.

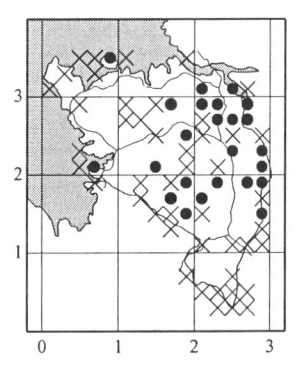

24 tetrads. Widespread but sparse on flushed ground in the higher hills in the east of the parish and very rare elsewhere.

*Calliergonella cuspidata* (Hedw.) Loeske
(*Calliergon cuspidatum* (Hedw.) Kindb.)

80 tetrads. Common and often abundant in a variety of wet habitats including damp grassland, grassy flushes, loch margins, tracksides, base of crags and ditches.

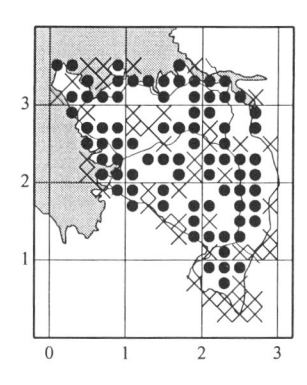

*Isothecium myosuroides* Brid. var. *myosuroides*

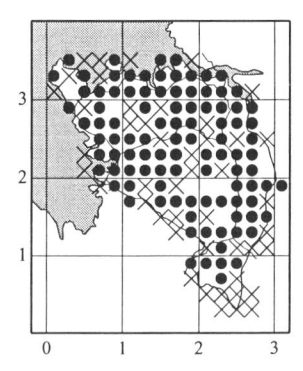

100 tetrads. Very common and often locally dominant on rocks and trees in woodland, scree, on crags, stone walls and coastal rocks. The deep mats of this moss, often with *Plagiochila spinulosa* and *Hymenophyllum wilsonii* are the characteristic community of the best oceanic, rocky woodland.

*Isothecium myosuroides* var. *brachythecioides* (Dixon) Braithw.

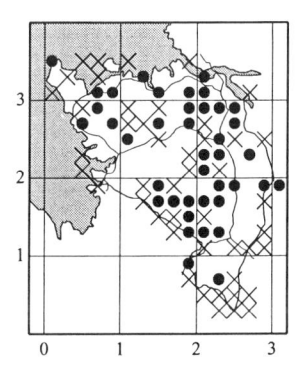

40 tetrads. A frequent moss on sheltered, acid rocks in woodland and on crags, the lax mats looking very different to var. *myosuroides* with which it often grows.

*Isothecium alopecuroides* (Dubois) Isov.
(*I. myurum* Brid.)

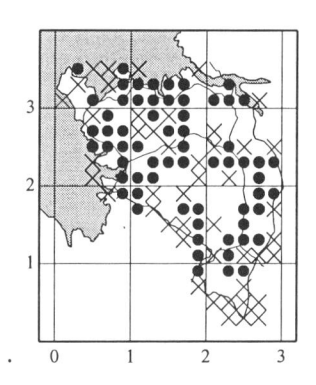

65 tetrads. Common and locally abundant on at least moderately basic rocks and tree bases; often a frequent feature of base-rich boulders on the margin of lochs.

*Homalothecium sericeum* (Hedw.) Bruch, Schimp. & W.Gümbel

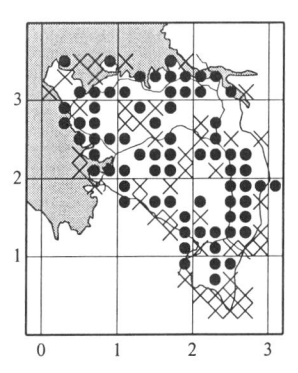

72 tetrads. This attractive moss is common on dry, base-rich rocks and also occurs on large elms and ash trees; it is particularly abundant on the limestone walls in the Inchnadamph area.

*Homalothecium lutescens* (Hedw.) H.Rob.

6 tetrads. This is an uncommon species, absent from much of Assynt but locally dominant on

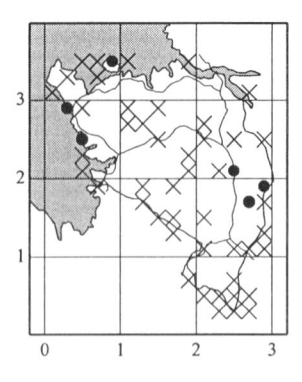

shell-sand in dunes and also occurs in dry limestone grassland.

*Brachythecium albicans* (Hedw.) Bruch, Schimp. & W.Gümbel

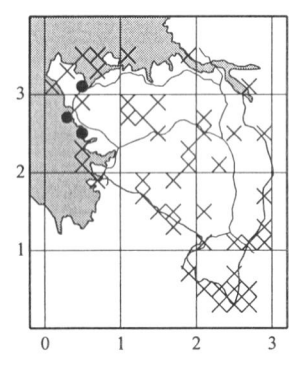

3 tetrads. Apparently limited to three areas of shell-sand dunes, at Achmelvich, Clachtoll and Clashnessie, where it is locally abundant.

*Brachythecium glareosum* (Spruce) Bruch, Schimp. & W.Gümbel

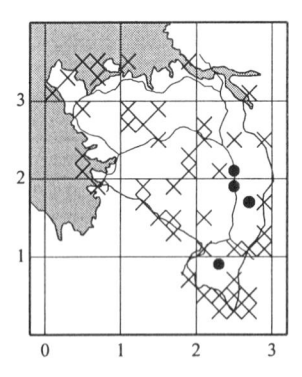

4 tetrads. A rare plant in Assynt, limited to limestone grassland in the Allt nan Uamh valley, at Stronechrubie and the limestone at Knockan.

*Brachythecium rutabulum* (Hedw.) Bruch, Schimp. & W.Gümbel

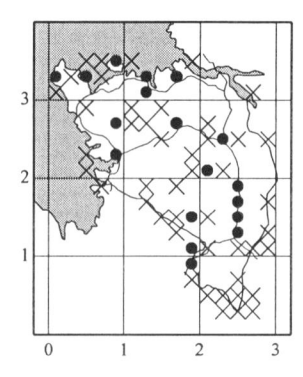

18 tetrads. An uncommon species in Assynt though it can be locally frequent in disturbed, eutrophic places by roads and in gardens or where livestock gather.

*Brachythecium rivulare* Bruch, Schimp. & W.Gümbel

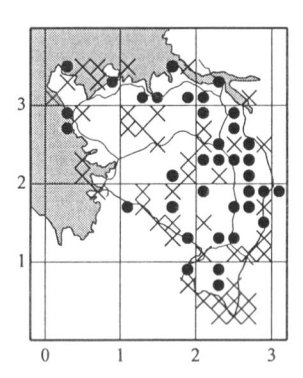

36 tetrads. A widespread species, particularly frequent on the limestone, growing on wet rocks by burns, in flushes, wet woodland or in marshy grassland.

*Brachythecium populeum* (Hedw.) Bruch, Schimp. & W.Gümbel

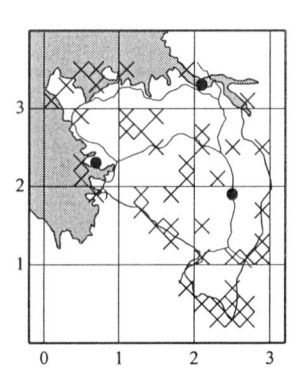

3 tetrads. An uncommon species in the north of Scotland with three records in Assynt from dry, sloping, moderately basic rocks at Tòrr a'Ghamhna, Loch Dubh (Ardroe) and Allt Mór (Stronechrubie).

*Brachythecium plumosum* (Hedw.) Bruch, Schimp. & W.Gümbel

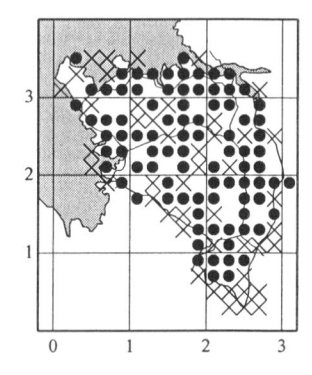

88 tetrads. Very common and often abundant, as part of a characteristic community of acid rocks in burns, with *Racomitrium aciculare*, *Scapania undulata* and *Marsupella emarginata*, but also occurring on loch margins and on wet crags.

*Scleropodium purum* (Hedw.) Limpr.
(Pseudoscleropodium purum (Hedw.) Fleisch. ex Broth.)

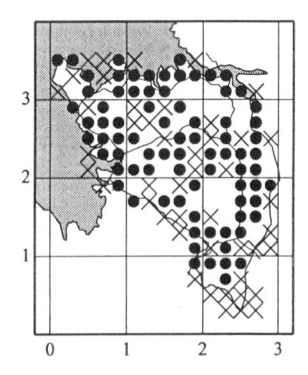

80 tetrads. Very common and locally abundant in grassland on soils that are not too acid; it can form large stands under bracken and is probably most frequent on the limestone.

*Cirriphyllum piliferum* (Hedw.) Grout

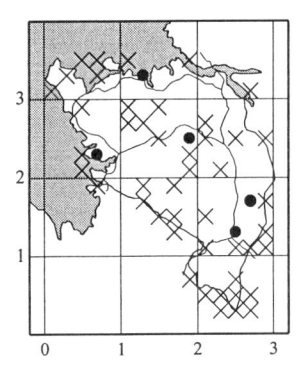

5 tetrads. A rare plant in Assynt, growing on richer soils in sheltered sites, often with *Eurhynchium striatum*: Duart (Nedd), Allt nan Uamh, Loch Dubh (Ardroe), Ledbeg River and Easter Tubeg.

*Rhynchostegium riparioides* (Hedw.) Cardot

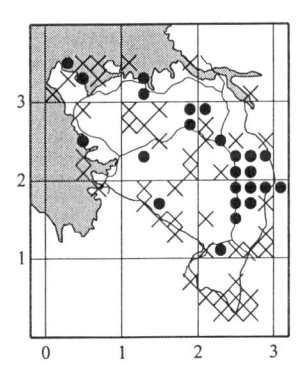

24 tetrads. A robust species of rocks in burns, which is surprisingly uncommon over much of Assynt, where it seems to avoid the gneiss, but it is abundant on the limestone.

*Rhynchostegium alopecuroides* (Brid.) A.J.E.Sm.
(R. lusitanicum (Schimp.) A.J.E.Sm.)

4 tetrads. This nationally scarce species has its most northerly localities in the world in Assynt where it occurs on Torridonian rocks in waterfalls in the Bealach Leireag and in and below the Sanctuary, both on the slopes of Quinag.

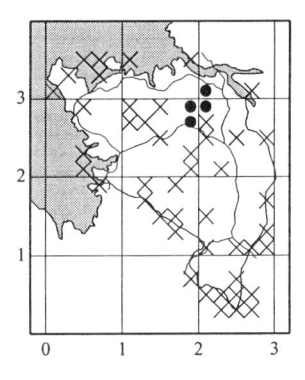

*Eurhynchium striatum* (Hedw.) Schimp.

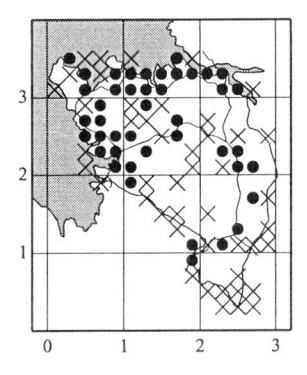

43 tetrads. Locally common on at least moderately basic soils and rocks in woodland, particularly under hazel, and also in sheltered sites in ravines and in limestone grassland.

*Eurhynchium pumilum* (Wilson) Schimp.

There are two old records for this species, which is very rare in the north of Scotland: limestone rocks by caves near Inchnadamph by Wallace in 1946 and from the Allt nan Uamh by Birks et al. in 1966.

*Eurhynchium praelongum* (Hedw.) Bruch, Schimp. & W.Gümbel

65 tetrads. Widespread but only locally frequent in a variety of grassy places in woodland, on dunes, tree bases and banks in ravines.

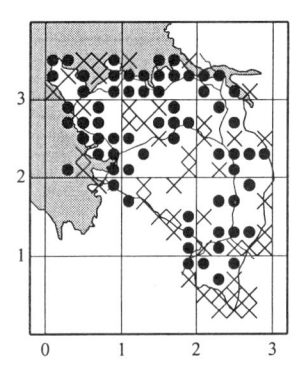

*Eurhynchium hians* (Hedw.) Sande Lac.
(E. swartzii (Turn.) Curn. in Rabenh.)

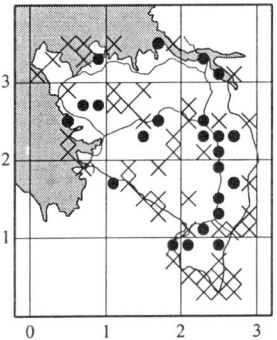

23 tetrads. Widespread but only at all frequent on the limestone and then usually in small stands; usually in deeply shaded crevices or on soil of sheltered ledges on crags.

*Eurhynchium crassinervium* (Wilson) Schimp.
(Cirriphyllum crassinervium (Wils.) Loeske & M.Fleisch.)

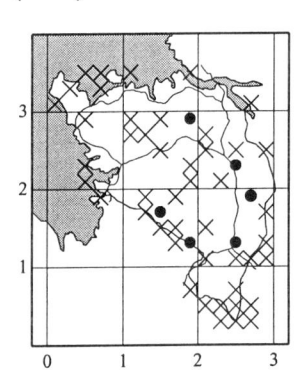

6 tetrads. A rare species in Assynt

251

with scattered sites on basic gneiss or limestone boulders in shaded sites, often near burns.

*Rhynchostegiella tenella* (Dicks.) Limpr.

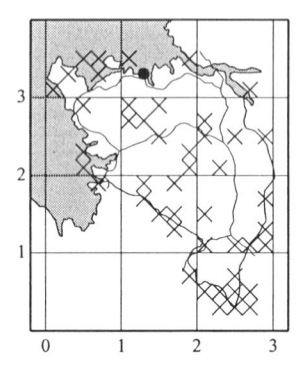

1 tetrad. Only recorded from one site in Assynt, a mortared wall in Nedd and seemingly absent from the limestone.

*Rhynchstegiella teneriffae* (Mont.) Dirkse & Bouman
(*R. teesdalei* (Bruch, Schimp. & W.Gümbel) Limpr.)

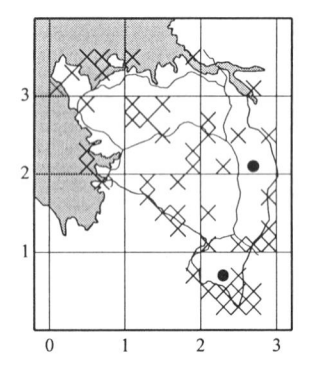

2 tetrads. Very rare with only two records, on wet limestone rocks, by the Traligill and the Allt a'Phollain behind Knockan.

*Entodon concinnus* (De Not.) Paris

15 tetrads. Quite common and locally abundant on the shell-sand on the coast and in dry limestone grassland, but absent elsewhere.

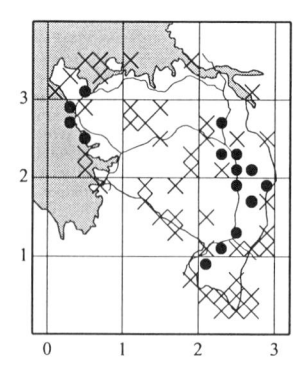

*Pleurozium schreberi* (Brid.) Mitt.

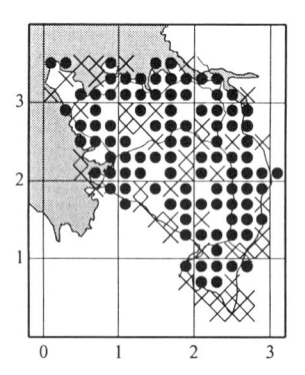

99 tetrads. Very common and often abundant, particularly on heathy ground with *Hylocomium splendens*, but also in scree, woodland, drier parts of mires and acid grassland.

*Platydictya jungermannioides* (Brid.) H.A.Crum

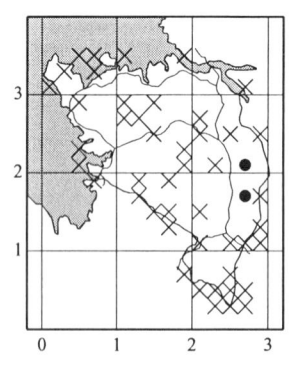

2 tetrads. Very rare, growing as fine, intricate wefts on soil on shaded ledges of limestone crags on Creag nan Uamh and in the Traligill valley.

*Orthothecium rufescens* (Brid.) Bruch, Schimp. & W.Gümbel

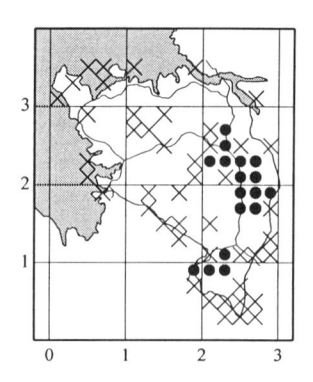

17 tetrads. The robust, red patches over the grey-white limestone make this one of our most beautiful mosses; frequent in damp places on the limestone with just one site on other basic rocks, on the north side of Beinn Gharbh.

*Orthothecium intricatum* (Hartm.) Bruch, Schimp. & W.Gümbel

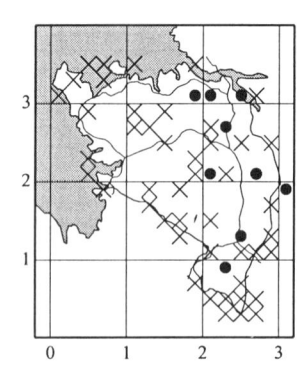

9 tetrads. Much less frequent than *Orthothecium rufescens*, occurring in similar sites but on the basic gneiss as well as on limestone.

*Plagiothecium denticulatum* var. *denticulatum* (Hedw.) Bruch, Schimp. & W.Gümbel

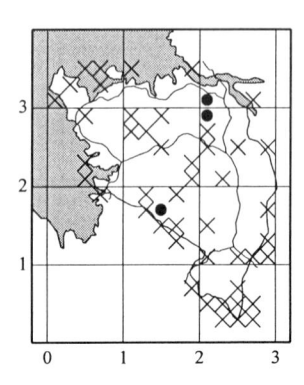

3 tetrads. A rare moss in Assynt with just three records from sheltered, damp earthy crevices on Torridonian sandstone crags.

*Plagiothecium denticulatum* var. *obtusifolium* (Turner) Moore

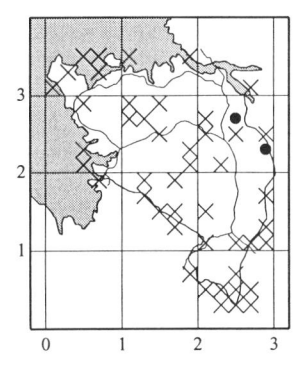

2 tetrads. Limited to a few sites in the bigger hills where it occurs in scree and crevices on crags: Imir Fada and in Coire Gorm on Glas Bheinn.

*Plagiothecium platyphyllum* Mönk.

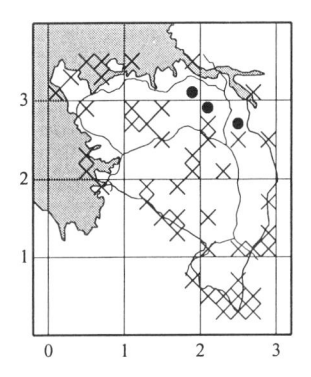

3 tetrads. A conspicuous but rare moss of wet, acid flushes or flushed ledges on crags in the higher hills: Sàil Gorm and the Sanctuary on Quinag and Coire Gorm on Glas Bheinn.

*Plagiothecium cavifolium* (Brid.) Z.Iwats.

1 tetrad. This unsatisfactory taxon is tentatively recorded from a ledge on a Torridonian sandstone crag below Saobhaidh Mhór on Quinag.

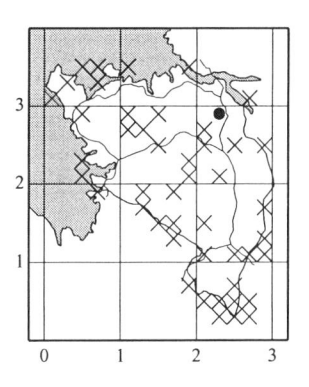

*Plagiothecium succulentum* (Wils.) Lindb.

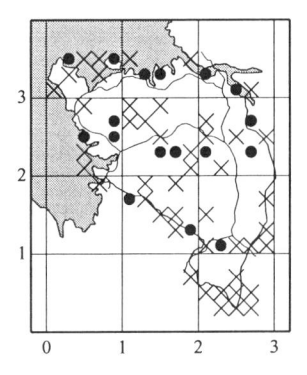

17 tetrads. Widespread but uncommon in sheltered sites in rocky woodland, ravines or below sheltered crags.

*Plagiothecium nemorale* (Mitt.) Jaeg.

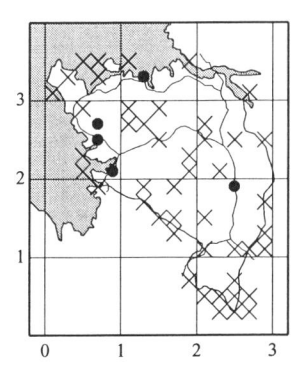

5 tetrads. In similar sites to *Plagiothecium succulentum* (from which it is not always easily distinguished), but much less frequent.

*Plagiothecium undulatum* (Hedw.) Bruch, Schimp. & W.Gümbel

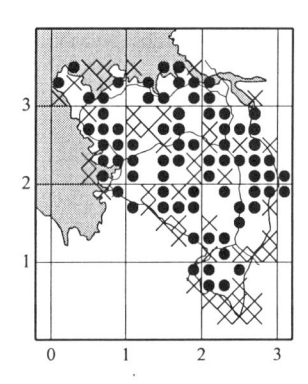

75 tetrads. A common and easily recognisable moss of sheltered damp sites, particularly wet banks in woodland, under heather on rocky slopes and in block scree.

*Isopterygiopsis muelleriana* (Schimp.) Z.Iwats.

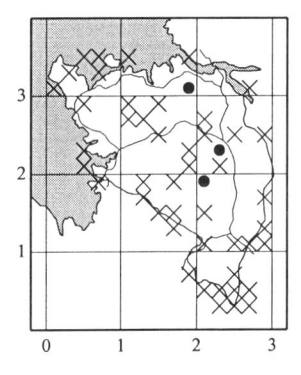

3 tetrads. This moss forms small, wispy mats in wet crevices at the base of montane crags; recorded from the north-facing crags on Beinn Garbh, Canisp and Sàil Gorm.

*Isopterygiopsis pulchella* (Hedw.) Z.Iwats.
(Isopterygium pulchellum (Hedw.) Jaeg.)

14 tetrads. Sparse but easily overlooked as it often grows through cushions of *Amphidium mougeottii* or *Anoectangium aestivum* on sheltered ledges on

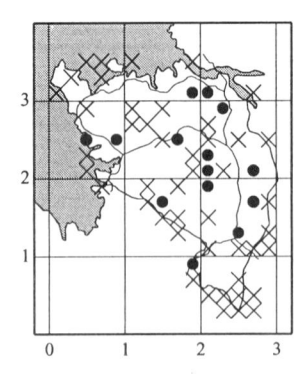

moderately basic crags; often only the sporophytes are visible.

*Pseudotaxiphyllum elegans* (Brid.) Z.Iwats.
(Isopterygium elegans (Brid.) Lindb.)

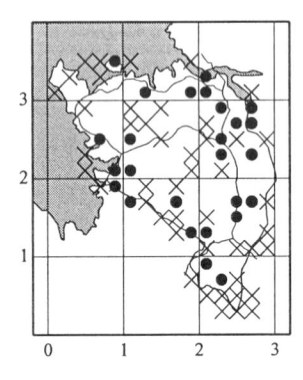

26 tetrads. Widespread but surprisingly infrequent compared with areas further south and limited to heavily shaded, acid soils in woodland, scree or at the base of crags.

*Herzogiella striatella* (Brid.) Z.Iwats.

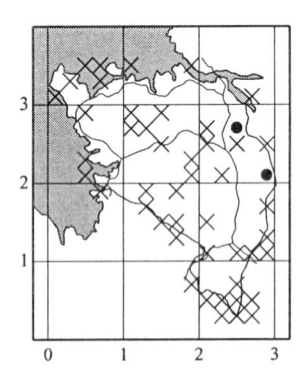

2 tetrads. A pretty species which is nearly always in fruit, but hides away in deep crevices in scree or under rocks in the higher hills: Breabag and Coire Gorm on Glas Bheinn, with an older record from Conival by Dixon and Nicholson in 1899.

*Taxiphyllum wissgrillii* (Garov.) Wijk & Margad.

There is a single, old record for this moss from the limestone at Knockan by Wallace in 1957.

*Rhytidium rugosum* (Hedw.) Kindb.

There is an old record from 'short turf on limestone cliffs near Inchnadamph' by C.D.Pigott in 1950.

*Hypnum cupressiforme* Hedw.

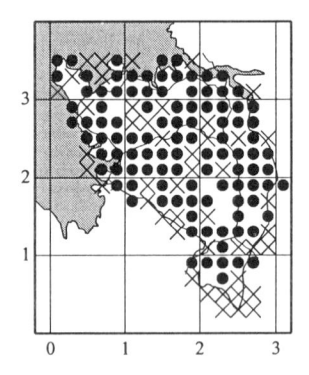

105 tetrads. Ubiquitous and often abundant on all kinds of dry, firm substrates, particularly trees and open rocks.

*Hypnum lacunosum* var. *lacunosum* (Brid.) Hoffm. ex Brid. (H. cupressiforme var. lacunosum Brid.)

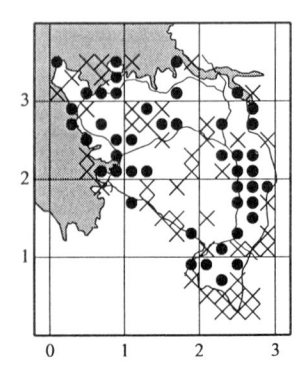

45 tetrads. A common moss of dry, moderately basic, firm habitats, particularly coastal rocks and boulders on loch margins, but also occurring in calcareous grassland.

*Hypnum resupinatum* Taylor (H. cupressiforme var. resupinatum (Tayl.) Schimp.)

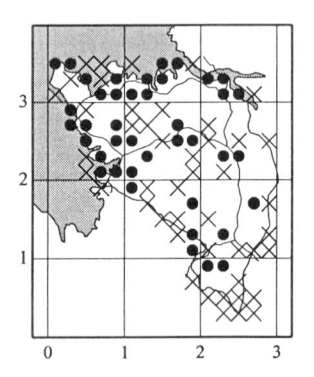

41 tetrads. In similar places to *Hypnum lacunosum* var. *lacunosum* but even more strongly coastal in its distribution.

*Hypnum andoi* A.J.E. Sm. (H. mammillatum (Brid.) Loeske)

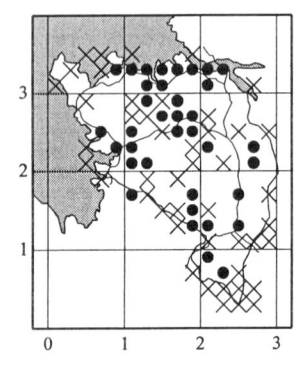

36 tetrads. Widespread and reasonably frequent on the limbs and trunks of larger trees, particularly birch, but also occurring on logs and, more rarely, on rocks.

*Hypnum jutlandicum* Holmen & E.Warncke

70 tetrads. A common and widespread moss in all kinds of heathy ground, particularly where

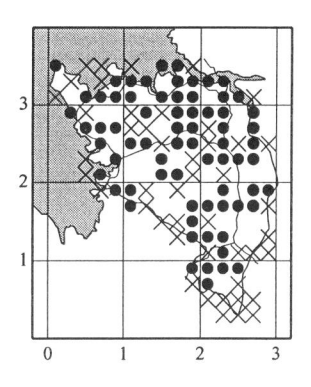

the dwarf shrub layer is well-developed.

*Hypnum bambergeri* Schimp.

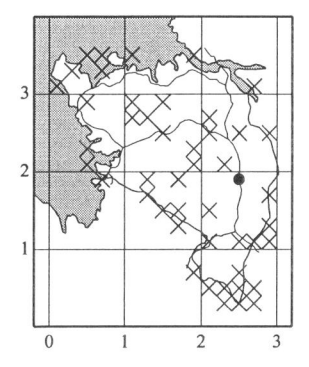

1 tetrad. Primarily a plant of the rich, calcareous mountains in the Central Highlands, this moss was a surprise discovery at relatively low altitude on the low limestone crags behind Stronechrubie where it occurs on a wet ledge next to a small waterfall, the only record north of the Great Glen.

*Hypnum lindbergii* Mitt.

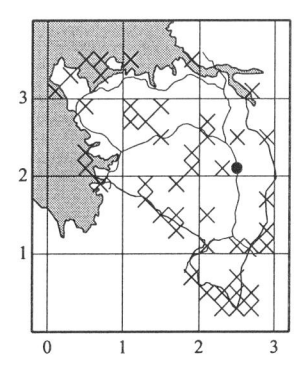

1 tetrad. Recorded from a damp roadside verge near the Inchnadamph Hotel by Long in 1981.

*Hypnum callichroum* Brid.

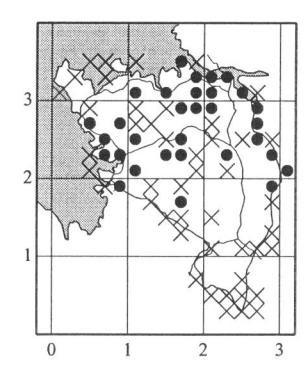

31 tetrads. Widespread but nowhere common, growing in rocky woodland and in sheltered sites on rocky ground in the hills, usually with other large pleurocarpous mosses like *Rhytidiadelphus loreus* or *Hylocomium splendens*.

*Hypnum hamulosum* Bruch, Schimp. & W.Gümbel

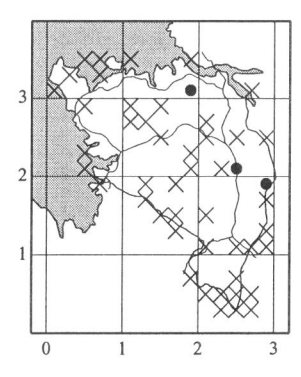

3 tetrads. Rare and always in small quantity, this delicate moss, somewhat similar to forms of *Ctenidium molluscum*, occurs on dry, basic crags in the hills: Bealach Traligill, Sàil Gorm and the Stronechrubie crags. There are older records from Quinag by Dixon and Nicholson in 1899 and 'near the caves, Inchnadamph' by Wallace in 1939.

*Ptilium crista-castrensis* (Hedw.) De Not.

21 tetrads. The unmistakable plumes of this moss are occasional but widespread and sometimes abundant in rocky woodland, within open stands of bracken and on steep heathy ground.

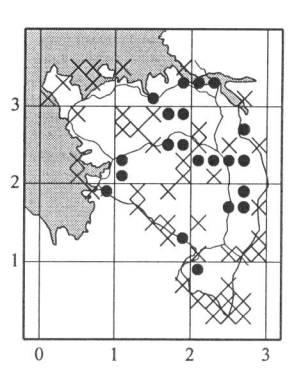

*Ctenidium molluscum* var. *molluscum* (Hedw.) Mitt.

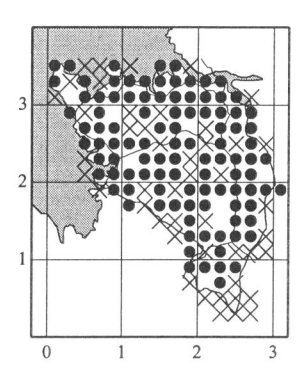

103 tetrads. A very common plant of flushed grassland, soil and rocks and particularly abundant on the limestone and Fucoid Beds where it can form large, pure cushions.

*Ctenidium molluscum* var. *condensatum* (Schimp.) E.Britton

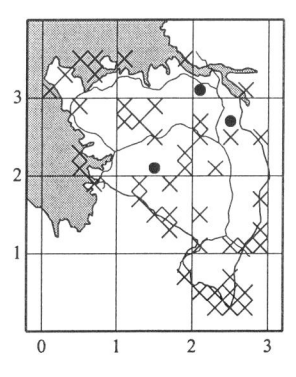

3 tetrads. A rare taxon, but possibly overlooked, growing at the bottom of damp, base-rich crags in the hills: Coire Dearg (Glas Bheinn), Suileag and Sàil Gharbh.

*Ctenidium molluscum* var. *robustum* Boulay

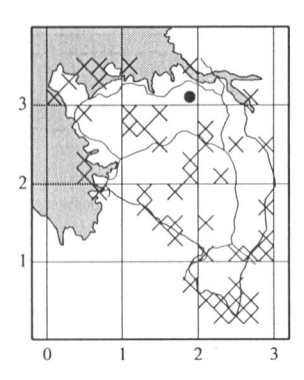

1 tetrad. This impressive moss, looking very different from var. *molluscum*, occurs on wet, base-rich gneiss at the foot of Sàil Gorm on Quinag where it was found by Blockeel on the B.B.S meeting in 1992.

*Hyocomium armoricum* (Brid.) Wijk & Marg.

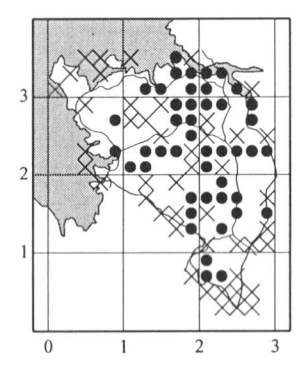

45 tetrads. A common plant of wet, acid rocks by burns and on wet crags but surprisingly absent from many suitable-looking sites.

*Rhytidiadelphus triquetrus* (Hedw.) Warnst.

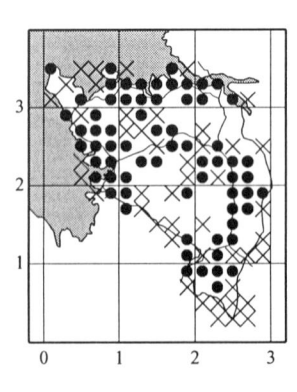

69 tetrads. A very common plant on the better soils and over base-rich scree, particularly abundant on the dune systems and on grassy banks on the limestone.

*Rhytidiadelphus squarrosus* (Hedw.) Warnst.

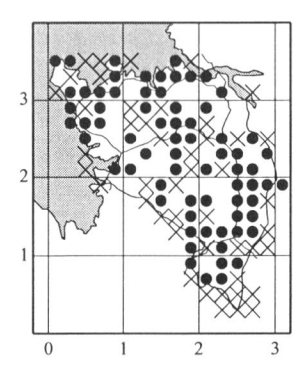

71 tetrads. Common and often abundant wherever there is damp grassland but absent from large areas of wet heath.

*Rhytidiadelphus loreus* (Hedw.) Warnst.

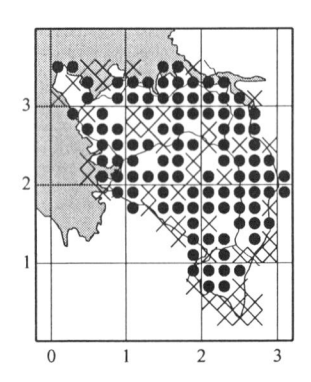

102 tetrads. A typical species of acid, rocky woodland, damp heath and of scree and crags in the hills where it can form thick wefts with other large pleurocarpous mosses.

*Hylocomium brevirostre* (Brid.) Bruch, Schimp. & W.Gümbel

49 tetrads. A widespread but often quite local species of base-rich rocks, probably most frequent over basic gneiss rocks in wooded areas near the coast.

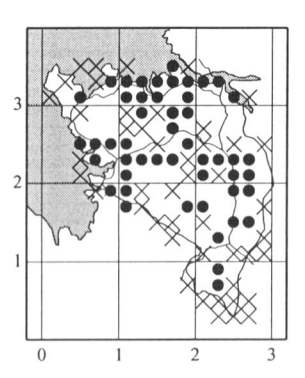

*Hylocomium umbratum* (Hedw.) Bruch, Schimp. & W.Gümbel

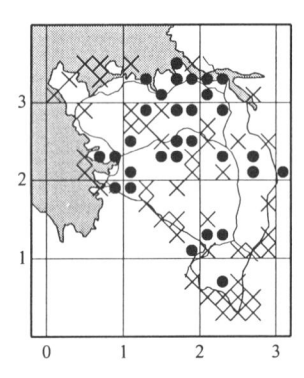

30 tetrads. A fairly consistent species of the better areas of shaded, rocky woodland and larger ravines, sometimes forming fine, large hummocks over low rocks.

*Hylocomium splendens* (Hedw.) Bruch, Schimp. & W.Gümbel

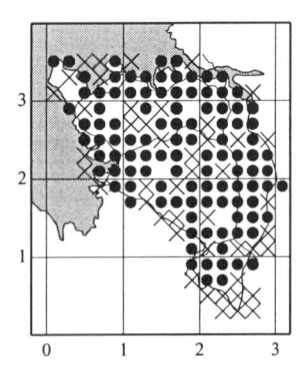

107 tetrads. A candidate for the most abundant moss in Assynt, occurring in quantity in both dry and wet heath, on sand dunes and fell-field and in both blanket bog and limestone grassland.

# BIBLIOGRAPHY

Adam, R.J. (ed.), 1960. *John Home's Survey of Assynt.* Edinburgh: Scottish History Society.

Anderson, G. and P., 1834. *Guide to the Highlands and Islands of Scotland.*

Anthony, J., 1959. Contribution to the flora of Sutherland. Bettyhill Region. *Trans. Bot. Soc. Edinb.* **38**, 7-15.

Anthony, J., 1967. *Sagina saginoides* (L.) Karst. in West Sutherland (v.c.108). *Trans. Bot. Soc. Edinb.* **40**, 335.

Averis, A.B.G., 1991. *A survey of the bryophytes of 448 woodlands in the Scottish Highlands.* Unpublished report for the Nature Conservancy Council, Edinburgh.

Balfour, I.B., 1907. A Catalogue of British Plants in Dr.Hope's Hortus Siccus, 1768. *Notes Roy. Bot. Gard. Edinb.* **4**, 147-192.

Balfour, J.H., 1865. Presidential Address. *Trans. Bot. Soc. Edinb.* **8**, 206-227.

Bangor-Jones, M., 1998. *The Assynt Clearances.* Dundee: The Assynt Press.

Bangor-Jones, M., 2000. *Historic Assynt.* Dundee: The Assynt Press.

Blockeel, T.L. and Long, D.G., 1998. *A checklist and census catalogue of British and Irish bryophytes.* Cardiff: British Bryological Society.

Blom, H.H., 1996. A revision of the *Schistidium apocarpum* complex in Norway and Sweden. *Bryophytorum Bibliotheca* **49**. Berlin: J.Cramer.

Bowden, J.K., 1989. *John Lightfoot, his work and travels.* Kew: Royal Botanic Gardens.

Burnett, J.M. (ed.), 1964. *The vegetation of Scotland.* Edinburgh: Oliver and Boyd.

Church, J.M., Hodgetts, N.G., Preston, C.D. and Stewart, N.F., 2001. *British Red Data Books: Mosses and Liverworts.* Peterborough: J.N.C.C.

Clark, J.W. and MacDonald, I., 1999. *Ainmean Gaidhlig Lusan. Gaelic names of plants.* North Ballachulish: J.W.Clark.

Coulson, M.G. and B.W.H., 1969. Plants found in S.W. Sutherland, July 20th to August 1st, 1969. *Rept.Oundle School N.H.S. for 1969*, 33-40.

Crampton, C.B. and MacGregor, M., 1913. The plant ecology of Ben Armine, Sutherlandshire. *Scot. Geogr. Mag.* **29**, 169-192, 256-266.

Crundwell, A.C., 1961. The autumn meeting at Ullapool. *Trans. Brit. Bryol. Soc.* **4**, 180-183.

Davies, E.W., 1953. Notes on two rare Scottish sedges. *Watsonia* **2**, 300-302. [*Carex rupestris* at Inchnadamph]

Druce, G.C., 1895. Notes on the flora of Elphin and the rocks of Cnoc-an-t'-Sasunnaich in West Sutherlandshire. *Ann. Scott. Nat. Hist. for 1895*, 35-38.

Druce, G.C., 1903. Notes on the flora of Western Ross-shire. *Ann. Scott. Nat. Hist. for 1903*, 166-175. [Record of *Ceratocapnos claviculata* from Elphin]

Druce, G.C., 1908. Plants of West Sutherland and Caithness. *Ann. Scott. Nat. Hist. for 1908*, 39-44, 106-109.

Druce, G.C., 1924a. Report of the Secretary. *Rept. Bot. Exch. Club for 1923*, 17-23.

Druce, G.C., 1924b. New county and other records. *Rept. Bot. Exch. Club for 1923*, 217. [*Paris quadrifolia* at Loch Awe]

Druce, G.C., 1926. Report of the Secretary. *Rept. Bot. Exch. Club for 1925*, 759-760.

Edwards, K.J. and Ralston, I.B.M. (eds.), 1997. *Scotland: Environment and Archaeology, 8000 BC - AD 1000.* Chichester: John Wiley and Sons.

Ferreira, R.E.C., 1959. Scottish mountain vegetation in relation to geology. *Trans. Bot. Soc. Edinb.* **37**, 229-250.

Ferreira, R.E.C., 1963, 1964. Some distinctions between calciphilous and basiphilous plants. *Trans. Bot. Soc. Edinb.* **39**. Pt.1: Field data, pp.399-413. Pt.2: Experimental data, pp.512-524.

Ferreira, R.E.C., 1995. *Vegetation Survey of North West Sutherland.* Unpublished report for Scottish Natural Heritage. 4 vols.

Gordon, C., 1845. *Parish of Assynt* in the *New Statistical Account of Scotland.*

Graham, G.G. and Primavesi, A.L., 1993. *Roses of Great Britain and Ireland.* London: Botanical Society of the British Isles.

Graham, R.A., 1826. Rare Scottish Plants. *Edinb. Phil. Journ.* **14**, 179-180.

Graham, R.A., 1827. Botanical Excursion in Sutherlandshire. *Edinb. New Phil. Journ.* **4**, 193-194.

Graham, R.A., 1833. Notice of Botanical Excursions into the Highlands of Scotland from Edinburgh this season, 1833. *Edinb. New Phil. Journ.* **15**, 358-361.

Gray, A. and Hinxman, L.W., 1888. A list of plants observed in West Sutherland (108). *Trans. Bot. Soc. Edinb.* **17**, 220-237.

Hamilton, A., Legg, C. and Zhaohua, L., 1997. *Blanket mire research in north-west Scotland: a view from the front.* In Tallis, J.H., Meade, R. and Hulme, P.D., (eds.). *Blanket mire degradation; causes, consequences and challenges.* Proceedings of the Mires Research Group Meeting, Manchester, 9th-11th April 1997. Aberdeen: Macaulay Land Research Institute. pp. 47-53.

Henderson, D.M., 1991, 1992. *Annotated checklist of flora of West Ross.* Privately published.

Henderson, D.M. and Dickson, J.H., 1994. *A naturalist in the Highlands. James Robertson: his life and travels in Scotland, 1767-1771.* Edinburgh: Scottish Academic Press.

Hill, M.O., Preston, C.D. and Smith, A.J.E., 1992. *Atlas of the bryophytes of Great Britain and Ireland.* **2**: *Mosses (except Diplolepidae).* Colchester: Harley Books.

Hodgetts, N.G., 1993. Atlantic bryophytes of the western seaboard. *British Wildlife* **4**, 287-295.

Hooker, W.J., 1821. *Flora Scotica.* Edinburgh: Archibald Constable.

Johnstone, G.S. and Mykura, W., 1989. *British regional geology: the Northern Highlands of Scotland.* 4th edition. H.M.S.O.: British Geological Survey.

Kenneth, A.G., 1985. A hybrid club-moss, *Diphasiastrum* x *issleri* (Rouy) Holub in West Sutherland. *Glasg. Nat.* **21:1**, 101.

Kent, D.H., 1992. *List of vascular plants of the British Isles.* London: Botanical Society of the British Isles.

Kenworthy, J.B. (ed.), 1976. *John Anthony's Flora of Sutherland.* Edinburgh: Botanical Society of Edinburgh.

# Bibliography

Lawson, T.J. (ed.)., 1995. *The Quaternary of Assynt and Coigach: Field Guide.* Cambridge: Quaternary Research Association.

Lightfoot, J., 1777. *Flora Scotica.* London.

Lipscomb, J. and David, R.W., 1981. *John Raven, by his friends.* Privately published.

Lowe, J., 1899. Note on the discovery of *Gentiana nivalis*, Linn., in Sutherlandshire. *Trans. Bot. Soc. Edinb.* **21**, 217.

Lousley, J.E. and Kent, D.H., 1981. *Docks and Knotweeds of the British Isles.* London: B.S.B.I.

Mackenzie, W., 1794. *Parish of Assint* in the *Old Statistical Account of Scotland.*

Macvicar, S.M., 1910. The distribution of Hepaticae in Scotland. *Trans. Proc. Bot. Soc Edinb.* **25**, 1-336.

Marshall, E.S., 1885. *Pinguicula alpina* in Sutherlandshire. *J. Bot. Lond.* **23**, 311.

Marshall, E.S., 1888. Notes on Highland Plants. *J. Bot. Lond.* **26**, 149-156.

Marshall, E.S., 1892. On an apparently endemic British *Ranunculus*. *J. Bot. Lond.* **30**, 289-290.

Marshall, E.S., 1899. *Epipactis atrorubens* Schultes. *J. Bot. Lond.* **37**, 328.

Marshall, E.S., 1910. *Callitriche intermedia* Hoffm. var. *tenuifolia*. *J. Bot. Lond.* **48**, 111.

Marshall, E.S., 1913. Two new Scottish hawkweeds. *J. Bot. Lond.* **51**, 119-122.

Marshall, E.S. and Hanbury, F.J., 1891. Notes on Highland plants, 1890. *J. Bot. Lond.* **29**, 108-118.

Marshall, E.S. and Shoolbred, W.A., 1909. Some Sutherland Plants. *J. Bot. Lond.* **47**, 220-223.

McVean, D.N., 1958. Island vegetation of some West Highland fresh-water lochs. *Trans. Bot. Soc. Edinb.* **37**, 200-208.

McVean, D.N. and Ratcliffe, D.A., 1962. *Plant communities of the Scottish Highlands.* London: H.M.S.O.

Mendum, J., Merritt, J., and McKirdy, A., 2001. *North West Highlands. A landscape fashioned by geology.* Perth: Scottish Natural Heritage.

Miller, R., 1967. Land use by summer sheilings. *Scottish Studies* **11**, 194-221.

Murray, A., 1836. *The Northern Flora.* Edinburgh: Adam and Charles Black.

Nicholson, W.E., 1900. Sutherlandshire mosses. *J. Bot. Lond.* **38**, 410-420.

Nicholson, W.E., 1923. Hepatics from West Sutherlandshire. *J. Bot. Lond.* **61**, 229-234.

Noble, R., 1997. *Changes in native woodland in Assynt, Sutherland, since 1774.* In Smout,T.C. (ed.) *Scottish woodland history.* Edinburgh: Scottish Cultural Press.

Noble, R., 2000. *The woods of Assynt.* Unpublished report for the Assynt Crofters' Trust.

Omand, D. (ed.), 1982. *The Sutherland Book.* Golspie: The Northern Times.

Page, C.N., 1982. *The ferns of Britain and Ireland.* Cambridge: C.U.P.

Paton, J.A., 1999. *The liverwort flora of the British Isles.* Colchester: Harley Books.

Pennant, T., 1774. *A tour in Scotland.* Chester.

Pennie, I.D., 1967. *The influence of man on the vegetation of Sutherland.* M.Sc.thesis. University of Aberdeen: Department of Botany.

Pennington, W., 1995. *Vegetation history of Assynt and Coigach.* In Lawson,T.J. (ed.). *The Quaternary of Assynt and Coigach: Field Guide.* pp. 104-131.

Pennington, W., Haworth, E.Y., Bonny, A.P. and Lishman, J.P., 1972. Lake sediments in Northern Scotland. *Phil. trans. Roy. Soc. (Lond.) Ser. B. (Biol Sci.)* **264**, 191-294.

Perring, F.H., (ed.), 1968. *Critical Supplement to the Atlas of the British Flora.* London: Thomas Nelson and Sons.

Perring, F.H. and Walters, S.M. (eds.), 1962. *Atlas of the British Flora.* London: Thomas Nelson and Sons.

Pigott, C.D., 1952. Plant Records. *Watsonia* **2**, 191. [*Thalictrum alpinum*]

Poore, M.E.D. and McVean, D.N., 1957. A new approach to Scottish mountain vegetation. *J. Ecol.* **45**, 401-439.

Preston, C.D. and Croft, J.M., 1997. *Aquatic plants in Britain and Ireland.* Colchester: Harley Books.

Ratcliffe, D.A., 1968. An ecological account of Atlantic bryophytes in the British Isles. *New Phytologist* **67**, 365-439.

Raven, J.E., 1952. *Roegneria doniana* (F.B.White) Meld. in Britain. *Watsonia* **2**, 180-185.

Raven, J.E. and Walters, S.M., 1956. *Mountain Flowers.* London: Collins.

Roberts, R.H., 1964. *Mimulus* hybrids in Britain. *Watsonia* **6**, 70-75.

Rodwell, J.S. (ed.), 1991. *British plant communities.* **2**: *Mires and heaths.* Cambridge: C.U.P.

Ross, S., 1982. *The geology of Sutherland.* In Omand, D. (ed.) *The Sutherland Book.* Golspie: the Northern Times.

Rothero, G.P., 1993. Summer field meeting, 1992, first week, Lochinver. *Bull. Brit. Bryol. Soc.* **61**, 5-8.

Salmon, C.E., 1900. Plant notes from Sutherland and Cantire. *J. Bot. Lond.* **38**, 299-303.

Scouller, C.E.K., 1988. *An introduction to the flowering plants and ferns of Lochbroom and Assynt.* Ullapool: Lochbroom Field Club.

Sell, P.D., West, C., and Tennant, D.J., 1995. Eleven new British species of *Hieracium* L. Section Alpina (Fries) F.N.Williams. *Watsonia* **20**, 351-365.

Smith, A.J.E., 1978. *The Moss Flora of Britain and Ireland.* Cambridge: C.U.P.

Spence, D.H.N., 1964. *The macrophytic vegetation of freshwater lochs, swamps and associated fens.* In Burnett, J.M. (ed.) *The vegetation of Scotland.* Edinburgh: Oliver and Boyd.

Spence, D.H.N. and Allen, E.D., 1979. The macrophytic vegetation of Loch Urigill and other lochs of the Ullapool area. *Trans. Bot. Soc. Edinb.* **43**, 131-144.

Stace, C., 1997. *New Flora of the British Isles.* 2nd edition. Cambridge: C.U.P.

Stewart, A., Pearman, D.A. and Preston, C.D., 1994. *Scarce plants in Britain.* Peterborough: J.N.C.C.

Stirling, A.McG., 1974. A fern new to Scotland - *Polystichum* x *illyricum* in West Sutherland. *Watsonia* **10**, 231.

Stirling, A.McG., 1990. A.G.Kenneth - an appreciation. *Watsonia* **18**, 241-242.

Taschereau, P.M., 1985. Taxonomy of *Atriplex* species indigenous to the British Isles. *Watsonia* **15**, 183-209.

Watson, H.C., 1883. *Topographical Botany.* 2nd edition. London

## Bibliography

Webster, M.McC., 1951. Plant Records. *Watsonia* **2**, 46. [*Carduus nutans* at Achmelvich]

Whitehouse, H.L.K., 1963. *Bryum riparium* Hagen in the British Isles. *Trans. Brit. Bryol. Soc.* **4**, 389-403.

Wigston, D.L., 1975. The distribution of *Quercus robur* L., *Q.petraea* (Matt.)Leibl. and their hybrids in south-western England. 1. The assessment of the taxonomic status of populations from leaf characters. *Watsonia* **10**, 345-369.

Wilmott, A.J. and Campbell, M.S., 1946. Autumn Botanising at Lochinver (West Sutherland). *Rept. Bot. Exch. Club for 1943-44*, 820-833.

Yapp, W.B., 1961. Oaks in Scotland. *Scott. Nat.* **70**, 2-6.

# GAZETTEER

All grid references fall within 100km. square NC (29). The grid reference given is generally that of the 1km. square containing the name of the feature on the 1:50,000 map or, failing that, the 1:25,000. Exceptions are hills, where the grid reference is that of the summit, and a few other features, whose name has been displaced for cartographic reasons, where it is that of the feature itself. Historic names are in *italics*. Entries have been computer-sorted, which means that, in compound names, spaces take preference over letters; thus Loch Beannach comes before Lochan Fada.

| Name | Grid |
|---|---|
| A' Chleit | 0220 |
| Abhainn a' Chnocain | 2209 |
| Abhainn Bad na h-Achlaise | 1221 |
| Abhainn na Clach Airigh | 1420 |
| Achadh Mór | 1422 |
| Achadh' an Ruighe Chòinich | 1422 |
| Achadhantuir | 0824 |
| Achiltibuie (West Ross) | 0208 |
| Achins Bookshop | 0819 |
| Achmelvich | 0524 |
| Achmelvich Bay | 0525 |
| Achmore | 2424 |
| Achnacarnin | 0432 |
| *Achumore* (Achmore) | 2424 |
| Airigh Bheag | 0627 |
| Airigh na Beinne | 2130 |
| Allt a' Bhealaich | 2819 |
| Allt a' Chalda Beag | 2424 |
| Allt a' Chalda Mór | 2523 |
| Allt a' Phollain | 2207 |
| Allt an Achaidh | 2307 |
| Allt an Tiaghaich | 1524 |
| Allt Caoruinn | 1821 |
| Allt Mhic Mhurchaidh Ghèir | 2317 |
| Allt Mór | 2518 |
| Allt na Braclaich | 1815 |
| Allt na Doire Cuilinn | 2025 |
| Allt na Ghamhna | 2132 |
| Allt na Glaic Móire | 2620 |
| Allt nam Meur | 2508 |
| Allt nan Damph | 1515 |
| Allt nan Uamh | 2617 |
| Allt Poll an Droighinn | 2722 |
| Allt Sgiathaig | 2226 |
| Alltan Beithe | 2219 |
| Alltana' bradhan | 0525 |
| Altnacealgach | 2610 |
| *An Coilean* | 2322 |
| An Coimhleum | 2322 |
| An Sgònnan | 1534 |
| Ardroe | 0623 |
| Ardvar | 1733 |
| Ardvreck (*Ardvrick*) | 2323 |
| Assynt School | 2112 |
| Bad a' Bhainne | 1024 |
| Bad an Dìoboirich | 1631 |
| Baddidarach | 0822 |
| Badnaban | 0720 |
| Bàgh an t-Srathain | 0721 |
| Balchladich | 0230 |
| Balchladich Bay | 0230 |
| Bathaich Cuinneige | 2029 |
| Bay of Stoer | 0328 |
| Bealach a' Chornaidh | 2028 |
| Bealach na h-Uidhe | 2626 |
| Bealach Traligill | 2919 |
| Beinn an Fhuarain | 2615 |
| Beinn an Fhurain | 2921 |
| Beinn Gharbh | 2122 |
| Beinn nan Cnaimhseag | 2717 |
| Beinn Reidh | 2121 |
| Beinn Uidhe | 2825 |
| Ben More Assynt | 3120 |
| Blar nam Fiadhag | 2520 |
| Bone Caves | 2616 |
| Bracklach | 1815 |
| Brackloch | 1123 |
| Breabag | 2917 |
| Caisteal Liath | 1518 |
| Cam Alltan | 2209 |
| Cam Loch (*Cama Loch*) | 2113 |
| Camas nam Bad | 1534 |
| Camasnafriaraich | 0623 |
| Canisp | 2018 |
| Canisp Road | 0922 |
| Cìrean Geardail | 0134 |
| Clach Airigh | 1720 |
| Clachtoll | 0427 |
| Clashmore | 0331 |
| Clashnessie | 0530 |
| Cnoc a' Bhainne | 1527 |
| Cnoc a' Phollain Bheithe | 0932 |
| Cnoc Aird na Seilge | 1725 |
| Cnoc an Each | 1021 |
| Cnoc an Leathaid Bhig | 2214 |
| Cnoc an Leathaid Bhuidhe | 2315 |
| Cnoc an t-Sasunnaich | 1908 |
| Cnoc Bad a'Bhainne | 1024 |
| Cnoc Beag | 0426 |
| Cnoc Eilid Mhathain | 2718 |
| Cnoc Gorm | 2508 |
| Cnoc na Creige | 2628 |
| Cnoc na Doire Daraich | 0921 |
| Cnoc na Sròine | 2612 |
| *Coineval/Coinne mheall* (Conival) | 3019 |
| Coir' an Lochain | 2006 |

| | |
|---|---|
| Coire Dearg | 2527 |
| Coire Dubh | 1518 |
| Coire Gorm | 2526 |
| Con a' Chreag | 2431 |
| Conival | 3019 |
| Corrag Ghorm | 2824 |
| Creag a' Choimhleum | 2423 |
| *Creag a-chnocaen* | 1909 |
| Creag an Spardain | 2232 |
| Creag Clais nan Cruineachd | 0627 |
| Creag Dharaich | 1431 |
| Creag Liath (Breabag) | 2715 |
| Creag Liath (Clachtoll) | 0526 |
| Creag Mhór | 2026 |
| Creag na h-Iolaire | 2617 |
| Creag nan Uamh | 2616 |
| Creag Ruigh a' Chàirn | 0825 |
| Creag Sròn Chrùbaidh | 2521 |
| Creagan a' Chait | 2122 |
| Creagan Beag | 2013 |
| Creagan Breaca (Inchnadamph) | 2619 |
| Creagan Mór | 1913 |
| *Creg-achnocean* | 1909 |
| Crom Allt | 2406 |
| Cromalt | 2407 |
| Cromalt Hills | 2106 |
| Cùil Dhubh | 2819 |
| Culag Hotel | 0922 |
| Culag River | 0921 |
| Culag Wood | 0921 |
| Culkein Drumbeg | 1133 |
| Culkein Stoer | 0332 |
| Doire Dhubh | 2309 |
| Druim nan Cnaimhseag | 2114 |
| Drumbeg | 1232 |
| Duart | 1333 |
| Duart Loch | 1333 |
| Duartmore | 1937 |
| Dubharlainn | 1229 |
| Eadar a' Chalda | 2423 |
| Eas Crom | 2307 |
| Eilean a'Ghamhna | 2033 |
| Eilean an Tuim | 2630 |
| Eilean Assynt | 1925 |
| Eilean Chrona | 0633 |
| Eilean Mór (West Ross) | 0517 |
| Eilean na Gartaig | 2112 |
| Elphin | 2111 |
| Feadan | 0724 |
| Fionn Loch | 1217 |
| Fuarain Ghlasa (Breabag) | 2816 |
| Fuaran nan Each | 2921 |
| Garbh Dhoire | 2123 |
| Garlic Island | 2112 |
| Geodh' Dearg | 0234 |
| Gillaroo Loch | 2719 |
| Glac Mhór | 2620 |
| Glas Bheinn | 2526 |
| Gleann Ardbhair | 1733 |
| Gleann Dorcha | 1817 |
| Gleann Dubh | 2720 |
| Gleann Leireag | 1630 |
| Gleannan a' Mhadaidh | 1416 |
| Glenbain | 2621 |
| Glencanisp Lodge | 1122 |
| Glenleraig | 1431 |
| Gorm Chnoc | 1528 |
| Gorm Cnoc | 2508 |
| Imir Fada | 2922 |
| Imirfada | 0631 |
| Inchnadamph | 2521 |
| *Inver* (Lochinver) | 0922 |
| Inver Park | 0922 |
| Inver Woods | 1023 |
| Inverkirkaig | 0719 |
| Kerrachar | 1734 |
| Knockan (*Knockain*) | 2110 |
| Knockan Crag | 1909 |
| *Knockneach* (Cnoc an Each) | 1021 |
| Kylesku (*Kylescu*) | 2333 |
| Kylestrome | 2134 |
| Lady Constance Bay | 0821 |
| Lairig Unapool | 2327 |
| Leathad Lianach | 2320 |
| Ledbeg | 2413 |
| Ledbeg River | 2413 |
| Ledmore | 2412 |
| Ledmore Junction | 2412 |
| Ledmore River | 2411 |
| Liath Bhad | 2530 |
| Little Assynt | 1525 |
| Loanan (River Loanan) | 2418 |
| Loch a' Choire Dheirg | 2527 |
| Loch a' Choire Dhuibh | 2528 |
| Loch a' Chroisg | 2215 |
| Loch a' Ghlinnein | 1723 |
| Loch a' Phollain Drisich | 0726 |
| Loch a' Phris | 2006 |
| Loch an Achaidh | 0233 |
| Loch an Aigeil | 0428 |
| Loch an Aon Aite | 0828 |
| Loch an Leothaid | 1729 |
| Loch an Ordain (Achmelvich) | 0625 |
| Loch an Ordain (Torbreck) | 0924 |
| Loch Ardbhair | 1633 |
| Loch Assynt | 1925 |
| Loch Assynt Lodge | 1726 |
| Loch Awe | 2415 |
| Loch Bad a' Chigean | 1627 |
| Loch Bealach a' Bhuirich | 2628 |
| Loch Beannach | 1326 |
| Loch Borralan | 2610 |
| Loch Cròcach | 1027 |
| Loch Culag | 0921 |
| Loch Doirean Rairidh | 1527 |
| Loch Druim Suardalain | 1121 |
| Loch Dubh | 0724 |
| Loch Eileanach | 0931 |
| Loch Fasg an t-Seana Chlaidh | 0724 |
| Loch Gleannan a' Mhadaidh | 1516 |

263

| | | | |
|---|---|---|---|
| Loch Glencoul | 2531 | Old Man of Stoer | 0135 |
| Loch Leathad an Lochain | 0720 | Oldany Farmhouse | 0932 |
| Loch Leathad nan Aighean | 0527 | Oldany Island | 0834 |
| Loch Leitir Easaidh (*Loch Letteressie*) | 1626 | Oldany River | 1031 |
| Loch Lurgainn (West Ross) | 1108 | Point of Stoer | 0235 |
| Loch Meall a' Chuna Beag | 0827 | Poll an Droighinn (Inchnadamph) | 2723 |
| Loch Mhaolach-coire | 2719 | Poll an Droighinn (Stoer) | 0428 |
| Loch na Barrach | 1518 | Poll Bhuidhe | 1332 |
| Loch na Bruthaich | 0731 | Pollachapuill | 1033 |
| Loch na Claise | 0330 | Pollan Beithe | 0932 |
| Loch na Creige Léithe | 0526 | Port a' Ghleannain | 0833 |
| Loch na Doire Daraich | 0921 | Port Achnacarnin | 0532 |
| Loch na Gainmhich | 2428 | Port Alltan na Bradhan | 0526 |
| Loch na h-Airigh Fraoich | 1321 | Port Dhrombaig | 1233 |
| Loch na h-Innse Fraoich | 1626 | Quinag | 2029 |
| Loch na h-Uidhe Doimhne | 0628 | Raffin | 0132 |
| Loch na Loinne | 1229 | Rhicarn | 0825 |
| Loch nam Meallan Liatha | 2220 | Rientraid | 1933 |
| Loch nan Cuaran | 2923 | River Inver | 1223 |
| Loch nan Eun | 1023 | River Kirkaig | 0918 |
| Loch Nedd | 1332 | River Loanan | 2418 |
| Loch Poll an Nigheidh | 1322 | River Traligill | 2720 |
| Loch Poll Dhàidh | 0729 | Ru(bha) Stoer | 0232 |
| Loch Roe | 0624 | Rubh' an Dùnain | 0434 |
| Loch Ruighean an Aitinn | 1232 | Rubh' Dhubhard | 1333 |
| Loch Sionascaig (West Ross) | 1113 | Ruigh Chnoc | 2120 |
| Loch Urigill | 2409 | Sàil Gharbh | 2129 |
| Loch Veyatie | 1813 | Sàil Gorm | 1930 |
| Loch(an) nan Caorach | 2923 | Saobhaidh Mhór | 2129 |
| Lochan Bealach Cornaidh | 2028 | Sgeir Liath | 1234 |
| Lochan a' Choire Ghuirm | 2626 | Sìdhean Mór | 0134 |
| Lochan Fada | 2016 | Skiag Bridge (*Skaig Bridge*) | 2324 |
| Lochan Fearna (Nedd) | 1231 | Soyea Island | 0421 |
| Lochan Fearna (Brackloch) | 1224 | Speicein Còinnich (West Ross) | 1004 |
| Lochan Feòir | 2225 | Spidean Còinich | 2027 |
| Lochan na Leobaig | 1033 | Sròn Chrùbaidh | 2519 |
| Lochan Sàile | 0724 | Star Pool | 1023 |
| Lochinver | 0922 | Stoer Lighthouse | 0032 |
| Lon Ruadh | 0932 | Stoer Peninsula | 0.3. |
| Luban Croma | 2812 | Stoer Point | 0235 |
| Lyne | 2514 | Stoer village | 0328 |
| Manse Loch | 0924 | Strathan | 0821 |
| Meall a' Bhuirich Rapaig | 2502 | Strathcroy | 0831 |
| Meall a' Chaoruinn | 2604 | Strone Brae | 0725 |
| Meall Beag | 1617 | Stronechrubie | 2419 |
| Meall Dearg | 1134 | Suilven | 1518 |
| Meall Diamhain (Breabag) | 2914 | The Sanctuary | 2029 |
| Meall Diamhain (Canisp) | 2117 | Torbreck | 0824 |
| Meall Meadhonach | 1617 | Torgawn | 2132 |
| Meall na Braclaich | 1916 | Tòrr a' Ghamhna | 2132 |
| Meall nam Imrichean | 2606 | Traligill River/Burn | 2720 |
| Meallan Liath Beag | 2220 | Tubeg | 1924 |
| Meallan Liath Mór | 2218 | Tumore | 1826 |
| Mòinteach na Dubha Chlaise | 1322 | Tumore Lodge | 1826 |
| Mòinteach na Totaig | 2510 | Uamh an Tartair | 2109 |
| Mol Ban | 0934 | Unapool | 2332 |
| Na Luirgean | 2211 | Unapool Burn | 2330 |
| Na Tuadhan | 3021 | Wood of Tumore | 1826 |
| Nedd | 1331 | | |
| Newton | 2331 | | |

# INDEX

# FLOWERING PLANTS AND FERNS

Entries are for the species accounts only. Scientific names are in *italics*, family names and synonyms in normal type and English names in **bold**.

*Abies procera* 74
*Acer pseudoplatanus* 116
ACERACEAE 116
*Achillea millefolium* 148
   *ptarmica* 148
**Adder's-tongue** 68
ADIANTACEAE 69
*Adoxa moschatellina* 137
ADOXACEAE 137
*Aegopodium podagraria* 119
*Aesculus hippocastanum* 116
*Agropyron caninum* 171
   *junceum* 171
   *repens* 171
*Agrostis canina* 169
   *capillaris* 168
   *castellana* 168
   *stolonifera* 169
   *tenuis* 168
   *vinealis* 169
*Aira caryophyllea* 168
   *praecox* 169
*Ajuga pyramidalis* 125
   *reptans* 124
*Alchemilla alpina* 105
   *filicaulis* 105
   *filicaulis* ssp. *filicaulis* 106
   *filicaulis* ssp. *vestita* 106
   *glabra* 106
   *glaucescens* 105
   *glomerulans* 106
   *mollis* 106
   *vulgaris* agg 105
   *wichurae* 106
**Alder** 81
*Alliaria petiolata* 94
*Allium ursinum* 173
*Alnus glutinosa* 81
*Alopecurus geniculatus* 169
   *pratensis* 169
**Alpine Bistort** 86
**Alpine Butterwort** 134
**Alpine Campion** 86
**Alpine Cinquefoil** 104
**Alpine Clubmoss** 65
**Alpine Enchanter's-nightshade** 115
**Alpine Gentian** 120
**Alpine Hybrid Shield-fern** 73
**Alpine Lady's-mantle** 105
**Alpine Lady-fern** 72
**Alpine Meadow-rue** 78

**Alpine Pearlwort** 85
**Alpine Rush** 153
**Alpine Saw-wort** 138
**Alpine Saxifrage** 101
**Alpine Willowherb** 114
**Alternate Water-milfoil** 113
**American Willowherb** 114
*Ammophila arenaria* 169
**Amphibious Bistort** 87
*Anagallis minima* 100
   *tenella* 100
*Anchusa arvensis* 121
*Anemone nemorosa* 76
*Angelica sylvestris* 119
*Anisantha sterilis* 170
**Annual Meadow-grass** 165
**Annual Pearlwort** 85
**Annual Wall-rocket** 97
*Antennaria dioica* 146
*Anthoxanthum odoratum* 168
*Anthriscus sylvestris* 118
*Anthyllis vulneraria* 110
*Aphanes arvensis* 106
   *australis* 107
   *microcarpa* 107
APIACEAE 118
**Apple-mint** 126
AQUIFOLIACEAE 115
*Arabidopsis thaliana* 94
*Arabis hirsuta* 95
ARALIACEAE 118
**Arctic Bearberry** 98
**Arctic Mouse-ear** 84
**Arctic Sandwort** 83
*Arctium minus* 138
*Arctostaphylos alpinus* 98
   *uva-ursi* 97
*Arenaria norvegica* ssp. *norvegica* 83
   *serpyllifolia* 82
*Armeria maritima* 89
*Arrhenatherum elatius* 167
*Artemisia absinthium* 148
   *vulgaris* 148
**Ash** 128
**Aspen** 91
ASPLENIACEAE 70
*Asplenium adiantum-nigrum* 70
   *marinum* 71
   *obovatum* ssp. *lanceolatum* 71
   *ruta-muraria* 72
   *septentrionale* 72

*trichomanes* 71
*trichomanes* ssp. *quadrivalens* 71
*viride* 71
*Aster tripolium* 147
ASTERACEAE 138
*Astragalus danicus* 110
*Athyrium distentifolium* 72
*filix-femina* 72
*Atriplex glabriuscula* 82
hastata 82
*patula* 82
*praecox* 82
*prostrata* 82
*Atriplex* spp. 82
Autumn Gentian 120
Autumn Hawkbit 140
Awlwort 96
Babington's Orache 82
BALSAMINACEAE 118
Barren Brome 170
Barren Strawberry 104
Bay Willow 91
Beaked Tasselweed 152
Bearberry 97
Bearded Couch 171
Beech 80
Beech Fern 70
Bell Heather 98
*Bellis perennis* 147
*Betula nana* 81
*pendula* 80
*pubescens* 80
BETULACEAE 80
Bifid Hemp-nettle 124
Bilberry 98
Bird Cherry 109
Bird's-nest Orchid 174
Biting Stonecrop 101
Bittersweet 120
Bitter-vetch 111
Black Bindweed 88
Black Bog-rush 157
Black Currant 100
Black Spleenwort 70
Black-poplar 91
Blackthorn 108
Bladder-Sedge 159
Bladderworts 134
Blaeberry 98
BLECHNACEAE 74
*Blechnum spicant* 74
Blinks 82
Blue Water-speedwell 130
Bluebell 173
*Blysmus rufus* 157
Bog Asphodel 173
Bog Blaeberry 98
Bog Hair-grass 167
Bog Myrtle 80
Bog Orchid 175

Bog Pimpernel 100
Bog Pondweed 150
Bog Stitchwort 84
Bogbean 121
Bog-sedge 162
BORAGINACEAE 121
*Botrychium lunaria* 68
Bottle Sedge 159
*Brachypodium sylvaticum* 170
Bracken 70
Brackish Water-crowfoot 77
Bramble 103
Branched Bur-reed 172
*Brassica rapa* 97
BRASSICACEAE 94
Bright-leaved Pondweed 151
Bristle Club-rush 156
Brittle Bladder-fern 72
*Briza media* 165
Broad Buckler-fern 74
Broad-leaved Cottongrass 155
Broad-leaved Dock 89
Broad-leaved Helleborine 174
Broad-leaved Osier 92
Broad-leaved Pondweed 150
Broad-leaved Willowherb 113
*Bromopsis ramosa* 170
*Bromus commutatus* 170
hordeaceus 170
hordeaceus ssp. hordeaceus 170
lepidus 170
Bromus ramosus 170
Brooklime 130
Broom 112
Brown Bent 169
Brown Sedge 158
Buck's-horn Plantain 127
Bugle 124
Bugloss 121
Bulbous Buttercup 77
Bulbous Rush 154
Burdock 138
Burnet Rose 107
Bush Vetch 111
Bushy Mint 126
*Cakile maritima* 97
*Calamagrostis epigejos* 169
CALLITRICHACEAE 127
*Callitriche hamulata* 127
*platycarpa* 127
*stagnalis* 127
*Calluna vulgaris* 98
*Caltha palustris* 76
*Calystegia pulchra* 121
*sepium* 120
*silvatica* 121
*Campanula cochleariifolia* 135
*latifolia* 135
*rotundifolia* 135
CAMPANULACEAE 135

# Flowering plant index

CAPRIFOLIACEAE  137
*Capsella bursa-pastoris*  96
*Cardamine flexuosa*  95
  *hirsuta*  95
  *pratensis*  95
*Carduus nutans*  138
*Carex aquatilis*  162
  *arenaria*  158
  *bigelowii*  162
  *binervis*  160
  *capillaris*  159
  *caryophyllea*  161
  *curta*  159
  *demissa*  161
  *dioica*  158
  *distans*  160
  *disticha*  158
  *echinata*  158
  *extensa*  160
  *flacca*  160
  *hostiana*  160
  *laevigata*  160
  *lasiocarpa*  159
  *lepidocarpa*  161
  *limosa*  162
  *nigra*  162
  *otrubae*  158
  *ovalis*  158
  *pallescens*  161
  *panicea*  160
  *paniculata*  157
  *pauciflora*  162
  *pilulifera*  161
  *pulicaris*  162
  *remota*  158
  *rostrata*  159
  *rupestris*  162
  *serotina*  161
  *sylvatica*  159
  *vesicaria*  159
  *viridula*  161
  *viridula* ssp. *brachyrrhyncha*  161
  *viridula* ssp. *oedocarpa*  161
  *viridula* ssp. *viridula*  161
  x *fulva*  160
**Carnation Sedge**  160
CARYOPHYLLACEAE  82
**Cat's-ear**  139
*Catabrosa aquatica*  166
*Catapodium marinum*  166
*Centaurea cyanus*  139
  *nigra*  139
*Cephalanthera longifolia*  174
*Cerastium arcticum*  84
  *arvense*  84
  *atrovirens*  84
  *diffusum*  84
  *fontanum*  84
  *glomeratum*  84
  *holosteoides*  84

  *semidecandrum*  84
*Ceratocapnos claviculata*  78
**Chaffweed**  100
*Chamaenerion angustifolium*  114
*Chamaepericlymenum suecicum*  115
*Chamerion angustifolium*  114
**Changing Forget-me-not**  122
CHENOPODIACEAE  81
*Chenopodium album*  81
  *murale*  81
  *rubrum*  81
Cherleria sedoides  83
**Chickweed Willowherb**  114
**Chickweeed Wintergreen**  99
*Chrysanthemum leucanthemum*  148
  parthenium  147
  *segetum*  148
  vulgare  147
*Chrysosplenium oppositifolium*  102
*Circaea alpina*  115
  *lutetiana*  115
  spp.  115
  x *intermedia*  115
*Cirsium arvense*  139
  *heterophyllum*  139
  *palustre*  139
  *vulgare*  138
  x *celakovskianum*  139
*Cladium mariscus*  157
*Claytonia sibirica*  82
**Cleavers**  137
**Climbing Corydalis**  78
**Cloudberry**  103
CLUSIACEAE  89
Cochlearia alpina  96
  *officinalis*  96
  *officinalis* ssp. *scotica*  96
  *pyrenaica* ssp. *alpina*  96
  scotica  96
**Cock's-foot**  166
*Coeloglossum viride*  176
**Colt's-foot**  149
**Common Bent**  168
**Common Bird's-foot-trefoil**  110
**Common Butterwort**  134
**Common Chickweed**  83
**Common Club-rush**  156
**Common Comfrey**  121
**Common Cornsalad**  137
**Common Cottongrass**  155
**Common Couch**  171
**Common Cow-wheat**  131
**Common Dog-violet**  90
**Common Duckweed**  152
**Common Field-speedwell**  131
**Common Figwort**  128
**Common Fumitory**  79
**Common Hemp-nettle**  124
**Common Knapweed**  139
**Common Marsh Bedstraw**  136

267

Common Milkwort 116
Common Mouse-ear 84
Common Nettle 79
Common Orache 82
Common Polypody 69
Common Ragwort 149
Common Reed 172
Common Saltmarsh-grass 164
Common Sedge 162
Common Sorrel 88
Common Spike-rush 156
Common Spotted Orchid 176
Common Stork's-bill 117
Common Twayblade 175
Common Valerian 138
Common Vetch 111
Common Water-starwort 127
Common Whitlowgrass 95
Common Yellow-sedge 161
Compact Rush 154
*Conopodium majus* 118
CONVOLVULACEAE 120
*Convolvulus arvensis* 120
Coppery Monkeyflower 129
Corn Buttercup 77
Corn Marigold 148
Corn Mint 125
Corn Spurrey 85
CORNACEAE 115
Cornflower 139
*Cornus suecica* 115
*Corydalis claviculata* 78
*Corylus avellana* 81
*Cotoneaster simonsii* 109
Cow Parsley 118
Cowberry 98
Cowslip 99
Crack Willow 92
CRASSULACEAE 101
*Crataegus monogyna* 109
Creeping Bent 169
Creeping Buttercup 77
Creeping Cinquefoil 104
Creeping Forget-me-not 122
Creeping Soft-grass 168
Creeping Thistle 139
Creeping Willow 93
*Crepis capillaris* 141
   *paludosa* 141
Crested Dog's-tail 164
Crested Hair-grass 167
*Crocosmia* x *crocosmiflora* 174
Cross-leaved Heath 98
Crosswort 137
Crowberry 97
*Cruciata laevipes* 137
*Cryptogramma crispa* 69
Cuckooflower 95
Curled Dock 88
Curved Woodrush 155

Cut-leaved Crane's-bill 117
Cut-leaved Dead-nettle 123
*Cynosurus cristatus* 164
CYPERACEAE 155
Cyphel 83
*Cystopteris fragilis* 72
*Cytisus scoparius* 112
*Dactylis glomerata* 166
Dactylorchis fuchsii 176
   incarnata 176
   maculata 176
   purpurella 177
*Dactylorhiza fuchsii* 176
   *incarnata* 176
   *incarnata* ssp. *coccinea* 176
   *incarnata* ssp. *incarnata* 176
   *incarnata* ssp. *pulchella* 177
   *incarnata* ssp. *cruenta* 177
   *lapponica* 177
   *maculata* ssp. *ericetorum* 176
   *purpurella* 177
   x *formosa* 177
Daffodil 173
Daisy 147
Dame's Violet 94
Dandelions 140
*Danthonia decumbens* 171
Dark-red Helleborine 174
*Datura stramonium* 120
*Daucus carota* 119
Deergrass 156
DENNSTAEDTIACEAE 70
*Deschampsia cespitosa* 167
   *cespitosa* ssp. *alpina* 167
   *cespitosa* ssp. *cespitosa* 167
   *flexuosa* 167
   *setacea* 167
Devil's-bit Scabious 138
*Digitalis purpurea* 129
Dioecious Sedge 158
*Diphasiastrum alpinum* 65
   complanatum ssp. issleri 65
   *issleri* 65
*Diplotaxis muralis* 97
DIPSACACEAE 138
Distant sedge 160
Dog-rose 107
Don's Twitch 171
Douglas Fir 75
Dove's-foot Crane's-bill 117
Downy Birch 80
Downy Oat-grass 166
Downy Willow 93
*Draba incana* 95
   *anglica* 90
   *intermedia* 90
   *rotundifolia* 89
   x *obovata* 90
DROSERACEAE 89
*Dryas octopetala* 105

DRYOPTERIDACEAE  73
Dryopteris abbreviata  73
   *aemula*  74
   *affinis*  73
   assimilis  74
   borreri  73
   *carthusiana*  74
   *dilatata*  74
   *expansa*  74
   *filix-mas*  72
   lanceolatacristata  74
   *oreades*  73
**Dwarf Birch**  81
**Dwarf Cornel**  115
**Dwarf Cudweed**  146
**Dwarf Willow**  94
**Eared Willow**  93
**Early Hair-grass**  169
**Early Orache**  82
**Early-purple Orchid**  177
*Echium vulgare*  121
**Eelgrass**  152
**Elder**  137
*Eleocharis acicularis*  156
   *multicaulis*  156
   *palustris*  156
   *quinqueflora*  156
   *uniglumis*  156
*Eleogiton fluitans*  157
*Elymus arenarius*  171
   *caninus*  171
   *caninus* var. *donianus*  171
*Elytrigia juncea*  171
   *repens*  171
EMPETRACEAE  97
*Empetrum nigrum*  97
   *nigrum* ssp. *hermaphroditum*  97
**Enchanter's-nightshade**  115
Endymion non-scriptus  173
**English Stonecrop**  101
Epilobium adnatum  113
   *alsinifolium*  114
   *anagallidifolium*  114
   *brunnescens*  114
   *ciliatum*  114
   *hirsutum*  113
   *montanum*  113
   *nerterioides*  114
   *obscurum*  113
   *palustre*  114
   *parviflorum*  113
   *tetragonum*  113
   x *boissieri*  114
   x *marshallianum*  113
   x *rivulicola*  114
*Epipactis atrorubens*  174
   *helleborine*  174
**Equal-leaved Knotgrass**  87
EQUISETACEAE  66
*Equisetum arvense*  67

*fluviatile*  67
*hyemale*  66
*palustre*  68
*pratense*  67
*sylvaticum*  67
*variegatum*  67
x *trachyodon*  67
*Erica cinerea*  98
   *tetralix*  98
ERICACEAE  97
*Erinus alpinus*  129
*Eriophorum angustifolium*  155
   *latifolium*  155
   *vaginatum*  155
*Erodium cicutarium*  117
*Erophila glabrescens*  96
   *verna*  95
*Eupatorium cannabinum*  150
*Euphorbia helioscopia*  115
   *peplus*  116
EUPHORBIACEAE  115
   *arctica* ssp. *borealis*  131
   *confusa*  131
   '*fharaidensis*'  132
   *foulaensis*  132
   *frigida*  132
   *marshallii*  132
   *micrantha*  132
   *nemorosa*  131
   *officinalis* agg.  131
   *ostenfeldii*  132
   *scottica*  132
**European Larch**  75
**Eyebright**  131
FABACEAE  110
FAGACEAE  80
*Fagus sylvatica*  80
**Fairy Flax**  116
**Fairy Foxglove**  129
**Fairy's-thimble**  135
*Fallopia baldschuanica*  88
   *convolvulus*  88
   *japonica*  88
**False Apple-mint**  127
**False Brome**  170
**False Fox-sedge**  158
**False Oat-grass**  167
**Fat-hen**  81
**Fen Bedstraw**  136
**Fennel Pondweed**  152
*Festuca altissima*  163
   *arundinacea*  163
   *filiformis*  164
   *ovina*  163
   *rubra*  163
   *rubra* ssp. *arctica*  163
   *rubra* ssp. *juncea*  163
   tenuifolia  164
   *vivipara*  164
**Feverfew**  147

**Few-flowered Sedge** 162
**Few-flowered Spike-rush** 156
**Field Bindweed** 120
**Field Forget-me-not** 122
**Field Gentian** 120
**Field Horsetail** 67
**Field Madder** 136
**Field Mouse-ear** 84
**Field Pansy** 91
**Field Penny-cress** 96
**Field Woodrush** 154
*Filipendula ulmaria* 103
**Fine-leaved Sheep's-fescue** 164
**Fir Clubmoss** 65
**Flame Nasturtium** 117
**Flax** 116
**Flea Sedge** 162
**Floating Bur-reed** 172
**Floating Club-rush** 157
**Floating Sweet-grass** 166
**Forked Spleenwort** 72
**Foxglove** 129
*Fragaria vesca* 104
**Fragrant Orchid** 176
*Fraxinus excelsior* 128
**Frog Orchid** 176
**Frog Rush** 153
*Fumaria bastardii* 79
   *officinalis* ssp. *officinalis* 79
FUMARIACEAE 78
*Galeopsis bifida* 124
   *tetrahit* 124
*Galium aparine* 137
   *boreale* 136
   *cruciata* 137
   *odoratum* 136
   *palustre* 136
   *saxatile* 136
   *sterneri* 136
   *uliginosum* 136
   *verum* 136
**Garlic Mustard** 94
*Gentiana nivalis* 120
   *verna* 120
GENTIANACEAE 120
*Gentianella amarella* ssp. *septentrionalis* 120
   *campestris* 120
GERANIACEAE 117
*Geranium dissectum* 117
   *lucidum* 117
   *molle* 117
   *robertianum* 117
**Germander Speedwell** 130
*Geum rivale* 104
   *urbanum* 104
**Giant Bellflower** 135
**Glabrous Whitlowgrass** 96
**Glaucous Meadow-grass** 165
**Glaucous Northern Dog-rose** 107
**Glaucous Sedge** 160

*Glaux maritima* 100
*Glechoma hederacea* 125
**Globeflower** 76
*Glyceria fluitans* 166
*Gnaphalium supinum* 146
   *sylvaticum* 146
   *uliginosum* 147
**Goat Willow** 92
**Goat's-beard** 140
**Goldenrod** 147
**Golden-scaled Male-fern** 73
**Goldilocks Buttercup** 77
**Gooseberry** 101
**Gorse** 112
**Grass-of-Parnassus** 102
**Great Fen-sedge** 157
**Great Mullein** 128
**Great Sundew** 90
**Great Willowherb** 113
**Great Woodrush** 154
**Greater Bird's-foot-trefoil** 110
**Greater Butterfly-orchid** 175
**Greater Chickweed** 83
**Greater Plantain** 128
**Greater Sea-spurrey** 85
**Greater Stitchwort** 83
**Greater Tussock-sedge** 157
**Green Alkanet** 122
**Green Field-speedwell** 131
**Green Spleenwort** 71
**Green-ribbed Sedge** 160
**Grey Willow** 92
GROSSULARIACEAE 100
**Ground-elder** 119
**Ground-ivy** 125
**Groundsel** 149
**Guelder-rose** 137
*Gymnadenia conopsea* 176
*Gymnadenia* x *Dactylorhiza* 176
*Gymnadenia* x *Platanthera* 176
*Gymnocarpium dryopteris* 72
   *robertianum* 72
**Gypsywort** 125
**Hair Sedge** 159
**Hairy Bindweed** 121
**Hairy Bitter-cress** 95
**Hairy Brome** 170
**Hairy Northern Dog-rose** 107
**Hairy Rock-cress** 95
**Hairy Woodrush** 154
HALORAGACEAE 112
*Hammarbya paludosa* 175
**Hard Fern** 74
**Hard Shield-fern** 73
**Hare's-tail Cottongrass** 155
**Harebell** 135
**Hart's-tongue** 70
**Hawkweeds** 141
**Hawthorn** 109
**Hay-scented Buckler-fern** 74

## Flowering plant index

**Hazel** 81
**Heath Bedstraw** 136
**Heath Cudweed** 146
**Heath Dog-violet** 90
**Heath Groundsel** 149
**Heath Milkwort** 116
**Heath Pearlwort** 84
**Heath Rush** 152
**Heath Speedwell** 130
**Heath Spotted Orchid** 176
**Heath Woodrush** 155
**Heather** 98
**Heath-grass** 171
*Hedera helix* 118
**Hedge Bindweed** 120
**Hedge Mustard** 94
**Hedge Woundwort** 123
*Helictotrichon pratense* 166
  *pubescens* 166
**Hemlock Water-dropwort** 119
**Hemp-agrimony** 150
**Henbit Dead-nettle** 124
*Heracleum sphondylium* 119
**Herb-Paris** 173
**Herb-Robert** 117
*Hesperis matronalis* 94
*Hieracium alpinum* 146
  *ampliatum* 145
  *anglicum* 144
  *argenteum* 144
  *caesiomurorum* 142
  *caledonicum* 143
  *camptopetalum* 143
  *dasythrix* 145
  *dicella* 144
  *duriceps* 143
  *eucallum* 144
  *flocculosum* 145
  *glandulidens* 146
  *hebridense* 145
  *holosericeum* 146
  *hyparcticoides* 145
  *iricum* 144
  *jovimontis* 144
  *kennethii* 146
  *langwellense* 145
  *latobrigorum* 142
  *lingulatum* 146
  *marginatum* 146
  *nitidum* 144
  *orcadense* 142
  *orimeles* 143
  *perscitum* 146
  *pictorum* 143
  *pilosella* 141
  *pollinarioides* 143
  *reticulatum* 142
  *rivale* 143
  *rubiginosum* 142
  *sarcophylloides* 144

  *saxorum* 144
  *shoolbredii* 145
  *sparsifolium* 142
  spp. 141
  *strictiforme* 142
  *subcrocatum* 142
  *subglobosum* 146
  *subtenue* 143
  *vulgatum* 142
**Highland Bent** 168
**Himalayan Cotoneaster** 109
HIPPOCASTANACEAE 116
HIPPURIDACEAE 127
*Hippuris vulgaris* 127
**Hoary Whitlowgrass** 95
**Hoary Willowherb** 113
**Hogweed** 119
*Holcus lanatus* 167
  *mollis* 168
**Holly** 115
**Holly Fern** 73
*Honckenya peploides* 83
**Honeysuckle** 137
**Hop Trefoil** 112
**Horse-chestnut** 116
*Huperzia selago* 65
*Hyacinthoides non-scripta* 173
**Hybrid Monkeyflower** 129
**Hybrid Woundwort** 123
*Hydrocotyle vulgaris* 118
HYDROPHYLLACEAE 121
HYMENOPHYLLACEAE 69
*Hymenophyllum wilsonii* 69
*Hypericum androsaemum* 89
  *pulchrum* 89
  *tetrapterum* 89
*Hypochaeris radicata* 139
*Ilex aquifolium* 115
*Impatiens glandulifera* 118
**Indian Balsam** 118
**Intermediate Water-starwort** 127
**Intermediate Wintergreen** 99
**Interrupted Clubmoss** 65
IRIDACEAE 173
*Iris pseudacorus* 173
ISOETACEAE 66
*Isoetes echinospora* 66
  *lacustris* 66
  x *hickeyi* 66
*Isolepis setacea* 156
**Italian Rye-grass** 164
**Ivy** 118
**Ivy-leaved Speedwell** 131
**Ivy-leaved Water-crowfoot** 77
**Japanese Knotweed** 88
**Jointed Rush** 153
JUNCACEAE 152
JUNCAGINACEAE 150
*Juncus acutiflorus* 153
  *alpinoarticulatus* 153

271

*ambiguus* 153
*articulatus* 153
*bufonius* 153
*bulbosus* 154
*conglomeratus* 154
*effusus* 154
*gerardii* 153
*squarrosus* 152
*tenuis* 152
*trifidus* 153
*triglumis* 154
x *surrejanus* 153
**Juniper** 75
*Juniperus communis* 75
**Keel-fruited Cornsalad** 138
**Kidney Vetch** 110
**Killarney Fern** 69
**Knotgrass** 87
**Knotted Pearlwort** 84
*Koeleria cristata* 167
*macrantha* 167
**Lady's Bedstraw** 136
**Lady-fern** 72
**Lady's-mantles** 105-106
LAMIACEAE 123
*Lamium album* 123
*amplexicaule* 124
*confertum* 123
*hybridum* 123
*molucellifolium* 123
*purpureum* 123
**Lanceolate Spleenwort** 71
**Lapland Orchid** 177
*Lapsana communis* 139
**Large Bindweed** 121
**Large Thyme** 125
*Larix decidua* 75
*Lathyrus linifolius* 111
*montanus* 111
*pratensis* 111
**Laurel-leaved Willow** 93
**Least Bur-reed** 172
*Lemna minor* 152
LEMNACEAE 152
**Lemon-scented Fern** 70
LENTIBULARIACEAE 134
*Leontodon autumnalis* 140
**Lesser Bladderwort** 135
**Lesser Butterfly-orchid** 175
**Lesser Celandine** 77
**Lesser Clubmoss** 66
**Lesser Meadow-rue** 78
**Lesser Sea-spurrey** 85
**Lesser Spearwort** 77
**Lesser Stitchwort** 83
**Lesser Trefoil** 112
**Lesser Twayblade** 175
*Leucanthemum vulgare* 148
*Leucorchis albida* 175
*Leymus arenarius* 171

*Ligusticum scoticum* 119
LILIACEAE 172
**Limestone Bedstraw** 136
**Limestone Fern** 72
LINACEAE 116
*Linum catharticum* 116
*usitatissimum* 116
*Listera cordata* 175
*ovata* 175
**Little Mouse-ear** 84
*Littorella uniflora* 128
*Lobelia dortmanna* 135
**Lodgepole Pine** 75
*Loiseleuria procumbens* 97
*Lolium multiflorum* 164
*perenne* 164
**Londonpride** 101
**Long-bracted Sedge** 160
**Long-headed Poppy** 78
**Long-stalked Pondweed** 151
**Long-stalked Yellow-sedge** 161
*Lonicera periclymenum* 137
*Lotus corniculatus* 110
*pedunculatus* 110
*uliginosus* 110
**Lousewort** 134
*Luzula arcuata* 155
*campestris* 154
*multiflora* 155
*pilosa* 154
*spicata* 155
*sylvatica* 154
*Lychnis flos-cuculi* 86
LYCOPODIACEAE 65
*Lycopodiella inundata* 65
*Lycopodium alpinum* 65
*annotinum* 65
*clavatum* 65
*inundatum* 65
*selago* 65
*Lycopsis arvensis* 121
*Lycopus europaeus* 125
**Lyme-grass** 171
*Lysimachia nemorum* 99
LYTHRACEAE 113
*Lythrum salicaria* 113
**Mackay's Horsetail** 67
**Maidenhair Spleenwort** 71
**Male-fern** 73
**Many-stalked Spike-rush** 156
**Mare's-tail** 127
**Marram** 169
**Marsh Arrowgrass** 150
**Marsh Cinquefoil** 103
**Marsh Clubmoss** 65
**Marsh Cudweed** 147
**Marsh Foxtail** 169
**Marsh Hawk's-beard** 141
**Marsh Horsetail** 68
**Marsh Lousewort** 133

## Flowering plant index

**Marsh Pennywort** 118
**Marsh Ragwort** 149
**Marsh Speedwell** 130
**Marsh Thistle** 139
**Marsh Violet** 90
**Marsh Willowherb** 114
**Marsh Woundwort** 123
**Marsh-marigold** 76
MARSILIACEAE 69
**Masterwort** 119
**Mat-grass** 163
*Matricaria discoidea* 149
    *recutita* 148
**Meadow Brome** 170
**Meadow Buttercup** 76
**Meadow Foxtail** 169
**Meadow Oat-grass** 166
**Meadow Vetchling** 111
**Meadowsweet** 103
*Meconopsis cambrica* 78
*Melampyrum pratense* 131
**Melancholy Thistle** 139
*Melica nutans* 166
*Mentha aquatica* 126
    *arvensis* 125
    *spicata* 126
    x *cordifolia* 126
    x *gentilis* 126
    x *gracilis* 126
    x *niliaca* 127
    x *piperita* 126
    x *rotundifolia* 127
    x *villosa* 126
    x *villosonervata* 126
MENYANTHACEAE 121
*Menyanthes trifoliata* 121
*Mertensia maritima* 122
*Mimulus guttatus* 129
    x *burnetii* 129
    x *maculosus* 129
    x *robertsii* 129
*Minuartia sedoides* 83
*Molinia caerulea* 171
**Monkeyflower** 129
**Montbretia** 174
*Montia fontana* 82
    *sibirica* 82
**Moonwort** 68
**Moschatel** 137
**Moss Campion** 86
**Mossy Saxifrage** 102
**Mountain Avens** 105
**Mountain Crowberry** 97
**Mountain Everlasting** 146
**Mountain Male-fern** 73
**Mountain Melick** 166
**Mountain Pansy** 91
**Mountain Sorrel** 89
**Mouse-ear Hawkweed** 141
**Mugwort** 148

**Musk Thistle** 138
*Myosotis arvensis* 122
    *caespitosa* 122
    *discolor* 122
    *laxa* 122
    *scorpioides* 122
    *secunda* 122
*Myrica gale* 80
MYRICACEAE 80
*Myriophyllum alternifolium* 113
    *spicatum* 112
*Myrrhis odorata* 118
*Narcissus pseudo-narcissus* 173
*Nardus stricta* 163
**Narrow Buckler-fern** 74
**Narrow-fruited Water-cress** 95
**Narrow-leaved Helleborine** 174
**Narrow-leaved Meadow-grass** 165
*Narthecium ossifragum* 173
**Needle Spike-rush** 156
*Neottia nidus-avis* 174
**Nettle-leaved Goosefoot** 81
**New Zealand Willowherb** 114
**Nipplewort** 139
**Noble Fir** 74
**Nordic Bladderwort** 134
**Northern Buckler-fern** 74
**Northern Bedstraw** 136
**Northern Dead-nettle** 123
**Northern Dock** 88
**Northern Knotgrass** 87
**Northern Marsh-orchid** 177
**Norway Spruce** 75
NYMPHACEAE 76
*Nymphaea alba* 76
**Oak Fern** 72
**Oaks** 80
**Oblong-leaved Sundew** 90
*Odontites vernus* 133
*Oenanthe crocata* 119
OLEACEAE 128
ONAGRACEAE 113
OPHIOGLOSSACEAE 68
*Ophioglossum azoricum* 68
    *vulgatum* 68
**Opium Poppy** 78
**Opposite-leaved Golden-saxifrage** 102
**Orache** 82
ORCHIDACEAE 174
*Orchis mascula* 177
*Oreopteris limbosperma* 70
*Orthilia secunda* 99
**Osier** 92
*Osmunda regalis* 68
**Oval Sedge** 158
OXALIDACEAE 117
*Oxalis acetosella* 117
**Ox-eye Daisy** 148
**Oxford Rampion** 135
*Oxyria digyna* 89

**Oysterplant**  122
**Pale Bladderwort**  134
**Pale Butterwort**  133
**Pale Sedge**  161
*Papaver dubium*  78
  *somniferum*  78
PAPAVERACEAE  78
*Paris quadrifolia*  173
*Parnassia palustris*  102
**Parsley Fern**  69
**Parsley-piert**  106
*Pedicularis palustris*  133
  *sylvatica*  134
*Pentaglottis sempervirens*  122
**Peppermint**  126
**Perennial Rye-grass**  164
**Perennial Sow-thistle**  140
**Perfoliate Pondweed**  151
*Persicaria amphibia*  87
  *hydropiper*  87
  *lapathifolium*  87
  *maculosa*  87
  *vivipara*  86
**Petty Spurge**  116
*Peucedanum ostruthium*  119
**Phacelia**  121
*Phacelia tanacetifolia*  121
*Phalaris arundinacea*  168
*Phegopteris connectilis*  70
*Phleum bertolonii*  170
  *pratense*  170
*Phragmites australis*  172
  *communis*  172
*Phyllitis scolopendrium*  70
*Phyteuma scheuchzeri*  135
*Picea abies*  75
  *sitchensis*  75
**Pick-a-back-plant**  102
**Pignut**  118
**Pill Sedge**  161
**Pillwort**  69
*Pilosella officinarum*  141
*Pilularia globulifera*  69
PINACEAE  74
**Pineappleweed**  149
*Pinguicula alpina*  134
  *lusitanica*  134
  *vulgaris*  134
**Pink Purslane**  82
*Pinus contorta*  75
  *sylvestris*  75
PLANTAGINACEAE  127
*Plantago coronopus*  127
  *lanceolata*  128
  *major*  128
  *maritima*  128
*Platanthera bifolia*  175
  *chlorantha*  175
PLUMBAGINACEAE  89
*Poa angustifolia*  165

*annua*  165
*glauca*  165
*humilis*  165
*nemoralis*  165
*pratensis*  165
*subcaerulea*  165
*trivialis*  165
POACEAE  162
*Polygala serpyllifolia*  116
  *vulgaris*  116
POLYGALACEAE  116
POLYGONACEAE  86
Polygonum amphibium  87
  *arenastrum*  87
  *aviculare*  87
  *boreale*  87
  convolvulus  88
  cuspidatum  88
  hydropiper  87
  lapathifolium  87
  *oxyspermum* ssp. *raii*  87
  persicaria  87
  viviparum  86
POLYPODIACEAE  69
*Polypodium interjectum*  69
  *vulgare*  69
  *vulgare* ssp. *prionodes*  69
*Polystichum aculeatum*  73
  *lonchitis*  73
  x *illyricum*  73
*Populus nigra*  91
  *tremula*  91
  *trichocarpa*  91
PORTULACEAE  82
*Potamogeton perfoliatus*  151
  *alpinus*  151
  *berchtoldii*  151
  *filiformis*  151
  *gramineus*  150
  *lucens*  150
  *natans*  150
  *pectinatus*  152
  *polygonifolius*  150
  *praelongus*  151
  *rutilus*  151
  x *nitens*  151
  x *sparganifolius*  150
POTAMOGETONACEAE  150
*Potentilla anserina*  103
  *crantzii*  104
  *erecta*  104
  *palustris*  103
  *reptans*  104
  *sterilis*  104
**Prickly Sow-thistle**  140
**Primrose**  99
*Primula veris*  99
  *vulgaris*  99
PRIMULACEAE  99
**Procumbent Pearlwort**  85

# Flowering plant index

*Prunella vulgaris* 125
*Prunus avium* 109
*Prunus padus* 109
   *spinosa* 108
*Pseudorchis albida* 175
*Pseudotsuga menziesii* 75
*Pteridium aquilinum* 70
*Puccinellia maritima* 164
**Purple Loosestrife** 113
**Purple Milk-vetch** 110
**Purple Moor-grass** 171
**Purple Saxifrage** 102
**Purple Willow** 92
**Pyramidal Bugle** 125
**Pyrenean Scurvygrass** 96
*Pyrola media* 99
PYROLACEAE 99
**Quaking-grass** 165
*Quercus* spp. 80
**Quillwort** 66
**Ragged-Robin** 86
**Ramsons** 173
RANUNCULACEAE 76
*Ranunculus acris* 76
   *arvensis* 77
   *auricomus* 77
   *baudotii* 77
   *bulbosus* 77
   *ficaria* 77
   *flammula* 77
   *hederaceus* 77
   *repens* 77
   *trichophyllus* 78
*Raphanus raphanistrum* 97
   raphanistrum var. *aureum* 97
**Raspberry** 103
**Ray's Knotgrass** 87
**Red Bartsia** 133
**Red Campion** 86
**Red Clover** 112
**Red Currant** 100
**Red Dead-nettle** 123
**Red Fescue** 163
**Red Goosefoot** 81
**Red Pondweed** 151
**Redshank** 87
**Reed Canary-grass** 168
**Remote Sedge** 158
*Rhinanthus minor* 133
   minor ssp. *borealis* 133
   minor ssp. *lintonii* 133
   minor ssp. *minor* 133
   minor ssp. *monticola* 133
   minor ssp. *stenophyllus* 133
*Rhynchospora alba* 157
**Ribbon-leaved Pondweed** 150
*Ribes nigrum* 100
   *rubrum* 100
   *sylvestre* 100
   *uva-crispa* 101

**Ribwort Plantain** 128
**Rock Sedge** 162
**Rock Whitebeam** 109
*Rorippa microphylla* 95
   *nasturtium-aquaticum* 94
   *nasturtium-aquaticum* agg. 95
*Rosa caesia* ssp. *caesia* 107
   caesia ssp. *glauca* 107
   caesia x *R. sherardii* 108
   *canina* 107
   *mollis* 108
   *pimpinellifolia* 107
   *rubiginosa* 108
   *sherardii* 108
   spp. 107
   x *dumalis* 107
   x *involuta* 108
   x *rothschildii* 108
   x *sabinii* 108
   x *suberecta* 108
ROSACEAE 103
**Rosebay Willowherb** 114
**Roseroot** 101
**Roses** 107
**Rough Horsetail** 66
**Rough Meadow-grass** 165
**Round-leaved Sundew** 89
**Rowan** 109
**Royal Fern** 68
RUBIACEAE 136
*Rubus chamaemorus* 103
   *fruticosus* agg. 103
   *idaeus* 103
   *saxatilis* 103
*Rumex acetosa* 88
   *acetosella* 88
   *crispus* 88
   *longifolius* 88
   *obtusifolius* 89
*Ruppia maritima* 152
RUPPIACEAE 152
**Russian Comfrey** 121
**Russian-vine** 88
**Rusty Willow** 92
*Sagina apetala* 85
   *maritima* 85
   *nodosa* 84
   *procumbens* 85
   *saginoides* 85
   *subulata* 84
**Salad Burnet** 105
SALICACEAE 91
*Salix alba* 92
   *aurita* 93
   *caprea* 92
   *cinerea* ssp. *oleifolia* 92
   *fragilis* 92
   *herbacea* 94
   *lapponum* 93
   *myrsinites* 94

*pentandra* 91
*purpurea* 92
*repens* 93
*viminalis* 92
x *ambigua* 93
x *cernua* 93
x *laurina* 93
x *ludificans* 93
x *multinervis* 92
x *sericans* 92
x *smithiana* 92
**Saltmarsh Flat-sedge** 157
**Saltmarsh Rush** 153
*Sambucus nigra* 137
**Sand Couch** 171
**Sand Sedge** 158
**Sand-spurrey** 85
*Sanguisorba minor* 105
**Sanicle** 118
*Sanicula europaea* 118
*Sarothamnus scoparius* 112
*Saussurea alpina* 138
*Saxifraga aizoides* 102
 *hypnoides* 102
 *nivalis* 101
 *oppositifolia* 102
 *stellaris* 101
 x *urbium* 101
SAXIFRAGACEAE 101
**Scented Mayweed** 148
**Scentless Mayweed** 149
*Schoenoplectus lacustris* 156
*Schoenus nigricans* 157
*Scilla verna* 173
Scirpus fluitans 157
 lacustris 156
 setacea 156
**Scots Lovage** 119
**Scots Pine** 75
**Scottish Asphodel** 172
**Scottish Monkeyflower** 129
**Scottish Scurvygrass** 96
*Scrophularia nodosa* 128
SCROPHULARIACEAE 128
**Scurvygrass** 96
*Scutellaria galericulata* 124
**Sea Arrowgrass** 150
**Sea Aster** 147
**Sea Campion** 86
**Sea Fern-grass** 166
**Sea Mayweed** 149
**Sea Mouse-ear** 84
**Sea Pearlwort** 85
**Sea Plantain** 128
**Sea Rocket** 97
**Sea Sandwort** 83
**Sea Spleenwort** 71
**Sea-milkwort** 100
*Sedum acre* 101
 *album* 101

 *anglicum* 101
 *rosea* 101
*Selaginella selaginoides* 66
SELAGINELLACEAE 66
**Selfheal** 125
*Senecio aquaticus* 149
 *jacobaea* 149
 *sylvaticus* 149
 *viscosus* 149
 *vulgaris* 149
**Serrated Wintergreen** 99
**Shade Horsetail** 67
**Sharp-flowered Rush** 153
**Sharp-toothed Mint** 126
**Sheep's Sorrel** 88
**Sheep's-fescue** 163
**Shepherd's-purse** 96
**Sherard's Downy-rose** 108
*Sherardia arvensis* 136
**Shetland Pondweed** 151
**Shining Crane's-bill** 117
**Shining Pondweed** 150
**Shoreweed** 128
**Short-fruited Willowherb** 113
**Sibbaldia** 104
*Sibbaldia procumbens* 104
*Sieglingia decumbens* 171
*Silene acaulis* 86
 *alba* 86
 *dioica* 86
 *latifolia* 86
 *maritima* 86
 *quadrifida* 86
 *uniflora* 86
 x *hampeana* 86
**Silky-leaved Osier** 92
**Silver Birch** 80
**Silver Hair-grass** 168
**Silverweed** 103
*Sisymbrium altissimum* 94
 *officinale* 94
**Sitka Spruce** 75
**Skullcap** 124
**Slender Parsley-piert** 107
**Slender Rush** 152
**Slender Sedge** 159
**Slender Soft-brome** 170
**Slender Speedwell** 131
**Slender Spike-rush** 156
**Slender St John's-wort** 89
**Slender-leaved Pondweed** 151
**Small Adder's-tongue** 68
**Small Nettle** 79
**Small Pondweed** 151
**Smaller Cat's-tail** 170
**Small-fruited Yellow-sedge** 161
**Small-white Orchid** 175
**Smooth Hawk's-beard** 141
**Smooth Meadow-grass** 165
**Smooth Sow-thistle** 140

# Flowering plant index

**Smooth-stalked Sedge** 160
**Sneezewort** 148
**Snowberry** 137
**Soft Downy-rose** 108
**Soft Rush** 154
**Soft-brome** 170
SOLANACEAE 120
*Solanum dulcamara* 120
*Solidago virgaurea* 147
*Sonchus arvensis* 140
   *asper* 140
   *oleraceus* 140
*Sorbus aucuparia* 109
   *rupicola* 109
SPARGANIACEAE 172
*Sparganium angustifolium* 172
   *emersum* 172
   *erectum* 172
   *minimum* 172
   *natans* 172
   x *diversifolium* 172
**Spear Thistle** 138
**Spear-leaved Orache** 82
**Spearmint** 126
*Spergula arvensis* 85
*Spergularia marina* 85
   *media* 85
   *rubra* 85
**Spiked Water-milfoil** 112
**Spiked Woodrush** 155
**Spreading Meadow-grass** 165
**Spring Gentian** 120
**Spring Quillwort** 66
**Spring Squill** 173
**Spring-sedge** 161
**Square-stalked St John's-wort** 89
**Square-stalked Willowherb** 113
**Squirreltail Fescue** 164
*Stachys palustris* 123
   *sylvatica* 123
   x *ambigua* 123
**Stag's-horn Clubmoss** 65
**Star Sedge** 158
**Starry Saxifrage** 101
*Stellaria alsine* 84
   *graminea* 83
   *holostea* 83
   *media* 83
   *neglecta* 83
   *uliginosa* 84
**Sticky Groundsel** 149
**Sticky Mouse-ear** 84
**Stiff Sedge** 162
**Stone Bramble** 103
*Subularia aquatica* 96
*Succisa pratensis* 138
**Sun Spurge** 115
**Sweet Cicely** 118
**Sweet Vernal-grass** 168
**Sweet-briar** 108

**Sycamore** 116
*Symphoricarpos albus* 137
*Symphytum officinale* 121
   x *uplandicum* 121
**Tall Fescue** 163
**Tall Ramping-fumitory** 79
**Tall Rocket** 94
*Tanacetum parthenium* 147
   *vulgare* 147
**Tansy** 147
*Taraxacum* spp. 140
   *faeroense* 140
   *unguilobum* 140
**Tawny Sedge** 160
*Teucrium scorodonia* 124
**Thale Cress** 94
*Thalictrum alpinum* 78
   *minus* 78
THELYPTERIDACEAE 70
*Thelypteris dryopteris* 72
   *oreopteris* 70
   *phegopteris* 70
   *robertiana* 72
*Thlaspi arvense* 96
**Thorn-apple** 120
**Thread-leaved Water-crowfoot** 78
**Three-flowered Rush** 154
**Three-leaved Rush** 153
**Thrift** 89
**Thyme-leaved Sandwort** 82
**Thyme-leaved Speedwell** 129
*Thymus drucei* 125
*Thymus polytrichus* 125
   *pulegioides* 125
**Timothy** 170
**Toad Rush** 153
*Tofieldia pusilla* 172
*Tolmiea menziesii* 102
**Tormentil** 104
*Tragopogon pratensis* 140
**Trailing Azalea** 97
*Trichomanes speciosum* 69
*Trichophorum cespitosum* 156
*Trientalis europaea* 99
*Trifolium campestre* 112
   *dubium* 112
   *pratense* 112
   *repens* 111
*Triglochin maritimum* 150
   *palustre* 150
*Tripleurospermum inodorum* 149
   *maritimum* 149
   maritimum ssp. *inodorum* 149
*Trisetum flavescens* 167
*Trollius europaeus* 76
TROPAEOLACEAE 117
*Tropaeolum speciosum* 117
**Tufted Forget-me-not** 122
**Tufted Hair-grass** 167

**Tufted Vetch** 110
**Turnip** 97
*Tussilago farfara* 149
**Tutsan** 89
*Ulex europaeus* 112
　*gallii* 112
ULMACEAE 79
*Ulmus glabra* 79
**Unbranched Bur-reed** 172
**Upland Enchanter's-nightshade** 115
*Urtica dioica* 79
　*urens* 79
URTICACEAE 79
*Utricularia australis* 134
　*minor* 135
　*ochroleuca* 134
　spp. 134
　*stygia* 134
*Vaccinium myrtillus* 98
　*uliginosum* 98
　*vitis-idaea* 98
*Valeriana officinalis* 138
VALERIANACEAE 137
*Valerianella carinata* 138
　*locusta* 137
**Variegated Horsetail** 67
**Various-leaved Pondweed** 150
**Various-leaved Water-starwort** 127
**Velvet Bent** 169
*Verbascum thapsus* 128
*Veronica agrestis* 131
　*anagallis-aquatica* 130
　*arvensis* 130
　*beccabunga* 130
　*chamaedrys* 130
　*filiformis* 131
　*hederifolia* 131
　*officinalis* 130
　*persica* 131
　*scutellata* 130
　*serpyllifolia* 129
　*serpyllifolia* ssp. *humifusa* 130
*Viburnum opulus* 137
*Vicia cracca* 110
　*orobus* 110
　*sativa* 111
　*sepium* 111
　*sylvatica* 111
*Viola arvensis* 91
　*canina* 90
　*lutea* 91
　*palustris* 90
　*riviniana* 90
　*tricolor* 91
VIOLACEAE 90
**Viper's-bugloss** 121
**Viviparous Sheep's-fescue** 164
*Vulpia bromoides* 164
**Wall Speedwell** 130
**Wall-rue** 72

**Water Avens** 104
**Water Forget-me-not** 122
**Water Horsetail** 67
**Water Lobelia** 135
**Water Mint** 126
**Water Sedge** 162
**Water-cress** 94-95
**Water-pepper** 87
**Wavy Bitter-cress** 95
**Wavy Hair-grass** 167
**Welsh Poppy** 78
**Western Balsam-poplar** 91
**Western Gorse** 112
**Western Polypody** 69
**White Beak-sedge** 157
**White Campion** 86
**White Clover** 111
**White Dead-nettle** 123
**White Sedge** 159
**White Stonecrop** 101
**White Water-lily** 76
**White Willow** 92
**Whorl-grass** 166
**Whortle-leaved Willow** 94
**Wild Angelica** 119
**Wild Carrot** 119
**Wild Cherry** 109
**Wild Pansy** 91
**Wild Radish** 97
**Wild Strawberry** 104
**Wild Thyme** 125
**Wilson's Filmy-fern** 69
**Wood Anemone** 76
**Wood Avens** 104
**Wood Bitter-vetch** 110
**Wood Fescue** 163
**Wood Horsetail** 67
**Wood Meadow-grass** 165
**Wood Sage** 124
**Wood Small-reed** 169
**Wood Vetch** 111
**Woodruff** 136
**Wood-sedge** 159
WOODSIACEAE 72
**Wood-sorrel** 117
**Wormwood** 148
**Wych Elm** 79
**Yarrow** 148
**Yellow Cypress Clubmoss** 65
**Yellow Iris** 173
**Yellow Oat-grass** 167
**Yellow Pimpernel** 99
**Yellow Saxifrage** 102
**Yellow-rattle** 133
**Yellow-sedge** 161
**Yorkshire-fog** 167
*Zostera marina* 152
ZOSTERACEAE 152

# INDEX

# BRYOPHYTES

Entries are for the species accounts only. Scientific names are in *italics* and synonyms are in normal type.

*Amblyodon dealbatus* 239
Amblystegium compactum 246
   *serpens* 246
   *tenax* 246
*Amphidium lapponicum* 240
   *mougeotii* 240
*Anastrepta orcadensis* 196
*Anastrophyllum donnianum* 198
   *hellerianum* 198
   *joergensenii* 198
   *minutum* 198
   *saxicola* 198
*Andreaea megistospora* 215
   *alpina* 215
   *rothii* 215
   *rupestris* 215
*Aneura pinguis* 209
*Anoectangium aestivum* 226
   *warburgii* 226
Anomobryum filiforme 235
   *julaceum* 235
*Anomodon viticulosus* 244
*Anthelia julacea* 195
   *juratzkana* 195
*Antitrichia curtipendula* 243
*Aphanolejeunea microscopica* 208
*Apometzgeria pubescens* 210
*Archidium alternifolium* 217
*Arctoa fulvella* 219
*Atrichum undulatum* 216
*Aulocomnium androgynum* 239
   *palustre* 238
   *turgidum* 239
*Barbilophozia atlantica* 195
   *attenuata* 195
   *barbata* 196
   *floerkei* 195
   *kunzeana* 195
   *lycopodioides* 196
Barbula ferruginascens 225
   convoluta 226
   cylindrica 227
   fallax 227
   icmadophila 226
   recurvirostra 225
   reflexa 227
   revoluta 225
   rigidula 226
   spadicea 227
   tophacea 227
   unguiculata 226
*Bartramia pomiformis* 239
   *ithyphylla* 239
*Bazzania pearsonii* 192
   *tricrenata* 192
   *trilobata* 192
*Blasia pusilla* 209
*Blepharostoma trichophyllum* 191
*Blindia acuta* 232
*Brachythecium albicans* 250
   *glareosum* 250
   *plumosum* 250
   *populeum* 250
   *rivulare* 250
   *rutabulum* 250
*Breutelia chrysocoma* 240
*Bryoerythrophyllum ferruginascens* 225
   *recurvirostrum* 225
*Bryum alpinum* 237
   *argenteum* 236
   *bicolor* 236
   *capillare* 236
   *dixonii* 236
   *elegans* 236
   *mildeanum* 236
   *muehlenbeckii* 237
   *pallens* 235
   *pseudotriquetrum* 236
   *radiculosum* 236
   *riparium* 236
   *weigelii* 235
*Calliergon cordifolium* 248
   cuspidatum 249
   giganteum 249
   sarmentosum 249
   stramineum 248
   trifarium 248
*Calliergonella cuspidata* 249
*Calypogeia arguta* 193
   *azurea* 193
   *fissa* 193
   *muelleriana* 193
   *sphagnicola* 193
   *suecica* 193
*Campyliadelphus chrysophyllus* 246
   *elodes* 246
Campylium chrysophyllum 246
   elodes 246
   polygamum 247
   *stellatum* var. protensum 246
   *stellatum* var. *stellatum* 246
*Campylopus atrovirens* var. *atrovirens* 221
   *atrovirens* var. *falcatus* 221
   *brevipilus* 222

*flexuosus* 221
*fragilis* 221
*gracilis* 220
*introflexus* 222
paradoxus 221
*pyriformis* 221
*schimperi* 220
schwarzii 220
*setifolius* 221
shawii 221
Cephalozia bicuspidata 193
   *connivens* 194
   *leucantha* 193
   *loitlesbergeri* 194
   *lunulifolia* 193
Cephaloziella divaricata 195
   hampeana 195
Ceratodon purpureus 217
Chiloscyphus polyanthos 204
Cinclidium stygium 237
Cinclidotus fontinaloides 228
Cirriphyllum crassinervium 251
   *piliferum* 251
Cladipodiella fluitans 194
Climacium dendroides 243
Cololejeunea calcarea 208
Colura calyptrifolia 208
Conardia compacta 246
Conocephalum conicum 211
Cratoneuron commutatum 245
   *filicinum* 246
Ctenidium molluscum var. condensatum 255
   var. *molluscum* 255
   var. *robustum* 256
Cynodontium jenneri 218
Dichodontium pellucidum 218
Dicranella grevilleana 218
   *heteromalla* 219
   *palustris* 218
   *schreberiana* 218
   *subulata* 219
   *varia* 219
Dicranodontium asperulum 220
   *denudatum* 220
   *uncinatum* 220
Dicranoweisia crispula 219
Dicranum bonjeanii 219
   *fuscescens* 220
   *majus* 220
   *scoparium* 219
   *scottianum* 220
Didymodon fallax 227
   *ferrugineus* 227
   *icmadophilus* 226
   *insulanus* 227
   *rigidulus* 226
   *spadiceus* 227
   *tophaceus* 227
   *vinealis* 227
Diphyscium foliosum 216

*Diplophyllum albicans* 202
   *obtusifolium* 202
Distichium capillaceum 217
   *inclinatum* 217
Ditrichum crispatissimum 217
   *flexicaule* 217
   *gracile* 217
   *heteromallum* 217
   *zonatum* var. *zonatum* 217
*Douinia ovata* 201
*Drepanocladus aduncus* 247
   *cossonii* 247
   exannulatus 247
   fluitans 247
   *polygamus* 247
   *revolvens* 247
   uncinatus 247
*Drepanolejeunia hamatifolia* 207
Dryptodon patens 230
*Encalypta alpina* 223
   *ciliata* 223
   *streptocarpa* 223
   *vulgaris* 223
*Entodon concinnus* 252
*Entosthodon attenuatus* 233
   *obtusus* 233
*Ephemerum serratum* var. *minutissimum* 233
*Eucladium verticillatum* 223
*Eurhynchium crassinervium* 251
   hians 251
   *praelongum* 251
   *pumilum* 251
   *striatum* 251
   swartzii 251
*Fissidens adianthoides* 223
   *bryoides* 222
   cristatus 222
   *dubius* 222
   *osmundoides* 222
   *pusillus* 222
   *taxifolius* 222
*Fontinalis antipyretica* 242
   *squamosa* 242
*Fossombronia foveolata* 208
*Frullania dilatata* 207
   *fragilifolia* 207
   *microphylla* 207
   *tamarisci* 207
   *teneriffae* 207
Funaria attenuata 233
   *hygrometrica* 233
   obtusa 233
*Geocalyx graveolens* 204
*Glyphomitrium daviesii* 232
Grimmia affinis 229
   *curvata* 230
   *decipiens* 230
   *funalis* 230
   *hartmanii* 230
   *longirostris* 229

*orbicularis* 229
*ovalis* 229
*pulvinata* var. *pulvinata* 229
*torquata* 230
*trichophylla* 230
*Gymnocolea inflata* 198
*Gymnomitrion concinnatum* 201
   *crenulatum* 201
   *obtusum* 201
*Gymnostomum aeruginosum* 226
   *calcareum* 226
   insigne 225
   recurvirostrum 225
*Gyroweisia tenuis* 226
*Haplomitrium hookeri* 191
*Harpalejeunea molleri* 207
*Harpanthus scutatus* 204
   *flotovianus* 204
*Hedwigia ciliata* var. *ciliata* 242
   *stellata* 242
*Herbertus aduncus* ssp. *hutchinsiae* 191
   *stramineus* 191
*Herzogiella striatella* 254
*Heterocladium heteropterum* var. *flaccidum* 245
   *heteropterum* var. *heteropterum* 244
*Homalia trichomanoides* 243
*Homalothecium lutescens* 249
   *sericeum* 249
*Hookeria lucens* 244
*Hygrobiella laxifolia* 194
Hygrohypnum dilatatum 248
   *duriusculum* 248
   *eugyrium* 248
   *luridum* var. *luridum* 248
   *ochraceum* 247
*Hylocomium brevirostre* 256
   *splendens* 256
   *umbratum* 256
*Hymenostylium insigne* 225
   *recurvirostrum* 225
*Hyocomium armoricum* 256
Hypnum andoi 254
   *bambergeri* 255
   *callichroum* 255
   *cupressiforme* 254
   cupressiforme var. lacunosum 254
   cupressiforme var. resupinatum 254
   *hamulosum* 255
   *jutlandicum* 254
   *lacunosum* var. *lacunosum* 254
   *lindbergii* 255
   mammillatum 254
   resupinatum 254
*Isopterygiopsis muelleriana* 253
   *pulchella* 253
Isopterygium elegans 254
   pulchellum 253
*Isothecium alopecurioides* 249
   *myosuroides* var. *brachythecioides* 249
   *myosuroides* var. *myosuroides* 249

   *myurum* 249
*Jungermannia atrovirens* 199
   *exsertifolia* ssp. *cordifolia* 199
   *gracillima* 200
   *hyalina* 200
   *obovata* 200
   *paroica* 200
   *pumila* 199
   *sphaerocarpa* 199
   *subelliptica* 200
*Kiaeria blyttii* 219
   *falcata* 219
   *starkei* 219
*Kurzia pauciflora* 191
   *sylvatica* 192
   *trichoclados* 192
*Leiocolea alpestris* 197
   *badensis* 197
   *bantriensis* 197
   *fitzgeraldiae* 197
   *gillmanii* 197
   *heterocolpos* 197
   *turbinata* 197
*Lejeunea cavifolia* 208
   *lamacerina* 208
   *patens* 208
   *ulicina* 207
*Lepidozia cupressina* 192
   *pearsonii* 192
   *reptans* 192
*Leptodontium flexifolium* 225
   recurvifolium 225
*Leptoscyphus cuneifolius* 204
*Leucobryum glaucum* 222
*Leucodon sciuroides* var. *sciuroides* 243
*Lophocolea bidentata* 204
*Lophozia bicrenata* 197
   *excisa* 196
   *incisa* 196
   *obtusa* 196
   *opacifolia* 196
   *sudetica* 196
   *ventricosa* 196
*Lunularia cruciata* 211
*Marchantia polymorpha* 211
*Marsupella adusta* 201
   *alpina* 201
   *emarginata* 200
   *funckii* 201
   *sphacelata* 201
   *sprucei* 201
*Mastigophora woodsii* 191
*Meesia uliginosa* 239
*Metzgeria conjugata* 210
   *fruticulosa* 210
   *furcata* 210
   *leptoneura* 210
*Microlejeunea ulicina* 207
*Mnium hornum* 237
   *stellare* 237

*thomsonii* 237
*Moerckia hibernica* 209
*Molendoa warburgii* 226
*Mylia anomala* 199
   *taylori* 199
*Myurium hochstetteri* 243
*Nardia compressa* 200
   *geoscyphus* 200
   *scalaris* 200
*Neckera complanata* 243
   *crispa* 243
*Nowellia curvifolia* 194
*Odontoschisma denudatum* 194
   *elongatum* 194
   *sphagni* 194
*Oligotrichum hercynicum* 216
*Oncophorus virens* 218
*Orthothecium intricatum* 252
   *rufescens* 252
*Orthotrichum affine* 241
   *anomalum* 241
   *cupulatum* 241
   *diaphanum* 241
   *pulchellum* 241
   *rupestre* 241
   *stramineum* 241
Oxystegus *hibernicus* 224
   *tenuirostris* 224
*Palustriella commutata* var. *commutata* 245
   *commutata* var. *falcata* 245
   *commutata* var. *virescens* 245
*Paraleptodontium recurvifolium* 225
*Pellia endiviifolia* 209
   *epiphylla* 209
   *neesiana* 209
*Philonotis calcarea* 240
   *fontana* 239
   *seriata* 239
*Plagiobryum zieri* 235
*Plagiochila asplenioides* 205
   *carringtonii* 204
   *exigua* 205
   *killarniensis* 205
   *porelloides* 205
   *punctata* 205
   *spinulosa* 205
*Plagiomnium cuspidatum* 238
   *elatum* 238
   *ellipticum* 238
   *rostratum* 238
   *undulatum* 238
Plagiopus oederi 239
*Plagiopus oederianus* 239
*Plagiothecium cavifolium* 253
   *denticulatum* var. *denticulatum* 252
   *denticulatum* var. *obtusifolium* 253
   *nemorale* 253
   *platyphyllum* 253
   *succulentum* 253
   *undulatum* 253

*Platydictya jungermannioides* 252
*Pleurozia purpurea* 205
*Pleurozium schreberi* 252
*Pogonatum aloides* 215
   *urnigerum* 215
*Pohlia annotina* 234
   *bulbifera* 234
   *camptotrachela* 234
   carnea 235
   *cruda* 234
   *drummondii* 234
   *elongata* 234
   *filum* 234
   *flexuosa* 235
   *ludwigii* 235
   *melanodon* 235
   *muyldermansii* 235
   *nutans* 234
   *scotica* 234
   *wahlenbergii* var. *wahlenbergii* 235
   *wahlenbergii* var. glacialis 235
*Polytrichum alpinum* 215
   *commune* var. *commune* 216
   *formosum* 215
   *juniperinum* 216
   *piliferum* 216
   *strictum* 216
*Porella arboris-vitae* 206
   *cordeana* 206
   *obtusata* 206
   *platyphylla* 206
*Preissia quadrata* 211
*Pseudobryum cinclidioides* 238
*Pseudocrossidium revolutum* 225
*Pseudoleskeella catenulata* 244
   *rupestris* 244
Pseudoscleropodium purum 250
*Pseudotaxiphyllum elegans* 254
*Pterigynandrum filiforme* 244
*Pterogonium gracile* 243
*Ptilidium ciliare* 206
   *pulcherrimum* 206
*Ptilium crista-castrensis* 255
*Ptychomitrium polyphyllum* 232
*Racomitrium aciculare* 230
   *aquaticum* 231
   *ellipticum* 230
   *elongatum* 231
   *ericoides* 231
   *fasciculare* 231
   *heterostichum* 231
   *lanuginosum* 231
   *macounii* ssp. *alpinum* 231
   *sudeticum* 231
*Radula aquilegia* 206
   *complanata* 206
   *lindenbergiana* 206
*Reboulia hemisphaerica* 211
*Rhabdoweisia crispata* 218
   *fugax* 218

## Bryophyte index

*Rhizomnium pseudopunctatum* 238
   *punctatum* 237
*Rhodobryum roseum* 237
*Rhynchostegiella teesdalei* 252
   *tenella* 252
   *teneriffae* 252
*Rhynchostegium alopecurioides* 251
   lusitanicum 251
   *riparioides* 251
*Rhytidiadelphus loreus* 256
   *squarrosus* 256
   *triquetrus* 256
*Rhytidium rugosum* 254
*Riccardia chamedryfolia* 209
   *incurvata* 210
   *latifrons* 210
   *multifida* 209
   *palmata* 210
*Riccia sorocarpa* 211
   *subbifurca* 211
*Saccogyna viticulosa* 204
*Sanionia uncinata* 247
*Scapania aequiloba* 203
   *aspera* 203
   *compacta* 202
   *cuspiduligera* 202
   *degenii* 202
   *gracilis* 203
   *irrigua* 202
   *nemorea* 202
   *nimbosa* 203
   *ornithopodioides* 203
   *scandica* 202
   *subalpina* 203
   *uliginosa* 203
   *umbrosa* 202
   *undulata* 203
*Schistidium agassizii* 228
   alpicola var. rivulare 228
   *apocarpum* 228
   *maritimum* 228
   *rivulare* 228
   *strictum* 229
   *trichodon* 229
*Scleropodium purum* 250
*Scorpidium scorpioides* 248
*Seligeria donniana* 232
   *pusilla* 232
   *recurvata* 232
   *trifaria* 232
*Sphagnum angustifolium* 214
   auriculatum 214
   auriculatum var. inundatum 214
   *austinii* 212
   *capillifolium* 213
   *compactum* 213
   *contortum* 214
   *cuspidatum* 214
   *denticulatum* 214
   *fallax* ssp. *fallax* 214

   *fimbriatum* 212
   *fuscum* 213
   *girgensohnii* 212
   imbricatum ssp. austinii 212
   *inundatum* 214
   *magellanicum* 212
   *molle* 213
   *palustre* var. *palustre* 212
   *papillosum* 212
   *platyphyllum* 214
   *quinquefarium* 212
   recurvum var. mucronatum 214
   recurvum var. tenue 214
   *russowii* 212
   *squarrosum* 212
   *strictum* 213
   *subnitens* var. *subnitens* 213
   *subsecundum* 213
   *tenellum* 214
   *teres* 212
   *warnstorfii* 213
*Sphenolobopsis pearsonii* 198
*Splachnum ampullaceum* 233
   *sphaericum* 233
*Syntrichia princeps* 228
   *ruraliformis* 228
   *ruralis* 228
*Taxiphyllum wissgrillii* 254
*Tetraphis pellucida* 216
*Tetraplodon angustatus* 233
   *mnioides* 233
*Thamnobryum alopecurum* 244
*Thuidium delicatulum* 245
   *philibertii* 245
   *recognitum* 245
   *tamariscinum* 245
*Tortella densa* 224
   *flavovirens* var. *flavovirens* 224
   *tortuosa* 224
*Tortula muralis* var. *muralis* 227
   princeps 228
   ruralis 228
   ruralis ssp. ruraliformis 228
   *subulata* 227
*Trichocolea tomentella* 191
*Trichostomum brachydontium* 224
   *crispulum* 224
   *hibernicum* 224
   *tenuirostre* var. *tenuirostre* 224
*Tritomaria exsectiformis* 198
   *polita* 199
   *quinquedentata* 199
*Ulota bruchii* 242
   *calvescens* 242
   *crispa* 241
   *drummondii* 241
   *hutchinsiae* 242
   *phyllantha* 242
*Warnstorfia exannulata* 247
   *fluitans* 247

*Weissia brachycarpa* 224
   *controversa* var. *controversa* 223
   microstoma 224
   *perssonii* 223
   *rutilans* 224

Zygodon baumgartneri 240
   *conoideus* 240
   *rupestris* 240
   *viridissimus* var. *stirtonii* 240
   *viridissimus* var. *viridissimus* 240